MODELING OF MICROSCALE TRANSPORT IN BIOLOGICAL PROCESSES

Edited by

SID M. BECKER

ELSEVIER

AMSTERDAM • BOSTON • HEIDELBERG • LONDON • NEW YORK
OXFORD • PARIS • SAN DIEGO • SAN FRANCISCO • SINGAPORE
SYDNEY • TOKYO
Academic Press is an imprint of Elsevier

AP

Academic Press is an imprint of Elsevier
125 London Wall, London EC2Y 5AS, United Kingdom
525 B Street, Suite 1800, San Diego, CA 92101-4495, United States
50 Hampshire Street, 5th Floor, Cambridge, MA 02139, United States
The Boulevard, Langford Lane, Kidlington, Oxford OX5 1GB, United Kingdom

Library of Congress Cataloging-in-Publication Data
A catalog record for this book is available from the Library of Congress

British Library Cataloguing-in-Publication Data
A catalogue record for this book is available from the British Library

ISBN: 978-0-12-804595-4

For information on all Academic Press publications
visit our website at https://www.elsevier.com

Publisher: Joe Hayton
Acquisition Editor: Fiona Geraghty
Editorial Project Manager: Maria Convey
Production Project Manager: Caroline Johnson
Designer: Greg Harris

Typeset by VTeX

Contents

Contributors ix
Preface xi

1. Molecular Simulations of Complex Membrane Models

D. JEFFERIES AND S. KHALID

1.1 Introduction 1
1.2 Unsaturated Carbon Chains 3
1.3 Membrane Proteins 4
1.4 Sterols 6
1.5 Eukaryotic Membranes 6
1.6 Prokaryotic Membranes 8
1.7 Viral Membranes 10
1.8 Membrane Fusion 11
1.9 Graphitic Nanomaterials 12
1.10 Nanoparticles 13
1.11 On-Going Work 13
1.12 Outlook and Conclusion 14

2. Microbial Strategies for Oil Biodegradation

G.E. KAPELLOS

2.1 Introduction 19
2.2 Overview of the Biodegradation Process 21
2.3 Microbial Growth Modes on Oily Substrates 25
2.4 Microscale Modeling Considerations 33
2.5 Summary and Outlook 34

3. Modeling and Measurement of Biomolecular Transport and Sensing in Microfluidic Cell Culture and Analysis Systems

J.F. WONG, C.A. SIMMONS, AND E.W.K. YOUNG

3.1 Introduction 41
3.2 Basic Principles of Microscale Cell Culture 42
3.3 Theory and Equations: Fluid Flow, Mass Transport, and Biochemical Reactions 44
3.4 Review of Microfluidic Transport Models 52
3.5 Review of Theoretical Model Experimental Validation 68
3.6 Summary and Conclusions 71

4. Coupling Microscale Transport and Tissue Mechanics: Modeling Strategies for Arterial Multiphysics

M. MARINO, G. PONTRELLI, G. VAIRO, AND P. WRIGGERS

4.1 Introduction 77
4.2 Brief on Arterial Tissues 79
4.3 Arterial Multiphysics Modeling 84
4.4 An Axisymmetric Case Study 94
4.5 Conclusions 105
Appendix A Along-the-Chord Collagen Fiber Tangent Modulus 106
Appendix B Microstructure of Aortic Media Layer 108

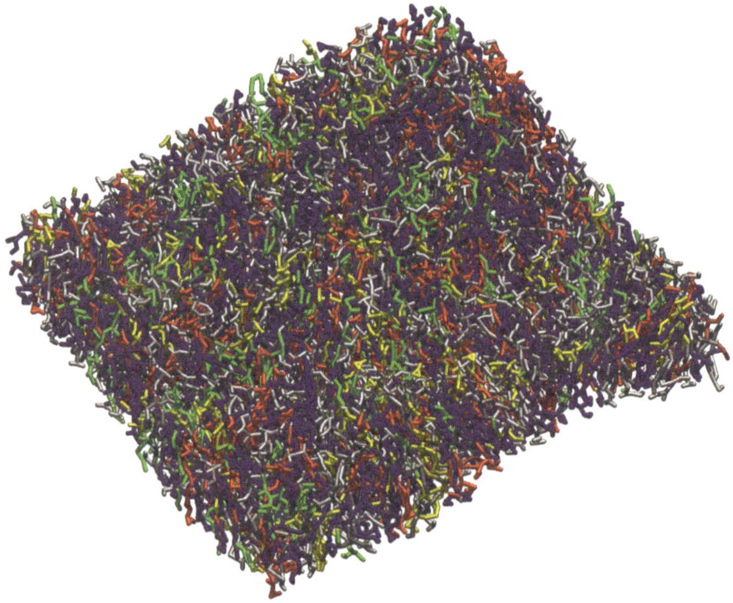

FIGURE 1.3 Coarse-grain (Martini) model of a thylakoid membrane from cyanobacteria. Phosphatidylglycerol lipids are colored green, digalactosyldiacylglycerol lipids are colored white, monogalactosyldiacylglycerol lipids are colored blue, and the two types of sulfoquinovosyldiacylglycerol lipids are colored yellow and red.

1.6 PROKARYOTIC MEMBRANES

Some of the most structurally diverse membranes are found within the family of single-celled prokaryotic microorganisms. Extreme diversity in lipid composition and cell wall structure has helped these primordial life-forms to thrive in the most inhospitable of environments from freezing arctic tundra, to heated hydrothermal vents, and even symbiotically, within the bodies of other organisms (Lauro and Bartlett, 2008; Madigan and Orent, 1999; Nap and Bisseling, 1990). Molecular simulation studies have thus far provided important insights into the characteristics of prokaryotic membranes, and on-going simulations are being used to address fundamental questions relating to the growth, and resilience of these microbes.

Gram-negative bacteria have one of the most complex cell envelope structures among all of the prokaryotes. The bacteria have a cell envelope that contains an inner cytoplasmic bilayer, and an outer membrane, which contains proteins, phospholipids, and lipopolysaccharides (Fig. 1.4). The two membranes are separated by the periplasm or periplasmic space, which contains peptidoglycan, a sugar-peptide polymer, this provides the cell with rigidity and mechanical strength (Beveridge, 1999). The periplasm is also host to a range of proteins. Atomistic simulations have shown that the outer membrane has many peculiar properties when compared with eukaryotic lipid systems. For one, the structural integrity of the membrane depends acutely on the availability of divalent metal ions. In the absence of metal ion

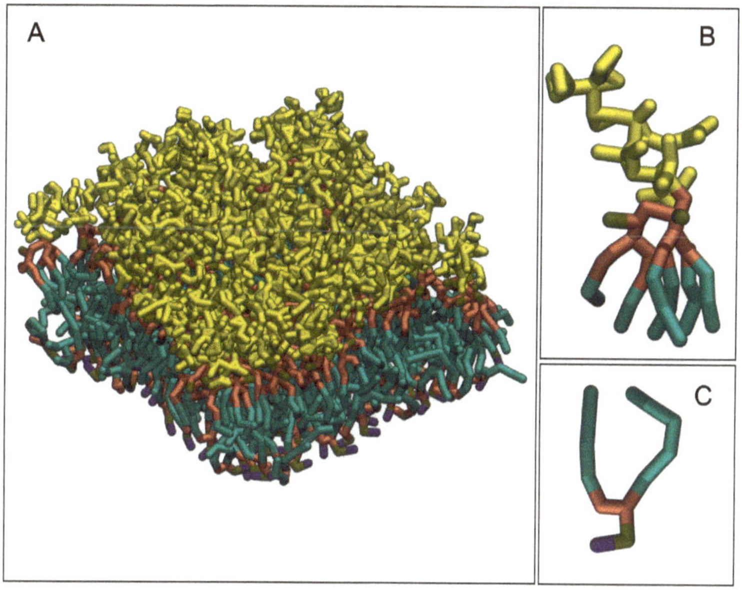

FIGURE 1.4 Panel A shows a coarse-grain (Martini) model of the *E. coli* outer membrane. The outer leaflet is composed of LPS and the inner leaflet contains a mixture of phospholipids. Carbon tails are colored cyan, phosphate groups are colored brown, glucosamine and glycerol groups are colored pink, choline head groups are colored blue, and the remaining core saccharide sections of LPS are colored yellow. Panel B shows a CG LPS molecule and panel C shows a POPE phospholipid, color schemes as panel A.

coordination, the membrane completely disaggregates on a picosecond timescale. Simulations also indicated that the outer membrane is less fluid than conventional phospholipid systems. The lipopolysaccharide molecules achieved unusually tight packing due to multiple stabilizing electrostatic interactions with metal ions, which immobilized the lipids and dramatically decreased surface accessibility to solvent (Wu et al., 2013; Soares et al., 2008; Soares and Straatsma, 2008).

In the case of Gram-positive membranes, molecular simulations have recently addressed fundamental questions relating to the growth and organization of the thick (20–35 nm) peptidoglycan network, which encases their single phospholipid membranes. By elucidating the mechanisms behind a distinct curling effect observed in three-dimensional electron cryotomography images, atomistic simulations were used to show that peptidoglycan polymers run circumferentially around Gram-positive cell bodies, providing support for a disorganized, layered model of the cell wall structure (Beeby et al., 2013). Coarse-grain simulations subsequently provided an explanation for how peptidoglycan layers are remodeled during the growth of cells. Local coordination of peptidoglycan remodeling enzymes sufficed to maintain the shape and structural integrity of the bacterial exoskeleton in molecular simulations, providing insight into the basic morphogenesis of Gram-positive and Gram-negative bacteria (Nguyen et al., 2015).

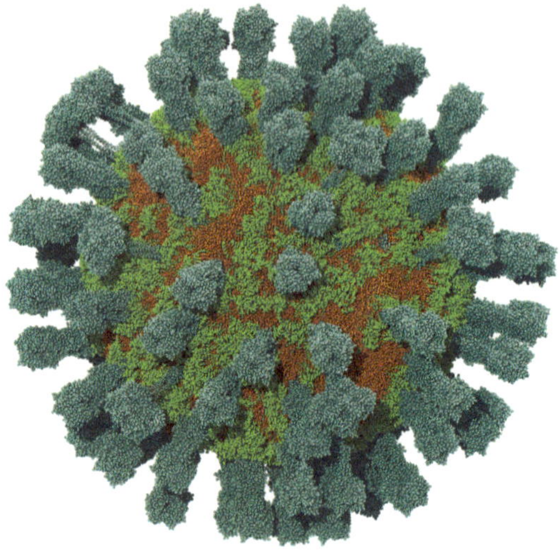

FIGURE 1.5 Snapshot of influenza A in water after ~5 μs of simulation. The spike glycoproteins are colored blue, lipids are colored orange, and the glycans of the Forssman glycolipids are colored green. The diameter is ~84 nm.

Extremophiles are some of the most durable life-forms owing to the intercalation of atypical tetraether lipids in their outer membranes. The lipids consist of two polar head groups, joined by two long hydrocarbon chains. The hydrocarbon chains can have a relatively simple saturated alkyl structure in-line with conventional phospholipid tails, but other mimetics with branching modifications are quite common (Longworthy et al., 1982). Molecular dynamics simulations revealed that the low permeability and unique durability of archaeal membranes could be ascribed to the tight packing of the tetraether lipid tails, which suppress fast-fluctuating void space formation (Chugunov et al., 2014; Shinoda et al., 2005). More recently, simulations showed that modifying the links between the tetraether lipid tails could drastically alter archaeal membrane properties, making tetraether lipid systems suitable for the design of artificial membranes and vesicles with increased durability and carrying capacity (Bulacu et al., 2011).

1.7 VIRAL MEMBRANES

In general, viruses are much smaller than prokaryote and eukaryote life forms. The infectious agents are so small (20–1500 nm scale) that molecular dynamics studies regularly consider all of the atoms that comprise entire viral particles, providing a view of hierarchal processes and large-scale emergent phenomena that are inaccessible in simulations of smaller viral subunits and fragments. For example, recent simulations of a complete Influenza A virion revealed that the overarching interactions of component Forssman glycolipids are responsible for the even spacing of spike proteins on the viral surface (Fig. 1.5) (Reddy et al., 2015).

When the glycolipids are included into a model for the virus, the spike proteins were suitably spaced to facilitate their crosslinking by immunoglobulin G (IgG) antibodies. Aside from affecting antibody-mediated immunity, the overarching interactions of Forssman glycolipids were speculated to confer physical robustness to the viral particles by reducing the mobility of proteins and lipids.

Moving forward, it will be interesting to simulate the interaction of Influenza A viruses with complex mammalian membrane models. Simulations exploring the pathogenicity of component viral fragments have already indicated that surface level hemagglutinin proteins can promote positive membrane curvature (Fuhrmans and Marrink, 2012). Simulations involving entire viral particles interacting with host membranes will resolve the full effects of the multiple affinities formed between surface level viral proteins and mammalian membranes at the host–pathogen interface.

Coarse-grain simulations were additionally used to model the complete dengue virion, including its complex lipid bilayer and peripheral E and M envelope proteins. The simulations revealed that protein–lipid interactions confer "raft-like" robustness to the cholesterol-free bilayer, providing resilience from environmental pressures and perturbations (Ayton and Voth, 2010). Other than this, a combination of coarse-grain and fine-grain modeling techniques were used to explore the structure of the immature HIV-1 virion. The multiscale simulations revealed the existence of an incomplete hexameric lattice formed from hexameric bundles of Gag polypeptide domains.

1.8 MEMBRANE FUSION

Membrane fusion was one of the first applications of coarse-grain molecular modeling given its pertinence to fundamental processes in cell biology, such as viral infection, endocytosis and exocytosis, and fertilization. From the outset, the coarse-grain simulation method successfully unraveled the molecular details of the process of membrane fusion, including the initial merging of apposed membranes, and the form of fusion intermediates (Marrink and Mark, 2003, 2004; Kasson and Pande, 2007). But as time progressed and fusion pathways became well established, efforts were directed at calculating the molecular free energy barriers involved in stalk formation, and investigating the role of fusion promoting peptides and polymers (Smirnova et al., 2010; Raudino et al., 2012).

The simulations showed that the prevailing notions of protein mediated membrane fusion were incomplete. Fusion promoting molecules like the SNARE complex did more than merely trigger fusion events by forcing opposing membranes into close proximity, the complexes overcame subsequent fusion barriers, and played an important role in actively guiding the membrane fusion reaction. It is now realized that the fusion process occurs as the SNARE complex releases mechanical stress to drive the expansion of a small stalk intermediate, which bridges the apposed membranes. The elucidation of the SNARE complex's full functional diversity helped to explain why so many molecular components are recruited in its functioning (Risselada and Grubmüller, 2012).

1.9 GRAPHITIC NANOMATERIALS

Molecular dynamics proved useful for elucidating important information about lipid bilayers as membrane models became increasingly complex, but the simulation technique had still more to offer. The simplicity and versatility of molecular dynamics encouraged studies that explored the interaction of complex membranes with a range of different organic and inorganic compounds.

Some of the most noteworthy simulations sought to decipher how novel nanomaterials interact with cellular membranes, given that many of these compounds are becoming more common, despite their cytotoxic effects being unknown. Graphitic nanomaterials, including zero-dimensional fullerenes, one-dimensional nanotubes, and two-dimensional nanosheets were placed in close apposition with eukaryotic and prokaryotic membrane models to elucidate their impact on human health or bacterial communities.

Starting with the molecular dynamics simulations of graphene and its different derivatives, it was discovered that the carbon nanosheets had the capacity to degrade the inner and outer membranes of Gram-negative bacteria. This degradation was mediated by the destructive extraction of phospholipids, once the nanosheets had intercalated among the inner and outer lipid membranes (Tu et al., 2013). The disruptive extraction of phospholipid molecules led not only to a sparser membrane, but also to a deformation of the basic bilayer structure. Additional simulation studies revealed that the specific orientation adopted by nanosheets in phospholipid membranes depends on their on their specific size and oxidation state, suggesting that the degree of membrane deformation may be sensitive to these physical and chemical properties (Wang et al., 2013).

Molecular simulation techniques were used in a similar way to evaluate the effects of graphitic nanomaterials on eukaryotic membrane models, and thereby human health. One-dimensional nanotubes were found to approach and penetrate the membranes in a way that depends on their exact lengths, widths, and the nature of their end cap groups (Shi et al., 2011). More specifically, nanotubes of different sizes were internalized into the membranes at different rates, while uncapped nanotubes were unable to enter the membrane core whatsoever. This inability to penetrate the membrane model was ascribed to the lack of carbon atoms at the nanotube tip, which prevents the receptor adhesion needed for tip wrapping and rotation. The receptors in the contact zone can only adhere to the carbon atoms on the sidewall of the nanotube, which preferentially pulls the nanotube parallel onto the membrane surface.

For C_{60} fullerenes, it was found that pristine nanoparticles readily permeate mammalian membrane models (Wong-Ekkabut et al., 2008), but polar derivatives are more frequently trapped at the water–lipid interface (D'Rozario et al., 2009; Qiao et al., 2007). This is interesting because C_{60} fullerenes were found to affect the structural and elastic properties of membranes more vigorously when they were embedded in the bilayer core. The simulations reveal how specific chemical features of graphitic nanomaterials relate to their cytotoxic effects, and in so doing provide an approach for achieving material safety by design.

Bock, L.V., 1
2013. Er
(12), 139
Bond, P.J., S
Biol. 21
Brown, M.F
(1), 159–
Bulacu, M.,
196–205.
Chiu, S.W.,
simulati
2462–24(
Chng, C.P.,
grained
Chugunov,
dynamic
Cornell, W.1
J.W., Kol
molecule
Daily, M.D.
choleste
de Jong, D.F
topologi
Domański, J
model m
Dong, J., Zh
Macrom
D'Rozario, I
bilayer v
Fotiadis, D.
rhodopsi
Fuhrmans, M
curvatur
Goss, R., La
anthin sc
67–75.
Hedger, G.,
sights fr
j.bbamen
Hezaveh, S.,
copolym
(49), 143
Hille, B., 197
(2), 283.
Ingólfsson, F
Tieleman
14554–14
Ingólfsson, F
Sci. 129 (
Izvekov, S., Y
(7), 2469–
Jahns, P., Lat
proteins a

1.10 NANOPARTICLES

Much like their carbon-based counterparts, metallic, branched, and coated nanoparticles have been the subject of a series of simulation studies (Lin et al., 2011; Nangia and Sureshkumar, 2012; Dong and Zhou, 2013; Prates Ramalho et al., 2011; Song et al., 2011). For example, the interaction of functionalized gold nanoparticles with phospholipid membranes was investigated, showing that penetration, pore formation, and membrane wrapping depends sensitively on the surface charge of both the nanoparticle and the membrane model. When the number of cationic ligands on the nanoparticle was reduced, there was a significant reduction in membrane penetration, and the complete cessation of pore formation.

In a related study on the interaction of dendrimer nanoparticles with phospholipid membranes, it was shown that minor modifications of surface charge could profoundly influence the nanoparticle–membrane interaction (Lee and Larson, 2006, 2008). When initially near the bilayer surface, neutral dendrimers clustered and deformed membrane models, in contrast to their positively charged analogues. Other polymer based delivery systems that have been studied with the Martini model include the semiflexible polymer-drug constructs: poly(gamma-glutamyl-glutamate) paclitaxel (PGG-PTX) and ABA type triblock copolymers (Peng et al., 2010, 2011a, 2011b; Hezaveh et al., 2012). In each case, the simulations provide an unprecedented view of nanoparticle–membrane interactions that is unavailable with conventional experimental methods alone.

1.11 ON-GOING WORK

Molecular dynamics simulations are currently being used to explain biological phenomena whose mechanistic description had previously been elusive. One particularly interesting application is the use of atomistic and coarse-grain force fields to help describe the molecular level mechanisms that underpin the bactericidal effects of antimicrobial peptides. The simulations have already revealed how many of these broad-spectrum antibiotics disrupt bacterial membranes, and confer their antibacterial effects. The helical antimicrobial peptides have primarily been found to mediate their pernicious influence through the formation of transmembrane pores, while the antimicrobial cyclotides and lipopeptides tend to alter membrane properties such as membrane curvature, thickness, and lateral organization (Berglund et al., 2015; Nawae et al., 2014; Santo and Berkowitz, 2012; Rzepiela et al., 2010). Continued efforts in this area promise to provide the pharmaceutical insights needed for the design of novel antibiotics, and thus quell the problems of multiple drug resistance among infectious bacteria.

Another area of active research is the formation, organization, and energetics of lipid rafts. The plasma membrane contains a mix of proteins and lipids that spontaneously self-assembly into transient microdomains, which modulate membrane fluidity and regulate protein trafficking. But characterizing these lateral segregation events *in vivo* has proved challenging given that experimental methods are poorly suited to studying the growth of fluctuating nanoscale assemblies on microsecond timescales. Some progress has been made with coarse-grain molecular dynamics, which can provide the required spatio-temporal resolution for

for the nutrition of animals and humans. Two mechanisms were then proposed to explain the ability of several microorganisms to access oily substrates with very low aqueous solubility. The first mechanism relied on the significant potential of surface active agents of biological origin (*biosurfactants*) to modify the rheological behavior of oil–water interfaces, primarily by reducing the interfacial tension. In shearing flows, this modification promotes the dispersion and solubilization of the oily phase into the aqueous one (Reddy et al., 1982). The second mechanism was based on the direct microbial attack to the oil–water interface (Rosenberg and Rosenberg, 1981). At that time, the biosurfactant-based mechanism received widespread recognition as a generic ability of microbes and boosted research on biosurfactants (Georgiou et al., 1992; Zajic et al., 1983), while the interfacial growth mechanism was deemed applicable to certain microorganisms exhibiting specific traits for adhesion and *direct uptake* from oil surfaces. Thereafter, the capacity of microbes to alter the mobility and the concentration of oily phases set the basis for extensive research in the context of the microbially enhanced oil recovery from subsurface reservoirs (Armstrong and Wildenschild, 2012; Lazar et al., 2007; Banat, 1995), the bioremediation of soil contaminated with non-aqueous phase liquids (Johnsen et al., 2005; Volkering et al., 1998), the ex situ biodegradation of hydrophobic contaminants (Déziel et al., 1999), and the confinement of oil spills in marine ecosystems (Hassanshahian and Cappello, 2013; McGenity et al., 2012). Over the last two decades, non-destructive visualization experiments revealed several cases where microbial cells responded in a coordinated manner to the presence of a nutritive oily substrate and actively formed complex *biofilms* (i.e., microbial communities residing in an extracellular matrix of biopolymers) over the oil–water interface (Macedo et al., 2005). More recently, the massive Deepwater Horizon blowout attracted significant public concern and scientific interest about the fate of oil in deep sea releases (Atlas and Hazen, 2011), thus causing research on oil–microbe interactions to bloom once again.

The scope of this chapter is to discuss the current understanding of the fundamental microscale mechanisms that underlie the biodegradation of an oily substrate by microbial communities. The majority of available information comes from bench-scale experiments with diverse oil–microbe–water systems and varying experimental approaches. An attempt was made to filter the most pertinent information, to identify firm evidence (but also knowledge gaps), and ultimately to put forth theoretical considerations for modeling the process at the microscale.

In order to fix ideas with respect to the scale of interest, consider, for example, Fig. 2.1 that schematically presents the key physicochemical and biological processes whose interplay across multiple spatial and temporal levels determines the fate of crude oil after a natural or accidental release in the deep sea. The *microscale* is defined by an observation window in space with extent up to several millimeters and resolution down to a few micrometers (inset in Fig. 2.1). Accordingly, for the previously mentioned applications of microbially-mediated oil recovery and bioremediation in porous media, the microscale coincides with the pore-scale (e.g., Fig. 8.3 in Kapellos et al., 2015). Microscale experiments involve an elementary volume up to circa 1 ml, while bench-scale experiments typically involve volumes from 100 ml to several liters. Focusing at the microscale enables the analysis of physical interactions between individual microbes and solitary oil droplets and, thus, lays the foundations for improved understanding and quantification of oil biodegradation (Stocker, 2012). Furthermore, throughout the chapter, the generic term *oil* is used to denote a distinct non-aqueous

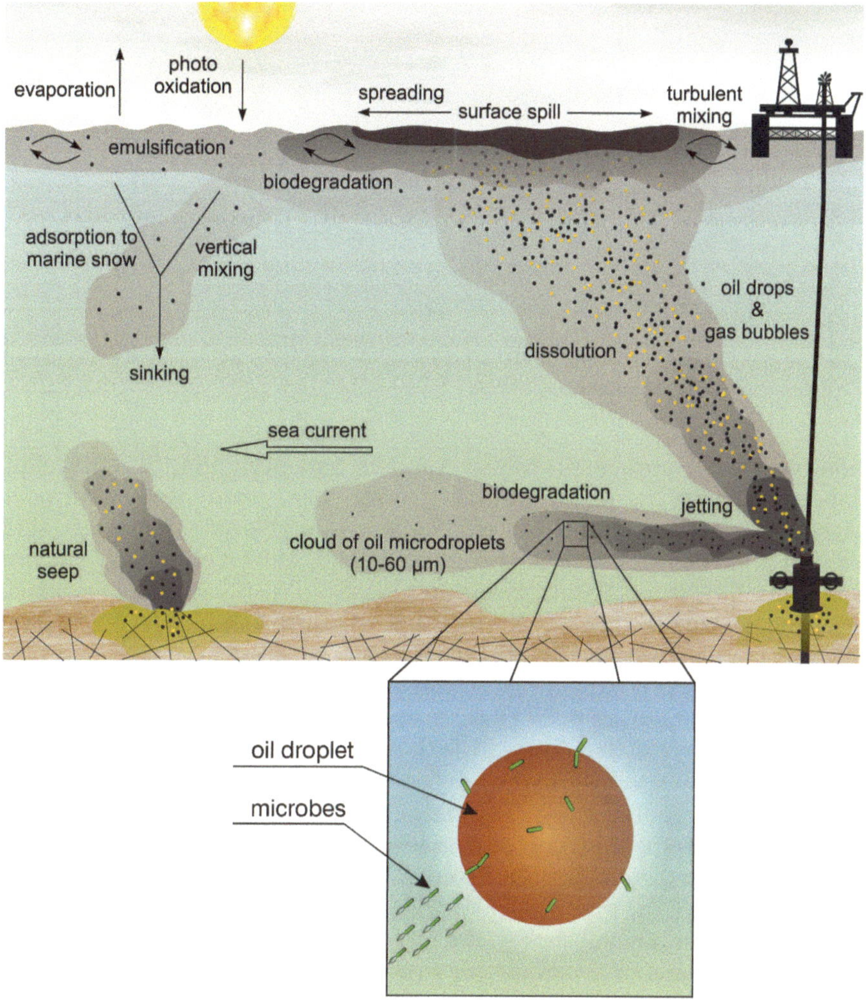

FIGURE 2.1 Schematic illustration of the spatial scales and key processes determining the fate of crude oil in subsea releases (adapted from Hazen et al., 2016).

phase which is formed by either a single compound or a mixture of organic compounds with extremely low affinity for water and the tendency to spontaneously coalesce when released in an aqueous solution. Besides the emphasis placed on crude oil so far, the considerations presented in this chapter may also apply to other natural organic oils extracted from plants, animals and even microorganisms as well as synthetic oils.

2.2 OVERVIEW OF THE BIODEGRADATION PROCESS

The microbial degradation of a chemical substrate involves three main steps: (i) physical contact between the substrate, or a substrate carrier particle, and the proper reception area

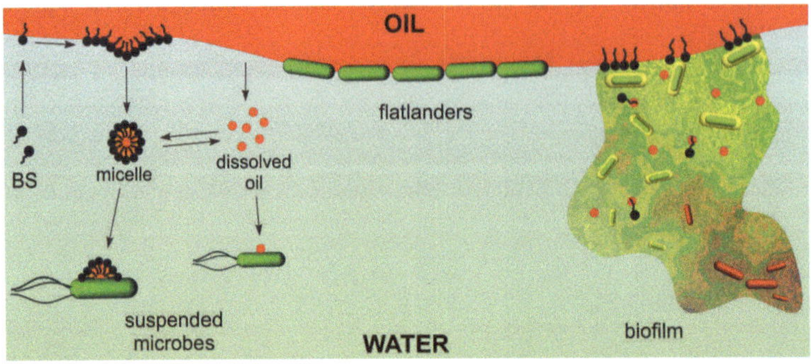

FIGURE 2.2 Schematic illustration of microbial tactics for substrate uptake and cell growth during oil biodegradation (BS = biosurfactant; see description in the text).

for ingestion on the cell surface; (ii) transmembrane transport; and (iii) intracellular enzymatic metabolism. In the literature, the first two steps are often jointly referred to as the *uptake mode* of a substrate by a biological cell. For water-soluble substrates (e.g., glucose), the physical contact is carried out with dissolved molecules. However, for oily substrates with low aqueous solubility, three distinct cases are usually considered (Bouchez-Naïtali et al., 1999; Déziel et al., 1999): (a) contact with organic molecules dissolved in the aqueous phase; (b) biosurfactant-mediated contact with oil containing micelles and emulsified oil droplets; and (c) direct contact with the surface of the (non-emulsified) oily phase (Fig. 2.2). Once physical contact is achieved, there are three possible routes for transferring the organic molecules to the intracellular space (van den Berg, 2010): (i) passive diffusion across the cell membrane; (ii) passive facilitated transport through membrane proteins; and (iii) active transport at the expenditure of cell energy. The enzymatic biotransformation of oil compounds takes place within the cell, although the first step of oxidation might also be performed by extracellular enzymes anchored to the cell surface. More information about pathways of hydrocarbon metabolism can be found, for instance, in Wentzel et al. (2007) and van Hamme et al. (2003). When microbes grow on oily substrates, the intracellular accumulation of lipophilic inclusions, which contain lipids or unaltered oil substrate, is often observed (Kim et al., 2002; Southam et al., 2001; Whyte et al., 1999; Alvarez et al., 1996). The interesting phenomenon of lipid storage is considered as a tactic developed for survival in oligotrophic environments, where the stored lipids are metabolized in later periods of famine.

In general, the detailed mechanisms (how it is done), the rate (how fast), and the extent (how much) of biodegradation of an oily substrate depend on a number of factors, including the genetic potential of microbes, the composition and physicochemical properties of the oily phase, the concentration of electron acceptors and mineral nutrients, as well as the hydrodynamic and other conditions (temperature, pH, salinity) in the aqueous phase (Fig. 2.3). Undoubtedly, the existence of microbes that are able to degrade specific oils is the primal (necessary and sufficient) condition for biodegradation to take place. Modern genetic analysis based on 16s rRNA gene sequences has led to the discovery of numerous microbial species that may use oil compounds as sole source of carbon and energy (Yakimov et al., 2007). Oil degrading species usually persist in small numbers in aquatic ecosystems, but rapidly dominate within the microbial community whenever a nutritive oily substrate becomes available

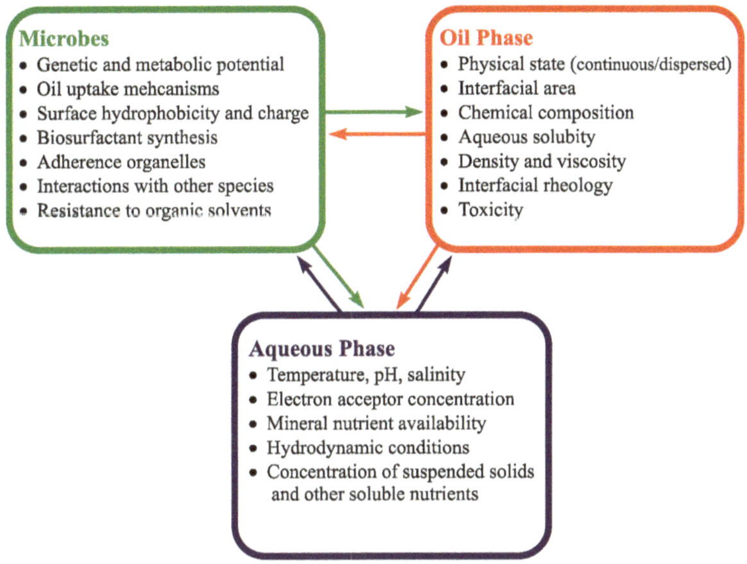

FIGURE 2.3 Key factors affecting oil biodegradation by microbes.

(Hazen et al., 2016; King et al., 2015; Head et al., 2006). Given that oil degrading microbes are available, the physical state of the oil, as a continuous or dispersed phase, is the next most important factor because it determines the interfacial area for oil dissolution and oil–microbe interactions. Among the other factors, the flow regime and the microbial synthesis of biosurfactants have a pronounced impact on the oil's physical state. The flow regime depends strongly on the specific environment, either it is a bioreactor (Déziel et al., 1999), a water column (Johansen et al., 2003), or porous sediments and soil (Joekar-Niasar and Hassanizadeh, 2012), and has received rather limited attention so far. Biosurfactants, either free or bound to the cell surface or in the form of a biofilm's extracellular matrix, represent the main microbial tactic for tackling the limitations posed by the low aqueous solubility of oil compounds. In the following subsections, basic aspects of biosurfactants and transmembrane transport are briefly introduced.

2.2.1 Biosurfactant-Mediated Uptake of Oil Compounds

Biosurfactant molecules consist of distinct *hydrophilic* (i.e., with high affinity for water) and *hydrophobic* (i.e., with low affinity for water) segments. The hydrophilic segments might be carbohydrates, polar peptides or acids, while the hydrophobic segments are usually lipids. Accordingly, biosurfactants are classified in glycolipids (e.g., rhamnolipids, trehalolipids, sophorolipids), lipopeptides and lipoproteins, phospholipids, protein–polysaccharide complexes, and membranous vesicles (Desai and Banat, 1997). Amphiphilic biomolecules of high molecular weight are often referred to as bioemulsifiers (Perfumo et al., 2010), but the term biosurfactant is used here for them as well. Biosurfactant molecules may be either secreted to the environment or anchored to the cell surface (Neu, 1996). In oil biodegradation, the primary role of biosurfactants is to increase the collision frequency between microbes and oil (Hua and Wang, 2014). The amphiphilic nature of biosurfactants drives them to partition at material interfaces, alter the interfacial composition and rheology, reduce the interfacial tensions, and increase the interfacial area (Zajic et al., 1983). For oil–water systems, biosurfactants *in synergy with flow* break down and disperse one phase into the other; this process is known as *emulsification*. Depending on the biosurfactant affinity for each phase (hydrophilic–lipophilic balance) as well as the volume and composition of each phase, the biosurfactants might disperse oil droplets in a continuous aqueous phase (oil-in-water emulsion) or vice versa (water-in-oil emulsion). Oil emulsification bears crucial importance for microbial biodegradation as it increases the oil–water interfacial area and the dissolution rate of oil compounds. For a given amount of oil, the average biodegradation rate increases as the size of emulsified oil droplets decreases (Brakstad et al., 2015).

If the biosurfactant concentration in the aqueous phase exceeds a threshold value, known as critical micelle concentration, then biosurfactant molecules spontaneously cluster together and form supramolecular aggregates with a hydrophobic core and a hydrophilic surface exposed to water. These nanometer-sized aggregates are known as micelles. The exact shape of a micelle depends on the relative size of the hydrophilic and hydrophobic segments of the biosurfactant (Langevin, 1992). Oil compounds might partition into the hydrophobic core of micelles, thus increasing their apparent concentration (molecular and micellar) into the aqueous phase; this process is known as *pseudo-solubilization*. There are two possible mechanisms for the formation of oil containing micelles (Fig. 2.2). In one mechanism, the creation of the micelle and the reversible absorption of oil compounds take place within the aqueous phase. This mechanism applies mainly to oil compounds with substantial aqueous solubility, such as polyaromatic hydrocarbons (PAHs: naphthalene, phenanthrene, etc.). In a second mechanism, the micelle is created at the oil–water interface and directly engulfs an oil patch (Perfumo et al., 2010). This mechanism works also for sparingly water-soluble compounds like *n*-alkanes. Oil containing micelles might affect biodegradation in two ways. First, the micelles might act as oil reservoirs by replenishing soluble oil compounds in the aqueous phase. Second, micelles might also serve as oil carriers by delivering their oily carriage to the microbial cell surface (Guha and Jaffé, 1996a, 1996b). The second effect depends on the specific microbe-surfactant compatibility and, in several cases, micellar oil might not be directly available to microbes (Li and Chen, 2009; Brown, 2007; Volkering et al., 1998). However, under favorable conditions, complete pseudo-solubilization of oil might result in faster biodegradation by one or two orders of magnitude (Miller and Bartha, 1989). Immense research has been devoted on biosurfactants, and interested readers can find more detailed information in excellent reviews from the literature (e.g., Perfumo et al., 2010; Ron and Rosenberg, 2001; Desai and Banat, 1997; Neu, 1996). Among the issues that remain obscure is the stability of micelles and their behavior close to interfaces.

2.2.2 Transmembrane Transport of Oil Compounds

In order for the intracellular enzymatic oxidation of organic oil compounds to take place, the hydrophobic molecules must somehow cross the cell envelope. Depending on the molecular structure of their envelope, microbes fall in two prime categories. Gram-positive microbes have a plasma membrane surrounded by a massive cell wall that consists mainly of peptidoglycan and lipoteichoic acid, with the latter being responsible for the degree of hydrophobicity of the cell surface (Miörner et al., 1983). Some Gram-positive microbes, such as *Mycobacterium* and *Rhodococcus* species, also have mycolic acids that render the cell surface superhydrophobic (Bendinger et al., 1993). The cell envelope of Gram-positive cells is not considered to pose any serious hindrance to passive transport of hydrophobic molecules across it. On the other hand, Gram-negative microbes have a second outer membrane whose external layer contains mainly lipopolysaccharides, instead of the regular phospholipids. Hydrophilic oligosaccharides, extending from the outer membrane surface, hinder the spontaneous passage of hydrophobic molecules which must, therefore, enter through some protein-mediated transport pathway. In the last few years, van den Berg and coworkers discovered a family of membrane proteins that facilitate hydrophobic molecules in bypassing the lipopolysaccharide barrier of the outer membrane, without the use of cellular energy (Lepore et al., 2011; van den Berg, 2010; Hearn et al., 2009). These proteins, exemplified by the long chain fatty acid – *FadL* – transporter in *Escherichia coli*, form a narrow hydrophobic channel which is normally plugged but opens up once the substrate binds at a specific receptor site within the channel. Nonetheless, several studies have presented evidence that transmembrane transport of hydrocarbons is, at least partially, energy-dependent (Hua and Wang, 2012; Noordman and Janssen, 2002; Kim et al., 2002). It has also been suggested that microbes might uptake entire oil nanodroplets through pinocytosis (Cameotra and Singh, 2009). These complicated membrane transport mechanisms merit further research.

2.3 MICROBIAL GROWTH MODES ON OILY SUBSTRATES

A microbial strategy for utilizing oily substrates depends on the microbe's tactics for substrate uptake and cell growth. As mentioned in the previous section, the uptake mode relates to the microbe's preference for the physical state of the oil compound, which might be dissolved in the aqueous phase, complexed with surfactant, or in the oily phase. In this section, the possible modes of microbial growth (i.e., where and how microbes grow) in a two phase oil–water system are examined. Depending on a microbe's *affinity* for the oily phase and its ability to proliferate in multicellular structures, such as biofilms and flocs, three growth modes can be distinguished:

(a) suspended growth either as single cells or flocs in the bulk aqueous phase,
(b) interfacial growth *at* the oil–water interface, and
(c) biofilm formation *over* the oil–water interface.

The affinity of a microbe for an oily phase is ideally quantified by the three-phase contact angle between the microbe and the oil–water interface. However, the determination of this contact angle is exceptionally challenging because of the small size of microbes and the existence of exocellular structures like porous capsule, fimbriae and flagella (Abbasnezhad et al., 2011). Experimental methods that indirectly quantify a microbe's affinity for an oily phase include the microbial adhesion to hydrocarbons (MATH) test, the contact angle between an oil droplet and a lawn of cells, and the mapping of cell surface hydrophobicity using atomic force microscopy (Rosenberg, 2013; Dorobantu et al., 2008).

Before entering into a growth mode, the microbes must first somehow detect the immiscible oily substrate in the physical system. For most non-motile microbes that perform their Brownian walk or are convected by fluid flow, the oil discovery is expected to occur after a random encounter with an oil–water interface. On the other hand, motile cells are able to locate a surface, or material interface, by design. The hydrodynamic interactions between a surface and an approaching motile cell strongly affect the cell's swimming behavior (Lopez and Lauga, 2014; Molaei et al., 2014; Lauga et al., 2006), mainly its orientation, and might alarm the cell so as to explore for a potentially nutritive surface. As the microbe moves closer to the surface, molecular interactions also come into play. These physicochemical interactions depend strongly on the charge and hydrophobicity of the cell surface, the charge and hydrophobicity of the oily surface, as well as the pH and ionic strength of the aqueous phase and can be quantified by the extended DLVO theory that includes the van der Waals, electrostatic and Lewis acid/base forces (Hermansson, 1999).

Another important microbial trait for survival in oligotrophic aquatic ecosystems is the *chemotactic* motility, that is, the ability of certain motile microbes to sense and move along chemical gradients. Chemotactic cells might perceive microscale variations (i.e., imperceptible at coarser spatial scales) in the concentration of dissolved oil compounds, or even metabolic products by other cells already feeding on the oil–water interface, and set their swimming direction towards the oily substrate (Parales and Ditty, 2010; Stocker et al., 2008; Pandey and Jain, 2002). For example, the chemotactic species *Pseudomonas putida* F1 has shown significantly higher accumulation in the vicinity of toluene droplets as compared to other non-chemotactic species (Wang et al., 2016).

The initial contact between microbes and the oily surface is of paramount importance, as it is considered to trigger the expression of microbial functions that facilitate oil uptake (van Hamme et al., 2003; Neu, 1996). In particular, two major adaptations have been widely reported: (a) the onset of biosurfactant synthesis and (b) the increase of cell surface hydrophobicity. The latter phenomenon may be attributed to: (i) active modulation of the cell envelope composition (Whyte et al., 1999); (ii) biosurfactants either masking or removing hydrophilic biopolymers from the cell surface (Al-Tahhan et al., 2000); or (iii) incorporation of oil nanodroplets in fibrillar structures of the cell surface (Baldi et al., 1999). The time required for these adaptations to take place on the population level is usually observed as a lag phase in the microbial growth curve (Kulkarni and Barnett, 1986). Nonetheless, the biological mechanisms responsible for recognizing the nature of the interface and triggering these adaptations remain unknown. One possibility is that a chemosensory system, analogous to those used in chemotaxis, could detect chemical compounds in the oil side of the interface. Another possibility, on a purely speculative basis, could be a mechanosensory system detecting one or more of the unique rheological characteristics of a liquid–liquid interface, such as interfacial currents and strain, tension at the oil–microbe interface or along the oil–microbe–water common line, buoyancy, and so on. Upon adaptation to feeding on an oily substrate, each microbe may enter one of the three basic growth modes analyzed in the following subsections.

2.3.1 Suspended Growth in the Aqueous Phase

In a first strategy, microbes grow either as individual planktonic cells or within flocs suspended in the bulk aqueous phase uptaking dissolved and micellar oil compounds. This strategy is suitable for cells with low affinity for the oily phase or cells experiencing inhibitory effects by the concentrated oil compounds. The oil dissolution rate is the limiting factor for microbial growth and oil biodegradation. Considering also that relatively few oil compounds (e.g., PAHs with low molecular weight) have substantial aqueous solubility, the synthesis of biosurfactants is of crucial importance for this growth mode. As the biosurfactants are released in the aqueous phase, some depletion is inevitable in *open* systems (i.e., exchanging mass with their surroundings). Therefore, this strategy would be efficient either in closed systems (e.g., batch bioreactors), or in physical systems characterized by slow flow rates, high retention times and, at best, immobilized oily phase. Such conditions are typically met in soil and marine sediments. In the sea or freshwater streams, severe depletion of biosurfactants by fluid flow would render this strategy rather ineffective for individual planktonic cells; only by assembling in flocs, could the microbes benefit in two important respects: (a) mitigate biosurfactant depletion and (b) captivate and emulsify oil droplets.

In line with the above analysis, the suspended growth mode has been observed in batch cultures of several Gram-negative hydrophilic *Pseudomonas* species originating from soil (see Table 2.1). The supernatant liquid from those cultures usually contained rhamnolipids and reduced at about half the air–water surface tension and the oil–water interfacial tension. *Exogenous* addition of surfactants above the critical micelle concentration is often beneficial to oil biodegradation by suspended microbes, as it increases both the dissolution rate and the apparent aqueous concentration of oil compounds. For example, the non-ionic surfactant Igepal CO-630 was found to enhance the extent of crude oil biodegradation by *Pseudomonas* JA-B45 (van Hamme and Ward, 2001), and hexadecane trapped in rhamnolipid micelles was directly available to *Pseudomonas aeruginosa* UG2 (Noordman and Janssen, 2002).

A couple of final remarks are in order. First and foremost, caution should be exercised in the extrapolation of laboratory observations to represent the behavior of microbes in their natural habitat. Soil microbes might grow in planktonic form in batch cultures while growing in biofilms within the confined pore space of porous media (Kapellos et al., 2015). Second, cell-free biosurfactants represent a great strategy for solubilizing oil compounds sorbed in the solid matrix of soil and sediments (Volkering et al., 1998), albeit an amount of biosurfactants is inevitably wasted by partitioning also to non-nutritive pore surfaces. In this regard, microbes from those ecosystems might be better candidates for the commercial production of biosurfactants. As an indicative example, marine microbes isolated from oil polluted seawater in Greece produced about 50 mg/l of rhamnolipids (Antoniou et al., 2015), while soil microbes isolated from an oil field in China produced about 500 mg/l of rhamnolipids (Hua and Wang, 2012). Of course, this trend should be taken with a grain of salt until confirmed from systematic research.

2.3.2 Flatlander's (Interfacial) Growth at the Oil–Water Interface

In a second strategy, microbes with very high affinity for the oily phase firmly adhere and grow at the oil–water interface. Depending on the degree of hydrophobicity and the existence

TABLE 2.1 Selected case studies of microbial strategies for oil biodegradation, using the oily substrate as sole source of carbon and energy. The species of *Acinetobacter*, *Marinobacter*, and *Pseudomonas* are Gram-negative, whereas *Arthrobacter*, *Mycobacterium* and *Rhodococcus* are Gram-positive. Abbreviations: BS = Biosurfactant; P. = Pseudomonas; IFT = interfacial tension (oil–water); ST = surface tension (air–water); CO = crude oil; EO = engine oil; HO = heating oil; DF = diesel fuel; ALK = alkanes; HEX = hexadecane; OCT = octadecane; PAH = polycyclic aromatic hydrocarbons; PCB = polychlorinated biphenyl

Study	Microbe	Oil	Observations
Strategy-I: ***Suspended growth mode***			
van Hamme and Ward (2001)	*Pseudomonas* JA-B45	CO	no BS synthesis; Igepal CO-630 enhanced degradation; formed biofilm in coculture; isolated from soil
Noordman and Janssen (2002)	*P. aeruginosa* UG2	HEX	cell-free BS: rhamnolipids; active uptake; rhamnolipids enhanced degradation; isolated from soil
Bouchez-Naïtali and Vandecasteele (2008)	*P. aeruginosa* GL1	HEX	cell-free BS: rhamnolipids; remained hydrophilic; reduced IFT; reduced ST (<37 mN/m); enhanced growth with BS addition; isolated from soil
Cameotra and Singh (2009)	*P. aeruginosa* SSC2	HEX	cell-free BS: rhamnolipids; reduced IFT; reduced ST: 65 → 30 mN/m; pinocytosis of micelles; intracellular oily inclusions; isolated from soil
Hua and Wang (2012, 2013)	*Pseudomonas* DG-17	OCT, CO	cell-free BS: rhamnolipids; strong emulsification ability of supernatant; reduced ST: 72 → 33 mN/m; active uptake; isolated from soil
Strategy-II: ***Flatlander's interfacial growth mode***			
Rosenberg et al. (1982)	*Acinetobacter* RAG-1	HEX	cell-bound BS: emulsan; eat 'n' go behavior; fimbriae-mediated adhesion; detachment via capsule release
Efroymson and Alexander (1991)	*Arthrobacter* sp. R1	HEX, PAH	no BS in the supernatant; Triton-X completely inhibited adhesion to oil and HEX degradation
Baldi et al. (1999)	*Acinetobacter* RAG-1	DF	cell-bound BS; always hydrophobic; emulsified oil
Whyte et al. (1999)	*Rhodococcus* Q15	DF, HEX	cell-bound BS; cells reduced ST: 70 → 36 mN/m; substrate-induced membrane alteration; cell flocculation
van Hamme and Ward (2001)	*Rhodococcus* F9-D79	CO	cell-bound BS; reduced ST; emulsified oil; Igepal CO-630 inhibited adhesion to oil; mycolic acid capsule
MacLeod and Daugulis (2005)	*Mycobacterium* PYR-1	PAH	super-hydrophobic; cells located in the oil side of the interface; partition in the oil phase even in the presence of water-soluble nutrients
Bouchez-Naïtali and Vandecasteele (2008)	*Rhodococcus equi* Ou2	HEX	cell-bound BS; substrate-enhanced hydrophobicity; BS release at a late stage of growth; ST reduction: 66 → 27 mN/m

continued on next page

TABLE 2.1 (*continued*)

Study	Microbe	Oil	Observations
Strategy-III: ***Biofilm growth mode***			
Marin et al. (1996)	*Acinetobacter* MM5	CO, HO	substrate-dependent BS synthesis; strong emulsification ability of the supernatant liquid
Baldi et al. (1999)	*Acinetobacter* VE-C3	DF	substrate-induced hydrophobicity; cell flocculation
Southam et al. (2001)	soil consortium	EO	BS-covered droplets; oil nanodroplets; intracellular oily inclusions
Wick et al. (2002)	*Mycobacterium* LB501T	PAH	no BS in the supernatant; crater formation on solid anthracene; substrate modulation of cell surface charge and hydrophobicity
Deppe et al. (2005)	marine consortium	CO	the consortium contained uncultivable cells and had greater biodegradation potential than individual species
Rodrigues et al. (2005)	*P. putida* ATCC 17514	PAH	substrate-modulation of cell surface charge and extracellular matrix
Macedo et al. (2005)	soil consortium	PCB	three stages of biofilm formation (see text); oil depressions
Klein et al. (2008, 2010)	*Marinobacter* SP17	ALK	cell-bound or interface-bound BS; IFT reduction: $\sim 38 \rightarrow 18$ mN/m for HEX; IFT reduction is related to active protein synthesis; intracellular oily inclusions
Dmitriev et al. (2011)	*Rhodococcus* S67, *P.* putida BS3701	CO	oil sorption to the extracellular matrix; extracellular oxidative enzymes; BS-covered oil droplets

of fimbriae on the cell surface (Rosenberg, 2013; Dorobantu et al., 2008), the cell positioning might vary from floating (similarly to water striders) to being fully submerged in the oily phase. Microbes whose overall existence (growth, proliferation, spreading, etc.) is confined on the two-dimensional space of an oil–water interface are here referred to as *flatlanders* (inspired by Abbott, 1992) in distinction to planktonic (suspended in a bulk phase), sessile (immobile, surface attached), and biofilm (matrix enmeshed) cells. Flatlanders may form *thin* films with thickness of a few cell diameters at the oil–water interface; with their direction of growth being always *tangential* to the interface. They may uptake dissolved and micellar oil but, because of their privileged position at the interface, they are expected to favor sipping oil compounds directly from the oily phase.

The oil–water interfacial area is the limiting factor for microbial growth and oil degradation in this second strategy. Typically, flatlanders do not release biosurfactants into the bulk aqueous phase and, thus, the supernatant liquid of their cultures does not reduce the surface or interfacial tensions. Oil emulsification and, consequently, increase of the interfacial area is usually achieved by cell-anchored biosurfactants (van Hamme and Ward, 2001; Whyte et al., 1999) and the specific partitioning of cell bodies at the oil–water interface (Dorobantu et al., 2004). Exogenous supply of surfactants is often detrimental for oil biodegradation by flatlanders because the surfactants alter the hydrophobicity of the oil surface. For example, in the study of Efroymson and Alexander (1991), the flatlander *Arthrobacter* sp. R1 was able to uptake hexadecane dissolved in non-biodegradable organic solvents directly from the oil–water interface. Hexadecane has an exiguous molecular solubility of about 1 ng/ml

TABLE 2.2 Recent studies on the rheological behavior of an oil–water interface with adhered microbes. Abbreviations for the Method: Capillary Pressure Tensiometry with micropipette (μCPT), Drop Profile Tensiometry (DPT), Disc Shear Viscometry (DSV), Particle Tracking (PT); for the Microbe: A. = *Acinetobacter*, R. = *Rhodococcus*, M. = *Marinobacter*, P. = *Pseudomonas*; and for the Oil: HEX = hexadecane, HMN = heptamethylnonane, MO = mineral oil, MCT = medium chain triglycerides. The "Age" of the interface refers to the time since the oily phase and the microbial suspension were brought to contact. The interfacial (IF) parameters are the initial interfacial tension (γ_0), the apparent interfacial tension (γ_{app}) of the aged interface, the interfacial *dilatational* elasticity (E') and *shear* elasticity (G') of the aged interface. (Unfortunately no information is provided in these studies about the thickness of the interfacial microbial film.)

					IF rheological parameters [mN/m]			
Study	**Method**	**Age**	**Microbe**	**Oil**	γ_0	γ_{app}	E'	G'
Kang et al. (2008)	μCPT	30 min	*A.* RAG-1	HEX	52	46	3.3	–
			R. 20S-E1-c	HEX	52	43	5.6	–
Klein et al. (2010)	DPT	250 min	*M.* SP17	HEX	38	18	45	–
				HMN	42	18	30	–
Rühs et al. (2014)	DSV/DPT	16 h	*P.* KT2442	MO	50	14	–	30
				MCT	26	10	–	10
Vaccari et al. (2015)	PT/DPT	22 h	*P.* P62	HEX	30	21	21	<< 1

in water. The addition of the non-ionic surfactant Triton-X profoundly increased the apparent aqueous concentration of hexadecane up to 700 ng/ml (presumably trapped within micelles), but also completely inhibited both the attachment of *Arthrobacter* to the oil–water interface and the biodegradation of hexadecane. In another recent study (Rahsepar et al., 2016), the surfactant Corexit EC9500A, which was used in the Deepwater Horizon spill, was found to inhibit degradation of crude oil by *Rhodococcus* sp. TUHH-12 (most probably a flatlander). In that case, the inhibition was related to the initial swift dissolution of toxic aromatics but prevention of adherence to the oil surface might have also played some role.

Flatlanders transform a simple nanometer-sized dividing surface between oil and water to a complex micrometer-sized transition zone. The mere presence of a layer of adhered microbes over oil droplets has been shown to prevent droplet coalescence in several occasions (Kang et al., 2008; Dorobantu et al., 2004). The stabilization of an emulsion by particles, either inanimate or living, is usually referred to as Pickering effect in the literature and is attributed to the steric hindrance brought about by the particles that cover the droplets (Binks, 2002). Although the size of most microbes is not sufficiently small to induce reduction of the interfacial tension by cell bodies alone, all microbes synthesize some biomolecules of amphiphilic nature (even for reasons not directly related to oil biodegradation) which, in turn, reduce the interfacial tension over time. Maturation of the interfacial microbial layer with cell–cell linkage via exocellular biopolymers, also leads to more complicated viscoelastic or elastic rheological behavior (Table 2.2).

The flatlander's growth mode has been observed, for example, in batch cultures of super-hydrophobic Gram-positive *Arthrobacter*, *Mycobacterium* and *Rhodococcus* species, as well as for the Gram-negative *Acinetobacter* RAG-1 cells (Table 2.1). Of particular interest is the extraordinary case of *Acinetobacter* RAG-1 which has developed an *eat 'n' go* strategy

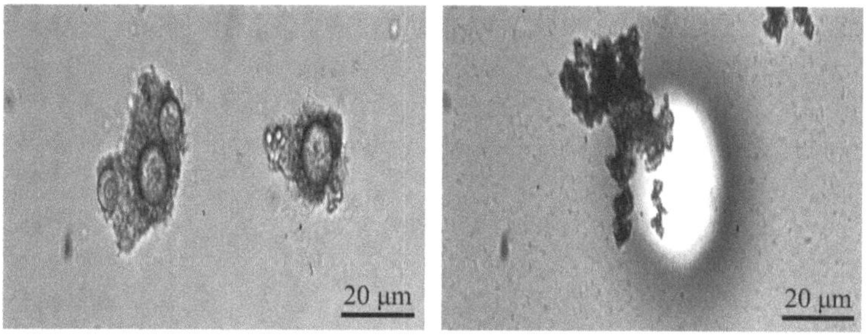

FIGURE 2.4 Micrograph of *Marinobacter hydrocarbonoclasticus* biofilms over crude oil droplets.

(Rosenberg, 2013). This microbe uses its fimbriae to adhere to a nutritive oil surface and the cell-bound biosurfactant *emulsan* to disperse the oily phase and increase the interfacial area. As soon as the biodegradable compounds within an oil droplet are consumed, the microbe detaches from the interface by dropping off its emulsan capsule (similarly to the tail autotomy in lizards). The remaining capsule layer has its hydrophilic side exposed to water and thus prevents adherence of other cells to the nutrient-depleted droplet.

2.3.3 Biofilm Growth Over the Oil–Water Interface

In a third strategy, microbes form complex biofilms over the oily surface (Fig. 2.4). Biofilm formation is an active process (O'Toole et al., 2000; Costerton et al., 1995). It occurs when microbes adhere to a material interface and secrete excessive amounts of biopolymers with high molecular weight. The biopolymers, mainly polysaccharides and proteins, do not dissolve into the bulk aqueous phase, but instead accumulate in the extracellular space and spontaneously assemble to form a matrix enmeshing the cells. Besides providing the necessary mechanical support for building a three-dimensional composite biomaterial, the extracellular matrix might also serve other important functions, such as retaining moisture and nutrients as well as protecting from predation and antimicrobials. The extracellular matrix is also the chief difference between the biofilm and flatlander's growth modes. The lack of an extensive matrix restricts flatlander's growth only along directions *tangential* to the interface, whereas biofilm growth occurs along the *normal* direction as well. At the microscale, a flatlander's film can be modeled as a reactive interface, while a biofilm will constitute a distinct phase.

In spite of the significant progress achieved over the last two decades in the microscale analysis of biofilm formation at water–solid interfaces (see, for instance, the review in Kapellos et al., 2015), the corresponding understanding for slippery and nutritive oil–water interfaces remains largely unclear. Using laser scanning microscopy and live cell staining methods, Macedo et al. (2005) visualized the dynamics of biofilm formation over an oil droplet (polychlorinated biphenyl). The oil droplet hanged from a hydrophobic solid surface, while being immersed in an aqueous solution. They observed three stages of biofilm development: (i) initial accumulation of the microbes at the oil–water–solid common line; (ii) spreading of microbes over the oil surface; and (iii) invasion of microbes into depressions formed in the oily phase. The formation of depressions might be related to hot spots of microbial activity very close to the oil surface that result in locally accelerated oil dissolution rate (Abbasnezhad et al., 2011). Furthermore, the prevention of coalescence of the remaining oily phase may be attributed to the extracellular polymeric matrix and implies that the oil–biofilm tension is probably low.

In several cases, high resolution electron microscopy has revealed the presence of a thin layer (25–50 nm) of biosurfactants at the oil–biofilm interface, which mediates any physical interactions between biofilm cells and the oily substrate (Dmitriev et al., 2011; Southam et al., 2001). The possible lack of direct contact between biofilm cells and the oil surface is another essential difference with the flatlander's growth mode. Another intriguing aspect of biofilm formation over oily substrates concerns the transport of hydrophobic oil compounds to biofilm cells located far away from the interface. Current knowledge on this issue is rather obscure. Two plausible mechanisms have been postulated. One mechanism relies on the formation of surfactant-covered ultrafine droplets at the oil–biofilm interface and, thereafter, the micellar transport of those droplets into the biofilm (Southam et al., 2001). Another mechanism is based on enhanced solubility of the oil compounds in the biofilm. For example, the solubility of the resin α-pinene was found to be two orders of magnitude higher in biofilms than in water (Miller and Grant Allen, 2004), and the solubility of PAHs was up to ten times higher in aqueous solutions of sphingans (components of the extracellular matrix produced by *Sphingomonas*) than in water (Johnsen and Karlson, 2004). The increased solubility of oil compounds in biofilms may be caused by sorption to hydrophobic moieties of the extracellular matrix or, perhaps, by capillary suction into the interstitial pore space of biofilms.

The biofilm growth mode over oily substrates has been reported for several pure cultures and mixed microbial consortia (Table 2.1). Microbes that independently employ the suspended or the flatlander's growth modes have been found to form biofilms when *in consortium*. For example, van Hamme and Ward (2001) observed that in pure cultures with crude oil as sole carbon source, *Pseudomonas* JA-B45 cells grew in the bulk aqueous phase while *Rhodococcus* F9-D79 cells adhered to the oil–water interface. However, in mixed culture, *Pseudomonas* cells formed a biofilm over the *Rhodococcus* cell layer. A final subtle point is with regard to the intrinsic relationship between flocs and biofilms. For example, Baldi et al. (1999) observed that *Acinetobacter* VE-C3 cells formed small aggregates in the aqueous phase which then adhered to large oil droplets and developed into complex oil–biofilm formations. This behavior has been recently proposed as a more general microbial tactic with small flocs representing an effective means of 'group traveling' for non-motile cells in aquatic ecosystems and contributing to the inoculation of interfaces for subsequent biofilm formation (Kragh et al., 2016; Valentine et al., 2012). Nonetheless, for the complex oil–microbe formations, which are routinely observed in batch cultures with oily substrates, it remains difficult to tell whether cells and small emulsified droplets passively flocculate in the bulk phase or the cells actively form biofilms on large droplets that break up within the biofilm over time. Probably, both processes occur simultaneously. The best way to reach sound conclusions about the relative importance of each process is through visualization experiments in combination with microscale simulations.

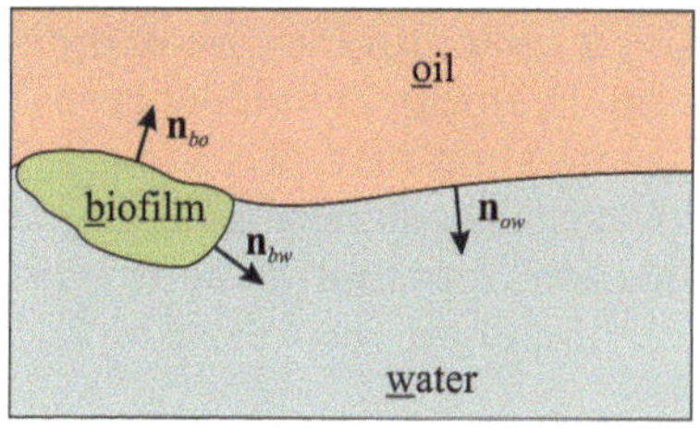

FIGURE 2.5 Schematic of an oil–biofilm–water system at the microscale.

2.4 MICROSCALE MODELING CONSIDERATIONS

Theoretical modeling and simulation of microbial growth over oily substrates, at the microscale, currently goes through an embryonic stage. With reference to Fig. 2.5, a minimal continuum-based formulation is presented here for the mathematical description of physical interactions between biofilms (b-phase), oil (o-phase), and an aqueous solution (w-phase). The mass and momentum balances for each one of the three continua can be expressed as follows:

$$\frac{\partial \rho_\alpha}{\partial t} + \nabla \cdot (\rho_\alpha \mathbf{v}_\alpha) = q_\alpha, \tag{2.1}$$

$$\frac{\partial \rho_\alpha \mathbf{v}_\alpha}{\partial t} + \nabla \cdot (\rho_\alpha \mathbf{v}_\alpha \mathbf{v}_\alpha) = \nabla \cdot \boldsymbol{\sigma}_\alpha + \mathbf{f}_\alpha \tag{2.2}$$

with $\alpha = w, o, b$. Here, ρ_α is the density of the αth phase, $\mathbf{v}_\alpha$ is the velocity of the αth phase, q_α is the rate of mass production or consumption per unit volume of the αth phase, $\boldsymbol{\sigma}_\alpha$ is the Cauchy stress tensor for the αth phase, and $\mathbf{f}_\alpha$ represents body forces (e.g., gravity) for the αth phase. Assuming *simple* massless interfaces (i.e., the interfacial tension being their only rheological feature), the mass and momentum balances at each one of the oil–water (o/w), water–biofilm (w/b), and biofilm–oil (b/o) interfaces can be expressed as follows (Edwards et al., 1991):

$$[\![\rho(\mathbf{v} - \mathbf{v}_{\alpha\beta})]\!]_\alpha^\beta \cdot \mathbf{n}_{\alpha\beta} = 0, \tag{2.3}$$

$$[\![\rho\mathbf{v}(\mathbf{v} - \mathbf{v}_{\alpha\beta}) - \boldsymbol{\sigma}]\!]_\alpha^\beta \cdot \mathbf{n}_{\alpha\beta} = 2\mathcal{H}_{\alpha\beta}\gamma_{\alpha\beta}\mathbf{n}_{\alpha\beta} + \nabla_S \gamma_{\alpha\beta} \tag{2.4}$$

with $\alpha, \beta = w, o, b$ and $\alpha \neq \beta$. Here, $\mathbf{n}_{\alpha\beta}$ is the unit normal vector on the $\alpha\beta$-interface pointing from the αth phase toward the βth phase, $\mathbf{v}_{\alpha\beta}$ is the interfacial velocity, $\gamma_{\alpha\beta}$ is the interfacial tension at the $\alpha\beta$-interface, $\mathcal{H}_{\alpha\beta}$ is the mean surface curvature of the $\alpha\beta$-interface, and ∇_S is the surface gradient. The surface mean curvature and gradient are defined as follows:

$$\mathcal{H}_{\alpha\beta} = -\frac{1}{2}\nabla_S \cdot \mathbf{n}_{\alpha\beta}, \qquad \nabla_S = \mathbf{I}_S \cdot \nabla \tag{2.5}$$

where $\mathbf{I}_S = \mathbf{I} - \mathbf{n}_{\alpha\beta}\mathbf{n}_{\alpha\beta}$ is the surface projection tensor, and $\mathbf{I}$ is the second-order unit tensor. The double brackets in Eqs. (2.3) and (2.4) denote the difference in the bracketed quantity between the two phases, e.g., $[\![\boldsymbol{\sigma}]\!]_{\alpha}^{\beta} = \boldsymbol{\sigma}_{\beta} - \boldsymbol{\sigma}_{\alpha}$. In the case of immiscible phases, with negligible interphase mass transfer, the normal velocity components are set equal: $\mathbf{v}_{\alpha} \cdot \mathbf{n}_{\alpha\beta} = \mathbf{v}_{\beta} \cdot \mathbf{n}_{\alpha\beta} = \mathbf{v}_{\alpha\beta} \cdot \mathbf{n}_{\alpha\beta}$, and Eqs. (2.3) and (2.4) are simplified accordingly. To keep the analysis tractable, the effects of line tension and its variation along the oil–biofilm–water contact line may not be considered in a first approximation. Transport of at least one nutritive oil compound that dissolves into the aqueous and biofilm phases must also be included in the formulation. Considering dilute solutions, we have:

$$\frac{\partial C}{\partial t} + \nabla \cdot (\mathbf{v}_{\alpha} C) - \nabla \cdot [D_{\alpha} \nabla C] + R_{\alpha} = 0 \tag{2.6}$$

with $\alpha = w, b$. Here, C is the concentration of the oil compound (molecular and micellar), D_{α} is the diffusion coefficient in the αth phase, and R_{α} is the reaction rate in the αth phase. At the oil–biofilm and oil–water interfaces, equilibrium partition may be considered.

In order to complete the formulation, constitutive relations for the stress tensors and the reaction rate must be defined. To this end, it is reasonable to consider that both the oily and the aqueous phases behave as Newtonian fluids. On the other hand, the rheological behavior of the biofilm is more complicated and might admit either a monophasic or multiphasic description (Kapellos et al., 2007, 2010, 2015; Kapellos and Alexiou, 2013). The absorption of water during volumetric biofilm growth is usually ignored in open systems, but must be modeled explicitly in closed systems as, for example, in the case of biofilm formation within an aqueous droplet floating on oil (Asadi et al., 2015). For poorly water-soluble oil compounds, the reaction rate is usually taken to be linear, but can also be modeled with Monod or Haldane kinetics. Reliable kinetic data for several compounds of crude oil have been recently reported in the literature (e.g., Nikolopoulou et al., 2013).

The parameters that raise some concern are the interfacial tensions at the oil–biofilm and water–biofilm interfaces, as there are no available data in the literature, yet. At the oil–biofilm interface, a reasonable hypothesis would be to assign a value in the range of data for the reduced interfacial tension by culture supernatants, or lower. At the water–biofilm interface, it is usually assumed that the interfacial tension is nil because biofilm contains a large amount of extracellular water. However, it has been recently shown that *Bacillus subtilis* biofilms are coated by a hydrophobic protein layer (Stanley-Wall and MacPhee, 2015), which could possibly give rise to tension with aqueous solutions. One step further, as most biofilms are heterogeneous by nature, spatial variation in the concentration or chemical type of the amphiphilic components of the extracellular matrix might result in interfacial tension gradient at the oil–biofilm interface and concomitantly induce flow within the adjacent bulk phases (Marangoni effect).

2.5 SUMMARY AND OUTLOOK

The strategies for microbial growth on nutritive oily substrates have been concisely reviewed in this chapter. Depending on a microbe's affinity for the oily phase and its ability to proliferate in multicellular structures, three fundamental growth modes have been identified: (a) suspended growth; (b) interfacial (flatlander's) growth; and (c) biofilm growth. Each growth mode might involve one or more of the three major modes for uptaking hydrophobic compounds: (i) uptake of dissolved compounds in the aqueous phase; (ii) uptake of compounds complexed with surfactants in micelles or emulsified fine droplets; and (iii) direct uptake from the oily phase.

In the literature, the emphasis on which microbial strategy might prevail has shifted over time from surfactant-supported suspended growth (including floc formation) in the 1970s, to direct interfacial uptake in the 1990s, and to biofilms (and flocs) at present. As a rule of thumb, Gram-negative microbes, like many *Pseudomonas* species, that produce biosurfactants of low molecular weight and have hydrophilic cell surface are prone to suspended growth in the bulk aqueous phase. However, in soil or sediments, those microbes are expected to form biofilms as well, but on the solid–water interface rather than the oil–water interface. On the other end, Gram-positive microbes, like *Mycobacterium* and *Rhodococcus* species, equipped with biosurfactants anchored to the hydrophobic cell surface are ideal candidates for flatlander's growth on the oil–water interface. Yet, in consortium with other species that are able to produce biopolymers of high molecular weight, those are also found to participate in biofilms. All in all, the biofilm growth mode appears to be the dominant strategy for oil biodegradation in nature by autochthonous microbial consortia.

The interactions between individual microbes and micelles, fine oil droplets, and large oil–water interfaces, as well as the microstructure–rheology correlation for flatlander's films, the transport of oil compounds in thick biofilms, and the impact of fluid flow on all these processes are among the topics that deserve further research. The next leap in our understanding of the complicated oil–microbe interactions that underpin oil biodegradation is expected to come from microscale experiments hand in hand with computer-aided simulations.

Acknowledgements

I am grateful to the Institute of Chemical Engineering Sciences (FORTH/ICE-HT) for providing financial and other support. This research was partially funded by the European Union (European Social Fund-ESF) and Greek National Funds through the Operational program "Education and Lifelong Learning" under the action Aristeia II (Code No. 4420). Many thanks are expressed to Professors Jian Sheng, Siva Vanapalli, Fazle Hussain, and their students, as I have greatly benefited from invaluable discussions with them on oil biodegradation during my research visit at Texas Tech University in 2013–2014.

References

Abbasnezhad, H., Gray, M., Foght, J.M., 2011. Influence of adhesion on aerobic biodegradation and bioremediation of liquid hydrocarbons. Appl. Microbiol. Biotechnol. 92, 653–675.

Abbott, E.A., 1992. Flatland. A Romance of Many Dimensions. Dover Publications, New York.

Al-Tahhan, R.A., Sandrin, T.R., Bodour, A.A., Maier, R.M., 2000. Rhamnolipid-induced removal of lipopolysaccharide from *Pseudomonas aeruginosa*: effect on cell surface properties and interaction with hydrophobic substrates. Appl. Environ. Microbiol. 66 (8), 3262–3268.

Alvarez, H.M., Mayer, F., Fabritius, D., Steinbüchel, A., 1996. Formation of intracytoplasmic lipid inclusions by *Rhodococcus opacus* strain PD630. Arch. Microbiol. 165, 377–386.

Antoniou, E., Fodelianakis, S., Korkakaki, E., Kalogerakis, N., 2015. Biosurfactant production from marine hydrocarbon-degrading consortia and pure bacterial strains using crude oil as carbon source. Front. Microbiol. 6, 274.

Armstrong, R.T., Wildenschild, D., 2012. Investigating the pore-scale mechanisms of microbial enhanced oil recovery. J. Pet. Sci. Eng. 94–95, 155–164.

Asadi, F., Phan, C.M., Obanijesu, E.O., 2015. Dynamic behavior of a floating water droplet with hydrocarbon-degrading bacteria. J. Environ. Chem. Eng. 3, 2311–2316.

Atlas, R.M., Hazen, T.C., 2011. Oil biodegradation and bioremediation: a tale of the two worst spills in U.S. history. Environ. Sci. Technol. 45, 6709–6715.

Baldi, F., Ivošević , N., Minacci, A., Pepi, M., Fani, R., Svetličić, V., Žutić, V., 1999. Adhesion of *Acinetobacter venetianus* to diesel fuel droplets studied with in situ electrochemical and molecular probes. Appl. Environ. Microbiol. 65 (5), 2041–2048.

Banat, I.M., 1995. Biosurfactants production and possible uses in microbial enhanced oil recovery and oil pollution remediation: a review. Bioresour. Technol. 51, 1–12.

Bendinger, B., Rijnaarts, H.H.M., Altendorf, K., Zehnder, A.J.B., 1993. Physicochemical cell surface and adhesive properties of coryneform bacteria related to the presence and chain length of mycolic acids. Appl. Environ. Microbiol. 59 (11), 3973–3977.

Binks, B.P., 2002. Particles as surfactants – similarities and differences. Curr. Opin. Colloid Interface Sci. 7, 21–41.

Bouchez-Naïtali, M., Rakatozafy, H., Marchal, R., Leveau, J.-Y., Vandecasteele, J.-P., 1999. Diversity of bacterial strains degrading hexadecane in relation to the mode of substrate uptake. J. Appl. Microbiol. 86, 421–428.

Bouchez-Naïtali, M., Vandecasteele, J.-P., 2008. Biosurfactants, an help in the biodegradation of hexadecane? The case of *Rhodococcus* and *Pseudomonas* strains. World J. Microbiol. Biotechnol. 24, 1901–1907.

Brakstad, O.G., Nordtug, T., Throne-Holst, M., 2015. Biodegradation of dispersed Macondo oil in seawater at low temperature and different oil droplet sizes. Mar. Pollut. Bull. 93, 144–152.

Brown, D.G., 2007. Relationship between micellar and hemi-micellar processes and the bioavailability of surfactant-solubilized hydrophobic organic compounds. Environ. Sci. Technol. 41, 1194–1199.

Cameotra, S.S., Singh, P., 2009. Synthesis of rhamnolipid biosurfactant and mode of hexadecane uptake by *Pseudomonas* species. Microb. Cell Fact. 8, 16.

Costerton, J.W., Lewandowski, Z., Caldwell, D.E., Korber, D.R., Lappin-Scott, H.M., 1995. Microbial biofilms. Annu. Rev. Microbiol. 49, 711–745.

Deppe, U., Richnow, H.-H., Michaelis, W., Antranikian, G., 2005. Degradation of crude oil by an arctic microbial consortium. Extremophiles 9, 461–470.

Desai, J.D., Banat, I.M., 1997. Microbial production of surfactants and their commercial potential. Microbiol. Mol. Biol. Rev. 61 (1), 47–64.

Déziel, E., Comeau, Y., Villemur, R., 1999. Two-liquid-phase bioreactors for enhanced degradation of hydrophobic/toxic compounds. Biodegradation 10, 219–233.

Dmitriev, V.V., Crowley, D., Rogachevsky, V.V., Negri, C.M., Rusakova, T.G., Kolesnikova, S.A., Akhmetov, L.I., 2011. Microorganisms form exocellular structures, trophosomes, to facilitate biodegradation of oil in aqueous media. FEMS Microbiol. Lett. 315, 134–140.

Dorobantu, L.S., Yeung, A.K.C., Foght, J.M., Gray, M.R., 2004. Stabilization of oil–water emulsions by hydrophobic bacteria. Appl. Environ. Microbiol. 70 (10), 6333–6336.

Dorobantu, L.S., Bhattacharjee, S., Foght, J.M., Gray, M.R., 2008. Atomic force microscopy measurement of heterogeneity in bacterial surface hydrophobicity. Langmuir 24 (9), 4944–4951.

Edwards, D.A., Brenner, H., Wasan, D.T., 1991. Interfacial Transport Processes and Rheology. Butterworth-Heinemann.

Efroymson, R.A., Alexander, M., 1991. Biodegradation by an *Arthrobacter* species of hydrocarbons partitioned into an organic solvent. Appl. Environ. Microbiol. 57 (5), 1441–1447.

Georgiou, G., Lin, S.-C., Sharma, M.M., 1992. Surface-active compounds from micro-organisms. Nat. Biotechnol. 10, 60–65.

Guha, S., Jaffé, P.R., 1996a. Biodegradation kinetics of phenanthrene partitioned into the micellar phase of nonionic surfactants. Environ. Sci. Technol. 30, 605–611.

Guha, S., Jaffé, P.R., 1996b. Bioavailability of hydrophobic compounds partitioned into the micellar phase of nonionic surfactants. Environ. Sci. Technol. 30, 1382–1391.

Hassanshahian, M., Cappello, S., 2013. Crude oil biodegradation in the marine environments. In: Chamy, R., Rosenkranz, F. (Eds.), Biodegradation – Engineering and Technology. InTech (ebook), Ch. 5.

Hazen, T.C., Prince, R.C., Mahmoudi, N., 2016. Marine oil biodegradation. Environ. Sci. Technol. 50, 2121–2129.

Head, I.M., Jones, D.M., Larter, S.R., 2003. Biological activity in the deep subsurface and the origin of heavy oil. Nature 426, 344–352.

Head, I.M., Jones, D.M., Röling, W.F.M., 2006. Marine microorganisms make a meal of oil. Nat. Rev. Microbiol. 4, 173–182.

Hearn, E.M., Patel, D.R., Lepore, B.W., Indic, M., Van Den Berg, B., 2009. Transmembrane passage of hydrophobic compounds through a protein channel wall. Nature 458, 367–370.

Hermansson, M., 1999. The DLVO theory in microbial adhesion. Colloids Surf. B, Biointerfaces 14, 105–119.

Hua, F., Wang, H.Q., 2012. Uptake modes of octadecane by *Pseudomonas* sp. DG17 and synthesis of biosurfactant. J. Appl. Microbiol. 112, 25–37.

Hua, F., Wang, H.Q., 2013. Selective pseudosolubilization capability of *Pseudomonas* sp. DG17 on *n*-alkanes and uptake mechanisms analysis. Front. Environ. Sci. Eng. 7 (4), 539–551.

Hua, F., Wang, H.Q., 2014. Uptake and transmembrane transport of petroleum hydrocarbons by microorganisms. Biotechnol. Biotechnol. Equip. 28 (2), 165–175.

Joekar-Niasar, V., Hassanizadeh, S.M., 2012. Analysis of fundamentals of two-phase flow in porous media using dynamic pore-network models: a review. Crit. Rev. Environ. Sci. Technol. 42 (18), 1895–1976.

Johansen, Ø., Rye, H., Cooper, C., 2003. DeepSpill – field study of a simulated oil and gas blowout in deep water. Spill Sci. Technol. Bull. 8 (5–6), 433–443.

Johnsen, A.R., Karlson, U., 2004. Evaluation of bacterial strategies to promote the bioavailability of polycyclic aromatic hydrocarbons. Appl. Microbiol. Biotechnol. 63, 452–459.

Johnsen, A.R., Wick, L.Y., Harms, H., 2005. Principles of microbial PAH-degradation in soil. Environ. Pollut. 133, 71–84.

Kang, Z., Yeung, A., Foght, J.M., Gray, M.R., 2008. Hydrophobic bacteria at the hexadecane–water interface: examination of micrometre-scale interfacial properties. Colloids Surf. B, Biointerfaces 67, 59–66.

Kapellos, G.E., Alexiou, T.S., Payatakes, A.C., 2007. Hierarchical simulator of biofilm growth and dynamics in granular porous materials. Adv. Water Resour. 30, 1648–1667.

Kapellos, G.E., Alexiou, T.S., Payatakes, A.C., 2010. Theoretical modeling of fluid flow through cellular biological media: an overview. Math. Biosci. 225 (2), 83–93.

Kapellos, G.E., Alexiou, T.S., 2013. Modeling momentum and mass transport in cellular biological media: from the molecular to the tissue scale. In: Becker, S.M., Kuznetsov, A.V. (Eds.), Transport in Biological Media. Academic Press (Elsevier). Ch. 1.

Kapellos, G.E., Alexiou, T.S., Pavlou, S., 2015. Fluid–biofilm interactions in porous media. In: Becker, S.M., Kuznetsov, A.V. (Eds.), Heat Transfer and Fluid Flow in Biological Processes. Academic Press (Elsevier). Ch. 8.

Kim, I.S., Foght, J.M., Gray, M.R., 2002. Selective transport and accumulation of alkanes by *Rhodococcus erythropolis* $S + {}^{14}He$. Biotechnol. Bioeng. 80 (6), 650–659.

King, G.M., Kostka, J.E., Hazen, T.C., Sobecky, P.A., 2015. Microbial responses to the Deepwater Horizon oil spill: from coastal wetlands to the deep sea. Ann. Rev. Marine Sci. 7, 377–401.

Klein, B., Grossi, V., Bouriat, P., Goulas, P., Grimaud, R., 2008. Cytoplasmic wax ester accumulation during biofilm-driven substrate assimilation at the alkane–water interface by *Marinobacter hydrocarbonoclasticus* SP17. Res. Microbiol. 159, 137–144.

Klein, B., Bouriat, P., Goulas, P., Grimaud, R., 2010. Behavior of *Marinobacter hydrocarbonoclasticus* SP17 cells during initiation of biofilm formation at the alkane–water interface. Biotechnol. Bioeng. 105 (3), 461–468.

Kragh, K.N., Hutchison, J.B., Melaugh, G., Rodesney, C., Roberts, A.E.L., Irie, Y., Jensen, P.Ø., Diggle, S.P., Allen, R.J., Gordon, V., Bjarnsholt, T., 2016. Role of multicellular aggregates in biofilm formation. mBio 7 (2), e00237-16.

Kulkarni, K., Barnett, S.M., 1986. A mechanistic model for microbial uptake of hydrocarbons in flocculating and non-flocculating systems. Chem. Eng. Commun. 45, 19–31.

Langevin, D., 1992. Micelles and microemulsions. Annu. Rev. Phys. Chem. 43, 341–369.

Lauga, E., DiLuzio, W.R., Whitesides, G.M., Stone, H.A., 2006. Swimming in circles: motion of bacteria near solid boundaries. Biophys. J. 90, 400–412.

Lazar, I., Petrisor, I.G., Yen, T.F., 2007. Microbial enhanced oil recovery (MEOR). Pet. Sci. Technol. 25, 1353–1366.

Lepore, B.W., Indic, M., Pham, H., Hearn, E.M., Patel, D.R., Van Den Berg, B., 2011. Ligand-gated diffusion across the bacterial outer membrane. Proc. Natl. Acad. Sci. USA 108 (25), 10121–10126.

Li, J.-L., Chen, B.-H., 2009. Surfactant-mediated biodegradation of polycyclic aromatic hydrocarbons. Materials 2, 76–94.

Lopez, D., Lauga, E., 2014. Dynamics of swimming bacteria at complex interfaces. Phys. Fluids 26, 071902.

Macedo, A.J., Kuhlicke, U., Neu, T.R., Timmis, K.N., Abraham, W.-R., 2005. Three stages of a biofilm community developing at the liquid–liquid interface between polychlorinated biphenyls and water. Appl. Environ. Microbiol. 71 (11), 7301–7309.

MacLeod, C.T., Daugulis, A.J., 2005. Interfacial effects in a two-phase partitioning bioreactor: degradation of polycyclic aromatic hydrocarbons (PAHs) by a hydrophobic *Mycobacterium*. Process Biochem. 40, 1799–1805.

Marin, M., Pedregosa, A., Laborda, F., 1996. Emulsifier production and microscopical study of emulsions and biofilms formed by the hydrocarbon-utilizing bacteria *Acinetobacter calcoaceticus* MM5. Appl. Microbiol. Biotechnol. 44, 660–667.

McCollom, T.M., 2013. Laboratory simulations of abiotic hydrocarbon formation in Earth's deep subsurface. Rev. Mineral. Geochem. 75, 467–494.

McGenity, T.J., Folwell, B.D., McKew, B.A., Sanni, G.O., 2012. Marine crude-oil biodegradation: a central role for interspecies interactions. Aq. Biosyst. 8, 10.

Miller, R.M., Bartha, R., 1989. Evidence from liposome encapsulation for transport-limited microbial metabolism of solid alkanes. Appl. Environ. Microbiol. 55 (2), 269–274.

Miller, M.J., Grant Allen, D., 2004. Transport of hydrophobic pollutants through biofilms in biofilters. Chem. Eng. Sci. 59, 3515–3525.

Miörner, H., Johansson, G., Kronvall, G., 1983. Lipoteichoic acid is the major cell wall component responsible for surface hydrophobicity of Group A *Streptococci*. Infect. Immun. 39 (1), 336–343.

Molaei, M., Barry, M., Stocker, R., Sheng, J., 2014. Failed escape: solid surfaces prevent tumbling of *Escherichia coli*. Phys. Rev. Lett. 113, 068103.

National Research Council, 2002. Oil in the Sea III: Inputs, Fates and Effects. National Academies Press, Washington D.C.

Neu, T.R., 1996. Significance of bacterial surface-active compounds in interaction of bacteria with interfaces. Microbiol. Rev. 60 (1), 151–166.

Nikolopoulou, M., Pasadakis, N., Kalogerakis, N., 2013. Evaluation of autochthonous bioaugmentation and biostimulation during microcosm-simulated oil spills. Mar. Pollut. Bull. 72, 165–173.

Noordman, W.H., Janssen, D.B., 2002. Rhamnolipid stimulates uptake of hydrophobic compounds by *Pseudomonas aeruginosa*. Appl. Environ. Microbiol. 68 (9), 4502–4508.

O'Toole, G.A., Kaplan, H.B., Kolter, R., 2000. Biofilm formation as microbial development. Annu. Rev. Microbiol. 54, 49–79.

Pandey, G., Jain, R.K., 2002. Bacterial chemotaxis toward environmental pollutants: role in bioremediation. Appl. Environ. Microbiol. 68 (12), 5789–5795.

Parales, R.E., Ditty, J.L., 2010. Chemotaxis. In: Timmis, K.N. (Ed.), Handbook of Hydrocarbon and Lipid Microbiology. Springer-Verlag, Berlin, Heidelberg. Ch. 49.

Perfumo, A., Smyth, T.J.P., Marchant, R., Banat, I.M., 2010. Production and roles of biosurfactants and bioemulsifiers in accessing hydrophobic substrates. In: Timmis, K.N. (Ed.), Handbook of Hydrocarbon and Lipid Microbiology. Springer-Verlag, Berlin, Heidelberg. Ch. 47.

Rahsepar, S., Smit, M.P.J., Murk, A.J., Rijnaarts, H.H.M., Langenhoff, A.A.M., 2016. Chemical dispersants: oil biodegradation friend or foe? Mar. Pollut. Bull. 108, 113–119.

Reddy, P.G., Singh, H.D., Roy, P.K., Baruah, J.N., 1982. Predominant role of hydrocarbon solubilization in the microbial uptake of hydrocarbons. Biotechnol. Bioeng. 24 (6), 1241–1269.

Rodrigues, A.C., Wuertz, S., Brito, A.G., Melo, L.F., 2005. Fluorene and phenanthrene uptake by *Pseudomonas putida* ATCC 17514: kinetics and physiological aspects. Biotechnol. Bioeng. 90 (3), 281–289.

Ron, E.Z., Rosenberg, E., 2001. Natural roles of biosurfactants. Environ. Microbiol. 3 (4), 229–236.

Rosenberg, M., Rosenberg, E., 1981. Role of adherence in growth of *Acinetobacter calcoaceticus* RAG-1 on hexadecane. J. Bacteriol. 148 (1), 51–57.

Rosenberg, M., Bayer, E.A., Delarea, J., Rosenberg, E., 1982. Role of thin fimbriae in adherence and growth of Acinetobacter calcoaceticus RAG-1 on hexadecane. Appl. Environ. Microbiol. 44 (4), 929–937.

Rosenberg, E., 2013. Hydrocarbon-oxidizing bacteria. In: Rosenberg, E., et al. (Eds.), The Prokaryotes. Springer-Verlag, Berlin, Heidelberg. Ch. 5.

Rühs, P.A., Böcker, L., Inglis, R.F., Fischer, P., 2014. Studying bacterial hydrophobicity and biofilm formation at liquid–liquid interfaces through interfacial rheology and pendant drop tensiometry. Colloids Surf. B, Biointerfaces 117, 174–184.

Southam, G., Whitney, M., Knickerbocker, C., 2001. Structural characterization of the hydrocarbon degrading bacteria–oil interface: implications for bioremediation. Int. Biodeterior. Biodegrad. 47, 197–201.

Stanley-Wall, N.R., MacPhee, C.E., 2015. Connecting the dots between bacterial biofilms and ice cream. Phys. Biol. 12, 063001.

Stocker, R., Seymour, J.R., Samadani, A., Hunt, D.E., Polz, M.F., 2008. Rapid chemotactic response enables marine bacteria to exploit ephemeral microscale nutrient patches. Proc. Natl. Acad. Sci. USA 105 (11), 4209–4214.

Stocker, R., 2012. Marine microbes see a sea of gradients. Science 338, 628–633.

Vaccari, L., Allan, D., Sharifi-Mood, N., Singh, A., Leheny, R., Stebe, K., 2015. Films of bacteria at interfaces: three stages of behavior. Soft Matter 11, 6062–6074.

Valentine, D.L., Mezić, I., Maćešić, S., Črnjarić-Žic, N., Ivić, S., Hogan, P.J., Fonoberov, V.A., Loire, S., 2012. Dynamic autoinoculation and the microbial ecology of a deep water hydrocarbon irruption. Proc. Natl. Acad. Sci. USA 109 (50), 20286–20291.

van den Berg, B., 2010. Going forward laterally: transmembrane passage of hydrophobic molecules through protein channel walls. ChemBioChem 11 (10), 1339–1343.

van Hamme, J.D., Ward, O.P., 2001. Physical and metabolic interactions of *Pseudomonas* sp. strain JA5-B45 and *Rhodococcus* sp. strain F9-D79 during growth on crude oil and effect of a chemical surfactant on them. Appl. Environ. Microbiol. 67 (10), 4874–4879.

van Hamme, J.D., Singh, A., Ward, O.P., 2003. Recent advances in petroleum microbiology. Microbiol. Mol. Biol. Rev. 67 (4), 503–549.

Volkering, F., Breure, A.M., Rulkens, W.H., 1998. Microbiological aspects of surfactant use for biological soil remediation. Biodegradation 8, 401–417.

Wang, X., Lanning, L.M., Ford, R.M., 2016. Enhanced retention of chemotactic bacteria in a pore network with residual NAPL contamination. Environ. Sci. Technol. 50 (1), 165–172.

Wentzel, A., Ellingsen, T.E., Kotlar, H.-K., Zotchev, S.B., Throne-Holst, M., 2007. Bacterial metabolism of long-chain *n*-alkanes. Appl. Microbiol. Biotechnol. 76, 1209–1221.

Whyte, L.G., Slagman, S.J., Pietrantonio, F., Bourbonnière, L., Koval, S.F., Lawrence, J.R., Inniss, W.E., Greer, C.W., 1999. Physiological adaptations involved in alkane assimilation at a low temperature by *Rhodococcus* sp. strain Q15. Appl. Environ. Microbiol. 65 (7), 2961–2968.

Wick, L.Y., Ruiz De Munain, A., Springael, D., Harms, H., 2002. Responses of *Mycobacterium* sp. LB501T to the low bioavailability of solid anthracene. Appl. Microbiol. Biotechnol. 58, 378–385.

Yakimov, M.M., Timmis, K.N., Golyshin, P.N., 2007. Obligate oil-degrading marine bacteria. Curr. Opin. Biotechnol. 18, 257–266.

Zajic, J.E., Seffens, W., Panchal, C., 1983. Biosurfactants. Crit. Rev. Biotechnol. 1 (2), 87–107.

CHAPTER

3

Modeling and Measurement of Biomolecular Transport and Sensing in Microfluidic Cell Culture and Analysis Systems

*J.F. Wong**[‡], C.A. Simmons**[†,‡], E.W.K. Young**[†,#]

*Institute of Biomaterials and Biomedical Engineering, University of Toronto, Toronto, ON, Canada [†]Department of Mechanical & Industrial Engineering, University of Toronto, Toronto, ON, Canada [‡]Translational Biology & Engineering Program, Ted Rogers Centre for Heart Research, Toronto, ON, Canada

3.1 INTRODUCTION

Microfluidic systems have become increasingly important in biology research and the life sciences. Over the past ten years, the principles of microscale fluid flow, defined as the mechanics of fluid motion within conduits and geometries ranging from 1–1000 microns, have been increasingly applied to develop innovative tools for studying and manipulating biological materials such as nucleic acids, proteins, and living cells, due to the fact that the geometries and dimensions of these systems are suitable for probing such biomolecules and biological entities. As a result, numerous microfluidic device designs have been generated, each with their own specific function.

Of the many microfluidic systems that have been developed, two major classes are: (i) microfluidic cell culture systems used for modeling physiological microenvironments, and (ii) microfluidic bioanalytical systems used to detect and measure the presence or activity of various biomolecular species. The engineering advances in both these device classes have progressed at a rapid pace. The complexity of microfluidic cell culture systems has increased steadily in complexity from two-dimensional (2D) monoculture platforms, to three-dimensional (3D) co-culture platforms, leading to recent demonstrations of biomimetic "organ-on-a-chip" systems, such as the lung-on-a-chip, gut-on-a-chip, and bone-marrow-on-

[#] Chapter coordinator.

a-chip devices (Bhatia and Ingber, 2014; Esch et al., 2015). In terms of microfluidic bioanalytical systems, the research field has undergone further development from basic biochemical assays (e.g., on-chip enzyme-linked immunosorbent assays, or ELISAs, in-cell Western blots, etc.) to more complex integrated functionalities such as electrochemical sensing and optical detection (Choi et al., 2011; Rackus et al., 2015; Wu et al., 2016). On both of these fronts, there remains room for much further research. Organ-on-a-chip systems remain a nascent field with remarkable potential for disease modeling and utility in the pharmaceutical drug development process. In comparison, bioanalytical systems have seen increased complexity in the form of aptamers and other surface chemistry approaches (Luka et al., 2015; Roper, 2016).

While the fields of microfluidic cell culture and bioanalytical systems will continue to chart their own course, a unique opportunity also exists for conducting research at the interface of these two fields, where microdevices that integrate cell culture together with on-chip detection and sensing would be of major benefit to our ability to study, probe, and measure (perhaps in real-time) the presence of certain molecular species, materials, and/or fluid properties. The goal of this integration is to be able to make measurements and perform analyses that make sense of the biological signals and responses from cells that have been cultured in the most physiologically relevant environment achievable while also avoiding the manual sample collection and processing steps required by traditional offline analysis methods such as ELISA.

The focus of work in the fields of microfluidic cell culture platforms and biochemical assays has primarily involved the development and experimental validation of different device concepts and functionality. By comparison, mathematical and computational modeling provides spatiotemporal characterization of physical processes within microfluidic platforms, enables exploration of design parameter effects, and facilitates the optimization of design parameters to produce desired outcomes for device design. Modeling therefore constitutes a valuable analysis tool that informs and guides the device design process.

The objective of this chapter is to present theoretical and computational modeling concepts and recent applications of those concepts toward the development of integrated microfluidic cell culture and bioanalytical systems. Understanding these modeling concepts will hopefully enable improved designs of microfluidic systems that aim to incorporate bioanalysis or biosensing into the on-chip functionality. This chapter includes a concise background on microscale cell culture, and then describes the governing constitutive equations of microscale fluid flow, mass transport, and reaction kinetics. Recent examples of modeling biomolecular transport in microfluidic cell culture and analysis systems are then provided. Finally, this chapter provides examples of experimental model validation and describes the latest advances in microfluidic systems for cell culture integrating bioanalysis or biosensing capabilities.

3.2 BASIC PRINCIPLES OF MICROSCALE CELL CULTURE

To facilitate the discussion of microfluidic cell culture and analytical systems, this section provides a basic background of key aspects of cell culture at the microscale, and describes the typical factors and system properties that are studied, monitored, or detected as part of biological analyses. Fundamental principles of microfluidic cell culture and approaches to controlling tissue microenvironments in microfluidic geometries have been discussed previously (Devadas and Young, 2016; Young and Beebe, 2010) and are summarized here for convenience.

Cell culture, in both traditional and microscale systems, involves establishing and maintaining biological functions of living cells in an engineered environment, which may consist of design features that differ from the natural physiological characteristics of the living organism from which the cells originated. Maintaining cell viability and function in culture requires essential elements in the surroundings of the cells that support their intrinsic cellular behavior. These essential elements include physical, chemical, and mechanical factors that all contribute to creating the proper culture environment for the cells. First, physicochemical properties such as temperature, pH, osmolality, and oxygen and carbon dioxide (CO_2) concentrations are all critical to cell viability, and engineered *in vitro* systems aim to replicate the conditions found *in vivo*. For cellular functions to maintain homeostasis, biochemical reactions need to occur at the proper temperature and pH. CO_2 is typically used as a buffer to maintain pH balance, and oxygen is necessary for proper cell metabolism.

Aside from these basic physicochemical properties in the surroundings, cell metabolism also depends on availability of nutrients that serve as an energy source for cells. In cell culture, these nutrients are provided in the ***culture medium***, an aqueous solution of amino acids, vitamins, salts, and other components within which the cells are immersed. Over time, nutrients are consumed and waste products are generated as a function of cell metabolism, and used culture medium must be replaced by fresh medium to replenish nutrients and remove accumulated waste. Rather than replacing the medium periodically, some systems (especially modern microfluidic systems) employ ***perfusion***, a continuous flow of medium that brings nutrients in and carries waste out via convective transport. Depending on the microscale geometries of the system, optimal perfusion conditions may be determined by estimating system parameters such as the critical perfusion rate (CPR) or the effective culture time (ECT) (Young and Beebe, 2010).

Cells may be either adherent or non-adherent in nature, and thus may prefer to attach and spread on 2D surfaces (e.g., epithelial or endothelial cells), within 3D scaffolds (e.g., fibroblasts), or may prefer to remain "suspended" in culture without any attachment at all (e.g., multiple myeloma cells). For adherent cells, adhesion occurs through focal adhesions that mechanically anchor the cells to the surface. Mechanobiological signals may be transmitted by the physical forces acting through these attachment sites, thereby triggering cellular responses that alter gene expression. Other mechanical forces may also be present around cells, such as shear stress generated by fluid flow over the cells when perfusion is applied. Thus, perfusion often must be carefully designed into the system to account for both convective transport of biomolecules as well as mechanobiological effects.

These various elements constitute the cellular microenvironment. Cells respond to these combined microenvironmental cues by expressing genes, resulting in production and release of various proteins and soluble factors into the microenvironment as signals that mediate cellular communication. Clearly, cells and their surroundings interact and are coupled via complex feedback loops, and deducing the underlying mechanisms of these interactions continues to drive basic biology research. To better study these mechanisms, new techniques and tools are essential for detecting and measuring the properties of the microenvironment.

As described above, microfluidic cell culture systems are becoming increasingly complex in modeling of cell and tissue microenvironments. At the same time, microfluidic on-chip analysis techniques have been developed over the last ten years, and are becoming more sophisticated in their functions and capabilities. The next stage in technological development is presumably the integration of cell culture and analysis on the same chip to enable automated real-time monitoring and analyses. A basic example of this integration is live-cell microscopy and the optical monitoring of cells within microchannels. However, further research is necessary to advance to the next phase of microfluidic cell culture system development involving the real-time monitoring of small secreted molecules.

3.3 THEORY AND EQUATIONS: FLUID FLOW, MASS TRANSPORT, AND BIOCHEMICAL REACTIONS

There are three core components to modeling the microenvironment within microfluidic cell culture platforms and microfluidic biochemical assays: (i) fluid motion, (ii) mass transport in fluids, and (iii) biochemical reactions. The coupled nature of these components typically results in a system of equations and boundary conditions that can be solved numerically using finite element (or finite volume) software packages, or analytically by incorporating additional assumptions into the problem formulation. In this section, the governing equations of fluid motion, mass transport, and biochemical reaction dynamics are described to provide the necessary background for modeling microfluidic platforms and assays that involve biomolecular transport.

3.3.1 Fluid Motion

In modeling of microfluidic culture systems and assays, the fluid is most often assumed to be incompressible and Newtonian, where density and viscosity of the fluid are considered constant throughout the flow. For an incompressible Newtonian fluid, the governing equations for fluid motion are:

$$\rho\left(\frac{\partial \mathbf{u}}{\partial t}+\mathbf{u}\cdot\nabla\mathbf{u}\right) = -\nabla p+\mu\nabla^2\mathbf{u}+\mathbf{F}, \tag{3.1}$$

$$\nabla\cdot\mathbf{u} = 0 \tag{3.2}$$

where ρ is the fluid density ($\mathrm{kg\,m^{-3}}$), μ is the viscosity (Pa s), $\mathbf{u}$ is the velocity field ($\mathrm{m\,s^{-1}}$), p is the pressure (Pa), and $\mathbf{F}$ are body forces per unit volume ($\mathrm{N\,m^{-3}}$). Typical values for ρ and μ are 1000 $\mathrm{kg\,m^{-3}}$ and 0.7–1 $\times$ 10^{-3} Pa s, respectively, for aqueous solutions in microfluidic cell culture devices and assays (where most of the variation in viscosity is due to temperature). Eq. (3.1) is a linear momentum equation called the Navier–Stokes equations, and relates the time-dependent velocity field to the pressure field. The terms on the left-hand side are the product of mass and acceleration per unit volume on a differential fluid element and represent inertial effects. The right-hand side lists the applied surface and body forces acting on the same fluid element. Examples of body forces **F** include gravity (typically not applicable at the microscale, Bruus, 2015) or an electric field (e.g., electroosmotic flow, Santiago, 2001). Eq. (3.2) is the continuity equation representing mass conservation.

Flow in a microfluidic channel is typically laminar, which consists of layers, or lamina, of fluid increasing in velocity from the channel walls to the channel center. The ratio of inertial to viscous forces in a fluid is given by a dimensionless parameter called the Reynolds number Re (derivation is found in Section 3.3.4.1):

$$\mathrm{Re} = \frac{\rho U \zeta}{\mu} \tag{3.3}$$

where U (m s^{-1}) and ζ (m) are the characteristic velocity (e.g., average velocity) and characteristic length (typically the hydraulic diameter), respectively. Laminar flow is characterized by low Re, meaning viscous forces dominate over inertial forces. For example, internal flow is characterized as laminar for Re < 2300 in a cylinder. When Re < 1, all inertial effects may be reasonably neglected and a special linearized case of the Navier–Stokes equation called Stokes flow (or creeping flow) arises:

$$-\nabla p + \mu \nabla^2 \mathbf{u} + \mathbf{F} = 0, \tag{3.4}$$

$$\nabla \cdot \mathbf{u} = 0. \tag{3.5}$$

Microfluidic channels typically have constant rectangular cross-sections due to the common methods of microfabrication employed. Flow is also often assumed to be fully developed. Steady fully developed laminar flow through these simple geometries yields analytical solutions to the Navier–Stokes equations that simplify the mathematical and computational model equations. For example, for low height-to-width aspect ratio (AR) microchannels, a common approximation is to consider the parallel plate approximation:

$$u(y) = \frac{H^2}{2\mu}\frac{\Delta p}{L}\frac{y}{H}\left(1 - \frac{y}{H}\right), \tag{3.6}$$

$$Q = \frac{WH^3}{12\mu}\frac{\Delta p}{L}. \tag{3.7}$$

Eqs. (3.6) and (3.7) describe the fully developed laminar parabolic velocity profile and volumetric flow rate, respectively, in a rectangular channel of height H (m), width W (m) and length L (m) where $H < W$, approximating an infinite parallel plate geometry (Fig. 3.1). For an aspect ratio of $H/W = 0.1$, the velocity profile is invariant across ~85% of the channel width (Stone, 2007), meaning that this two-dimensional approximation may be adequately used to model flow in the three-dimensional rectangular geometry (expect near the side walls). The form of Eq. (3.6) is analogous to that of Poiseuille flow in a cylindrical pipe. The derivation of these equations from the Navier–Stokes equation may be found in most introductory texts in fluid mechanics (Fox et al., 2004).

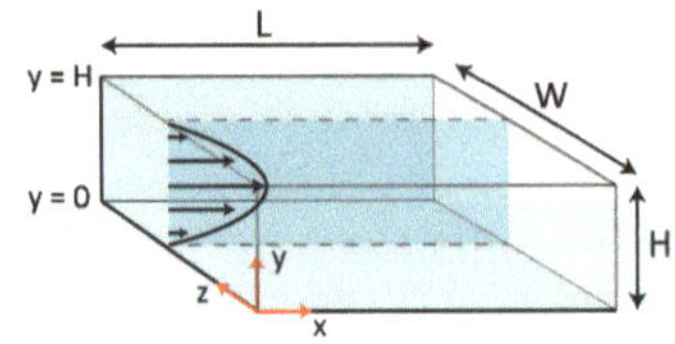

FIGURE 3.1 Schematic of a rectangular microchannel with laminar flow characterized by a parabolic velocity profile. The microchannel has a height H, width W, and length L.

3.3.2 Mass Transport in Fluids

Transport of dissolved gases and nutrients is a process commonly modeled in microfluidic devices, and is important when considering a microfluidic design. Examples of mass transport problems in microchannels include the transport of oxygen to cultured living cells, and the transport of an analyte such as a protein to a biosensor. The transport of a dilute solute species in solution is given by the convection–diffusion equation:

$$\frac{\partial c_i}{\partial t} + \mathbf{u} \cdot \nabla c_i = D_i \nabla^2 c_i + R_i \quad \text{for } i = 1, \ldots, N. \tag{3.8}$$

For a given species i, c_i is the concentration (mol m^{-3}), D_i is the diffusion coefficient (m^2 s^{-1}), R_i is a bulk source or sink reaction term (mol m^{-3} s^{-1}), and $\mathbf{u}$ is the velocity field (m s^{-1}). Eq. (3.8) assumes that the concentration of solute is sufficiently dilute such that it does not affect motion of the solvent, an assumption commonly and appropriately applied to species in culture medium and clinical samples. The left-hand side of the equation contains the convective transport term while the right-hand side contains the diffusive transport term. In the case of pure diffusion transport and neglecting any reactions, the equation reduces to Fick's second law of diffusion:

$$\frac{\partial c_i}{\partial t} = D_i \nabla^2 c_i. \tag{3.9}$$

The diffusion coefficient D in Eqs. (3.8) and (3.9) is the binary diffusion coefficient that describes the interdiffusion of one species in another (e.g., a solute in a solvent).

3.3.3 Biochemical Reactions

Biochemical reactions are an important component of microfluidic cell culture systems and quantitative assays. Oxygen consumed by a cell monolayer within a microchannel may be modeled as a biochemical reaction (Buchwald, 2009). The supply of oxygen within a cell culture platform is critical to maintaining cell viability, proliferation and function due to large surface area-to-volume ratios compared to traditional cell culture systems (Halldorsson et al., 2015; Meyvantsson and Beebe, 2008; Young and Beebe, 2010). Microfluidic quantitative assays involve immobilizing receptors onto a microchannel surface and injecting solution containing the analyte into the channel. Binding of the analyte to the receptor, represented as a biochemical reaction, is then transduced into a measurable signal using an optical, electronic or mass-based readout (Choi et al., 2011).

Reactions in microfluidic cell culture systems and quantitative assays typically occur at a surface or interface as opposed to the bulk fluid as in the case of a typical chemical reactor. The binding kinetics of receptors is a complex process that depends on many parameters. To control and optimize a microfluidic system, it is necessary to understand the role(s) of each parameter in altering binding kinetics within the system. The mathematical representations of biochemical reactions commonly encountered in microfluidic culture platforms and assays are described next.

3.3.3.1 *Ligand–Receptor Binding Kinetics*

The binding of a ligand to a surface-immobilized receptor is represented by the following chemical equation:

$$l_s + r \overset{k_{on}, k_{off}}{\longleftrightarrow} b \tag{3.10}$$

where l_s is the concentration of ligand at the reaction site (mol m^{-3}), b is the surface concentration of bound receptors (mol m^{-2}), r is the surface concentration of free unbound receptors (mol m^{-2}), k_{on} is the association rate constant ($\text{mol}^{-1}\,\text{m}^3\,\text{s}^{-1}$), and k_{off} is the dissociation rate constant (s^{-1}). The equation states that ligand l_s reversibly binds to receptor r to form the bound ligand–receptor complex b and describes a second order reversible reaction. A one-to-one binding ratio between ligand and receptor is assumed as well as a homogeneous surface where all binding sites are equivalent. This model of binding is used to represent processes such as antigen–antibody binding and nucleic acid hybridization. The quantity of receptors bound to a surface is often expressed in the form of number of surface receptors per unit area. The change in the concentration of bound surface receptors over time is then given by:

$$\frac{\partial b}{\partial t} = k_{on} l_s r - k_{off} b. \tag{3.11}$$

The first term on the right-hand side of the equation indicates that receptor binding is proportional to the association rate constant, the concentration of ligand in the binding region, and the concentration of free (unbound) surface receptors. The second term indicates that the reaction is reversible and that dissociation is proportional to the dissociation rate constant k_{off} and the concentration of bound receptors. Eq. (3.11) above may then be simplified, recognizing that the concentration of free unbound receptors is given by the difference, $b_{max} - b$, where b_{max} is the total surface concentration of receptors (mol m^{-2}). The rate equation for a second order reversible receptor–ligand binding reaction is then:

$$\frac{\partial b}{\partial t} = k_{on} l_s (b_{max} - b) - k_{off} b. \tag{3.12}$$

Receptor–ligand binding is occasionally modeled as an irreversible second order reaction, eliminating the dissociation term in Eq. (3.12), which results in the following equation:

$$\frac{\partial b}{\partial t} = k_{on} l_s (b_{max} - b). \tag{3.13}$$

Eq. (3.12) may also be formulated as a pseudo-first order reaction by assuming that the concentration of ligand is supplied in excess such that it remains constant. This is a valid assumption when the quantity of ligand that can be bound by the receptors is small compared to the quantity in the bulk solution (Myszka et al., 1998). The rate equation represented by Eq. (3.12) then only depends on the surface concentration of bound receptors, a first order reaction, where at equilibrium:

$$0 = k_{on} l_s \left(b_{max} - b_{eq}\right) - k_{off} b_{eq}. \tag{3.14}$$

The concentration of bound receptors at equilibrium in the equation above is represented by b_{eq} (mol m^{-2}). Rearranging the terms by isolating b_{max} and b_{eq} on the left side yields:

$$\frac{b_{eq}}{b_{max}} = \frac{K_A l_s}{1 + K_A l_s} \tag{3.15}$$

where K_A is the equilibrium association constant (m^3 mol^{-1}) and is a measure of the binding affinity of a given receptor towards its ligand. It is represented as

$$K_A = \frac{k_{on}}{k_{off}}. \tag{3.16}$$

The physical meaning of K_A is that half of the total number of receptors are bound at a concentration K_A^{-1} at equilibrium. A higher K_A therefore indicates a greater affinity. Eq. (3.15), called the Langmuir isotherm, describes the fraction of bound receptors at equilibrium for a given concentration of ligand. Quantitative assay readouts are proportional to this fraction and therefore the standard curve for a given assay often closely resembles the Langmuir isotherm.

3.3.3.2 *Michaelis–Menten Kinetics*

Michaelis–Menten kinetics were originally derived as a mathematical model of enzymatic reaction rates, and are frequently used to describe the uptake of nutrients like oxygen by cultured cells (Cho et al., 2007). The model describes a cell c forming a complex c_s with substrate s, consuming the substrate, and finally resulting in the production of a product p. The cell is then available to uptake more nutrients. The chemical equation representing this process is:

$$c + s \overset{k_f, k_r}{\longleftrightarrow} c_s \overset{k_c}{\longleftrightarrow} c + p \tag{3.17}$$

where k_f is the cell–substrate association rate constant (mol^{-1} m^3 s^{-1}), k_r is the cell–substrate dissociation rate constant (s^{-1}), and k_c is the dissociation rate constant resulting in the conversion of substrate to product (s^{-1}). The rate equation for the formation of product is given by:

$$\frac{\partial p}{\partial t} = k_c c_s. \tag{3.18}$$

Assuming that the formation and dissociation of complex c_s occurs initially at identical constant rates, the following rate equation is obtained:

$$k_f cs = k_r c_s + k_c c_s. \tag{3.19}$$

This equation may be rearranged to isolate the concentration terms on one side and rate constants on the other:

$$\frac{k_r + k_c}{k_f} = \frac{cs}{c_s}. \tag{3.20}$$

Isolating the term c_s in Eq. (3.20) above and then substituting into Eq. (3.19) yields the Michaelis–Menten equation:

$$\frac{\partial p}{\partial t} = \frac{V_{max} \gamma s}{K_m + s} \tag{3.21}$$

where p is the concentration of product formed (mol m^{-3}), s is the substrate concentration (mol m^{-3}), and V_{max} is the maximum species consumption rate per cell (mol s^{-1} cell^{-1}). Note that in order to express the reaction rate $\partial p/\partial t$ as the consumption rate for a monolayer of cells, Eq. (3.21) has been multiplied by γ, the surface density of cells (cells m^{-2}). The parameter K_m is the Michaelis–Menten constant of the cells (mol m^{-3}) and is represented as:

$$K_m = \frac{k_r + k_c}{k_f}. \tag{3.22}$$

Physically, K_m is the concentration of substrate at which the reaction rate is half of V_{max}.

3.3.4 Non-Dimensionalization

The non-dimensionalization of the governing equations can provide additional insight into the behavior of a given system by combining physical system parameters into naturally arising dimensionless numbers that reduce the overall number of equation parameters. Typically, dimensionless numbers describe competing physical phenomena that influence system behavior. Dimensionless numbers often help to simplify modeling equations by revealing phenomena (and its associated terms in the equation) that are dominant, and identifying terms that can be considered negligible. For example, if convective transport dominates over diffusive transport then diffusive equation terms may potentially be neglected.

In microfluidic device engineering, it is useful to obtain a qualitative understanding of system behavior using readily obtained device parameter values without resorting to complex calculations. The optimization of microfluidic cell culture devices and assays requires that many parameters be considered (e.g., geometry, cell properties, receptor properties, flow characteristics). Instead of restricting findings to a single set of parameter values, it is valuable to extend and generalize design optimization using dimensionless numbers that combine the effect of multiple design parameters. Indeed, dimensionless numbers describe systems that are dynamically similar despite having different physical parameters.

Non-dimensionalization involves scaling equation variables by characteristic values typically informed by the geometry, physical parameters, and system boundary conditions. In the case of a microfluidic device model incorporating the core components of fluid flow, mass transport and biochemical reactions, the Navier–Stokes equation, the convection–diffusion equation, and the biochemical reaction equations, as introduced in Sections 3.3.1 to 3.3.3, can be non-dimensionalized, as shown below.

3.3.4.1 *Navier–Stokes Equation*

In order to develop the non-dimensional representations of the governing equations, consider first the following non-dimensional variables:

$$x^* = \frac{x}{\zeta}, \quad y^* = \frac{y}{\zeta}, \quad z^* = \frac{z}{\zeta}, \quad \mathbf{u}^* = \frac{\mathbf{u}}{U},$$
$$t^* = t\left(\frac{\mu}{\rho\zeta^2}\right), \quad p^* = p\left(\frac{\zeta}{\mu U}\right), \quad \mathbf{F}^* = \mathbf{F}\left(\frac{\rho\zeta^2}{\mu U}\right)$$

where ζ and U are the characteristic length and characteristic velocity, respectively. Substituting these variables into Eq. (3.1) results in the non-dimensional Navier–Stokes equation:

$$\frac{\partial \mathbf{u}^*}{\partial t^*} + \mathrm{Re}\,\mathbf{u}^* \cdot \nabla^* \mathbf{u}^* = -\nabla^* p^* + \nabla^{*2} \mathbf{u}^* + \mathbf{F}^*, \tag{3.23}$$

$$\mathrm{Re} = \frac{\rho U \zeta}{\mu}. \tag{3.3}$$

The Reynolds number Re, introduced in Section 3.3.1, represents the ratio between inertial and viscous forces in the fluid. Flow at low Re is considered to be laminar, which is typically true for microfluidic devices. For internal flows specifically, a general rule is to consider flows to be laminar for $\mathrm{Re} < 2300$. Stokes flow arises when $\mathrm{Re} \ll 1$, and the non-linear inertial term in Eq. (3.23) may be neglected under all flow situations (i.e., internal or external flows, fully developed or not fully developed). The characteristic length ζ is often represented by the hydraulic diameter $D_H = 4A/P_w$ for internal flows, where A is the cross-sectional area and P_w is the wetted perimeter length, which depends on the channel geometry. For example, a filled rectangular channel has a hydraulic diameter of $D_H = 2HW/(H + W)$ where H and W are the channel height and width, respectively.

3.3.4.2 *Mass Transport Equation*

We define additional dimensionless variables:

$$c^* = \frac{c}{c_0}, \quad t^* = t\left(\frac{D}{\zeta^2}\right).$$

c_0 is typically a fixed inlet concentration of solute (e.g., analyte concentration), and t is scaled by the characteristic diffusion time in one dimension. Substituting these variables into the convection–diffusion equation shown in Eq. (3.8), we obtain:

$$\frac{\partial c_i^*}{\partial t^*} + \mathrm{Pe}\,\mathbf{u}^* \cdot \nabla c_i^* = \nabla^{*2} c_i^* + R_i^*, \tag{3.24}$$

$$\mathrm{Pe} = \frac{U\zeta}{D}. \tag{3.25}$$

Eq. (3.24) is the non-dimensional convection–diffusion equation. A single dimensionless parameter is contained in this equation, the Péclet number Pe, which is the ratio of convective and diffusive transport timescales. When $\mathrm{Pe} \ll 1$, diffusion dominates over convection. This may significantly impact the quantity of analyte transported to a sensor surface or oxygen available for cells to consume in the context of microfluidics. When $\mathrm{Pe} \gg 1$, convection dominates and it is often possible to neglect diffusion in the axial direction of convection, simplifying modeling equations. The length parameter ζ is typically taken to be one of the microchannel dimensions, e.g., the height H if diffusion in the vertical direction is of interest, or width W if lateral diffusion from one side wall to the other is of more concern.

Note that the non-dimensionalization described above has been independently performed for the Navier–Stokes and mass transport equations. The reason for this is to reveal the dimensionless parameter that is most pertinent to each equation, namely Re for the Navier–Stokes equations, and Pe for the mass transport equation. Indeed, if it were necessary to solve these equations simultaneously, the dimensionless variables would need to be selected differently to ensure that the time scale was consistent throughout the set of equations. However, for many transport problems, it is quite conceivable to solve the Navier–Stokes equations first, obtain the velocity field, and then subsequently solve the mass transport equation with the determined velocity field.

3.3.4.3 *Ligand–Receptor Kinetics*

The following non-dimensional variables are defined:

$$b^* = \frac{b}{b_{max}}, \quad l_s^* = \frac{l_s}{l_0}.$$

Here b is scaled by the total surface concentration of receptors b_{max} while ligand concentration l_s is scaled by the fixed inlet concentration l_0. Non-dimensionalizing Eq. (3.12), the second order reversible reaction, results in:

$$\frac{\partial b^*}{\partial t^*} = \xi\,\mathrm{Da}_l \left[l_s^*(1 - b^*) - \frac{1}{\overline{c}} b^* \right], \tag{3.26}$$

$$\xi = \frac{l_0 \zeta}{b_{max}}, \tag{3.27}$$

$$\mathrm{Da}_l = \frac{k_{on} b_{max} \zeta}{D}, \tag{3.28}$$

$$\overline{c} = \frac{l_0 k_{on}}{k_{off}} = \frac{l_0}{K_D}, \tag{3.29}$$

$$K_A = \frac{1}{K_D}. \tag{3.30}$$

The Damköhler number Da frequently arises in the non-dimensionalization of rate equations. The Damköhler number is the ratio of reaction rate timescale to transport rate timescale, which in this analysis is the diffusion timescale. A system with $Da_l \ll 1$ is reaction rate-limited, which is useful for determining binding kinetic constants (e.g., surface plasmon resonance binding analysis, Myszka, 1999) or designing rapid assays. When $Da_l \gg 1$, a system is mass transport-limited where the rate of diffusion (or convection depending on the Da number) is slower than the reaction rate and therefore species such as analyte or oxygen are bound or consumed before they can accumulate.

The adsorption capacity, ξ, is the ratio of ligand l_0 concentration along the characteristic length ζ associated with mass transfer (e.g., channel height) to the total surface concentration that may be bound b_{max}. It is a measure of the relative abundance of ligand to receptors. $\overline{c}$ is the ratio of the analyte concentration l_0 to the equilibrium dissociation constant K_D. Larger $\overline{c}$ results in faster binding and more receptors bound at equilibrium than smaller $\overline{c}$.

3.3.4.4 *Michaelis–Menten Kinetics*

Similar to the non-dimensionalization of ligand–receptor kinetics, the Michaelis–Menten equation may also be non-dimensionalized. Consider the dimensionless parameters:

$$p^* = \frac{p}{s_0}, \quad s^* = \frac{s}{s_0}$$

where both product concentration p and substrate concentration s are non-dimensionalized by s_0, typically a fixed initial inlet concentration of substrate. Substituting these variables into the Michaelis–Menten equation (3.21) yields:

$$\frac{\partial p^*}{\partial t^*} = \mathrm{Da}_m \frac{s^*}{\overline{K}_m + s^*}, \tag{3.31}$$

$$\mathrm{Da}_m = \frac{V_{max}\gamma\zeta}{Ds_0}, \tag{3.32}$$

$$\overline{K}_m = \frac{K_m}{s_0}. \tag{3.33}$$

$\overline{K}_m$ is the non-dimensionalized Michaelis–Menten constant. The Damköhler number Da_m, as in the case of Da_l (Eq. (3.28)), represents the ratio between reaction and diffusive transport timescales. For $\mathrm{Da}_m \gg 1$, substrate is consumed faster than diffusion can replenish it while for $\mathrm{Da}_m \ll 1$ the substrate is supplied by diffusion faster than its consumption.

3.4 REVIEW OF MICROFLUIDIC TRANSPORT MODELS

Modeling of microfluidic cell culture and biochemical assays typically involves modeling microchannels with rectangular cross-sections and reaction sites representing a cell monolayer or a biosensor surface. Simple device geometries allow for simplifications of the governing equations such as using a 2D model approximation of a 3D geometry. These simplifications significantly increase the computational efficiency of models. Microfluidic modeling therefore does not require significant computational resources compared to CFD studies involving complex 3D geometries. Moreover, computational models of microfluidic devices are typically implemented using finite element software such as COMSOL Multiphysics that is widely available, accessible and user-friendly. The barrier and resources required to perform useful and predictive modeling studies is therefore relatively low, which facilitates a thorough design process incorporating modeling prior to the prototyping and testing stages as opposed to a purely experimental validation approach.

There are a number of general outputs commonly studied in models of microfluidic cell culture and quantitative assays. A spatial description of fluid motion in the domain is obtained, which includes the fluid velocity, flow streamlines and pressure distribution. The time evolution of mass transport dynamics and nutrient consumption/binding kinetics for concentrated species are also computed. These parameters are often challenging or time-consuming to measure and quantify experimentally, requiring specialized instruments and techniques. Modeling therefore enables the study and visualization of quantities that may be difficult to study experimentally in a given device. The importance of physical, controllable design parameters on critical design outcomes may also be assessed in this manner. In this section, mathematical and computational models for microfluidic cell culture and microfluidic biochemical assays are reviewed, with reference to the fundamental governing equations presented in Section 3.3. Furthermore, these model geometries are described in the context of specific applications of microfluidic devices that have incorporated biosensing or the measurement of biochemical reactions into their system. The section is divided based on model geometry.

3.4.1 Straight Microchannel with Biochemical Assay Reaction Site(s)

Microfluidic biochemical assays such as microfluidic immunoassays (Ng et al., 2010), nucleic acid hybridization (Hui et al., 2007), surface plasmon resonance (Luo et al., 2008; Lee et al., 2007), and affinity-based sensors (Rogers, 2000) typically consist of a straight microchannel with a sensing area located on the bottom surface of the channel. The sensing area, commonly a glass or metal substrate, is functionalized with surface receptors (e.g., antibody, aptamer, complementary DNA strand, peptide nucleic acid (PNA)) that recognize and bind the analyte of interest (Fig. 3.2).

Fig. 3.2A illustrates a representative modeling domain for a microfluidic biochemical assay. The microchannel has a rectangular cross-section with height H, width W, and length L. At typical microfluidic channel fluid velocities, Re (Eq. (3.3)) is low and flow is assumed to be fully developed at the inlet. Therefore, fully developed laminar flow $\mathbf{u}$ may be imposed throughout the entire domain with the velocity profile being either parabolic in the case of pressure-driven flow or constant in the case of electrokinetic flow. The Navier–Stokes momentum equations therefore do not need to be solved for this geometry. A fixed concentration of analyte c_0 is also imposed at the inlet, describing a constant sample concentration. At the bottom surface of the microchannel, one or more planar reaction sites with length L_s are located downstream of the inlet where binding of analyte occurs in biochemical assays.

The microchannel aspect ratio (H/W) is frequently assumed to be small enough such that the fully developed velocity profile is invariant along the channel width. Therefore, models are often two-dimensional. Furthermore, convective flow is often dominant over diffusion

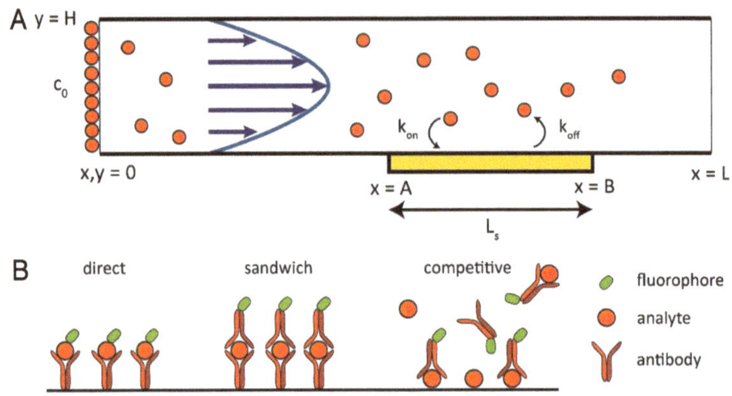

FIGURE 3.2 Representative microfluidic biochemical assay modeling domain. (A) Analyte concentration c_0 is fixed at the inlet ($x = 0$). Laminar flow in the straight microchannel transports analyte to a surface functionalized with analyte-specific receptors of length L_s. Binding and unbinding are proportional to rate constants k_{on} and k_{off}, respectively. (B) Schematics of direct, sandwich and competitive immunoassays.

in the axial direction, i.e., Pe $\gg 1$. Thus, convective fluid motion is restricted to the axial direction and diffusion to the y-direction, simplifying the mass transport equation (Eq. (3.8)). Note that in this model, the phenomena of interest are primarily located at the reaction sites and are essentially independent of the microchannel length. The length only needs to be large enough to capture the spatial changes in the local concentration upstream and downstream of the reaction sites (Squires et al., 2008). This leads to a more computationally efficient model.

The modeling domain described above leads to the following set of boundary conditions for a two-dimensional model where $\mathbf{n}$ is the outward unit normal vector to the surface:

$$\mathbf{u} = 0 \text{ (no-slip condition)} \quad \text{at the walls} \quad \begin{array}{l} 0 \le x \le L,\ y = 0, \\ 0 \le x \le L,\ y = H, \end{array} \tag{3.34}$$

$$u(y) = 6u_{avg}\frac{y}{H}\left(1 - \frac{y}{H}\right) \text{ (pressure-driven flow)} \quad \text{entire domain} \quad 0 \le x \le L,\ 0 \le y \le H, \tag{3.35}$$

$$\mathbf{n} \cdot (D\nabla c) = 0 \text{ (convective flux only)} \quad \text{at the outlet} \quad x = L,\ 0 \le y \le H, \tag{3.36}$$

$$\mathbf{n} \cdot (c\mathbf{u} - D\nabla c) = 0 \text{ (no flux)} \quad \text{at the walls} \quad \begin{array}{l} 0 \le x < A,\ y = 0, \\ B < x \le L,\ y = 0 \\ 0 \le x \le L,\ y = H, \end{array} \tag{3.37}$$

$$c = c_0 \text{ (fixed concentration)} \quad \text{at the inlet} \quad x = 0,\ 0 \le y \le H, \tag{3.38}$$

$$\mathbf{n} \cdot (D\nabla c) = k_{on}c_s(b_{max} - b) - k_{off}b \quad \text{at the reaction surface} \quad A \le x \le B,\ y = 0 \tag{3.39}$$

(analyte flux given by second order reversible reaction, Eq. (3.12)).

The values of constants that need to be specified include the diffusion coefficient D of the analyte, the association and dissociation constants k_{on} and k_{off}, and the surface receptor density b_{max}. The diffusion coefficient for a small protein in an aqueous solution is $\sim 10^{-10}\ m^2\ s^{-1}$. k_{on} and k_{off} values range from $\sim 10^2$–$10^3\ mol^{-1}\ m^3\ s^{-1}$ (Housden et al., 2004) and 10^{-2}–$10^{-3}\ s^{-1}$, respectively. b_{max} values typically span a wide range between 10^{-5}–$10^{-8}\ mol\ m^{-2}$. Average microchannel fluid velocities are generally between 10^{-6}–$10^{-2}\ m\ s^{-1}$. Inlet concentrations correspond to the concentration of analyte present in a sample, which varies between $\sim 10^{-12}$–$10^{-6}\ mol\ L^{-1}$ (~pM–μM).

The binding reaction sites are frequently used to represent microfluidic immunoassays. Fig. 3.2B illustrates different immunoassays that have been modeled. The direct immunoassay involves monitoring the direct binding of an analyte to a surface-immobilized antibody. In the case of fluorescence detection, the analyte must be labeled with a fluorescent reporter molecule. The sandwich immunoassay involves flow of an analyte solution over a surface functionalized with analyte-specific antibodies, leading to analyte binding at the surface. A labeled secondary antibody is subsequently introduced that then binds to a secondary binding site on the analyte and generates a measurable signal (e.g., fluorescence, colorimetric) that is correlated to the concentration of analyte in solution. The step involving the labeled secondary antibody is not usually modeled because high concentrations of antibody are used and rapid binding is assumed. Competitive immunoassays have also been modeled in microfluidic devices. This type of assay involves incubating a sample containing the analyte with a known concentration of analyte-specific antibodies. The equilibrated solution is then introduced into a microchannel containing a binding reaction site functionalized with the analyte. The rate of binding of the bound and unbound antibodies to the binding reaction site is monitored, yielding a signal correlated to the analyte concentration. These immunoassay approaches have been studied and modeled extensively in the microfluidics literature, and several of them are described in detail below.

Modeling of microfluidic biochemical assays initially began in studies from modeling surface reaction binding kinetics to determine binding affinity constants using systems such as the Biacore instrument. Myszka et al. (1998) developed, numerically simulated, and validated a two-compartment model describing the dependence of ligand–receptor binding kinetics at a surface on convective mass transport. This early model considered the binding reaction site as mass transfer between an outer bulk compartment with constant analyte concentration and an inner compartment where analyte concentration changed over time due to binding. Transport between the two compartments was governed by a diffusion-limited mass transfer coefficient, k_M. Physically, the assumption of constant analyte concentration in the outer bulk compartment was satisfied by replenishing analyte supply using flow. In the case where insufficient convective transport is present, Vijayendran et al. (1999) developed and validated a computational model of mass transport-limited binding reactions where transport occurred solely by diffusion. Computational models have gradually improved to operate under both reaction-limited and transport-limited regimes (Jenkins et al., 2004), and have since been extensively applied to the modeling of microfluidic quantitative assays where the full set of governing equations may be solved within a modeling domain.

Using the finite difference method, Zimmermann et al. (2005) modeled convective–diffusive transport coupled to antigen–antibody binding in a microfluidic immunoassay with capillary flow. Model parameters including channel dimensions, binding constants and flow velocities were based on their previous experimental work developing a microfluidic sandwich immunoassay targeting an inflammatory cytokine, tumor necrosis factor-α (TNF-α) (Benn et al., 2006). They determined that increased convective transport ($\sim 10^{-6}$–10^{-4} m s^{-1} average velocities) significantly increased the rate of binding and therefore reduced immunoassay times at the cost of incomplete collection of analyte molecules that pass over the reaction site without binding. This has important implications in cases where sample volumes and/or analyte concentrations are small (~ 5 µL, $c/K_D \ll 1$) where it is critical to maximize the number of molecules captured. They also found that large surface receptor densities increased the number of molecules bound at the cost of significantly increasing assay times, which in turn can be mitigated by increasing convective transport. These findings were used to optimize the TNF-α microfluidic immunoassay with respect to increasing the speed, decreasing the required volumes, and maximizing the assay dynamic range.

Hu et al. (2007a) investigated the same type of assay with the addition of multiple reaction sites and electrokinetically-driven flow. With four sequential binding reaction sites in the microfluidic channel, the equilibrium binding times for a given fluid velocity increase for sites that are further downstream due to the increasingly depleted supply of analyte. By alternating the binding sites between the upper and lower surfaces of the channel, it is possible to minimize depletion and differences between equilibrium binding times. For their given binding reaction parameters, they demonstrated that electrokinetically-driven flow can reduce the time required for the reaction to reach equilibrium relative to pressure-driven flow since uniform velocity profiles arising from electroosmotic flow avoid the gradient of small fluid velocities near the channel wall of parabolic flow profiles arising from pressure-driven flow.

Friedrich et al. (2008) developed a theoretical model of a microfluidic assay described by an irreversible second order binding reaction. Analyte solution flow was modeled as a fluid "plug", representing a limited sample volume. Three strategies were developed for maximizing the number of analyte molecules captured and ensuring uniformity of binding across the entire reaction site: (i) gradually decreasing the flow velocity as the analyte plug traveled across the binding surface, (ii) increasing the density of surface receptors along the reaction site, and (iii) using asymmetric tapered channel and reaction site geometries. Using fitting parameters and functions, the optimal rate of velocity and density change as a function of position were determined, as well as the optimal reaction site shape. The resulting velocity, density, and shape functions were then implemented in a computational model and verified numerically for complete analyte capture and uniform binding across the surface.

Computational modeling of competitive immunoassays has also been performed. Foley et al. (2007) modeled a competitive immunoassay implemented in a microfluidic T-sensor where a small analyte diffuses into a parallel stream containing antibodies that then bind to an analyte-functionalized surface. They studied the importance of flow rate, microchannel dimensions and antibody concentrations on assay sensitivity. More recently, the same group modeled a microfluidic competitive immunoassay (Fu et al., 2009) that eliminates the need for a *T* sensor and targets phenytoin, an important anti-seizure medication. The model was used to evaluate the sensitivity of the assay signal, i.e., the rate of analyte binding, to different parameters. An ideal quantitative assay output signal would only be sensitive to the analyte concentration while changes to other parameters such as surface receptor concentration and flow would produce minimal changes in the output signal. The assay signal was found to significantly increase as average flow velocities increased between $\sim 10^{-5}$–10^{-4} m s^{-1} with negligible signal increases beyond $>10^{-4}$ m s^{-1}, indicating sensitivity of the assay signal to average flow velocities between $\sim 10^{-5}$–10^{-4} m s^{-1} and minimal sensitivity at average flow velocities $>10^{-4}$ m s^{-1}. This demonstrates the role of convective transport in increasing binding rate until the reaction-limited kinetics regime is reached. The assay signal was also found to be sensitive to increasing phenytoin concentration and surface density of phenytoin binding sites. The significant sensitivity of the assay to parameters other than analyte concentration highlights the importance of replicating assay parameters for accurate, reproducible measurements.

Many of the previous examples involved theoretical models of microfluidic assays that had previously been developed where some model parameters were varied while others were assigned fixed values. While qualitative trends may be extracted from these studies, quantitative findings are dependent on the specific microfluidic device geometry being modeled, and are limited to the parameter values studied. Using order-of-magnitude estimates, Squires et al. (2008) developed a generalized analysis of biosensing assays for the straight microchannel geometry in Fig. 3.2. Critical biosensor performance characteristics such as equilibration times and sensor fluxes were approximated and predicted using dimensionless numbers. Their analysis centered on estimating the size of the depletion region that forms above a biosensor as analyte binds to the surface and is depleted from the bulk solution. Two Péclet numbers were derived: (i) Pe_H, which describes the size of the depletion layer relative to the channel height, and (ii) Pe_s, which describes the size of the depletion region relative to the size of the biosensor, and in contrast to Pe_H, accounts for microchannel geometry. Pe_H is identical to Eq. (3.25) while $\mathrm{Pe}_s = 6(\frac{L_s}{H})^2\,\mathrm{Pe}_H$. If $\mathrm{Pe}_H \gg 1$, the size of the depletion region is smaller than the channel height and convective transport dominates. If Pe_H and $\mathrm{Pe}_s \gg 1$, then the depletion region is also thin relative to the biosensor size. Pe_H was correlated to numerically-simulated dimensionless flux at the biosensor surface for different Pe_s, and compared to the flux corresponding to the full collection of analyte. A Damköhler number Da was also derived and correlated to the equilibration time where it was found that the equilibration time τ_{eq} is approximately equivalent to the reaction-limited equilibration time τ_R scaled by Da, i.e., $\tau_{eq} \sim \mathrm{Da}\,\tau_R$. Using these three dimensionless numbers Pe_H, Pe_s, and Da, it is possible to obtain estimates of the equilibration time and reaction site flux for a wide range of design parameters including flow velocity, sensor dimensions, surface receptor densities, and binding constants.

3.4.2 Straight Microchannel with Cell Monolayer

In a microfluidic cell culture, it is important to ensure that a proper supply of nutrients is available to the cells. Perfusion culture at low flow rates enables nutrient delivery to the cells while also removing waste products that would otherwise accumulate under static culture conditions (Young and Beebe, 2010). This is particularly important in microfluidic devices where low media volumes and high cell densities are common. Using this technique,

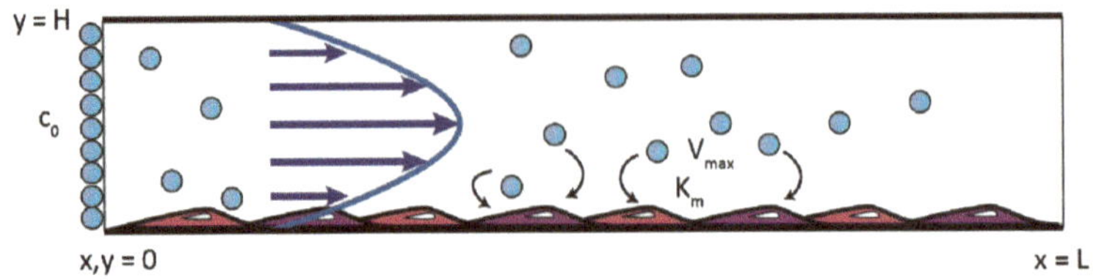

FIGURE 3.3 Representative microfluidic cell culture device modeling domain. A cell monolayer of length L is cultured at the base of a microchannel ($y = 0$). Nutrient concentration c_0 is fixed at the inlet ($x = 0$). Laminar flow of cell culture medium supplies nutrients to the cells, which are consumed at a rate described by V_{max} and K_m.

long term culture of cells within microchannels has been demonstrated (Kimura et al., 2008; Giulitti et al., 2013).

Mathematical and numerical modeling assist in the development of microfluidic cell culture models by enabling the study of critical design considerations without the need for extensive, costly and time-consuming experiments. These considerations include flow-induced shear stress, nutrient consumption by cells, and the spatiotemporal distribution of nutrient concentration gradients within the device. Flow-induced shear stress, an important regulator of cell function (Papaioannou and Stefanadis, 2005), must be controlled when cells are shear-sensitive. The transport of oxygen in microfluidic cell culture devices has been the most common design consideration examined (Mäki et al., 2015). Aside from its critical role in cell viability, oxygen has a low solubility in cell culture medium (Allen et al., 2001), which results in its limited availability and rapid depletion from the microenvironment. Therefore, adequate supply needs to be ensured to maintain stable culture conditions (Naciri et al., 2008). Gradual consumption of oxygen and nutrient supply along the channel length in perfusion culture may also result in significant concentration gradients and uneven distribution of nutrients that must be considered.

Fig. 3.3 illustrates a representative 2D modeling domain for a simple microfluidic cell culture device consisting of a microchannel with rectangular cross-section of height, width and length. Assuming fully developed flow at the inlet, low Re within the domain allow laminar flow with a parabolic velocity profile to be imposed throughout the domain, eliminating the need to solve the momentum equations. A fixed nutrient concentration is also present at the inlet. On the bottom surface of the microchannel, a cultured cell monolayer of length L consumes nutrients, the kinetics of which are modeled using Michaelis–Menten kinetics.

As in Section 3.4.1, the microchannel aspect ratio H/W is frequently assumed to be sufficiently small such that the fully developed velocity profile is invariant along the channel width, leading to a 2D model. Another common assumption is that the fluid entering the domain (and initially in the domain) is saturated with nutrients such as oxygen. Furthermore, the bottom surface is typically glass and impermeable to nutrients. The length of the modeling domain typically corresponds to the length of the cell monolayer. This results in the following set of boundary conditions for a 2D model where $\mathbf{n}$ is the outward unit normal vector to the surface:

$$\mathbf{u}=0 \text{ (no-slip condition)} \quad \text{at the walls} \quad \begin{array}{l} 0 \le x \le L,\ y=H, \\ 0 \le x \le L,\ y=0, \end{array} \tag{3.40}$$

$$u(y)=6u_{avg}\frac{y}{H}\left(1-\frac{y}{H}\right) \text{ (pressure-driven flow)} \quad \text{entire domain} \quad 0 \le x \le L,\ 0 \le y \le H, \tag{3.41}$$

$$\mathbf{n}\cdot(D\nabla c)=0 \text{ (convective flux only)} \quad \text{at the outlet} \quad x=L,\ 0 \le y \le H, \tag{3.42}$$

$$\mathbf{n}\cdot(c\mathbf{u}-D\nabla c)=0 \text{ (no flux)} \quad \text{at the walls} \quad 0 \le x \le L,\ y=H, \tag{3.43}$$

$$c=c_0 \text{ (fixed concentration)} \quad \text{at the inlet} \quad x=0,\ 0 \le y \le H, \tag{3.44}$$

$$\mathbf{n}\cdot(D\nabla c)=\frac{V_{max}\gamma c}{K_m+c} \quad \text{at the cell surface} \quad 0 \le x \le L,\ y=0 \tag{3.45}$$

(nutrient consumption flux given by Michaelis–Menten kinetics, Eq. (3.21)).

In the common scenario where oxygen concentrations are simulated, the fixed inlet concentration is set to the saturation concentration of oxygen in cell culture medium ($0.2\ \mathrm{mol\,m^{-3}}$), and the oxygen diffusion coefficient in culture medium is $2.5 \times 10^{-9}\ \mathrm{m^2\,s^{-1}}$ (Buchwald, 2009). The diffusion coefficient for a small protein in an aqueous solution is $\sim 10^{-10}\ \mathrm{m^2\,s^{-1}}$. Parameter values for V_{max} and K_m depend on the cell type and culture conditions. For hepatocytes, a commonly modeled cell type, V_{max} and K_m values in models are $3.8 \times 10^{-16}\ \mathrm{mol\,s^{-1}\,cell^{-1}}$ and $6.6 \times 10^{-3} - 7.39 \times 10^{-3}\ \mathrm{mol\,m^{-3}}$, respectively (Tilles et al., 2001; Tan et al., 2013). Cell densities vary significantly depending on the application, but values between $\sim 10^7$–$10^9\ \mathrm{cells\,m^{-2}}$ are standard. Microchannel average velocities in cell culture models range approximately between 10^{-6}–$10^{-3}\ \mathrm{m\,s^{-1}}$ (Zahorodny-Burke et al., 2011).

In designing microfluidic cell culture models, it is important to determine conditions where sufficient oxygen will be delivered to cells along the entire channel length, and to predict the presence of oxygen concentration gradients that may affect cell behavior. Zeng et al. (2006) studied the effect of mass transport and cell-related parameters on oxygen concentration at the cell monolayer along the length of a microchannel reactor. They derived two non-dimensional parameters, the effective distance L^* and the concentration–reaction parameter ε_k, which are respectively:

$$L^* = \frac{x}{H}\frac{1}{\mathrm{Pe}}, \tag{3.46}$$

$$\varepsilon_k = \frac{(\overline{c}_0-1)}{\mathrm{Da}_m}\left(\frac{1+3\overline{K}_m}{1+\overline{K}_m}\right) \tag{3.47}$$

where $\overline{c}_0$ is the non-dimensionalized concentration at the base of the microchannel, and Da_m and $\overline{K}_m$ are given in Eqs. (3.32) and (3.33). The effective distance parameter L^* contains channel height and flow velocity information (embedded in Pe), while the concentration–reaction parameter contains cell density and cell oxygen consumption rate information (Da_m, K_m). By correlating numerical simulation results to these two dimensionless parameters, the authors determined that these parameters were capable of collapsing data for different cell consumption rates and flow velocities onto nearly identical curves for Da <0.5 and Pe <1000 and

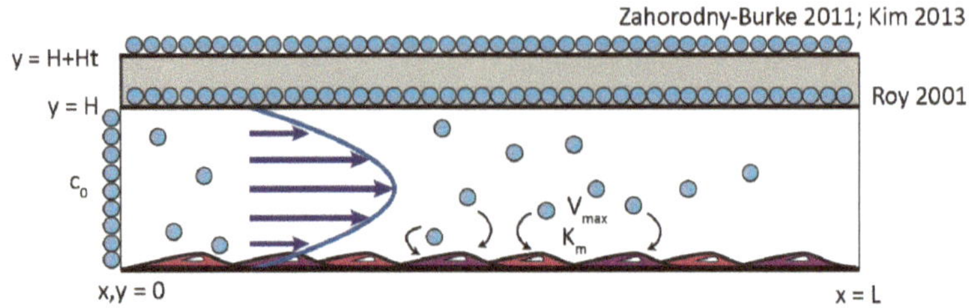

FIGURE 3.4 Representative microfluidic oxygenator modeling domain. The channel upper domain ($y \in [H, H_t]$) is permeable to oxygen (blue circles), which diffuses across the culture–upper domain interface as oxygen is depleted by the cell monolayer ($y = 0$). The location of fixed oxygen concentration in the upper domain is marked for different works by Roy et al. (2001) ($y = H$), Zahorodny-Burke et al. (2011) ($y = H_t$) and Kim et al. (2013) ($y = H_t$).

microchannel aspect ratios $H/W < 0.4$, where the 2D model approximation is valid. The effective distance and concentration–reaction parameters can therefore be used to describe the concentration of oxygen in a microchannel for a wide variety of reaction and mass transport conditions. Using these parameters, they determined critical channel lengths for the concentration–reaction parameter ε_k between 0–4 such that the outlet oxygen concentration would be equal to K_m, avoiding anoxic conditions in the device and thus ensuring sufficient oxygen supply to the cells.

3.4.2.1 *Oxygenators*

The previous model assumes that the microchannel material is impermeable to oxygen. It does not account for mass transfer that may occur from a permeable channel upper surface into the cell culture medium. Microfluidic devices for cell culture, however, are frequently fabricated in poly(dimethylsiloxane) (PDMS) due in part to its well-known property of permeability to gases such as oxygen and carbon dioxide (Mcdonald et al., 2000; Mehta et al., 2007). When a device is fabricated with a gas permeable material such as PDMS, the upper surface of the microchannel can no longer be modeled as an impermeable wall. Fig. 3.4 illustrates a representative modeling domain for a microfluidic cell culture device with an internal oxygenator, i.e., an oxygen permeable top layer.

Roy et al. (2001) developed a numerical model to assess and compare oxygen concentrations in flat-plate bioreactors with and without an internal oxygenator. This was motivated by previous work that demonstrated sensitivity of rat hepatocytes to oxygen levels in bioreactors (Tilles et al., 2001). For their given channel dimensions, numerical simulations of oxygen concentration profiles indicated that low Pe flows (<80) resulted in hypoxic conditions for rat hepatocytes if an internal oxygenator was not used. Conversely, with the use of an oxygenator, their simulations demonstrated that hypoxic conditions could be avoided. The oxygenator, a permeable membrane, was numerically modeled using a boundary condition that imposes a constant oxygen partial pressure at the interface between the upper surface of the channel and the culture medium (Fig. 3.4 – Roy et al., 2001):

$$D_{O2}\frac{\partial c}{\partial y} = \sigma[c_g - c_{O2}] \quad \text{at } y = H \tag{3.48}$$

where σ is the mass transfer coefficient (m s^{-1}), c_g is a fixed gas phase oxygen concentration, and c_{O2} is the concentration of oxygen. The oxygen permeability of the membrane was described using the Sherwood number Sh:

$$\text{Sh} = \frac{\sigma H}{D_{O2}}. \tag{3.49}$$

For a given cellular rate of oxygen consumption, i.e., Da_m, it is possible to avoid hypoxic conditions by selecting a membrane with an appropriate Sh where larger Da_m results in larger Sh being necessary. In combination with strategies of increasing c_g and the Pe number, the oxygenator was demonstrated to be an effective means of achieving sufficient oxygen concentrations in the bioreactor. In the case of the bioreactor with no internal oxygenator, strategies are limited to increasing Pe, typically by increasing flow velocity, which may subject cells to undesired levels of shear stress.

The previous oxygenator model assumes that the permeable membrane layer remains saturated with oxygen. While this is true for thin PDMS membranes, thicker layers of material may not remain saturated, leading to concentration gradients within the material and possible depletion of oxygen. Zahorodny-Burke et al. (2011) developed a computational model that incorporates the permeable layer as a separate domain (Fig. 3.4 – Zahorodny-Burke et al., 2011) with an oxygen flux throughout the permeable layer domain and at the culture medium-permeable layer interface. New parameters such as the permeable layer thickness, and permeable layer-specific diffusivity, and oxygen solubility become important. Using stiff-spring analysis, they described boundary conditions that govern the oxygen flux across the permeable layer-culture medium interface, and specify mass balance equations that enforce continuity where $\mathbf{n}$ is the outward unit normal vector to the surface:

$$\mathbf{n} \cdot \left(c\mathbf{u_x} - D_{O2}\frac{\partial c_{O2}}{\partial y} \right) = M\left(c_p - K_c\right) \qquad \text{at } y = H, \tag{3.50}$$

$$\mathbf{n} \cdot \left(-D_p\frac{\partial c_p}{\partial y} \right) = M\left(K_c - c_p\right) \qquad \text{at } y = H, \tag{3.51}$$

$$\nabla \cdot (c_{O2}\mathbf{u_x} - D_{O2}\nabla c_{O2}) = 0 \qquad \text{at } y = H, \tag{3.52}$$

$$\nabla \cdot (-D_p \nabla c_p) = 0 \qquad \text{at } y = H \tag{3.53}$$

where c_{O2} is the concentration of oxygen in culture medium, c_p is the concentration of oxygen in the permeable layer, D_{O2} is the diffusion coefficient of oxygen in medium, D_p is the diffusion coefficient of oxygen in the permeable layer, K_c is the partition coefficient (ratio of permeable layer and medium solubilities), and M is the spring velocity selected to ensure sufficient oxygen exchange across the interface. These boundary conditions ensure a continuous oxygen flux while allowing discontinuous concentrations across the interface resulting from differences in oxygen solubilities. Model validation was performed by comparing oxygen concentration distributions to the model previously described in Roy et al. (2001) for the case of a thin permeable layer. They determined that diffusive transport of oxygen through PDMS permeable layer thicknesses between 25 μm and 1.9 mm ensured sufficient and near-uniform oxygen supply at the cell monolayer even for low average flow velocities of 10^{-6} m s^{-1} in a 15-mm long channel. Lower channel heights also resulted in increased concentrations of oxygen over the channel length. Finally, they demonstrated that low oxygen diffusivities in other polymers used to fabricate microfluidic devices, such as cyclic olefin copolymer (COC) and poly(methylmethacrylate) (PMMA), result in complete depletion of oxygen over a short channel length unless elevated flow velocities are used. Material selection therefore has important ramifications on oxygen transport in microfluidic cell culture devices.

Kim et al. (2013) modeled a separating-type oxygenator, a configuration identical to Zahorodny-Burke et al. (2011) with the addition of an oxygen gas channel above the permeable top layer, allowing greater control of oxygen concentration. To model the interface between the PDMS layer and culture medium, continuous partial pressures of oxygen were enforced across the interface through application of Henry's law (Eq. (3.54)), and imposed mass balance equations (Eq. (3.55)):

$$\frac{c_{O2}}{S_m} = \frac{c_p}{S_p} \quad \text{at } y = H, \tag{3.54}$$

$$D_p \frac{\partial c_p}{\partial y} = D \frac{\partial c_{O2}}{\partial y} \quad \text{at } y = H, \tag{3.55}$$

$$c_p = S_p p_g \quad \text{at } y = H + H_t \tag{3.56}$$

where c_{O2} and c_p are the concentrations of oxygen in culture medium and polymer (PDMS), respectively. S_m, S_p, and p_g are the solubilities of oxygen in culture medium and PDMS (mM atm^{-1}), and the partial pressure of oxygen (Pa), respectively. Eq. (3.56) imposes a constant concentration at the top surface of the permeable layer. In this model, a theoretical "rise time" is defined as the time required for oxygen to reach 90% of its stabilized value in a microchannel with no cells. Theoretical rise times between $\sim 10^1$–10^4 s were found for different permeable layer and channel heights. Low rise times of $\sim 10^2$ s occurred at permeable layer thicknesses <1 mm. They also derived an analytical equation for the oxygen pressure drop (OPD) along the length of the cell monolayer. Large OPDs indicate that large concentration gradients are present along the length of the microchannel. For a fixed cell type, a dimensionless effective length parameter L^* was used to correlate numerically simulated oxygen partial pressure P^*_{cell} (i.e., concentration) along the microchannel length for different Pe, channel heights and PDMS membrane thicknesses. For $L^* > 1$, the oxygen concentration profile was fully-developed with P^*_{cell} approaching 1-OPD. For low channel heights and thicknesses, P^*_{cell} approached unity, meaning that OPDs were small and concentration gradients in the microchannel were minimal.

3.4.2.2 *Cell Signaling, Cell Proliferation, and Co-Culture*

Cells modulate their local microenvironment through autocrine and paracrine signaling. For example, growth factors such as epidermal growth factor (EGF) play important roles in promoting cell proliferation (Schlessinger et al., 1983). The resulting distribution of secreted factors in the microfluidic cell culture model may therefore induce different cell behaviors depending on the concentration of factors present (Kothapalli et al., 2011; Chung et al., 2005).

Fig. 3.5 shows a schematic of a microfluidic cell culture model with autocrine signaling at the cell monolayer. Autocrine signaling was computationally modeled by Mehta and Linder-

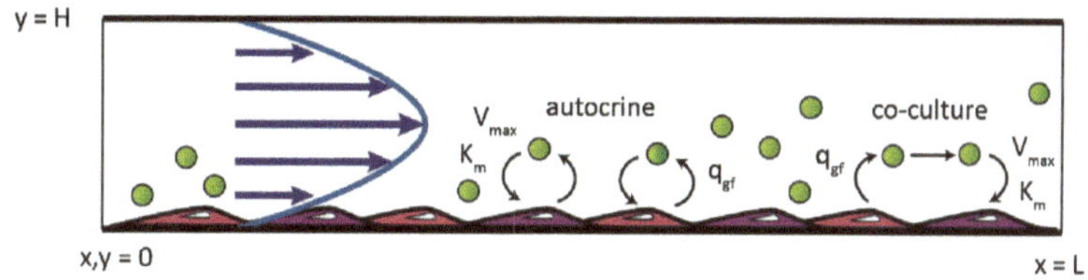

FIGURE 3.5 Representative microfluidic cell culture device modeling domain incorporating cell secretion and co-culture. Autocrine signaling involves the cell monolayer ($y = 0$) both releasing and absorbing a secreted factor at rates governed by q_{gf} and V_{max}/K_m, respectively. Co-culture of releasing and absorbing cell types involves one cell type releasing secreted factor at rate q_{gf} and absorbing at a rate governed by V_{max}/K_m.

man (2006) in a microfluidic cell culture device as a constant secretion rate combined with uptake rates governed by Michaelis–Menten kinetics. This leads to the following flux boundary condition at the cell monolayer:

$$D_{gf}\frac{\partial c_{gf}}{\partial y} = q_c\gamma - \frac{V_{max}c_{gf}}{K_m + c_{gf}}\gamma \quad \text{at } y = 0 \tag{3.57}$$

where D_{gf} is the growth factor diffusion coefficient, c_{gf} is the concentration of growth factor, q_c is the secretion rate (mol cell^{-1} s^{-1}), and γ is the cell surface density (cell m^{-2}). Growth factor concentrations at the microchannel outlet were found to decrease for lower cell secretion-to-uptake rate ratios as expected. Outlet concentrations exhibited an approximately linear dependence on secretion and uptake rate values. When the cell secretion-to-uptake rate ratio was $\sim$1, growth factor accumulation at the cell monolayer was limited in cases where increased convective flow or a lower value in the ratio H/L reduced the local growth factor concentrations in the device. The availability of growth factors in perfusion culture devices may therefore be reduced in comparison to static culture conditions, potentially inducing different cell behaviors. The fraction of growth factor lost was found to increase with higher Pe and lower values in the ratio H/L where a minimum of 55% was lost at a moderate $\text{Da}_m \sim 0.349$, increasing to over 90% as Pe increased and H/L decreased. Simulations involving recirculation of growth factors showed a decrease in the fraction lost, mitigating growth factor concentration reductions in the microenvironment.

Similarly, Zeng et al. (2007) developed a computational model of cell co-culture in a microfluidic device where one secretory cell type releases a growth factor that is then consumed by a second absorbing cell type. For example, hepatocyte function in culture may be improved when co-cultured with fibroblasts (Bhatia et al., 1998). A constant secretion flux is assumed while consumption is modeled using Michaelis–Menten kinetics:

$$D_{gf}\frac{\partial c_{gf}}{\partial y} = q_c\gamma \text{ (releasing cell type)} \quad \text{at } y = 0, \tag{3.58}$$

$$D_{gf}\frac{\partial c_{gf}}{\partial y} = -\frac{V_{max}c_{gf}}{K_m + c_{gf}}\gamma \text{ (absorbing cell type)} \quad \text{at } y = 0. \tag{3.59}$$

They modified their previously derived concentration–reaction parameter ε_k in Zeng et al. (2006) (Eq. (3.47)) to incorporate a term that accounts for the secretion flux:

$$\varepsilon_k = \frac{(\overline{c}_0 - 1)}{\mathrm{Da}_m \, \beta}, \tag{3.60}$$

$$\beta = \frac{1 + 3\overline{K}_m}{1 + \overline{K}_m} - \frac{\mathrm{Da}_s}{\mathrm{Da}_m} \tag{3.61}$$

where $\overline{c}_0$ is the dimensionless concentration along the microchannel base, Da_s is the secreting cell type Da number, Da_m is the absorbing cell type Da number, and $\overline{K}_m$ is the dimensionless Michaelis–Menten constant. Dimensionalizing the concentration–reaction parameter provides the distribution of secreted factor along the axial direction of the microchannel. Using the modified concentration–reaction parameter ε_k, they demonstrated a close correlation between ε_k and the effective distance parameter L^* (Eq. (3.46)) for different $\overline{K}_m$, β, and the condition of positive flux $\mathrm{Da}_m \, \beta > 0$ where absorption flux exceeds secretion flux. The correlation was not valid for $\mathrm{Da}_m \, \beta < 0$ where secretion flux exceeds absorption flux. The model may therefore be useful in cases where the absorbing cell type is present in higher densities than the secreting cells, which is a common scenario in hepatocyte culture. The model was used to determine the critical inlet concentration for a given effective channel length that results in a concentration K_m at the channel outlet, avoiding complete depletion of the secreted factor by the absorbing cells along the microchannel length.

Mehta and Linderman (2006) modeled the process of cell proliferation in a computational model of a microfluidic cell culture device. Under the proper culture conditions, cells grow and proliferate in culture, thereby increasing in density and resulting in increased nutrient demand over time. This cellular process was modeled using a continuous cell growth model based on work by Pathi et al. (2005) where cell density over time is described by the following equation:

$$\frac{\partial \gamma_j}{\partial t} = \mu_{g,j}\gamma_j - k_d \gamma_j \quad \text{for } j = 1, \ldots, N. \tag{3.62}$$

This equation describes the change in cell density for cell type j over time, which depends on proliferation rate parameter $\mu_{g,j}$ and death rate constant k_d. For a given cell type and nutrient, the proliferation rate parameter depends on nutrient concentration and cell density:

$$\mu_g = \mu_{max}\left(1 - \frac{\gamma}{\gamma_{max}}\right)\left(\frac{k_1 c}{1 + k_2 c + k_3 c^2}\right) \tag{3.63}$$

where μ_{max} is the maximum cell density, c is the concentration of nutrients, and k_1, k_2, and k_3 are constants that describe the effect of nutrient concentration on the proliferation rate. Large k_3 values imply that large nutrient concentrations reduce the cell proliferation rate. Eqs. (3.62) and (3.63) describe a logistic model of cell proliferation. The consumption of nutrients by Michaelis–Menten kinetics was coupled to the growth and proliferation of cells in the culture devices in order to predict cell densities over time. The presence of significant axial nutrient concentration gradients led to similar gradients in cell density along the microchannel length while uniform axial nutrient concentrations led to uniform cell densities over time.

Hu et al. (2007b) extended this work by applying different oxygen concentration gradients (e.g., linear, parabolic) across the inlet width in a 3D model of the culture device and studying the spatial variations in cell densities across the channel width at different points along the length.

3.4.3 Alternative Geometries

One of the many advantages of microfabrication and rapid prototyping capabilities is the freedom to design and explore different device geometries. Indeed, microfluidic cell culture platforms and biochemical assays are not limited to only straight microchannel geometries. This section describes mathematical and numerical modeling, based on the same principles described above, of microfluidic device designs that utilize alternative geometries to control fluid flow and mass transport.

3.4.3.1 *Upper and Lower Microchannels Separated by a Porous Membrane*

Inamdar et al. (2011) developed a computational model of a two layer microfluidic device in which upper and lower chambers are separated by a porous membrane. A cell monolayer is located in the lower chamber where fluid flow is minimized. Convective flow of culture medium in the upper chamber replenishes nutrients and oxygen that diffuse across the porous membrane to the lower chamber where they are consumed by the cell monolayer. This device configuration enables the continuous replenishment of nutrients while shielding the cells from elevated shear stress, which is important in applications such as mesenchymal stem cell (MSC) differentiation (Zhao et al., 2007).

Fig. 3.6 illustrates the modeling domain of the two layer microfluidic cell culture device. Flux of oxygen across the membrane is governed by the following boundary condition:

$$D\frac{\partial c_u}{\partial y} = \frac{D_m}{\delta_m}\left[c_l\left(x, H\right) - c_u\left(x, H_t\right)\right] \tag{3.64}$$

where D is the diffusion coefficient of oxygen in culture medium, c_u is the oxygen concentration in the upper channel, c_l is the oxygen concentration in the lower channel, D_m is the diffusion coefficient of oxygen in the porous membrane, and δ_m is the membrane thickness. This boundary condition assumes that membrane permeability is small enough such that only diffusive flux is present. The flux of oxygen across the membrane is equivalent to the flux of oxygen at the cell monolayer, which was assumed to be governed by first-order Michaelis–Menten kinetics (Eq. (3.21) where $s \ll 1$). No oxygen was assumed to be initially present in the lower chamber in order to isolate the effect of oxygen flux from the upper chamber.

For a given cell oxygen consumption rate, a higher flow velocity and thus shear stress was required for a single layer microchannel geometry to achieve a uniform oxygen concentration across the device length compared to the two layer geometry. This shear stress exceeded the 0.03 Pa threshold selected by the authors based on MSC sensitivity to shear stress. Increasing the upper chamber flow velocity increased the uniformity of the oxygen distribution in the lower chamber. The two-layer microfluidic membrane device may therefore be used to achieve uniform oxygen concentrations while minimizing the shear stress experienced by cultured cells.

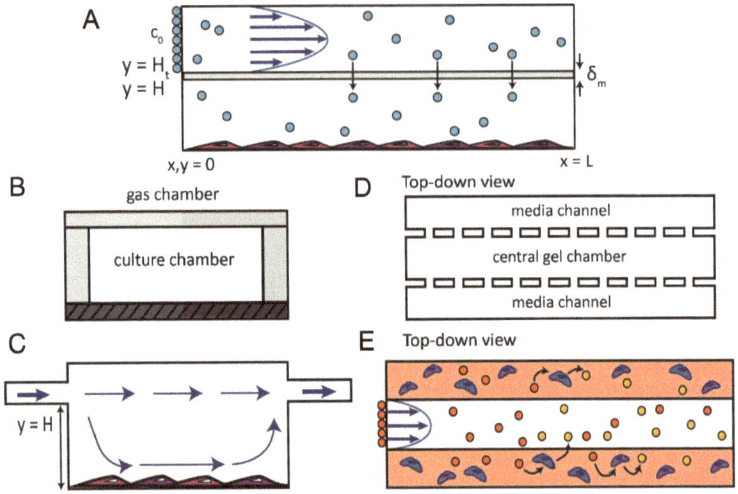

FIGURE 3.6 Schematics of alternative microfluidic device geometries. (A) Upper and lower microchannels separated by a porous membrane. Nutrients are transported from the upper to lower microchannel via diffusive flux. Cells in the lower channel are protected from elevated shear stresses. (B) Microchamber (2D cross-section). The chamber walls and top are permeable to gases. (C) Perfused microchamber (2D cross-section). Cells located at the base are shielded from elevated shear stresses. (D) Central channel flanked by two media channels (top-down view). (E) Straight microchannels with porous walls. Perfusion flow transports drug substrate (red circles) into the porous hydrogel walls where it is metabolized into a product (yellow circles) by embedded cells.

3.4.3.2 *Microchamber with Thin Membrane Top for Static Culture*

Most work in the literature on modeling microfluidic cell culture involves the study of oxygen transport. Mäki et al. (2015) developed the first model of carbon dioxide (CO_2) transport in silicone-based devices motivated by the fact that CO_2 concentrations are critical to maintaining cell culture medium pH *in vitro*. In practice, 5% CO_2 is typically dissolved in culture medium. A microchamber geometry was modeled where static cell culture conditions were assumed and a thin PDMS membrane covering the top of the microchamber enables efficient transfer of CO_2 between a gas chamber and culture medium chamber (Fig. 3.6B). The microchamber walls were also composed of PDMS, allowing for increased gas exchange. Experimental monitoring of % CO_2 over time was used to determine diffusion and mass transfer coefficients in and between the different material phases (gas, liquid and PDMS), which were previously unavailable in the literature. These parameters were then used in a computational model incorporating CO_2 transport in the different phases of the microchamber. Significant reductions in CO_2 transport rise times were found for thinner microchamber wall thicknesses. Rise times of ~35 min were computed for a 14-mm wall thickness, whereas prohibitively long rise times of >200 min were obtained for a 30-mm wall thickness.

Mazzei et al. (2010) computationally modeled a microfluidic cell culture device based on microwells found in standard multiwell plates. The device concept consisted of a perfused microchamber with inlet and outlet channels located near the upper portion of the microwell that provided continuous mass transfer and minimized shear stress at the base of the microwell where a hepatocyte monolayer was cultured (Fig. 3.6C). The effects of different inlet/outlet heights on the flow velocities, streamlines, shear stresses and oxygen concentrations in the microwell were studied. For simulated heights of 3, 4, and 6 mm above the cell surface, oxygen concentration at the cell monolayer remained above 0.04 $mol\,m^{-3}$, which is the assumed threshold for normoxic hepatocyte function based on previously reported data (Smith and Mooney, 2007). At 9 mm, however, the oxygen concentration at the monolayer was below this threshold. Maximum shear stresses on the order of 10^{-5} Pa were found for a representative inlet flow rate of 180 $\mu L\,min^{-1}$ at all simulated heights, indicating that hepatocytes would be shielded from shear stresses that would affect function, i.e., shear stresses above 3×10^{-2} Pa. Maximum velocities were on the order of 10^{-6} $m\,s^{-1}$. Increasing inlet/outlet heights led to increased departure from a laminar flow profile, with flow streamlines being no longer parallel to the cell culture surface.

Hsu et al. (2014) numerically simulated the same device configuration and cell type, i.e., hepatocytes, with additional nutrient consumption and protein secretion pathways in addition to oxygen consumption. This work added cell proliferation, albumin and ammonia production, as well as glucose and glutamine production to the computational model. Rate equations were defined for each concentrated species based on consumption/production rates found in the literature. Conditional statements were added to equations to specify minimum or maximum concentration thresholds that determined whether or not a given process occurred. For example, albumin production was given by the following equation:

$$F_{alb} = \gamma(k_1 \ln \tau + k_2 + k_3 c_{O2}) \tag{3.65}$$

where F_{alb} is albumin flux at the cell surface, γ is the cell density, τ is the shear stress (Pa), c_{O2} is the concentration of oxygen, c_{gluc} is the concentration of glucose, and c_{glut} is the concentration of glutamine. Eq. (3.65) applies only if $c_{O2} > 0.03\ mol\,m^{-3}$ and $c_{gluc}, c_{glut} > 0\ mol\,m^{-3}$. The parameters k_1 and k_2 were determined by using experimental data found in Tanaka et al. (2006) relating albumin production to shear stress, and fitting it to a logarithmic function. k_3 was determined by fitting a linear function to data in Anada et al. (2012) relating albumin production to oxygen partial pressure. Similar flux equations were implemented for the other secreted/nutrient species. Species concentrations were simulated over a three-day culture period, and their average concentrations within the microfluidic device analyzed. Experimentally measured albumin production over three days agreed closely with simulated production values.

3.4.3.3 *Central Channel Flanked by Two Media Channels*

Ochs et al. (2014) numerically simulated and experimentally measured oxygen concentrations in a three-channel microfluidic cell culture device where a central channel is flanked by two culture media channels (Fig. 3.6D). The central channel is typically filled with a hydrogel embedded with cells, which can be perfused on both sides by the media channels. Close agreement was found between simulated and experimental oxygen levels over 120 min for endothelial cells and hepatocytes cultured in devices made out of PDMS, COC and poly(methylpentene) (PMP). Initial oxygen levels were ~17% in all cases. Endothelial cells maintained a minimum of ~12% oxygen after 120 min for all materials. Hepatocytes, a more fastidious cell type to culture, maintained ~16% and ~11% for PDMS and PMP, respectively. Oxygen levels quickly dropped to 4% for COC, an oxygen impermeable material. To match experimental results, "hepatic hibernation" (Subramanian et al., 2007) was incorporated into numerical simulations where hepatocytes decreased their oxygen uptake rate to maintain viability in hypoxic conditions. All simulations and measurements were under static fluid conditions.

3.4.3.4 *Straight Microchannel with Porous Walls*

Tourlomousis and Chang (2015) developed a computational model of a straight microchannel domain with walls composed of a porous medium that represents a hydrogel. Instead of being located within the microchannel, cells were encapsulated into the hydrogel walls. This model was used to characterize drug metabolism rates in the microscale platform where cells, modeled as a uniform distribution of enzymes within the porous domain, metabolized a drug substrate transported by perfusion flow through the microchannel and subsequent permeation into the porous material (Fig. 3.6E).

Porous medium flow was governed by the Stokes–Brinkman equations:

$$0 = \nabla \cdot \left[-pI + \frac{\kappa}{\phi} \left(\nabla \mathbf{u} + (\nabla \mathbf{u})^T \right) - (\nabla \cdot \mathbf{u})\, I \right] \tag{3.66}$$

where I is the identity matrix, p is the pressure, κ is the permeability (m^2), and ϕ is the porosity. Flow in the microchannel was governed by the Navier–Stokes equations. They determined that flow velocity profiles were identical at the entrance and exit of the modeling domain, and were continuous across the channel–porous medium boundary, as expected. Shear stresses within the porous medium were very small, on the order of $\sim 10^{-7}$ Pa for a flow rate of 0.015 $mL\,h^{-1}$, with higher shear stresses near the entrance and exit of the channel. The hydrogel enabled cells to be protected from elevated shear stress, mimicking organ structures like the liver sinusoid. Average drug metabolite production was governed by a two site Michaelis–Menten model modified by mechanotransduction terms relating metabolite production to wall shear stress, similar to the modeling performed in Hsu et al. (2014). For a two-day period of simulated perfusion flow, the average drug metabolite concentration was predicted to stabilize around 11 μM for a 120 μM inflow concentration of drug substrate.

3.5 REVIEW OF THEORETICAL MODEL EXPERIMENTAL VALIDATION AND MICROFLUIDIC ON-CHIP ANALYSIS HIGHLIGHTS

Section 3.4 focused on summarizing the major results and findings of theoretical models describing transport in microfluidic cell culture and quantitative assays. In many cases, models were supplemented with experimental evidence supporting theoretical model outputs. This section provides examples of experimental model validation in order to emphasize their predictive nature. Furthermore, examples of technological advances enabling on-chip biomolecule quantitation assays in microfluidic culture models are discussed.

3.5.1 Experimental Validation of Microfluidic Biomolecular Transport and Sensing Models

Qualitative validation of microfluidic bioreactor models has been performed by assessing cell viability under different flow conditions. For example, Mazzei et al. (2010) selected bioreactor dimensions for their perfusion microchamber device based on numerically simulated velocity fields and shear stresses at the cell monolayer. They fabricated the bioreactor, cultured hepatocytes within the system, and then measured cell viability and albumin production, an important marker of hepatocyte function. They determined that no significant decreases in viability and albumin production were observed for flow rates up to 180–500 $\mu L\,min^{-1}$. At 1000 $\mu L\,min^{-1}$, both viability and albumin production decreased due to hepatocyte sensitivity to shear stress. The bioreactor design, guided by computational modeling, was therefore capable of culturing hepatocytes over a range of flow rates as intended. Roy et al. (2001) measured oxygen levels in microfluidic flat-plate bioreactors with and without an internal oxygenator, and determined that oxygen levels predicted by their model qualitatively agreed with their experimental measurements, i.e., the internal oxygenator resulted in significantly higher oxygen concentrations in the device exit region than a device without an oxygenator in which oxygen was nearly depleted.

Quantitative experimental validation of microfluidic models has also been demonstrated. Kim et al. (2013) measured oxygen rise times in an oxygenator with no cells for PDMS oxygenator membrane thicknesses between 10^1–10^4 μm and channel heights between 100–800 μm. They found good agreement between their experimental results and semi-analytical solution. They also measured the OPD across cell monolayers at different cultured cell densities and found that experimental OPD was proportional to surface cell coverage, which agrees with their analytical expression for OPD. Ochs et al. (2014) measured oxygen levels in a three-channel microfluidic device containing ECs or hepatocytes over a two-hour period. Three different materials were also tested: COC, PDMS, and PMP. In all cases, experimental and simulated oxygen levels closely matched. Mäki et al. (2015) constructed a cell culture microchamber with internal oxygenator and simulated the time evolution of CO_2 concentration in the chamber liquid. The average carbon dioxide concentrations were then converted to pH and compared to experimental pH values where a good correlation was found.

Fu et al. (2009) validated numerical modeling of their competitive immunoassay by performing surface plasmon resonance (SPR) imaging measurements of antiphenytoin antibodies binding to a surface functionalized with phenytoin. The rates of binding were then determined for different phenytoin concentrations and compared to the simulated rates where close agreement was found. Their numerical simulations therefore successfully predicted the immunoassay standard curve. Benn et al. (2006) developed an analytical model of microarray printing and DNA hybridization based on the derivation of a total mass transfer coefficient H_{tot}. Experimental validation was performed by demonstrating the ability of H_{tot} to predict the hybridization assay signal intensity over a range of experimental conditions. A linear correlation ($R^2 = 0.93$) between the signal intensity and H_{tot} was found as predicted.

3.5.2 Technological Advances in Microfluidic On-Chip Analysis

As presented above, models of transport within microfluidic cell culture devices and biochemical assays can serve as predictive models to guide the design process and optimize device performance based on specific criteria and requirements. Furthermore, models can support and validate experimental *in vitro* results by relating these findings back to the fundamental governing equations for fluid flow, mass transport, and reaction kinetics. In the context of modeling in microfluidics, the integration of sensing capabilities into microfluidic cell culture platforms has been an area of great interest in the research community. Here, technological advances that enable on-chip detection of different molecules in microfluidic cell culture and analysis platforms are briefly examined, with the objective of bridging theoretical analyses and technology development.

Experimental validation of oxygen concentrations in microfluidic devices is enabled by the integration of on-chip oxygen sensors. Early oxygen sensors utilized integrated electrodes and electrochemical detection, but were easily contaminated, and consumed oxygen as part of the measurement. Alternatively, optical oxygen sensors are highly sensitive, do not consume oxygen and are inherently isolated from surface contaminants. These sensors are based on luminescence intensity or excited-state lifetime quenching of oxygen-sensitive indicator molecules embedded in an oxygen permeable material. The indicator molecules are patterned on microfluidic device substrates where imaging of luminescence using an excitation light source and detector allows for complete 2D oxygen concentration gradients to be quantitated. Optical oxygen sensors have been extensively applied in microfluidic cell culture as reviewed by Grist et al. (2010), and have also been applied to validate theoretical modeling work. Kim et al. (2013) integrated a platinum (II) octaethyl-porphyrin ketone (PtOEPK)-polystyrene oxygen sensor into their microfluidic oxygenator to validate their diffusion time model. Ochs et al. (2014) employed a commercially available handheld microscope for oxygen measurements (Visisens, Regensburg, Germany) that uses oxygen-sensitive foils bonded to the bottom of their three-channel device.

The coupling of mass spectrometry and microfluidic devices has been explored in efforts to advance the "lab-on-a-chip" concept, where analytical tools are miniaturized and integrated onto a single chip, effectively "shrinking" the size of a lab. Mass spectrometry (MS) is a powerful bioanalytical tool due to its sensitivity in detecting a wide range of biochemical species (Gao et al., 2013a). The ability of lab-on-a-chip platforms to manipulate small volumes and incorporate sample handling steps such as extraction and separation complements this analytical tool because time-consuming and laborious sample preparation steps are normally required for MS analysis. While MS is most often applied to proteomics (Oedit et al., 2015), microfluidic cell culture platforms with online MS measurements have also been recently developed for drug testing purposes. For example, the permeation of the drug curcumin through a layer of Caco-2 epithelial cells in a bilayer microfluidic culture device was monitored by incorporating downstream micro solid-phase extraction prior to MS detection (Gao et al., 2013b). In a separate report, the cytotoxicity of products from acetaminophen metabolism on liver cells was assessed in a microfluidic culture system with simultaneous online detection of metabolism products using MS (Mao et al., 2012).

The integration of on-chip biochemical assays in microfluidic cell culture devices has also been demonstrated. Zhou et al. (2015) developed a co-culture device in which transforming growth factor (TGF)-β1 secretion from primary rat hepatocytes and hepatic stellate cells was monitored *in situ* using electrochemical sensors using redox reporter-labeled aptamers. Aptamers are oligonucleotides that undergo binding-induced conformation changes in the presence of target molecules. These conformation changes alter measured electrode currents, producing changes correlated to different analyte concentrations. A system of pressure-actuated "micro-cups" was used to isolate the aptasensors from the cells prior to induction of cell secretion and the subsequent detection period. A computational model of diffusion–reaction in the microfluidic device was used to extract experimental cell secretion rates and confirm that both hepatocytes and stellate cells contributed to TGF-β1 secretion. Dugan et al. (2014) developed a multilayer microfluidic device for the online multiplexed analysis of glycerol and non-esterified fatty acid (NEFA) secretion from cultured adipocytes. Near real-time measurements were achieved using flow-based fluorescence detection of analyte in adipocyte perfusate. The glycerol and NEFA assays required multiple reagent solutions and flow channels. Numerical modeling using COMSOL was used to select flow rates to ensure optimal detection sensitivity as well as proper mixing of reagents prior to detection.

Overall, the integration of on-chip sensing has the potential to enable real-time online monitoring of cellular secretion and other biomolecules of interest such as oxygen, which reduces the need for manual sample acquisition and time-consuming offline analysis while also providing valuable insight into the state of the cellular microenvironment.

3.6 SUMMARY AND CONCLUSIONS

In this chapter, microfluidic cell culture systems and their applications to biomedical research were briefly introduced as well as microfluidic biochemical assays that have the potential for downstream integration in cell culture systems. The fundamental theory and equations governing fluid motion, mass transport and biochemical reactions were then developed and explained in detail. Representative microfluidic biochemical assay and cell culture models from the literature were reviewed and categorized based on model geometry and boundary conditions. Finally, experimental validation of theoretical models as well as technological advances in microfluidic on-chip analysis were reviewed.

The utility of microfluidic cell culture platforms and biochemical assays stems largely from precise spatial and temporal control over fluid motion, mass transport and physical features in the device microenvironment. Theoretical modeling of processes within microfluidic devices offers an additional dimension to this control by providing a fundamental understanding of the physical phenomena that may be used to predict and design critical outputs. These outputs often determine whether or not a design conforms to specifications, and include nutrient distributions, flow-induced shear stress, analyte collection efficiency, and analyte binding times. Sensitivity analyses may also be performed to determine the relative importance of design parameters. Modeling biomolecular transport and sensing in microfluidic cell culture and analysis systems is therefore a valuable and useful analytical tool that facilitates design of microfluidic cell culture platforms and quantitative biochemical assays. Ultimately, use of theoretical modeling can guide device design, minimize costly and time-consuming experimentation, and increase the efficiency and effectiveness of the design process.

The next phase in the development of microfluidics technology will undoubtedly involve further advances in integration of on-chip biosensing with increasingly more sophisticated designs and functionalities. Achieving these advances will require innovation in all areas, including microfabrication methods, detection and measurement techniques, and microscale cell culture control. Advances in microfabrication will be required to integrate electrodes in devices that are not only fabricated with glass substrates, but also in thermoplastics that have more potential for mass manufacturing. Translation of these technologies to intended end users will additionally require that they be manufactured in standardized formats (e.g., multiwell plates) that are compatible with usual workflows. Recent excitement around organ-on-a-chip technology will also create new opportunities for developing new on-chip biosensing approaches at the tissue- and organ-on-chip level, creating microfluidic systems with added complexity in both physiologic mimicry and analysis capabilities. The ability to detect and measure biomolecular secretion in both 2D monolayer cultures and 3D organ-on-chip cultures will enable researchers to compare different culture arrangements, and assess the need for added complexity in these platforms. All of these future developments will likely benefit from theoretical modeling that combines fluid motion, mass transport, and biochemical reaction kinetics.

References

Allen, C.B., Schneider, B.K., White, C.W., 2001. Limitations to oxygen diffusion and equilibration in in vitro cell exposure systems in hyperoxia and hypoxia. Am. J. Physiol., Lung Cell. Mol. Physiol. 281, L1021–L1027.

Anada, T., Fukuda, J., Sai, Y., Suzuki, O., 2012. An oxygen-permeable spheroid culture system for the prevention of central hypoxia and necrosis of spheroids. Biomaterials 33, 8430–8441.

Benn, J.A., Hu, J., Hogan, B.J., Fry, R.C., Samson, L.D., Thorsen, T., 2006. Comparative modeling and analysis of microfluidic and conventional DNA microarrays. Anal. Biochem. 348, 284–293.

Bhatia, S.N., Balis, U.J., Yarmush, M.L., Toner, M., 1998. Microfabrication of hepatocyte/fibroblast co-cultures: role of homotypic cell interactions. Biotechnol. Prog. 14, 378–387.

Bhatia, S.N., Ingber, D.E., 2014. Microfluidic organs-on-chips. Nat. Biotechnol. 32, 760–772.

Bruus, H., 2015. Governing equations in microfluidics. In: Microscale Acoustofluidics, pp. 1–28.

Buchwald, P., 2009. FEM-based oxygen consumption and cell viability models for avascular pancreatic islets. Theor. Biol. Med. Model. 6, 5.

Cho, C.H., Park, J., Nagrath, D., Tilles, A.W., Berthiaume, F., Toner, M., Yarmush, M.L., 2007. Oxygen uptake rates and liver-specific functions of hepatocyte and 3T3 fibroblast co-cultures. Biotechnol. Bioeng. 97, 188–199.

Choi, S., Goryll, M., Sin, L.Y.M., Wong, P.K., Chae, J., 2011. Microfluidic-based biosensors toward point-of-care detection of nucleic acids and proteins. Microfluid. Nanofluid. 10, 231–247.

Chung, B.G., Flanagan, L.a, Rhee, S.W., Schwartz, P.H., Lee, A.P., Monuki, E.S., Jeon, N.L., 2005. Human neural stem cell growth and differentiation in a gradient-generating microfluidic device. Lab Chip 5, 401–406.

Devadas, D., Young, E.W.K., 2016. Microfluidic Methods for Molecular Biology. Springer, pp. 323–347.

Dugan, C.E., Cawthorn, W.P., MacDougald, O.A., Kennedy, R.T., 2014. Multiplexed microfluidic enzyme assays for simultaneous detection of lipolysis products from adipocytes. Anal. Bioanal. Chem. 406, 4851–4859.

Esch, E.W., Bahinski, A., Huh, D., 2015. Organs-on-chips at the frontiers of drug discovery. Nat. Rev. Drug Discov. 14, 248–260.

Foley, J.O., Nelson, K.E., Mashadi-Hossein, A., Finlayson, B.A., Yager, P., 2007. Concentration gradient immunoassay. 2. Computational modeling for analysis and optimization. Anal. Chem. 79, 3549–3553.

Fox, R.W., Macdonald, A.T., Pritchard, P.J., 2004. Introduction to Fluid Mechanics. Wiley.

Friedrich, D., Please, C., Melvin, T., 2008. Optimisation of analyte transport in integrated microfluidic affinity sensors for the quantification of low levels of analyte. Sens. Actuators B, Chem. 131, 323–332.

Fu, E., Nelson, K.E., Ramsey, S.A., Foley, J.O., Helton, K., Yager, P., 2009. Modeling of a competitive microfluidic heterogeneous immunoassay: sensitivity of the assay response to varying system parameters. Anal. Chem. 81, 3407–3413.

Gao, D., Liu, H., Jiang, Y., Lin, J.-M., 2013a. Recent advances in microfluidics combined with mass spectrometry: technologies and applications. Lab Chip 13, 3309–3322.

Gao, D., Liu, H., Lin, J.-M., Wang, Y., Jiang, Y., 2013b. Characterization of drug permeability in Caco-2 monolayers by mass spectrometry on a membrane-based microfluidic device. Lab Chip 13, 978–985.

Giulitti, S., Magrofuoco, E., Prevedello, L., Elvassore, N., 2013. Optimal periodic perfusion strategy for robust long-term microfluidic cell culture. Lab Chip 13, 4430–4441.

Grist, S.M., Chrostowski, L., Cheung, K.C., 2010. Optical oxygen sensors for applications in microfluidic cell culture. Sensors (Basel) 10, 9286–9316.

Halldorsson, S., Lucumi, E., Gómez-Sjöberg, R., Fleming, R.M.T., 2015. Advantages and challenges of microfluidic cell culture in polydimethylsiloxane devices. Biosens. Bioelectron. 63, 218–231.

Housden, N.G., Harrison, S., Housden, H.R., Thomas, K.A., Beckingham, J.A., Roberts, S.E., Bottomley, S.P., Graille, M., Stura, E., Gore, M.G., 2004. Observation and characterization of the interaction between a single immunoglobulin binding domain of protein L and two equivalents of human kappa light chains. J. Biol. Chem. 279, 9370–9378.

Hsu, M.N., Tan, G.D.S., Tania, M., Birgersson, E., Leo, H.L., 2014. Computational fluid model incorporating liver metabolic activities in perfusion bioreactor. Biotechnol. Bioeng. 111, 885–895.

Hu, G., Gao, Y., Li, D., 2007a. Modeling micropatterned antigen–antibody binding kinetics in a microfluidic chip. Biosens. Bioelectron. 22, 1403–1409.

Hu, G., Quaranta, V., Li, D., 2007b. Modeling of effects of nutrient gradients on cell proliferation in microfluidic bioreactor. Biotechnol. Prog. 23, 1347–1354.

Hui, W.C., Yobas, L., Samper, V.D., Heng, C.K., Liw, S., Ji, H., Chen, Y., Cong, L., Li, J., Lim, T.M., 2007. Microfluidic systems for extracting nucleic acids for DNA and RNA analysis. Sens. Actuators A, Phys. 133, 335–339.

Inamdar, N.K., Griffith, L.G., Borenstein, J.T., 2011. Transport and shear in a microfluidic membrane bilayer device for cell culture. Biomicrofluidics 5, 022213.

Jenkins, J., Prabhakarpandian, B., Lenghaus, K., Hickman, J., Sundaram, S., 2004. Fluidics-resolved estimation of protein adsorption kinetics in a biomicrofluidic system. Anal. Biochem. 331, 207–215.

Kim, M.C., Lam, R.H.W., Thorsen, T., Asada, H.H., 2013. Mathematical analysis of oxygen transfer through polydimethylsiloxane membrane between double layers of cell culture channel and gas chamber in microfluidic oxygenator. Microfluid. Nanofluid. 15, 285–296.

Kimura, H., Yamamoto, T., Sakai, H., Sakai, Y., Fujii, T., 2008. An integrated microfluidic system for long-term perfusion culture and on-line monitoring of intestinal tissue models. Lab Chip 8, 741–746.

Kothapalli, C.R., van Veen, E., de Valence, S., Chung, S., Zervantonakis, I.K., Gertler, F.B., Kamm, R.D., 2011. A high-throughput microfluidic assay to study neurite response to growth factor gradients. Lab Chip 11, 497–507.

Lee, K.-H., Su, Y.-D., Chen, S.-J., Tseng, F.-G., Lee, G.-B., 2007. Microfluidic systems integrated with two-dimensional surface plasmon resonance phase imaging systems for microarray immunoassay. Biosens. Bioelectron. 23, 466–472.

Luka, G., Ahmadi, A., Najjaran, H., Alocilja, E., Derosa, M., Wolthers, K., Malki, A., Aziz, H., Althani, A., Hoorfar, M., 2015. Microfluidics integrated biosensors: a leading technology towards lab-on-a-chip and sensing applications. Sensors (Switzerland) 15, 30011–30031.

Luo, Y., Yu, F., Zare, R.N., 2008. Microfluidic device for immunoassays based on surface plasmon resonance imaging. Lab Chip 8, 694–700.

Mäki, A.-J., Peltokangas, M., Kreutzer, J., Auvinen, S., Kallio, P., 2015. Modeling carbon dioxide transport in PDMS-based microfluidic cell culture devices. Chem. Eng. Sci. 137, 515–524.

Mao, S., Gao, D., Liu, W., Wei, H., Lin, J.M., 2012. Imitation of drug metabolism in human liver and cytotoxicity assay using a microfluidic device coupled to mass spectrometric detection. Lab Chip 12, 219–226.

Mazzei, D., Guzzardi, M.A., Giusti, S., Ahluwalia, A., 2010. A low shear stress modular bioreactor for connected cell culture under high flow rates. Biotechnol. Bioeng. 106, 127–137.

Mcdonald, J.C., Duffy, D.C., Anderson, J.R., Chiu, D.T., 2000. Review General Fabrication of microfluidic systems in poly(dimethylsiloxane). Electrophoresis 21, 27–40.

Mehta, G., Mehta, K., Sud, D., Song, J.W., Bersano-Begey, T., Futai, N., Heo, Y.S., Mycek, M.A., Linderman, J.J., Takayama, S., 2007. Quantitative measurement and control of oxygen levels in microfluidic poly(dimethylsiloxane) bioreactors during cell culture. Biomed. Microdevices 9, 123–134.

Mehta, K., Linderman, J.J., 2006. Model-based analysis and design of a microchannel reactor for tissue engineering. Biotechnol. Bioeng. 94, 596–609.

Meyvantsson, I., Beebe, D.J., 2008. Cell culture models in microfluidic systems. Annu. Rev. Anal. Chem. 1, 423–449.

Myszka, D.G., 1999. Improving biosensor analysis. J. Mol. Recognit. 12, 279–284.

Myszka, D.G., He, X., Dembo, M., Morton, T.A., Goldstein, B., 1998. Extending the range of rate constants available from BIACORE: interpreting mass transport-influenced binding data. Biophys. J. 75, 583–594.

Naciri, M., Kuystermans, D., Al-Rubeai, M., 2008. Monitoring pH and dissolved oxygen in mammalian cell culture using optical sensors. Cytotechnology 57, 245–250.

Ng, A.H.C., Uddayasankar, U., Wheeler, A.R., 2010. Immunoassays in microfluidic systems. Anal. Bioanal. Chem. 397, 991–1007.

Ochs, C.J., Kasuya, J., Pavesi, A., Kamm, R.D., 2014. Oxygen levels in thermoplastic microfluidic devices during cell culture. Lab Chip 14, 459–462.

Oedit, A., Vulto, P., Ramautar, R., Lindenburg, P.W., Hankemeier, T., 2015. Lab-on-a-chip hyphenation with mass spectrometry: strategies for bioanalytical applications. Curr. Opin. Biotechnol. 31, 79–85.

Papaioannou, T.G., Stefanadis, C., 2005. Vascular wall shear stress: basic principles and methods. Hellenic J. Cardiol. 46, 9–15.

Pathi, P., Ma, T., Locke, B.R., 2005. Role of nutrient supply on cell growth in bioreactor design for tissue engineering of hematopoietic cells. Biotechnol. Bioeng. 89, 743–758.

Rackus, D.G., Shamsi, M.H., Wheeler, A.R., 2015. Electrochemistry, biosensors and microfluidics: a convergence of fields. Chem. Soc. Rev. 44, 5320–5340.

Rogers, K.R., 2000. Principles of affinity-based biosensors. Mol. Biotechnol. 14, 109–129.

Roper, M.G., 2016. Cellular analysis using microfluidics. Anal. Chem. 88, 381–394.

Roy, P., Baskaran, H., Tilles, A.W., Yarmush, M.L., Toner, M., 2001. Analysis of oxygen transport to hepatocytes in a flat-plate microchannel bioreactor. Ann. Biomed. Eng. 29, 947–955.

Santiago, J.G., 2001. Electroosmotic flows in microchannels with finite inertial and pressure forces. Anal. Chem. 73, 2353–2365.

Schlessinger, J., Schreiber, A., Levi, A., Lax, I., Libermann, T., Yarden, Y., 1983. Regulation of cell proliferation by epidermal growth factor. Crit. Rev. Biochem. 14 (2), 93–111.

Smith, M.K., Mooney, D.J., 2007. Hypoxia leads to necrotic hepatocyte death. J. Biomed. Mater. Res., Part A 80A, 520–529.

Squires, T.M., Messinger, R.J., Manalis, S.R., 2008. Making it stick: convection, reaction and diffusion in surface-based biosensors. Nat. Biotechnol. 26, 417–426.

Stone, H.A., 2007. Introduction to fluid dynamics for microfluidic flows. In: CMOS Biotechnology, pp. 5–30.

Subramanian, R.M., Chandel, N., Scott Budinger, G.R., Schumacker, P.T., 2007. Hypoxic conformance of metabolism in primary rat hepatocytes: a model of hepatic hibernation. Hepatology 45, 455–464.

Tan, G.D.S., Toh, G.W., Birgersson, E., Robens, J., van Noort, D., Leo, H.L., 2013. A thin-walled polydimethylsiloxane bioreactor for high-density hepatocyte sandwich culture. Biotechnol. Bioeng. 110, 1663–1673.

Tanaka, Y., Yamato, M., Okano, T., Kitamori, T., Sato, K., 2006. Evaluation of effects of shear stress on hepatocytes by a microchip-based system. Meas. Sci. Technol. 17, 3167–3170.

Tilles, A.W., Baskaran, H., Roy, P., Yarmush, M.L., Toner, M., 2001. Effects of oxygenation and flow on the viability and function of rat hepatocytes cocultured in a microchannel flat-plate bioreactor. Biotechnol. Bioeng. 73, 379–389.

Tourlomousis, F., Chang, R.C., 2015. Numerical investigation of dynamic microorgan devices as drug screening platforms. Part I: macroscale modeling approach & validation. Biotechnol. Bioeng. 113, 612–622.

Vijayendran, R.A., Ligler, F.S., Leckband, D.E., 1999. A computational reaction–diffusion model for the analysis of transport-limited kinetics. Anal. Chem. 71, 5405–5412.

Wu, J., He, Z., Chen, Q., Lin, J.-M., 2016. Biochemical analysis on microfluidic chips. TrAC, Trends Anal. Chem. 80, 213–231.

Young, E.W.K., Beebe, D.J., 2010. Fundamentals of microfluidic cell culture in controlled microenvironments. Chem. Soc. Rev. 39, 1036–1048.

Zahorodny-Burke, M., Nearingburg, B., Elias, A.L., 2011. Finite element analysis of oxygen transport in microfluidic cell culture devices with varying channel architectures, perfusion rates, and materials. Chem. Eng. Sci. 66, 6244–6253.

Zeng, Y., Lee, T., Yu, P., Roy, P., Low, H.-T., 2006. Mass transport and shear stress in a microchannel bioreactor: numerical simulation and dynamic similarity. J. Biomech. Eng. 128, 185–193.

Zeng, Y., Lee, T.-S., Yu, P., Low, H.-T., 2007. Numerical simulation on mass transport in a microchannel bioreactor for co-culture applications. J. Biomech. Eng. 129, 365–373.

Zhao, F., Chella, R., Ma, T., 2007. Effects of shear stress on 3-D human mesenchymal stem cell construct development in a perfusion bioreactor system: experiments and hydrodynamic modeling. Biotechnol. Bioeng. 96, 584–595.

Zhou, Q., Patel, D., Kwa, T., Haque, A., Matharu, Z., Stybayeva, G., Gao, Y., Diehl, A.M., Revzin, A., 2015. Liver injury-on-a-chip: microfluidic co-cultures with integrated biosensors for monitoring liver cell signaling during injury. Lab Chip 15, 4467–4478.

Zimmermann, M., Delamarche, E., Wolf, M., Hunziker, P., 2005. Microfluidic-based surface immunoassays. Biomedica 7, 99–110.

CHAPTER

4

Coupling Microscale Transport and Tissue Mechanics: Modeling Strategies for Arterial Multiphysics

M. Marino*, G. Pontrelli†,#, G. Vairo‡, P. Wriggers*

*Institute of Continuum Mechanics, Leibniz Universität, Hannover, Germany †Istituto per le Applicazioni del Calcolo (IAC) - CNR, Rome, Italy ‡Department of Civil Engineering and Computer Science, Università di Roma "Tor Vergata", Rome, Italy

4.1 INTRODUCTION

The complete understanding of the physical mechanisms underlying the correct functioning of the human body, from the cellular to the organ level, stands out as one of the major challenges in modern medicine (Ethier and Simmons, 2007). Indeed, the ultimate goal is to turn heuristic knowledge into predictive capabilities, via the quantitative modeling of the fundamental interactions between basic biochemical processes and biomechanical mechanisms.

In this framework, although cardiovascular diseases are the leading cause of deaths worldwide, their etiology is still often debated and therapeutic approaches are usually driven by risk ranges deduced via clinical records (Fuster and Kelly, 2010). Nevertheless, when addressing, for instance, arterial diseases (such as atherogenesis and aneurysms), it is well-known that pathologies are driven by the biological activity of cells in response to both biomechanical and biochemical stimuli (Taylor and Humphrey, 2009). Unfortunately, the specific causative link between biomechanical/biochemical factors and arterial pathogenesis remains essentially unknown because of the substantial complexity of the biomechanical/biochemical environment affecting arterial tissues.

Many biochemical substances (such as oxygen, nutrients, hormones, enzymes, proteins, tissue inhibitor, regulators of growth, and sometimes drugs) are dispersed in a living body with concentrations controlled by complex regulatory mechanisms (Sagi and Gaffney, 2015; Lilly, 2014; Taylor and Humphrey, 2009). Alterations in transport phenomena play a crucial

Chapter coordinator.

role in tissue inflammatory states, mechanical dysfunctions and damage. As a matter of fact, for instance, the synthesis/degradation of the constituents of the extracellular matrix (ECM) is regulated by cell–cell signaling pathways involving matrix metalloproteinases (MMPs), transforming growth factor-β (TGF-β) and many other molecules (Deguchi et al., 2005; Freedman et al., 2015; García-Alvarez et al., 2006; Jones et al., 2009; Kucich et al., 2002; Rastogi et al., 2013; Sagi and Gaffney, 2015; van der Slot et al., 2005; Visse and Nagase, 2003). In turn, ECM has a major influence on tissue mechanical properties. As a consequence, molecular transport phenomena highly affect the macroscale mechanics of organs and macrobiological structures. Moreover, since biological cells respond to mechanical stimuli by altering the biochemical environment, mechanical quantities have a strong influence on transport phenomena (Ethier and Simmons, 2007; Freedman et al., 2015). Therefore, mechanical and transport mechanisms represent a closed-loop control system characterized by an internal feedback.

Under this perspective, the analysis of physio-pathological mechanisms in organs and macrobiological structures should be addressed via a multiphysics strategy taking into account the two-way interaction between transport processes and mechanics. This would surely open to a better understanding of the onset of many pathologies, as well as to develop novel therapeutic and clinical approaches. For instance, addressing the cardiovascular system, alteration in the concentration of MMPs have been shown to play important roles during the development of cardiovascular diseases (such as plaque formation and rupture, restenosis), highly affecting the mechanical functionalities of cardiovascular structures (Jones et al., 2009; Visse and Nagase, 2003).

This chapter aims to present a multiphysics computational strategy for modeling the previously-introduced coupled mechanical-transport system. This is achieved by accounting for the two-way interaction between mechanical tissue response and transport mechanisms, coupled via biochemically-motivated remodeling laws. The proposed approach opens to the analysis of pathological arterial behavior. It is worth observing that a number of well-established models are available in the specialized literature aiming to describe the evolution of arterial mechanics in disease (Baek et al., 2006; Figueroa et al., 2009; Humphrey and Rajagopal, 2002; Humphrey, 2009; Volokh and Vorp, 2008; Watton et al., 2009), and they are generally referred to as growth and remodeling approaches. Nevertheless, in existing literature, molecular transport mechanisms involved in cell–cell signaling pathways are not accounted for, despite of their well-documented importance. Moreover, the proposed strategy would allow to develop novel computational tools, able to gain effective insights into the physical factors that influence molecular transport processes in otherwise inaccessible locations. As a matter of fact, modeling and computational approaches in conjunction with appropriately designed experiments are helping researchers to better understand the link between biochemical environment and arterial diseases. This also aids in the design of novel devices and the development of novel therapies. For instance, numerical and analytical models have been developed to study the transport of drugs (D'Errico et al., 2015; Pontrelli and De Monte, 2007), of lipoproteins (Dabagh et al., 2009), or of chemoattractants (Leemasawatdigul and Gappa-Fahlenkamp, 2012) through vascular walls. Nevertheless, there are many important transport problems that have not been solved yet or require further attention, especially referring to those coupled with chemical reactions or integrated with tissue mechanics.

The mechanics of arterial tissues is highly nonlinear and anisotropic (Holzapfel et al., 2000; Humphrey, 2002; Taylor and Humphrey, 2009), due to the presence of crimped collagen fibers, whose mechanical response at the microscale is strongly affected by nonlinear molecular and inter-molecular mechanisms at the nanoscale (Fratzl, 2008). A number of constitutive formulations for arterial tissues have been proposed in the literature and remarkable overviews are provided by Holzapfel et al. (2000) and by Humphrey (2002). In the framework of structural approaches, aiming to link model parameters with structural properties of the tissue, the most widely known constitutive model is probably the one by Holzapfel et al. (2000), based on an orthotropic hyperelastic material description and characterized by a clear distinction among arterial principal constituents. Nevertheless, in that model, several collagen nonlinearities (e.g., nanomechanics, cross-link effects, crimped microstructure) are taken into account by choosing an exponential-like representation for the fiber strain-energy density (as in a phenomenological approach), disregarding any direct relationship with micro- and nanoscale mechanisms, and thus losing predictive capability of the model on these issues.

Alternatively, constitutive hyperelastic models of collagenous tissues based on multiscale homogenization techniques have been recently introduced (Maceri et al., 2013; Marino and Vairo, 2013, 2014). These models explicitly incorporate nano- and microscale mechanisms, giving a special insight on the deep link between histology, biochemistry and mechanical response of collagenous tissues. Such an approach allows us to straightforwardly incorporate histological and biochemical alterations and it has been also generalized for including damage evolution at different scales, induced by both mechanical and non-mechanical sources (Maceri et al., 2012; Marino, 2016).

With the aim of coupling transport processes with mechanical response, multiscale constitutive approaches are the most promising methodology because they allow introducing only model parameters with a clear physical meaning, avoiding any phenomenological description. Accordingly, the multiphysics framework proposed in the present work couples a multiscale constitutive description of arterial tissues with a mechanistic modeling of molecular transport, through the definition of remodeling laws based on biochemical evidence.

After a brief overview of some of the main histological–mechanical–biochemical features of arterial tissues in Section 4.2, the multiphysics model is described in Section 4.3. A case study is addressed in Section 4.4, analyzing the effects of ECM remodeling mediated by the transport of MMPs and TGF-β on the compliance of an axisymmetric aortic segment. In detail, proposed results clearly highlight that the alteration of arterial macroscale mechanical properties straight results from the alteration of the tissue biochemical environment. Finally, some concluding remarks are drawn in Section 4.5.

4.2 BRIEF ON ARTERIAL TISSUES

An overview of main histological and mechanical features is presented in Section 4.2.1. Moreover, main transport processes within arterial tissues are described in Section 4.2.2 and some mechanisms involved in tissue remodeling are introduced in Section 4.2.3.

4.2.1 Histology and Mechanics of Arterial Tissues

Arterial walls are made up of three different tissue layers which are named (from inner to outer) the tunica intima, the tunica media and the tunica adventitia (Humphrey, 2002). These three layers are sometimes referred to simply in their shortened version as the intima, the media, and the adventitia. In young and healthy arteries, the intima consists of a single layer of endothelial cells and its thickening is associated with ageing and onset of arteriosclerosis. The media is significantly thicker than the intima and it is made up of concentric layers (named medial lamellar units) consisting of smooth muscle cells embedded in an organized network of loose connective tissue (Clark and Glagov, 1985; O'Connell et al., 2008; Wolinsky and Glagov, 1967). A similar layered structure characterizes the adventitia which in general is thicker than the intima but thinner than the media (Chen et al., 2011).

The three-dimensional histological structure of a single lamellar unit has been recently investigated (O'Connell et al., 2008) and described as a thick sub-layer of elastin sheets, sided by an interlamellar substance made up of water, elastin, smooth muscle cells, and collagen. Collagen results in about 20–30% of the aortic wall dry-weight (Behmoaras et al., 2005) and is organized in crimped fibrils with radius varying from 25 to 50 nm (Merrilees et al., 1987). Fibrils are in turn arranged in both thick and thin bundles (namely, fibers). Electron microscopy scan reveals that the fiber period is on the order of 5 μm and the fiber amplitude-to-period ratio is about 0.2–0.5 (O'Connell et al., 2008).

In both media and adventitia, collagen fibers are arranged in circumferential sub-lamellae. In each of them, collagen fibers are helically wrapped around the vessel axis, so that the collagen fibers are approximately parallel to one another. Fiber orientation of each sub-lamella differs from the adjacent ones resulting in a variation of orientations throughout the mural thickness. Symmetric uni-modal (Chen et al., 2011; O'Connell et al., 2008) and bi-modal (Schriefl et al., 2012) distributions have been reported for the wrapping angle. In general, fiber main orientation is close to the circumferential one in the media and close to the axial one in the adventitia. A schematic representation of arterial histology is depicted in Fig. 4.1.

From a mechanical point of view, the pressure–radius relationship of aortic segments is characterized by a relatively high distensibility at low pressures (associated with the elastin content) and a stiffening response for high pressures (related to collagen) (Wolinsky and Glagov, 1964), see Fig. 4.1. Arterial compliance strongly depends on the highly nonlinear tissue constitutive response which, in turn, is mainly affected by collagen mechanics (Fratzl, 2008). As a matter of fact, in the case of tissues with collagen fibers mainly aligned along a single direction and subjected to an along-the-fiber uniaxial traction test, a progressive fiber straightening and the disappearance of nanoscale kinks within molecules are experienced, resulting in an increase of the overall tissue stiffness. Accordingly, the stress/strain curves are typically J-shaped (see Fig. 4.1) and can be subdivided into three main regions (Buehler and Wong, 2007; Fratzl, 2008; Sasaki and Odajima, 1996):

1. Toe region (strain range, 0–2% circa) is a low stiffness region associated with the removal of the microscopic crimp in collagen fibers.
2. Heel region (strain range, 2–4% circa) is a region associated with a significant stiffening response due to the straightening of labile domains of collagen triple-helices (denoted as molecular kinks) that counteract the entropic forces associated with thermal fluctuations.

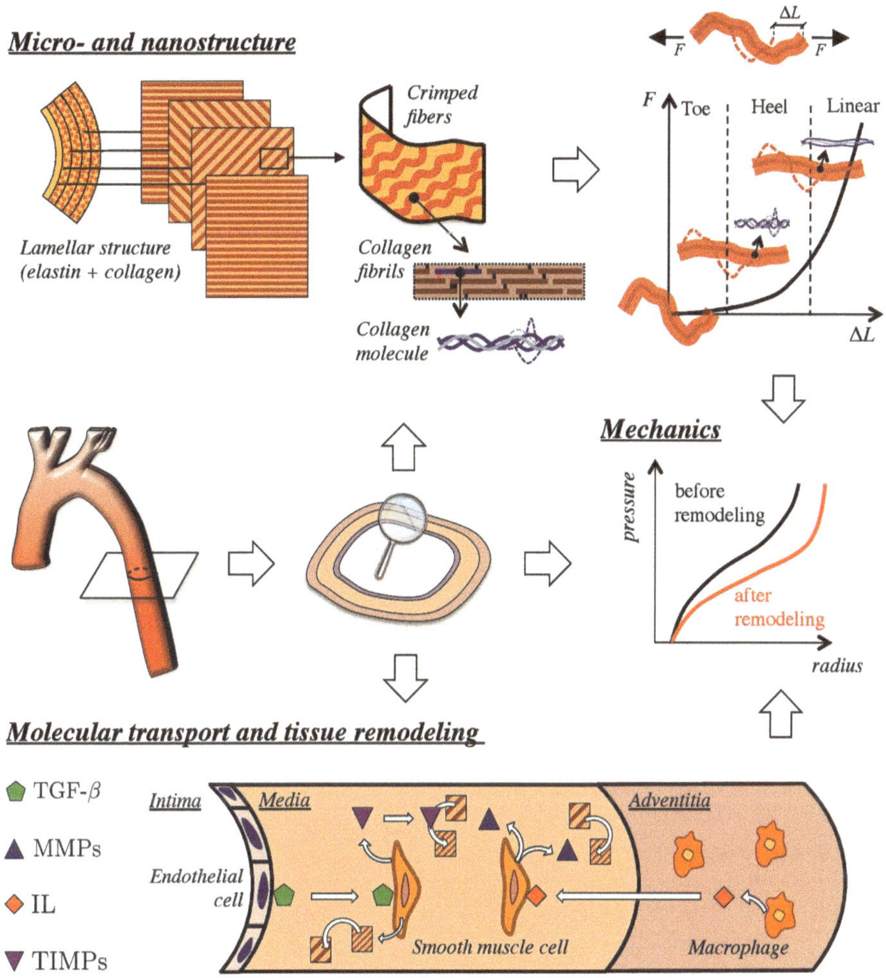

FIGURE 4.1 Multiphysics of arterial response: coupling of arterial micro/nanostructure, mechanics and molecular transport phenomena through the remodeling of tissue constituents: TGF-β, transforming growth factor beta; MMPs, matrix metalloproteinases; IL, Interleukin; TIMPs, tissue inhibitors of MMPs.

3. Linear region (strain range, greater than 4% circa) is a high stiffness region that is mainly related to the stretching of collagen triple-helices and to molecular rearrangement mechanisms (i.e., intermolecular sliding highly affected by covalent cross-links among collagen molecules (Bailey, 2001; Svensson et al., 2013).

Finally, it is worth pointing out that aortic segments in living bodies are pre-stressed and pre-stretched under zero loads: when the vessel is excised, the aortic segment shortens; when a ring cross-section of an artery free of external loads is cut radially, an open sector appears (Kassab, 2006; Rachev et al., 1996; Zhang et al., 2005).

4.2.2 Molecular Transport in Arterial Tissues

Transport mechanisms are involved in physio-pathological paths that are strictly associated with an alteration of arterial mechanical response. As a matter of fact, there is clear evidence that cell–cell communications in the vascular wall can be operated by means of signaling pathways mediated by diffusion of soluble factors, and that the latter contribute to vascular physiological activity, pathogenesis, and disease progression (Lilly, 2014). Some important transport phenomena within arterial tissues are (Lilly, 2014; Taylor and Humphrey, 2009; Sagi and Gaffney, 2015):

- *Oxygen transport.* Biological cells require oxygen and other nutrients for their biological activity which ensure the maintenance of tissue homeostasis. Oxygen diffuses from the luminal surface and, for large blood vessel beyond circa 30 lamellar units, from vasa vasorum in the outer medial and adventitial layers. Severe vascular diseases that thicken the wall (e.g., occlusive atherosclerotic plaques) or increase the distance between the flowing blood and intramural cells (e.g., intraluminal thrombus) affect these transport phenomena and thus the activity of biological cells. As a matter of fact, these diseases are generally related to the development of new blood vessels, or neovascularization, which likely occurs due to chemokines released in response to the local inflammation and/or hypoxic conditions;
- *Transport of vasoactive molecules.* The intimal endothelium produces vasoactive molecules, such as nitric oxide (NO), a potent vasodilator, and endothelin-1 (ET-1), a strong vasoconstrictor. Vasoactive molecules are produced strongly depending on local wall shear stress at the endothelium; they diffuse within aortic thickness and their rate of consumption depends on vascular smooth muscle cells. These molecules play an important role also on arterial growth and remodeling because NO is an inhibitor of smooth muscle cells proliferation and collagen synthesis whereas ET-1 promotes both proliferation and matrix synthesis;
- *Lipid transport.* The development of atherosclerotic process is due to the transport of atherogenic molecules (e.g., low density lipoproteins) between the blood stream and the vessel wall. In this mechanisms, shear-dependent changes in the endothelial permeability have a crucial role;
- *Transport of other molecules.* Cells (e.g., endothelial cells in the intima, smooth muscle cells in the media and macrophages in the adventitia) respond to alterations in mechanical loading via altered gene expression producing, among others, growth factors (such as the transforming growth factor-beta (TGF-β)), cytokines (inflammatory mediators such as interleukines (ILs)), matrix metalloproteinases (MMPs), and tissue inhibitors of MMPs (TIMPs). For instance, the local alteration of intramural stress induces the production and the activation of growth factors (e.g., TGF-β), as well as of MMPs. The latter are also upregulated by indirect effects driven by an altered wall shear stress on endothelial cells and activated by plasmin. In this framework, transport processes are of main importance. As a matter of fact, the media layer (the most important from a biomechanical point of view) is poorly accessible to inflammatory cells but remains accessible to soluble mediators (Michel et al., 2007). A detailed description of signaling pathways involving these molecules is beyond the scope of the present work. As illustrative examples, ILs (produced for instance by macrophages in the adventitia) diffuse up to smooth muscle cells (SMCs) in the media and induce the secretion of MMPs by SMCs. On the other hand, TGF-β (produced for instance by endothelial cells) activates SMCs to express TIMPs which, in turn, inhibit

MMPs. Both TGF-β and MMPs participate in a wide array of cellular responses including: proliferation, angiogenesis, differentiation, apoptosis, inflammation, and wound healing (Jones et al., 2009; Sagi and Gaffney, 2015). Moreover, non-homeostatic concentrations of these molecules affect both composition and organization of constituents of the extracellular matrix. The above-introduced mechanisms (schematically depicted in Fig. 4.1) are further addressed in Section 4.2.3 in terms of remodeling effects.

4.2.3 Extracellular Matrix Remodeling

Extracellular matrix (ECM) is an acellular component of all tissues and organs which provides the necessary physical scaffolding and serves as a source of crucial biochemical and biomechanical signals which, in turn, regulate tissue morphogenesis, differentiation, and homeostasis (Ethier and Simmons, 2007; Sagi and Gaffney, 2015). ECM remodeling is a change in tissue structure, achieved by the reorganization of existing constituents (e.g., altered orientation or cross-linking) or by the synthesis of new constituents. Remodeling may or may not alter the mass density, but it does change tissue stiffness and strength properties which are mainly conferred by the elastin network and collagen fibers, as recalled in Section 4.2.1.

The dysregulation of the degradation and the deposition of the extracellular matrix (ECM) leads to pathological remodeling of cardiovascular tissues which represents an open issue still under investigation (Deguchi et al., 2005; García-Alvarez et al., 2006; Jones et al., 2009; Kucich et al., 2002; Rastogi et al., 2013; Sagi and Gaffney, 2015; van der Slot et al., 2005; Visse and Nagase, 2003). For instance, alterations in collagen cross-linking are often related to connective tissue diseases (such as the Marfan's syndrome) and common cardiovascular pathologies (e.g., restenosis after angioplasty, aortic dilation, dissecting aneurysm, Brüel et al., 1998; Carmo et al., 2002).

Many biochemical pathways intervene *in vivo* when ECM remodels and most of them are open-issues still under investigations. In this framework, a major biochemical mechanisms involves the metabolism of TGF-β, MMPs, ILs and TIMPs. Growth factors TGF-β bind to a specific receptor in SMCs and activates a cascade of events that promote ECM deposition (e.g., deposition of collagen and elastin, Jones et al., 2009; Kucich et al., 2002), affect ECM biochemical features (e.g., inducing collagen cross-linking, van der Slot et al., 2005) and repress ECM degradation (e.g., promoting TIMPs expression, García-Alvarez et al., 2006). On the other hand, MMPs are proteinases that induce ECM degradation (e.g., proteolysis of collagen and elastin, Sagi and Gaffney, 2015; Visse and Nagase, 2003) and affect the disorganization of ECM constituents (e.g., misalignment of collagen fibers, Deguchi et al., 2005; Rastogi et al., 2013). Under normal physiological conditions, MMPs activity is down-regulated, among other mechanisms, by endogenous tissue inhibitors (TIMPs) (Visse and Nagase, 2003; Sagi and Gaffney, 2015). On the contrary, ILs induce the secretion of MMPs by SMC (Lee et al., 1995; Maiellaro and Taylor, 2007).

Under homeostatic conditions, a balance between these mechanisms is maintained in order to tightly control matrix degradation and matrix deposition. A loss of activity control may result in diseases such as arthritis, cancer, atherosclerosis, aneurysms, nephritis, tissue ulcers, and fibrosis (Sagi and Gaffney, 2015; Visse and Nagase, 2003). For instance, within the aneurysmal aorta, this balance is disrupted by an overproduction of MMPs or an underproduction of TIMPs, favoring an enhanced proteolytic state and driving matrix degradation

(Jones et al., 2009; Visse and Nagase, 2003). Thus, while the physiological remodeling process within the arterial wall operates to maintain the physiological arterial function, pathological dysregulation can result in an excessive degradation of critical ECM components, leading to loss of mechanical strength and integrity, and resulting in arterial dilatation, dissection, or rupture (Jones et al., 2009).

Accordingly, the concentration of these molecules within arterial walls are involved in the etiology of many cardiovascular disease, resulting associated with a coupled system made up by arterial biomechanics, transport phenomena and molecular biological activity. The afore-described balance between TGF-β, MMPs, ILs and TIMPs (schematically depicted in Fig. 4.1) will represent our study case, addressed in Section 4.4.

4.3 ARTERIAL MULTIPHYSICS MODELING

After a brief introduction of geometrical features and of the general notation (see Section 4.3.1), the general multiphysics rationale is introduced in Section 4.3.2. In the following, possible strategies for modeling each ingredient which builds up the proposed multiphysics approach are described. In detail, arterial mechanics is presented in Section 4.3.3. Thereafter, molecular transport problem is addressed in Section 4.3.4, and biochemically-motivated remodeling laws are defined in Section 4.3.5. Finally, a computational strategy for the coupled problems involved in arterial multiphysics is presented in Section 4.3.6, moving towards an analytical solution.

4.3.1 Geometric Description and General Notation

An arterial segment, $\mathcal{A}$, is regarded as a set of continuously distributed material points $X \in \mathcal{A}$, namely a continuum-body, undergoing a deformation mechanism along the time-path governed by the variable $t \in \mathbb{R}^+$, with $\mathbb{R}^+$ denoting the set of non-negative real numbers (and with $\mathbb{R}^{++} = \mathbb{R}^+ \setminus \{0\}$). Moreover, let $\mathbf{x} = \mathbf{x}(X, t) \in \mathbb{R}^3$ denote the position of a material point X in a three-dimensional Euclidean space at time t, with $\mathbb{R}^n$ denoting an n-dimensional real coordinate space.

For what follows, it is useful to introduce the following terms:

- The mechanical state is described by the current configuration $\Omega = \Omega(t)$ (namely, the region occupied by $\mathcal{A}$ at time t), the second-order right Cauchy–Green deformation tensor $\mathbf{C} = \mathbf{C}(\mathbf{x}, t)$ and the second-order Cauchy stress tensor $\boldsymbol{\sigma} = \boldsymbol{\sigma}(\mathbf{x}, t)$;
- The biochemical environment is defined by the occurrence of M active molecules in Ω and is described by the vector $\mathbf{c} = \mathbf{c}(\mathbf{x}, t) \in \mathbb{R}^M$ where the qth component of $\mathbf{c}$ (namely, $c_q = c_q(\mathbf{x}, t)$) represents the concentration of the qth biologically active molecule in Ω;
- The structural (histological, biochemical, and biophysical) features relevant to tissue constitutive response, s_j, are collected in the set $\mathcal{S} = \{s_1, \ldots, s_S\}$. Due to possible remodeling mechanisms, space-/time-dependency of the values of $s_j \in \mathcal{S}$ is accounted for, namely $s_j = s_j(\mathbf{x}, t)$.

As general notation rules, let $\Sigma = \partial\Omega$ be the boundary of Ω and $\mathbf{n}$ be the outward normal unit vector to Σ. Moreover, let $\mathbf{I}$ be the second-order identity tensor, $\mathrm{Div}(\cdot)$ and $\mathrm{Grad}(\cdot)$ (resp.,

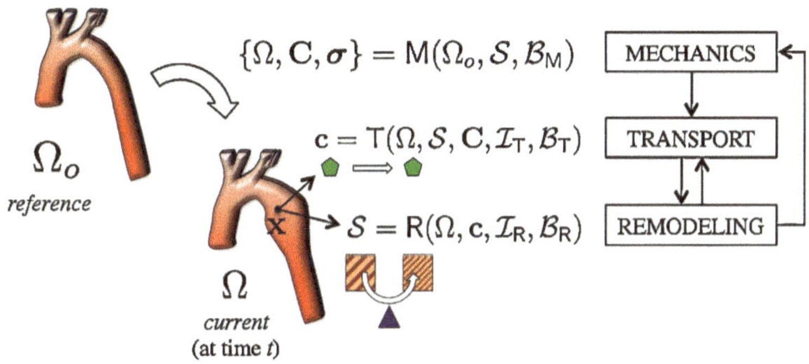

FIGURE 4.2 Multiphysics strategy for arterial behavior: coupling of mechanical M, transport T, and remodeling R problems for obtaining mechanical state $\{\Omega, \mathbf{C}, \boldsymbol{\sigma}\}$, biochemical environment $\mathbf{c}$, and structural features $\mathcal{S}$. Sets $\mathcal{I}_*$ and $\mathcal{B}_*$ respectively collect initial and boundary conditions for problem $* = \{\mathsf{M}, \mathsf{T}, \mathsf{R}\}$.

div(·) and grad(·)) be the divergence and gradient operators with respect to the reference (resp., current) configuration. As subscript/superscript notation, the subscript o indicates quantities in the reference configuration, the superscript T denotes the transpose operator, the subscript q implies values in $\{1, \ldots, M\}$, and the subscript j implies values in $\{1, \ldots, S\}$. Finally, $\langle \cdot \rangle$ denotes the Macaulay brackets, that is, $\langle x \rangle = (x + |x|)/2$.

4.3.2 Multiphysics Modeling Rationale

The multiphysics strategy is based on the following ingredients (schematically depicted in Fig. 4.2):

1. *Description of arterial mechanics.* Arterial mechanics is obtained by solving problem M (defined in Section 4.3.3) giving the mechanical state $\{\Omega, \mathbf{C}, \boldsymbol{\sigma}\}$ starting from reference configuration Ω_o, structural parameters $\mathcal{S}$ and the set of mechanical boundary conditions $\mathcal{B}_\mathsf{M}$, namely
$$\{\Omega(t), \mathbf{C}(\mathbf{x}, t), \boldsymbol{\sigma}(\mathbf{x}, t)\} = \mathsf{M}(\Omega_o, \mathcal{S}(\mathbf{x}, t), \mathcal{B}_\mathsf{M}) . \tag{4.1}$$
Tissue constitutive properties within problem M are defined from a strain-energy density Ψ_T which depends on deformation $\mathbf{C}$ and structural features $\mathcal{S}$, namely $\Psi_T = \Psi_T(\mathbf{C}, \mathcal{S})$;
2. *Description of molecular transport.* The biochemical environment $\mathbf{c}$ is obtained by solving the transport problem T (defined in Section 4.3.4) starting from current configuration Ω, structural features $\mathcal{S}$, strain $\mathbf{C}$ and the set of transport initial $\mathcal{I}_\mathsf{T}$ and boundary $\mathcal{B}_\mathsf{T}$ conditions, namely
$$\mathbf{c}(\mathbf{x}, t) = \mathsf{T}(\Omega(t), \mathcal{S}(\mathbf{x}, t), \mathbf{C}(\mathbf{x}, t), \mathcal{I}_\mathsf{T}, \mathcal{B}_\mathsf{T}) ; \tag{4.2}$$
3. *Description of remodeling laws.* Tissue remodeling is obtained by solving problem R (defined in Section 4.3.5) which gives the values of structural features in $\mathcal{S}$ starting from current configuration Ω, biochemical environment $\mathbf{c}$, and the set of initial $\mathcal{I}_\mathsf{R}$ and boundary $\mathcal{B}_\mathsf{R}$ conditions, namely
$$\mathcal{S}(\mathbf{x}, t) = \mathsf{R}(\Omega(t), \mathbf{c}(\mathbf{x}, t), \mathcal{I}_\mathsf{R}, \mathcal{B}_\mathsf{R}) . \tag{4.3}$$

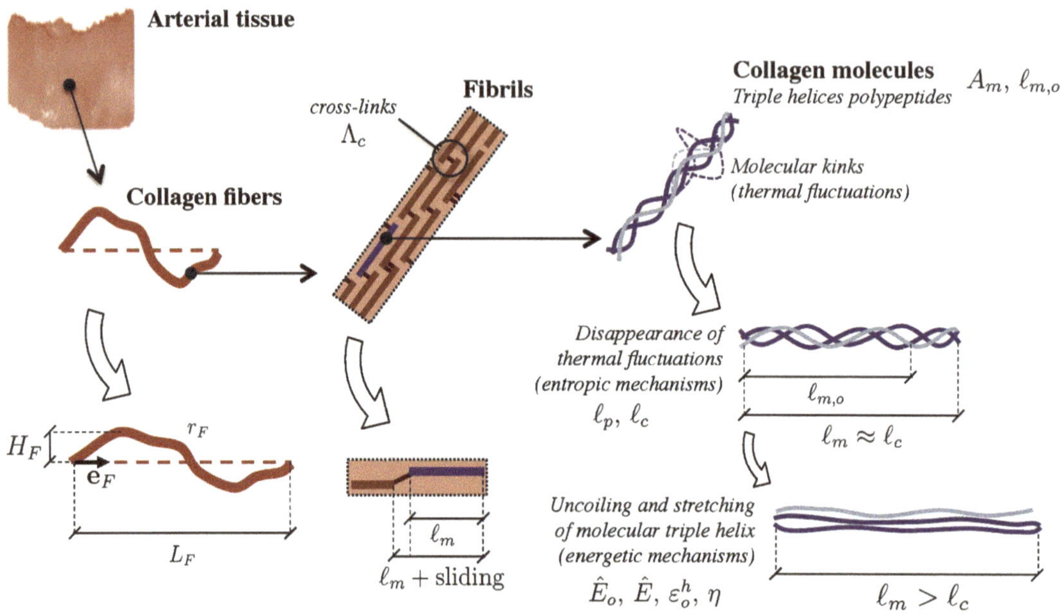

FIGURE 4.3 Collagen-related nonlinearities in arterial tissues – multiscale coupled mechanisms. Symbols are defined in Section 4.3.3.1.

4.3.3 Arterial Mechanical Problem

Arterial mechanics is here described by firstly introducing the nonlinear behavior of collagen fibers in Section 4.3.3.1 which is coupled with a macroscale constitutive description in Section 4.3.3.2, employed in the solving equations for arterial macroscopic response, presented in Section 4.3.3.3.

4.3.3.1 *Collagen Multiscale Nonlinearities*

Collagen fibers in arterial tissues are characterized by a nonlinear mechanical behavior, related to multiscale coupled mechanisms (see Fig. 4.3). Fibers are assumed to have a circular cross-section of radius r_F (and area measure $A_F = \pi r_F^2$). The crimped structure of collagen fibers is taken into account by considering locally periodic fibers of along-the-chord period length L_F and amplitude H_F in the current configuration (resp., $L_{F,o}$ and $H_{F,o}$ in the reference configuration).

In turn, collagen fibers are bundles of densely-packed fibrils wherein collagen triple-helical molecules (of cross-sectional area A_m and length ℓ_m) are mutually interconnected into a head-to-tail arrangement through intermolecular covalent cross-links, described by their average occurrence per molecule Λ_c. Therefore, fibril mechanics can be described as related to molecular sliding (counteracted by cross-links with stiffness k_c) *in series* with molecular elongation (Maceri et al., 2012; Marino, 2016).

At the nanoscale, the elongation of collagen molecules is governed by in-series entropic and energetic mechanisms (Buehler and Wong, 2007; Fratzl, 2008; Maceri et al., 2012; Marino,

2016). The former are associated with thermal fluctuations (depending on temperature T, molecular persistence ℓ_p and contour ℓ_c lengths, as well as on molecular length $\ell_{m,o}$ in the reference configuration), and the latter with the uncoiling and the stretching of collagen triple helices (depending on low-strain $\hat{E}_o$ and high-strain $\hat{E}$ tangent moduli, molecular uncoiling strain ε_o^h, and uncoiling resistance η).

Accordingly, the set $\mathcal{S}_F$ of histological, biochemical and biophysical features, describing collagen fiber mechanical behavior, results in

$$\mathcal{S}_F = \{\mathcal{S}_f,\ H_{F,o},\ L_{F,o},\ r_F\}, \tag{4.4a}$$

$$\mathcal{S}_f = \{\mathcal{S}_m,\ k_c,\ \Lambda_c\}, \tag{4.4b}$$

$$\mathcal{S}_m = \{\ell_{m,o},\ \ell_p,\ \ell_c,\ A_m,\ T,\ \hat{E}_o,\ \hat{E},\ \varepsilon_o^h\}, \tag{4.4c}$$

where the multiscale nature is highlighted by the inclusions $\mathcal{S}_m \subset \mathcal{S}_f \subset \mathcal{S}_F$.

By following the Maceri–Marino–Vairo multiscale rationale (Maceri et al., 2013; Marino and Vairo, 2013, 2014) (summarized in Appendix A for the sake of completeness), collagen fiber mechanics is defined in terms of their along-the-chord tangent modulus C_F, equal to (Marino and Wriggers, 2016)

$$C_F = E_f \frac{\ell_F^2 + H_F^2}{\sqrt{\ell_o^2 + H_o^2}} \left[\ell_F + \frac{4H_F^2}{3r_F^2\,\ell_F}\left(\ell_F^2 + H_F^2\right) \right]^{-1}, \tag{4.5}$$

where $\ell_F = L_F/4$ and E_f is the tangent modulus of collagen fibrils (see Eqs. (A.1) in Appendix A).

It is worth pointing out that, due to multiscale mechanisms, E_f, ℓ_F, and H_F are functions of the fiber along-the-chord stretch $\lambda_F = L_F/L_{F,o}$. In particular, it results in $\ell_F = \ell_F(\lambda_F) = \lambda_F L_{F,o}/4$, while functions $E_f = E_f(\lambda_F)$ and $H_F = H_F(\lambda_F)$ are determined from inter-scale compatibility relationships, expressed by the system of differential equations (A.6) in Appendix A.

Moreover, it is worth noticing that fibril modulus E_f in Eqs. (A.1) is fully defined in terms of structural parameters $\mathcal{S}_f$. In order to highlight the explicit dependence of fiber mechanical response by the set of structural parameters $\mathcal{S}_F$ in Eqs. (4.4), the functional form $C_F = C_F(\lambda_F, \mathcal{S}_F)$ is conveniently employed in what follows.

4.3.3.2 *Tissue Constitutive Model*

Arterial tissue is regarded as a three-phase substance comprising crimped collagen fibers with volume fraction V_C, a non-collagenous matrix with volume fraction V_M and other non-bearing-load constituents of volume fraction $V_0 = 1 - V_C - V_M$. It is remarked that V_M comprises the volume fraction V_{EL} of the elastin network that bears load (i.e., not including the fragmented elastin), and other constituents of similar stiffness (e.g., cells). Accordingly, V_0 represents the amount of fragmented elastin, degraded collagen as well as other possible non-bearing-load constituents (Maceri et al., 2013; Marino and Vairo, 2013; Rezakhaniha et al., 2011).

Accounting for the lamellar structure of the arterial tissue (see Section 4.2.1), the main direction of collagen fibers in a single sub-lamella is described by the unit vector $\mathbf{e}_F$, in the following referred to as the fiber-chord direction. Disregarding any curvature effect, collagen fibers are organized in thin planar sheets identified by the orthonormal basis $(\mathbf{e}_F, \mathbf{n}_F, \mathbf{k}_F)$ where $\mathbf{k}_F$ is the normal direction to the tissue plane identified by $(\mathbf{e}_F, \mathbf{n}_F)$. Due to arterial lamellar structure, both volume fractions and collagen orientation are space-dependent functions, even when remodeling mechanisms do not occur (Fig. 4.1).

In the context of nonlinear hyperelasticity, the passive mechanical behavior of the arterial tissue is described by introducing the tissue strain-energy function Ψ_T per unit reference volume from which the Cauchy stress tensor $\boldsymbol{\sigma}$ is obtained as:

$$J\boldsymbol{\sigma} = -p\mathbf{I} + 2\mathbf{F}\frac{\partial \Psi_T}{\partial \mathbf{C}}\mathbf{F}^T , \tag{4.6}$$

where $J = \sqrt{I_3}$ with $I_3 = \det(\mathbf{C})$, $\mathbf{F}$ is the deformation gradient, $\mathbf{C} = \mathbf{F}^T\mathbf{F}$ and p acts as a Lagrange multiplier introduced to enforce the incompressibility constraint (namely, $J = 1$) (Auricchio et al., 2013).

In agreement with a constrained mixture approach (Humphrey and Rajagopal, 2002; Humphrey, 2009), tissue strain-energy density Ψ_T is split in collagen-related Ψ_C and non-collagen-related Ψ_M contributions (Holzapfel et al., 2000; Wriggers, 2008). Accordingly, it results in

$$\Psi_T(\mathbf{C}, \mathcal{S}) = V_C\Psi_C(\mathbf{C}, \mathcal{S}) + V_M\Psi_M(\hat{\mathbf{C}}) , \tag{4.7a}$$

where $\hat{\mathbf{C}} = I_3^{-1/3}\mathbf{C}$ is the isochoric part of deformation. Collagen-related strain-energy contribution is defined as

$$\Psi_C(\mathbf{C}, \mathcal{S}) = \int_1^{1+\langle\lambda_F-1\rangle}\int_1^{1+\langle\xi-1\rangle}\frac{C_F(\eta, \mathcal{S}_F)}{\eta}d\eta d\xi , \tag{4.7b}$$

where C_F is given in Eq. (4.5) and fiber along-the-chord stretch λ_F results in

$$\lambda_F = \sqrt{\mathrm{Tr}(\mathbf{CM})} = \|\mathbf{F}\mathbf{e}_{F,o}\| . \tag{4.7c}$$

Furthermore, a standard neo-Hookean approach for incompressible materials is employed for non-collagen-related strain-energy contribution Ψ_M, resulting in

$$\Psi_M(\hat{\mathbf{C}}) = k_M(\hat{I}_1 - 3) , \tag{4.7d}$$

with k_M being a material constant and $\hat{I}_1 = I_3^{-1/3}\mathrm{Tr}(\mathbf{C})$.

Accordingly, arterial tissue mechanical response depends on set $\mathcal{S}$ of biophysical, biochemical, and histological parameters,

$$\mathcal{S} = \{\mathcal{S}_F, \mathbf{e}_{F,o}, V_C, V_M, k_M\} , \tag{4.8}$$

where $\mathcal{S}_F$ (defined in Eqs. (4.4)) accounts for collagen nonlinearities.

It is worth pointing out that cells in arterial tissues (i.e., SMCs) may contract and relax, affecting the *in vivo* tissue response. SMCs active contraction is here neglected, although present formulation can be straightforwardly generalized by following well-established approaches (e.g., Nardinocchi and Teresi, 2007).

4.3.3.3 *Mechanical Equilibrium*

Arterial mechanical equilibrium is associated with the solution of problem M (see Eq. (4.1)) which is here defined. In general, arterial mechanics is a fluid–structure interaction problem, with boundary conditions $\mathcal{B}_\mathsf{M}$ on the arterial luminal surface related to blood flow. In this work, the problem is simplified by considering stationary conditions for the fluid. Therefore, apart from possible kinematic constraints, a given uniform luminal pressure is considered, neglecting any unsteady fluid–structure interaction effects. Accordingly, boundary conditions are defined by:

$$\mathcal{B}_\mathsf{M} : \begin{cases} \text{a given surface traction } \hat{\mathbf{t}} \in \mathbb{R}^3 & \text{on } \Sigma_{\sigma,o}, \\ \text{a prescribed displacement } \hat{\mathbf{u}} \in \mathbb{R}^3 & \text{on } \Sigma_{u,o}, \end{cases} \tag{4.9}$$

with $\Sigma_{u,o} \cup \Sigma_{p,o} = \Sigma_o$ and with $\hat{\mathbf{t}}$ describing the assigned luminal pressure.

The Mechanical Problem M is defined as follows:

Mechanical Problem M. Find

$$\Omega = \{\mathbf{x}(X,t) \text{ with } X \in \mathcal{A},\ t \in \mathbb{R}^+\}, \qquad \mathbf{S} = \sqrt{I_3}\mathbf{F}^{-1}\boldsymbol{\sigma}\mathbf{F}^{-T},$$

such that

$$\mathrm{Div}(\mathbf{FS}) = \mathbf{0} \text{ in } \Omega_o, \text{ with } \begin{cases} \mathbf{FSn}_o = \hat{\mathbf{t}} & \text{on } \Sigma_{\sigma,o}, \\ \mathbf{u} = \hat{\mathbf{u}} & \text{on } \Sigma_{u,o}, \end{cases}$$

where $\mathbf{F} = \mathbf{I} + \mathrm{Grad}\,\mathbf{u}$ is the deformation gradient, $\mathbf{u} = \mathbf{x} - \mathbf{x}_o$ is the displacement field, and $\mathbf{S}$ is the second Piola–Kirchhoff stress tensor defined from the Cauchy stress tensor $\boldsymbol{\sigma}$ in Eq. (4.6), with $\Psi_T = \Psi_T(\mathbf{C}, \mathcal{S})$ in Eq. (4.7), $\mathbf{C} = \mathbf{F}^T\mathbf{F}$ and $I_3 = \det(\mathbf{C})$.

4.3.4 Molecular Transport Problem

The biochemical environment $\mathbf{c}$ in arterial tissue is obtained defining the Transport Problem T (see Eq. (4.2)) as an advection–diffusion–reaction problem for each active molecule q (Truskey et al., 2010). To this aim, let us introduce:

- The diffusivity $\mathbf{D}_q$ that generally results in a space dependent second-order tensor with non-negative elements and accounts for possible anisotropy effects. It can be a function of tissue microstructure (described by $\mathcal{S}$) and of the existing biochemical environment (represented by $\mathbf{c}$). Accordingly, it becomes $\mathbf{D}_q = \mathbf{D}_q(\mathbf{x}, \mathcal{S}, \mathbf{c})$;
- A reaction term $\mathfrak{R}_q$ that is split in a source term $\mathfrak{R}_q^+$ and a consumption term $\mathfrak{R}_q^-$, namely $\mathfrak{R}_q = \mathfrak{R}_q^+ - \mathfrak{R}_q^-$. These terms include possible sources, sinks, or interactions among the substances. For example, a substance can bound to cell receptor and can activate a signaling effect that acts as a source or sink for other species (see Section 4.2.3); an anomalously stretched tissue can promote or can inhibit microchannels formation, paving the way to chemical agent reactions. Accordingly, the reaction term generally depends on the existing

biochemical environment (described by $\mathbf{c}$) and on tissue strain[1] (namely, $\mathbf{C}$), resulting in $\mathfrak{R}_q^{\pm} = \mathfrak{R}_q^{\pm}(\mathbf{c}, \mathbf{C})$;

- The filtration velocity $\mathbf{v}$ that is associated with seepage. Vessel walls are indeed permeable to blood plasma, resulting in a filtration flux over the entire luminal surface, induced by the occurrence of transmural pressure gradients (Wada and Karino, 1999). As a consequence, the molecules are also transported in the vessel radial direction, with a filtration velocity that can be considered almost constant and, for physiological transmural pressure gradients, on the order of 10^{-6} cm/s (Pontrelli and De Monte, 2007; Vairo et al., 2010).

Boundary conditions are defined by

$$\mathcal{B}_{\mathsf{T}}: \begin{cases} \text{a given concentration } \hat{c}_q \in \mathbb{R}^+ & \text{on } \Sigma_c, \\ \text{a prescribed concentration flux } \hat{J}_q \in \mathbb{R} & \text{on } \Sigma_J, \\ \text{a permeable surface with permeability } \hat{\lambda}_q \in \mathbb{R}^+ & \text{on } \Sigma_r, \end{cases} \tag{4.10}$$

with $\Sigma_c \cup \Sigma_J \cup \Sigma_r = \Sigma$, and associated respectively to Dirichlet-, Neumann- and Robin-type conditions (Farlow, 1982; Tarbell, 2003). An initial concentration profile $c_q^o(\mathbf{x})$ at time $t = 0$ defines the initial condition $\mathcal{I}_{\mathsf{T}}$.

The Transport Problem T is defined as follows:

Transport Problem T. Find $c_q(\mathbf{x}, t)$ such that

$$\frac{\partial c_q}{\partial t} + \operatorname{div}(-\mathbf{D}_q(\mathbf{x}, \mathcal{S}, \mathbf{c}) \operatorname{grad} c_q + \mathbf{v} c_q) = \mathfrak{R}_q^+(\mathbf{c}, \mathbf{C}) - \mathfrak{R}_q^-(\mathbf{c}, \mathbf{C}) \text{ in } \Omega,$$

with $c_q(\mathbf{x}, 0) = c_q^o(\mathbf{x})$, and

$$\begin{cases} c_q(\mathbf{x}, t) = \hat{c}_q & \text{on } \Sigma_c, \\ -\mathbf{D}_q \operatorname{grad} c_q(\mathbf{x}, t) \cdot \mathbf{n} = \hat{J}_q & \text{on } \Sigma_J, \\ -\mathbf{D}_q \operatorname{grad} c_q(\mathbf{x}, t) \cdot \mathbf{n} = \hat{\lambda}_q c_q + \hat{J}_q & \text{on } \Sigma_r. \end{cases}$$

4.3.5 Remodeling Laws

The Remodeling Problem R (see Eq. (4.3)) is here presented as a non-local problem and by taking inspiration from the logistic function which finds application in a range of fields including biology, chemistry, medicine, and biomathematics (Vandermeer, 2010). To this aim, the following quantities are introduced:

- An homeostatic value for the jth structural feature, denoted by $\bar{s}_j = \bar{s}_j(\mathbf{x})$, corresponds to a set of values (collected in $\bar{\mathcal{C}}_j$) for the concentration profiles $\mathbf{c}$ not activating remodeling;

[1] Alternatively (or in conjunction), reaction terms might depend on tissue stress $\boldsymbol{\sigma}$. This is not considered in what follows, although the present approach can be straightforwardly generalized.

- The homeostatic imbalance, denoted by $\mathfrak{I}_j$, represents the imbalance from the homeostatic state induced by remodeling such that, roughly speaking, one has

$$s_j \to \bar{s}_j + \mathfrak{I}_j(\mathbf{c})\,. \tag{4.11a}$$

The quantity $\mathfrak{I}_j$ depends on biochemical mechanisms (described by $\mathbf{c}$), resulting in $\mathfrak{I}_j = \mathfrak{I}_j(\mathbf{c})$, and it activates for non-homeostatic concentrations, namely $\mathfrak{I}_j(\mathbf{c}) = 0$ if $\mathbf{c} \in \bar{\mathcal{C}}_j$ and $\mathfrak{I}_j(\mathbf{c}) \neq 0$ if $\mathbf{c} \notin \bar{\mathcal{C}}_j$. A general expression for $\mathfrak{I}_j(\mathbf{c})$ cannot be provided because of high case-by-case variability. In fact, experimental observations (see, for instance, the ones reported in Section 4.2.3) highlight that the same structural feature can be altered in a very different way by different molecules, cell–cell signaling pathways can be involved, and biochemistry-dependent up- and down-regulations often occur. As a general rule-of-thumb and assuming an additive decomposition of the remodeling mechanisms, $\mathfrak{I}_j(\mathbf{c})$ could be defined as

$$\mathfrak{I}_j(\mathbf{c}) = \sum_{q=1}^{M} \mathfrak{i}_{qj}(c_q)\,, \tag{4.11b}$$

with $\mathfrak{i}_{qj}(c_q)$ being the remodeling law which drives s_j as a function of c_q. For instance, a linear threshold-based activation law is obtained by defining $\mathfrak{i}_{qj}(c_q)$ as

$$\mathfrak{i}_{qj}(c_q) = \pm K_{qj}^{\mathsf{R}} \langle c_q - C_{qj}^{\mathsf{R}} \rangle \tag{4.11c}$$

where C_{qj}^{R} is the threshold value activating remodeling, K_{qj}^{R} is a positive remodeling constant, and the sign depends on the biochemical activity (namely, it is positive if c_q induces the synthesis of s_j and negative for its degradation);
- The set $\mathcal{K}_j$ of admissible values for s_j accounts for physical constraints;
- The remodeling diffusivity $\mathbf{A}_j$ accounts for non-local and possible anisotropic effects associated with the influence area of the remodeling stimulus. In general, $\mathbf{A}_j$ is a space-dependent second-order tensor with non-negative elements and can be a function of tissue microstructure (described by $\mathcal{S}$) and biochemical environment (described by $\mathbf{c}$), namely $\mathbf{A}_j = \mathbf{A}_j(\mathbf{x}, \mathcal{S}, \mathbf{c})$;
- The remodeling viscosity $\nu_j \in \mathbb{R}^{++}$ accounts for the biochemical resistance to remodeling.

In general, a prescribed flux $Q_j \in \mathbb{R}$ can be imposed as a boundary condition $\mathcal{B}_{\mathsf{R}}$ on Σ in order to account for possible external agents (e.g., chemical, electromagnetical, radioactive factors) acting on the boundary surface and inducing remodeling. An initial value $s_j^o(\mathbf{x})$ at time $t = 0$ defines the initial condition $\mathcal{I}_{\mathsf{R}}$.

The Remodeling Problem R is defined as follows:

Remodeling Problem R**.** Find $s_j = s_j(\mathbf{x}, t) \in \mathcal{K}_j$ such that

$$\frac{\partial s_j}{\partial t} - \mathrm{div}[\mathbf{A}_j(\mathbf{x}, \mathcal{S}, \mathbf{c})\mathrm{grad}(s_j - \bar{s}_j)] = -\left(\frac{s_j}{\bar{s}_j + \mathfrak{I}_j(\mathbf{c})} - 1\right)\frac{s_j}{\nu_j} \quad \text{in } \Omega\,,$$

with $s_j(\mathbf{x}, 0) = s_j^o(\mathbf{x})$ and $-\mathbf{A}_j\,\mathrm{grad}(s_j) \cdot \mathbf{n} = Q_j$ on Σ.

The solution of Problem R gives the implicit relationship $\mathcal{S} = \mathcal{S}(\mathbf{c})$. In particular, considering an initial homeostatic state (namely, $\mathbf{c} = \bar{\mathbf{c}}_j$ and $s_j^o = \bar{s}_j$), it is immediate to observe that R gives $s_j = \bar{s}_j$ (since $\mathfrak{I}_j(\bar{\mathbf{c}}_j) = 0$). On the other hand, when $\mathbf{c} \neq \bar{\mathbf{c}}_j$, the remodeling is activated and it induces a variation of s_j.

It is worth highlighting that the description of remodeling through problem R does not allow us to consider the natural turn-over of ECM constituents but it represents an effective strategy for describing tissue pathological remodeling associated with alterations in tissue biochemical environment.

4.3.6 Integrated Computational Strategy: Towards an Analytical Solution

From the complex interplay of the afore-described ingredients, arterial multiphysics is governed by a nonlinear and strongly-coupled system of partial differential equations, resulting in a closed-loop system (see Fig. 4.2). As a matter of fact, Transport Problem T (Section 4.3.4) affects the Remodeling Problem R (Section 4.3.5) and vice versa. Moreover, remodeling mechanisms alter $\mathcal{S}$ and thus the solution of the Mechanical Problem M (Section 4.3.3) which, in turn, affects the Transport Problem T because the latter is solved in the current configuration Ω and the reaction terms $\mathfrak{R}_q^{\pm}$ depend on $\mathbf{C}$ (namely, a mechanical feedback mechanism occurs).

Rigorously, the solution of Problems M, T and R should be faced via a monolithic approach. Accordingly, the development of a solution strategy for arterial multiphysics is not a trivial task. Nevertheless, it is worth highlighting that, due to arterial adaptation processes, remodeling laws are characterized by large time scales (i.e., months to years) which can be believed to be significantly higher than the ones of microscale transport (i.e., hours to days) and of arterial mechanics (i.e., seconds due to the characteristic times of the cardiac cycle).

Taking advantage of the separation of time scales, Problem i (with $i = \mathsf{M}, \mathsf{T}, \mathsf{R}$) can be solved by considering its own time-variable t_i and a characteristic time T_i in which the solution of Problem i attains a steady-state, with $T_{\mathsf{R}} \gg T_{\mathsf{T}} \gg T_{\mathsf{M}}$. Accordingly, the principle of separation of time-scales might be exploited for obtaining an effective solution strategy. Nevertheless, a solution strategy for the fully-coupled formulation is beyond the scope of present work. Here a number of simplifying assumptions are introduced in order to move towards a quasi-analytical solution strategy and to show the effectiveness of the present approach. In detail, the following assumptions are considered:

1. $\mathbf{D}_q$ in T does not depend on tissue microstructure $\mathcal{S}$;
2. $\mathfrak{R}_q^{\pm}$ in T does not depend on $\mathbf{C}$ (i.e., no mechanical feedback is accounted for);
3. T is solved on the arterial configuration $\bar{\Omega}$ corresponding to the homeostatic state, associated with $\bar{\mathcal{S}} = \{\bar{s}_1, \ldots, \bar{s}_S\}$;
4. R is solved assuming that remodeling is a local phenomenon (namely, $\mathbf{A}_j = \mathbf{0}$);
5. Since $T_{\mathsf{R}} \gg T_{\mathsf{T}}$, R is solved employing steady-state concentration profiles $\tilde{\mathbf{c}}(\mathbf{x}) = (\tilde{c}_1(\mathbf{x}) \ldots \tilde{c}_M(\mathbf{x}))$;

In agreement with previous assumptions, Reduced Transport $\tilde{\mathsf{T}}$ and Remodeling $\tilde{\mathsf{R}}$ Problems are introduced such that

$$\tilde{\mathbf{c}}(\mathbf{x}) = \tilde{\mathsf{T}}(\bar{\Omega}, \mathcal{B}_{\mathsf{T}}), \qquad \mathcal{S}(\mathbf{x}, t) = \tilde{\mathsf{R}}(\tilde{\mathbf{c}}(\mathbf{x}), \mathcal{I}_{\mathsf{R}}). \tag{4.12}$$

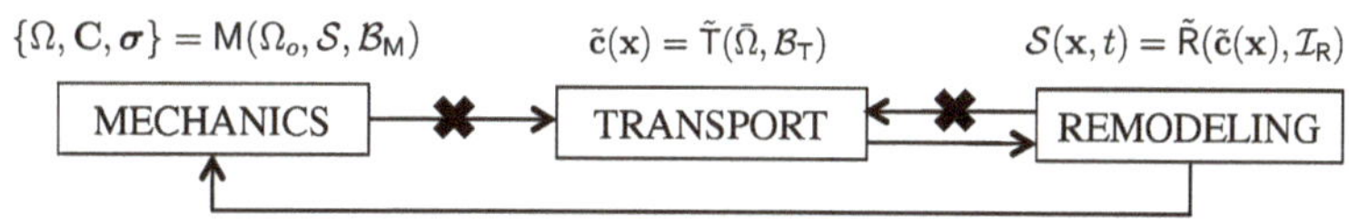

FIGURE 4.4 Arterial multiphysics modeling: the fully-coupled system reduces to the open-loop one by neglecting the mechanical feedback on molecular transport (crossed right-oriented arrow) and the dependence of the transport on the remodeling (crossed left-oriented arrow) obtained from the fully-coupled system.

Problems $\tilde{\mathsf{T}}$ and $\tilde{\mathsf{R}}$ are defined as follows:

Reduced Transport Problem $\tilde{\mathsf{T}}$. Find $\tilde{c}_q(\mathbf{x})$ such that

$$\operatorname{div}(-\mathbf{D}_q(\mathbf{x},\tilde{\mathbf{c}})\operatorname{grad}\tilde{c}_q + \mathbf{v}\tilde{c}_q) = \mathfrak{R}_q^+(\tilde{\mathbf{c}}) - \mathfrak{R}_q^-(\tilde{\mathbf{c}}) \quad \text{in } \bar{\Omega}$$

with

$$\begin{cases} \tilde{c}_q(\mathbf{x}) = \hat{c}_q & \text{on } \bar{\Sigma}_c, \\ -\mathbf{D}_q \operatorname{grad}\tilde{c}_q(\mathbf{x})\cdot\mathbf{n} = \hat{J}_q & \text{on } \bar{\Sigma}_J, \\ -\mathbf{D}_q \operatorname{grad}\tilde{c}_q(\mathbf{x})\cdot\mathbf{n} = \hat{\lambda}_q\tilde{c}_q(\mathbf{x}) + \hat{J}_q & \text{on } \bar{\Sigma}_r. \end{cases}$$

Reduced Remodeling Problem $\tilde{\mathsf{R}}$. Find $s_j = s_j(\mathbf{x}, t) \in \mathcal{K}_j$ such that

$$\frac{\partial s_j}{\partial t} = -\left(\frac{s_j}{\bar{s}_j + \mathfrak{I}_j(\tilde{\mathbf{c}})} - 1\right)\frac{s_j}{\nu_j},$$

with $s_j(\mathbf{x}, 0) = s_j^o(\mathbf{x})$.

It is worth pointing out that, assuming $\mathcal{K}_j \equiv \mathbb{R}$, problem $\tilde{\mathsf{R}}$ can be solved in an analytical way, obtaining

$$s_j(\mathbf{x}, t) = \frac{s_j^o(\mathbf{x})[\bar{s}_j(\mathbf{x}) + \mathfrak{I}_j(\tilde{\mathbf{c}})]\exp(t/\nu_j)}{\bar{s}_j(\mathbf{x}) + \mathfrak{I}_j(\tilde{\mathbf{c}}) + s_j^o(\mathbf{x})[\exp(t/\nu_j) - 1]}, \tag{4.13a}$$

where it is immediate to verify that

$$\lim_{t\to+\infty} s_j(\mathbf{x}, t) = \bar{s}_j(\mathbf{x}) + \mathfrak{I}_j(\tilde{\mathbf{c}}). \tag{4.13b}$$

Accordingly, under the afore-introduced assumptions, M does not affect the solution of T and R (apart from the computation of the homeostatic configuration). Clearly, the solution of T affects R which, in turn, affects M. As schematically depicted in Fig. 4.4, an open-loop multiphysics system is obtained, opening to the investigation of the effects of biochemical processes on mechanics but not vice versa.

In this case, open-loop arterial multiphysics can be obtained from the pseudocode in Table 4.1. The proposed solution strategy opens a way to develop analytical solutions, which allow the implementation of extensive parametric analyses aiming to furnish insights into the effects of pathological/healing biochemical stimuli on arterial mechanics. This is the aim of the case study addressed in the following Section 4.4.

TABLE 4.1 Pseudocode of the algorithm for open-loop arterial multiphysics

1. $\{\bar{\Omega}, \bar{\mathsf{C}}, \bar{\sigma}\} = \mathsf{M}(\Omega_o, \bar{\mathcal{S}}(\mathbf{x}), \mathcal{B}_\mathsf{M})$,
2. $\tilde{\mathsf{c}}(\mathbf{x}) = \tilde{\mathsf{T}}(\bar{\Omega}, \mathcal{B}_\mathsf{T})$,
3. $\mathcal{S}(\mathbf{x}, t) = \tilde{\mathsf{R}}(\tilde{\mathsf{c}}(\mathbf{x}), \mathcal{I}_\mathsf{R})$,
4. $\{\Omega(t), \mathsf{C}(\mathbf{x}, t), \sigma(\mathbf{x}, t)\} = \mathsf{M}(\Omega_o, \mathcal{S}(\mathbf{x}, t), \mathcal{B}_\mathsf{M})$.

4.4 AN AXISYMMETRIC CASE STUDY

In agreement with the evidence in Section 4.2.3, a model for the physiopathological control system involved on ECM remodeling in aortic tissues is described in what follows, by addressing the effects of microscale transport phenomena of MMPs (produced by SMCs in the media), TGF-β (produced by endothelial cells in the inner intima), and IL (produced by macrophages in the outer adventitia) on the aortic response.

Only the media layer is addressed, since it is the most important from the mechanical point of view. Under axisymmetrical geometric assumptions, the arterial segment (characterized by internal and external radii r_i and r_e, respectively) is loaded by a uniform internal pressure P_i. The effects of ECM remodeling mechanisms on arterial compliance is analyzed, by obtaining the pressure–radius relationship of arterial segments associated with the transport and the biochemical activity of MMPs, TGF-β and ILs which alter, as structural features: elastin and collagen volume fractions (V_{EL} and V_C), collagen fiber radius (r_F), inter-molecular cross-links density (Λ_c), and fiber orientation distribution (see Fig. 4.5). In the following, subscript a takes values in $\{2, 3\}$, q in $\{1, 2, 3\}$, and j in $\{1, \ldots, S\}$.

4.4.1 Arterial Geometry and Structure

A single-layer thick-walled right cylinder (L_a long) is addressed, modeling the media layer of an aortic segment. By following the opening angle method (see Fig. 4.5) and denoting by α the opening angle (Auricchio et al., 2014; Holzapfel et al., 2000), the material points in the reference configuration Ω_o (assumed to be load-free and stress-free) are identified by the cylindrical coordinates $\mathbf{x}_o = (r_o, \theta_o, z_o)$ where

$$r_{i,o} \leq r_o \leq r_{e,o}\,, \quad 0 \leq \theta_o \leq 2\pi - \alpha\,, \quad 0 \leq z_o \leq L_{a,o}\,, \tag{4.14a}$$

for the radial, circumferential, and axial coordinate, respectively. On the other hand, the current configuration Ω is described by the cylindrical coordinates $\mathbf{x} = (r, \theta, z)$ where

$$r_i \leq r \leq r_e\,, \qquad 0 \leq \theta \leq 2\pi\,, \qquad 0 \leq z \leq L_a\,. \tag{4.14b}$$

Introducing $\kappa = 2\pi/(2\pi - \alpha)$ and λ_z as a constant axial stretch, function $\mathbf{x} = \mathbf{x}(\mathbf{x}_o)$ results in

$$r = r(r_o) = \sqrt{\frac{r_o^2 - r_{i,o}^2}{\kappa\lambda_z} + r_i^2}\,, \quad \theta = \theta(\theta_o) = \kappa\theta_o\,, \quad z = z(z_o) = \lambda_z z_o\,, \tag{4.14c}$$

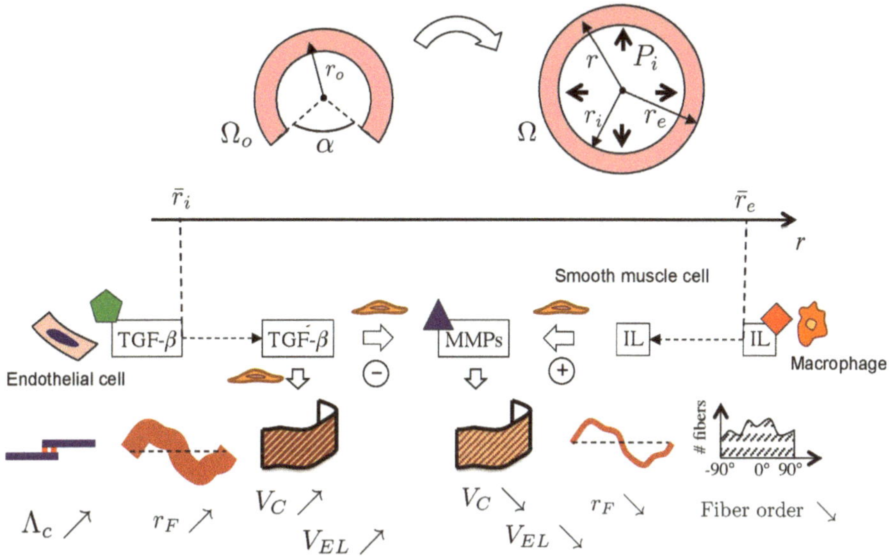

FIGURE 4.5 Axisymmetric case study for arterial multiphysics. Reference (with opening angle α and radial coordinate r_o) and current configuration (with radial coordinate r, internal and external radius r_i and r_e, and luminal pressure P_i). It is sketched the transport of matrix metalloproteinases (MMPs), transforming growth factor-beta (TGF-β), interleukines (ILs), the remodeling of elastin and collagen volume fractions (V_{EL} and V_C), collagen fiber radius (r_F), inter-molecular cross-links density (Λ_c), and fiber orientation distribution.

due to the incompressibility condition. The inverse function $r_o(r)$ can be also straightforwardly defined as

$$r_o = r_o(r) = \sqrt{\kappa\lambda_z(r^2 - r_i^2) + r_{i,o}^2}\,. \tag{4.14d}$$

Moreover, the deformation gradient expressed in the cylindrical coordinate system results in

$$\mathbf{F} = \frac{\partial \mathbf{x}}{\partial \mathbf{x}_o} = \mathrm{diag}[\lambda_r,\, \lambda_\theta,\, \lambda_z] = \mathrm{diag}\left[\frac{\partial r}{\partial r_o},\, \frac{r}{r_o}\frac{\partial \theta}{\partial \theta_o},\, \frac{\partial z}{\partial z_o}\right] \tag{4.15a}$$

with

$$\lambda_r = \lambda_r(r) = \frac{r_o(r)}{\kappa\lambda_z r}\,, \qquad \lambda_\theta = \lambda_\theta(r) = \frac{\kappa\, r}{r_o(r)}\,. \tag{4.15b}$$

Moreover, introducing $\boldsymbol{\theta}$ and $\mathbf{z}$ as the unit vectors in the circumferential and axial direction, respectively, the unit vector identifying the direction of collagen fibers is described by

$$\mathbf{e}_F = \cos\beta\,\boldsymbol{\theta} + \sin\beta\,\mathbf{z}\,, \tag{4.16a}$$

where β represents collagen fiber angle in the current configuration with respect to the circumferential direction. Accordingly, from Eq. (4.7c), one has

$$\lambda_F = \sqrt{\lambda_\theta^2 \cos^2 \beta_o + \lambda_z^2 \sin^2 \beta_o}\,, \tag{4.16b}$$

where β_o represents the angle of collagen fibers in the reference configuration.

Addressing arterial microstructure (see Section 4.2.1), media lamellar units (MLUs) are explicitly considered and each MLU is regarded as made up of a thick elastin sub-layer and a composite interlamellar substance with elastin and collagen fibers (see Appendix B for a detailed description). In order to account for the heterogeneous lamellar structure of aortic media layer and for remodeling mechanisms, structural features $\mathcal{S}$ in Eq. (4.8) are assumed to vary on the radial coordinate r only, namely $\mathcal{S} = \mathcal{S}(r,t)$.

As a model assumption, structural features $\mathcal{S}_{\mathsf{R}} = \{V_{EL}, V_C, r_F, \Lambda_c, \beta_o\} \subset \mathcal{S}$ are assumed to undergo remodeling, while the remaining model parameters $\mathcal{S}_{\mathsf{NR}} = \mathcal{S} \setminus \mathcal{S}_{\mathsf{R}}$ are assumed to be constant in time.

4.4.2 Quasi-Analytical Arterial Mechanics

Under the afore-introduced assumptions on arterial geometry and microstructure, the solution of Problem M in Section 4.3.3 becomes axisymmetric and it can be solved through simple analytical relationships. In particular, arterial configuration is fully described by the line segment connecting internal and external radii, namely $\Omega \equiv [r_i, r_e]$. Moreover, Cauchy stresses σ_r and σ_θ (respectively in the radial and circumferential directions) and the luminal pressure P_i have to satisfy the following equilibrium equation:

$$P_i = \int_{r_i}^{r_e} \frac{\sigma_\theta - \sigma_r}{r}\, dr = \int_{r_i}^{r_e} \left[\lambda_\theta \frac{\partial \Psi_T}{\partial \lambda_\theta} - \lambda_r \frac{\partial \Psi_T}{\partial \lambda_r} \right] \frac{dr}{r}\,. \tag{4.17a}$$

Here Eq. (4.6) has been accounted for, and the Lagrange multiplier p has been straightforwardly eliminated due to the appearance of stress difference $\sigma_\theta - \sigma_r$. Accounting for Eq. (4.16b) and referring to the constitutive model in Eqs. (4.7), we obtain

$$\frac{\partial \Psi_T}{\partial \lambda_r} = V_M \frac{\partial \Psi_M}{\partial \lambda_r} + \frac{\partial \lambda_F}{\partial \lambda_r} V_C \int_1^{1+\langle \lambda_F - 1 \rangle} \frac{C_F(\eta, \mathcal{S}_F)}{\eta}\, d\eta = V_M \frac{\partial \Psi_M}{\partial \lambda_r}\,, \tag{4.17b}$$

$$\begin{aligned} \frac{\partial \Psi_T}{\partial \lambda_\theta} &= V_M \frac{\partial \Psi_M}{\partial \lambda_\theta} + \frac{\partial \lambda_F}{\partial \lambda_\theta} V_C \int_1^{1+\langle \lambda_F - 1 \rangle} \frac{C_F(\eta, \mathcal{S}_F)}{\eta}\, d\eta = \\ &= V_M \frac{\partial \Psi_M}{\partial \lambda_\theta} + V_C \frac{\lambda_\theta \cos^2 \beta_o}{\lambda_F} \int_1^{1+\langle \lambda_F - 1 \rangle} \frac{C_F(\eta, \mathcal{S}_F)}{\eta}\, d\eta\,. \end{aligned} \tag{4.17c}$$

Accordingly, Eqs. (4.17) provide the quasi-analytical relationship

$$P_i = P_i(r_i, \mathcal{S}(r,t))\,. \tag{4.18}$$

Therefore, the homeostatic state can be computed as

$$\bar{\Omega} \equiv [\bar{r}_i, \bar{r}_e] \ \text{ s.t. } \ P_i(\bar{r}_i, \bar{\mathcal{S}}) = \bar{P}_i\,, \tag{4.19}$$

where $\bar{P}_i$ represent a stationary physiological pressure.

4.4.3 Analytical Arterial Molecular Transport

Let c_1, c_2, and c_3 respectively denote the concentrations of MMPs, TGF-β, and IL, where

- IL activates the synthesis of MMPs by SMCs;
- TGF-β inhibits the synthesis of MMPs by SMCs.[2]

The steady-state (axisymmetric) concentration profiles $\tilde{c}_q = \tilde{c}_q(r)$ are obtained from the Reduced Transport Problem $\tilde{\mathsf{T}}$ (Section 4.3.6) by assuming an isotropic diffusion with constant diffusivity (namely, $\mathbf{D}_q = \hat{D}_q\mathbf{I}$). Moreover, curvature effects and convective-terms are neglected. Furthermore, nontrivial reaction terms are defined as:

$$\mathfrak{R}_1^+(\tilde{\mathbf{c}}) = \eta_1 V_S \, R^+(\tilde{c}_3)\,(C_1 - \tilde{c}_1)\,, \tag{4.20a}$$

$$\mathfrak{R}_1^-(\tilde{\mathbf{c}}) = [\zeta_1 + \xi_1 V_S \, R^-(\tilde{c}_2)]\tilde{c}_1\,, \tag{4.20b}$$

$$\mathfrak{R}_a^-(\tilde{\mathbf{c}}) = \zeta_a \, \tilde{c}_a\,, \tag{4.20c}$$

where

- $V_S \in [0, 1]$ is the volume fraction of smooth muscle cells (SMCs), assumed to be constant in time and space;
- $R^+(\tilde{c}_3)$ (resp., $R^-(\tilde{c}_2)$) is MMPs source (resp., consumption) dimensionless function, activated by the molecular species IL (resp., TGF-β) through SMCs. Simple linear expressions are herein employed, by assuming a dependence on the average concentration within the media layer, namely $R^{\pm}(\tilde{c}_a) = \tilde{c}_a^{av}/\gamma_a$, $\tilde{c}_a^{av}$ being

$$\tilde{c}_a^{av} = \frac{1}{\bar{r}_e - \bar{r}_i} \int_{\bar{r}_i}^{\bar{r}_e} \tilde{c}_a(r)\, dr\,, \tag{4.21}$$

  and γ_a representing a reference concentration value corresponding to a unitary stimulus;
- C_1 is the saturation limit of MMPs production by SMCs;
- η_1 (resp., ξ_1) is MMPs source (resp., consumption) rate mediated by SMCs;
- ζ_q are MMPs/TGF-β/IL consumption rates associated with molecular natural decay.

Accordingly, introducing $\rho \in [0, 1]$ as the normalized radial coordinate $\rho = (r - \bar{r}_i)/\bar{\delta}$ with $\bar{\delta} = \bar{r}_e - \bar{r}_i$, the Reduced Transport Problem $\tilde{\mathsf{T}}$ (Section 4.3.6) reads: find $\tilde{c}_1$, $\tilde{c}_2$, and $\tilde{c}_3$ such that

$$\frac{\partial^2 \tilde{c}_1}{\partial \rho^2} + B - h_1^2 \tilde{c}_1 = 0\,, \tag{4.22a}$$

$$\frac{\partial^2 \tilde{c}_a}{\partial \rho^2} - h_a^2 \tilde{c}_a = 0\,, \tag{4.22b}$$

[2] It is worth pointing out that TGF-β promotes the synthesis of TIMPs which, in turn, reduce the activity of MMPs. In the present model, TIMPs are not explicitly modeled but their effect is taken into account by introducing a negative feedback of MMPs synthesis activated by TGF-β.

in $\rho \in [0, 1]$, with

$$B = \eta_1 \bar{\delta}^2 \frac{V_S\, \tilde{c}_3^{av}\, C_1}{\hat{D}_1\, \gamma_3}, \tag{4.22c}$$

$$h_1^2 = \frac{\bar{\delta}^2}{\hat{D}_1}\left(\zeta_1 + \xi_1 \frac{V_S\, \tilde{c}_2^{av}}{\gamma_2} + \eta_1 \frac{V_S\, \tilde{c}_3^{av}}{\gamma_3}\right), \tag{4.22d}$$

$$h_a^2 = \frac{\bar{\delta}^2 \zeta_a}{\hat{D}_a}. \tag{4.22e}$$

The solution of Eqs. (4.22) can be obtained analytically: firstly, the independent Eqs. (4.22b) are solved and $\tilde{c}_a^{av}$ are computed; then, average concentrations $\tilde{c}_a^{av}$ are employed for obtaining the solution of Eq. (4.22a). Accordingly, it results in

$$\tilde{c}_1(\rho) = k_{11} \exp(-h_1 \rho) + k_{21} \exp(h_1 \rho) + \frac{B}{h_1^2}, \tag{4.23a}$$

$$\tilde{c}_a(\rho) = k_{1a} \exp(-h_a \rho) + k_{2a} \exp(h_a \rho), \tag{4.23b}$$

where constants k_{1q} and k_{2q} can be determined from boundary conditions $\mathcal{B}_{\mathsf{T}}$. The latter are assumed to be of Robin-type, except for Dirichlet boundary conditions at the intima-media layer for $\tilde{c}_2$ and at the media-adventitia layer for $\tilde{c}_3$, in order to respectively account for (see Section 4.2.2 and Fig. 4.5):

- A constant source of TGF-β by endothelial cells in the intima;
- A constant source of IL by macrophages in the adventitia.

Accordingly,

$$\text{at } \rho = 0 \quad \begin{cases} \dfrac{\partial \tilde{c}_1}{\partial \rho} = \dfrac{\bar{\delta}\hat{\lambda}_{1i}}{\hat{D}_1} \tilde{c}_1, \\ \tilde{c}_2 = \hat{c}_2, \\ \dfrac{\partial \tilde{c}_3}{\partial \rho} = \dfrac{\bar{\delta}\hat{\lambda}_{3i}}{\hat{D}_3} \tilde{c}_3, \end{cases} \qquad \text{at } \rho = 1 \quad \begin{cases} \dfrac{\partial \tilde{c}_1}{\partial \rho} = -\dfrac{\bar{\delta}\hat{\lambda}_{1e}}{\hat{D}_1} \tilde{c}_1, \\ \dfrac{\partial \tilde{c}_2}{\partial \rho} = -\dfrac{\bar{\delta}\hat{\lambda}_{2e}}{\hat{D}_2} \tilde{c}_2, \\ \tilde{c}_3 = \hat{c}_3. \end{cases} \tag{4.23c}$$

4.4.4 Analytical Arterial Remodeling Induced by MMPs, TGF-β, and IL

The remodeling laws are formulated by defining the homeostatic imbalance $\mathfrak{I}_j$ in the Remodeling Problem R (Section 4.3.5). As general assumptions, local remodeling is addressed (namely, $\mathbf{A}_j = \mathbf{0}$) and an additive decomposition of remodeling mechanisms based on threshold-based activation laws is employed, analogously to the form of Eqs. (4.11b) and (4.11c).

Remodeling laws are defined in agreement with the evidence in Section 4.2.3 and as schematically depicted in Fig. 4.5, accounting for the activity by MMPs ($q = 1$) and TGF-β ($q = 2$). To this aim, let constants $C_{qj}^{\mathsf{R}} = C_{\mathsf{R}}$, $K_{1j}^{\mathsf{R}} = \bar{s}_j / C_{\mathsf{R}}$, and $K_{2j}^{\mathsf{R}} = \bar{s}_j V_S / C_{\mathsf{R}}$ be introduced, where $C_{\mathsf{R}} \in \mathbb{R}^{++}$ is a given concentration value. It is highlighted that volume fraction V_S is accounted for in K_{2j}^{R}, since TGF-β activity is mediated by SMCs. Accordingly,

- Elastin volume fraction $s_1 = V_{EL}$, collagen volume fraction $s_2 = V_C$ and collagen fiber radius $s_3 = r_F$ increase with TGF-β and decrease with MMPs:

$$\mathfrak{I}_1(c_1, c_2) = \bar{V}_{EL}\,[V_S\langle c_2 - C_{\mathsf{R}}\rangle - \langle c_1 - C_{\mathsf{R}}\rangle]/C_{\mathsf{R}}\,, \tag{4.24a}$$

$$\mathfrak{I}_2(c_1, c_2) = \bar{V}_C\,[V_S\langle c_2 - C_{\mathsf{R}}\rangle - \langle c_1 - C_{\mathsf{R}}\rangle]/C_{\mathsf{R}}\,, \tag{4.24b}$$

$$\mathfrak{I}_3(c_1, c_2) = \bar{r}_F\,[V_S\langle c_2 - C_{\mathsf{R}}\rangle - \langle c_1 - C_{\mathsf{R}}\rangle]/C_{\mathsf{R}}\,; \tag{4.24c}$$

- Inter-molecular cross-link density $s_4 = \Lambda_c$ increases with TGF-β:

$$\mathfrak{I}_4(c_2) = \bar{\Lambda}_c V_S\langle c_2 - C_{\mathsf{R}}\rangle/C_{\mathsf{R}}\,; \tag{4.24d}$$

- Collagen fiber angle $s_5 = \beta_o$ tends to an isotropically distributed pattern (namely, without any preferred direction) in the presence of MMPs:

$$\mathfrak{I}_5(c_1) = (\beta_{iso} - \bar{\beta}_o)\, f_h(c_1 - C_{\mathsf{R}})\,, \tag{4.24e}$$

  where $\beta_{iso} = \beta_{iso}(r)$ corresponds to an isotropic angle distribution (see Eq. (B.2) in Appendix B), and $f_h(x)$ is the Heaviside function (i.e., $f_h(x) = 0$ for $x \leq 0$ and $f_h(x) = 1$ for $x > 0$).

Let the homeostatic state be assumed as an initial condition and let ν be a remodeling viscosity, constant among different structural features (namely, $s_j^o = \bar{s}_j$ and $\nu_j = \nu$). Employing steady-state concentration profiles $\tilde{c}_1$ and $\tilde{c}_2$ in Eqs. (4.23), the Remodeling Problem corresponds to its Reduced form $\tilde{\mathsf{R}}$ (Section 4.3.6). Accounting for Eqs. (4.24), analytical solutions for ECM remodeling in arterial tissues of the form of Eqs. (4.13) directly follow as:

$$V_{EL}(r,t) = \frac{\bar{V}_{EL}\left[C_{\mathsf{R}} + V_S\langle \tilde{c}_2(r) - C_{\mathsf{R}}\rangle - \langle \tilde{c}_1(r) - C_{\mathsf{R}}\rangle\right]}{C_{\mathsf{R}} + \left[V_S\langle \tilde{c}_2(r) - C_{\mathsf{R}}\rangle - \langle \tilde{c}_1(r) - C_{\mathsf{R}}\rangle\right]\exp(-t/\nu)}, \tag{4.25a}$$

$$V_C(r,t) = \frac{\bar{V}_C\left[C_{\mathsf{R}} + V_S\langle \tilde{c}_2(r) - C_{\mathsf{R}}\rangle - \langle \tilde{c}_1(r) - C_{\mathsf{R}}\rangle\right]}{C_{\mathsf{R}} + \left[V_S\langle \tilde{c}_2(r) - C_{\mathsf{R}}\rangle - \langle \tilde{c}_1(r) - C_{\mathsf{R}}\rangle\right]\exp(-t/\nu)}, \tag{4.25b}$$

$$r_F(r,t) = \frac{\bar{r}_F\left[C_{\mathsf{R}} + V_S\langle \tilde{c}_2(r) - C_{\mathsf{R}}\rangle - \langle \tilde{c}_1(r) - C_{\mathsf{R}}\rangle\right]}{C_{\mathsf{R}} + \left[V_S\langle \tilde{c}_2(r) - C_{\mathsf{R}}\rangle - \langle \tilde{c}_1(r) - C_{\mathsf{R}}\rangle\right]\exp(-t/\nu)}, \tag{4.25c}$$

$$\Lambda_c(r,t) = \frac{\bar{\Lambda}_c\left[C_{\mathsf{R}} + V_S\langle \tilde{c}_2(r) - C_{\mathsf{R}}\rangle\right]}{C_{\mathsf{R}} + V_S\langle \tilde{c}_2(r) - C_{\mathsf{R}}\rangle\exp(-t/\nu)}, \tag{4.25d}$$

$$\beta_o(r,t) = \frac{\bar{\beta}_o(r)\left[\bar{\beta}_o(r) + (\beta_{iso}(r) - \bar{\beta}_o(r))\, f_h(\tilde{c}_1(r) - C_{\mathsf{R}})\right]}{\bar{\beta}_o(r) + [\beta_{iso}(r) - \bar{\beta}_o(r)]\, f_h(\tilde{c}_1(r) - C_{\mathsf{R}})\exp(-t/\nu)}. \tag{4.25e}$$

From Eq. (4.25a), the volume fraction of non-collagenous matrix $V_M(r,t) = V_{EL}(r,t) + V_S$ is obtained. Moreover, from both Eqs. (4.25a) and (4.25b), the volume fraction of non-bearing-load constituents (i.e., fragmented elastin and degraded collagen) is obtained as $V_0(r,t) = 1 - (V_C(r,t) + V_{EL}(r,t) + V_S)$. For defining the admissible set of values $\mathcal{K}_j$, physical constraints $V_C(r,t) \geq 0$, $V_{EL}(r,t) \geq 0$ and $V_0(r,t) \geq 0$ (or equivalently, $V_C(r,t) + V_{EL}(r,t) \leq 1 - V_S$) are employed as lower and upper bound limits for Eqs. (4.25a) and (4.25b), as well as $r_F(r,t) \geq 0$ for Eq. (4.25c) and $\Lambda_c(r,t) \geq 0$ for Eq. (4.25d).

TABLE 4.2 Pseudocode of the algorithm for quasi-analytical arterial multiphysics

1. Find $[\bar{r}_i, \bar{r}_e]$ from Eq. (4.19),
2. Find $\{\tilde{c}_1, \tilde{c}_2, \tilde{c}_3\}$ from Eqs. (4.23),
3. Find $\mathcal{S}(r, t)$ from Eqs. (4.25),
4. Find $P_i(r_i, \mathcal{S}(r, t))$ from Eq. (4.18).

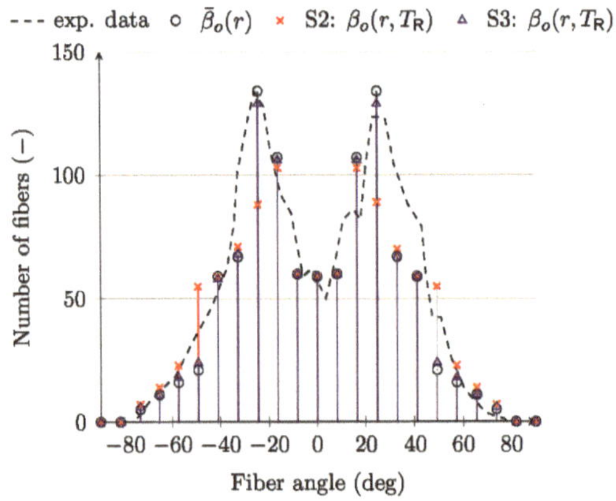

FIGURE 4.6 Total number of collagen fibers for different fiber angles: homeostatic $\bar{\beta}_o(r)$ and remodeled distributions at steady-state $\beta_o(r, T_R)$ for scenarios S2 and S3, compared with experimental data by Schriefl et al. (2012). Values of parameters in Tables 4.3 and 4.4.

To evaluate the characteristic remodeling time interval T_R (namely, the time interval to reach a steady state), let $\epsilon \ll 1$ be an arbitrary positive constant and, in agreement with Eq. (4.13b), define T_R such that $s_j(r, T_R) = \bar{s}_j + \mathfrak{I}_j \approx (1 - \epsilon)(\bar{s}_j + \mathfrak{I}_j)$. Accordingly, from Eqs. (4.25) and assuming $\mathfrak{I}_j \propto \bar{s}_j$, T_R can be estimated as:

$$T_R = \nu \log\left(\frac{(1-\epsilon)\mathfrak{I}_j}{\epsilon\,\bar{s}_j}\right) \approx \nu \log\left(\frac{1-\epsilon}{\epsilon}\right). \tag{4.26}$$

4.4.5 Results

In the line of the solution strategy introduced in Section 4.3.6 and Table 4.1, numerical results are obtained by adopting the procedure in the pseudocode of Table 4.2. Addressing arterial media layer only, the model is parameterized for a thoracic aortic segment. It is worth highlighting that well-established experimental data (Buehler and Wong, 2007; Clark and Glagov, 1985; O'Connell et al., 2008; Marino and Vairo, 2013, 2014; Marino, 2016; Schriefl et al., 2012) are available for setting the values of model parameters describing the homeostatic state $\bar{\mathcal{S}}$, which is assumed as corresponding to the initial state (namely, at $t = 0$). In the following simulations, values in Table 4.3 are employed for model parameters at homeostasis. For instance, addressing fiber orientation distribution, Fig. 4.6 shows a comparison between model setting and histological measurements.

TABLE 4.3 Values of structural parameters defining reference geometry ($r_{i,o}$, $r_{e,o}$, and α) and the homeostatic state $\bar{\mathcal{S}}$ in agreement with data by Auricchio et al. (2014), Buehler and Wong (2007), Clark and Glagov (1985), O'Connell et al. (2008), Holzapfel and Ogden (2010), Marino and Vairo (2013, 2014), Marino (2016), Schriefl et al. (2012) and with the microstructure model in Appendix B (the bar symbol is omitted for parameters not undergoing remodeling). Arterial tissue comprises N_M media lamellar units (MLUs) along its thickness, and each MLU is divided in n_M sub-layers. MLUs and sub-layers are assumed to be identical in thickness. Within each sub-layer, $s_j \in \mathcal{S}$ is piecewise constant and, if variable across arterial thickness, defined by the value at $r = r^{k,p}$, representing the middle position of the pth sub-layer in the kth MLU. In the table, $k \in \{1, \ldots, N_M\}$ and $c \in \{2, \ldots, n_M\}$

Parameter	Value (with unit of measure)	Definition
$r_{i,o}$	8.24 mm	Arterial internal radius (in Ω_o)
$r_{e,o}$	9.14 mm	Arterial external radius (in Ω_o)
α	80°	Arterial opening angle
N_M	60	Number of Media Lamellar Units (MLUs)
n_M	18	Number of sub-layers in each MLU
$\bar{V}_{EL} = \bar{V}_{EL}(r)$	$\begin{cases} 0.5 & \text{for } r = r^{k,1} \\ 0.36 & \text{for } r = r^{k,c} \end{cases}$	Elastin volume fraction
$\bar{V}_C = \bar{V}_C(r)$	$\begin{cases} 0 & \text{for } r = r^{k,1} \\ 0.14 & \text{for } r = r^{k,c} \end{cases}$	Collagen volume fraction
V_S	0.2	Volume fraction of smooth muscle cells
$\bar{r}_F$	0.5 μm	Collagen fiber radius
$H_{F,o}$	1.3 μm	Collagen fiber amplitude (in Ω_o)
$L_{F,o}$	5 μm	Collagen fiber period (in Ω_o)
$\bar{\beta}_o = \bar{\beta}_o(r)$	Eqs. (B.1) with $\hat{\beta} = 25°$	Collagen fiber angle (in Ω_o)
$\bar{\Lambda}_c$	1 mol/mol	Inter-molecular cross-links density
k_c	105 nN/μm	Inter-molecular cross-links stiffness
$\ell_{m,o}$	279 nm	Molecular end-to-end length (in Ω_o)
ℓ_c	287 nm	Molecular contour length
ℓ_p	14.5 nm	Molecular persistence length
T	37°C	Body temperature
A_m	1.41 nm^2	Molecular cross-sectional area
$\hat{E}_o$	1 GPa	Low-strain triple-helix tangent modulus
$\hat{E}$	80 GPa	High-strain triple-helix tangent modulus
η	22.5	Triple-helix uncoiling resistance
ε_o^h	0.1	Triple-helix uncoiling strain
k_M	2.2 MPa	Elastin stiffness

As shown in Fig. 4.7, the obtained pressure–radius relationship $P_i(r_i, \bar{\mathcal{S}})$ is reported, clearly showing that it fully agrees with available experimental data (Hallock and Benson, 1937) and proving the effectiveness of present approach in capturing a realistic mechanical behavior of thoracic aortic segments. Moreover, by choosing $\bar{P}_i = 100$ mm Hg as reference physiological pressure, the initialization step (step 1 in Table 4.2) gives:

$$\bar{r}_i = 9.1\,\text{mm}, \qquad \bar{r}_e = 9.74\,\text{mm}. \tag{4.27}$$

Addressing transport-remodeling phenomena, let $\hat{c}$ denote a basal molecular production (assumed to be constant among different molecules for the sake of simplicity) and let $\gamma_a =$

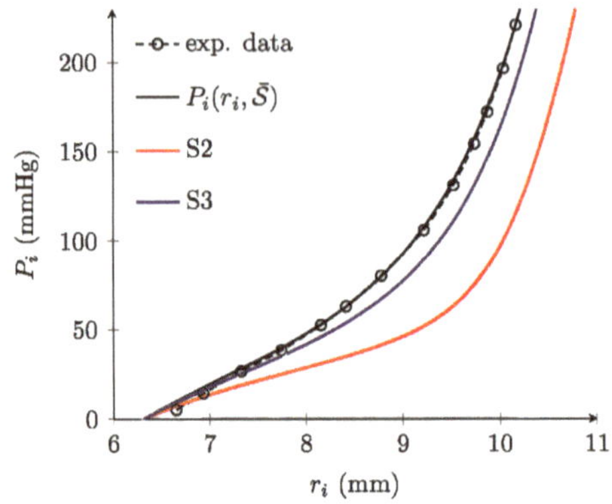

FIGURE 4.7 Pressure–radius relationship $P_i(r_i, \bar{\mathcal{S}})$ (corresponding to the one obtained in scenario S1) and $P_i(r_i, \mathcal{S}(r, T_R))$ for scenarios S2 and S3 (at steady-state), compared with experimental data by Hallock and Benson (1937). Values of parameters in Tables 4.3 and 4.4.

TABLE 4.4 Values of parameters for the transport-remodeling problems in the axisymmetric arterial study case, with $\hat{D}_q = \hat{D}$, $\zeta_q = \zeta$, and $\hat{\lambda}_{qi} = \hat{\lambda}_{qe} = \hat{\lambda}$ in Eqs. (4.23). With reference to Eq. (4.26), ratio $\nu/T_R = 0.1$ ensures $\epsilon < 10^{-4}$

Parameter	$\hat{D}$	ζ	η_1	ξ_1	$\hat{\lambda}$	ν	T_R
Unit of measurement	$mm^2\,s^{-1}$	s^{-1}	s^{-1}	s^{-1}	$mm\,s^{-1}$	years	years
Value	10^{-6}	10^{-5}	10^{-3}	10^{-3}	1	0.1	1

$C_R = C_1/2 = \hat{c}$ be addressed in what follows. Under these conditions, it is straightforward to observe that Eqs. (4.23) and (4.25) do not depend on the value of $\hat{c}$ but only on ratios $\hat{c}_a/\hat{c}$ that are herein varied in order to reproduce three modeling scenarios:

Scenario S1 Maintenance of the homeostatic state with $\hat{c}_2 = \hat{c}_3 = \hat{c}$;

Scenario S2 Increased macrophages activity with $\hat{c}_3 = 2\hat{c}$, coupled with a basal production of TGF-β, with $\hat{c}_2 = \hat{c}$;

Scenario S3 Increased macrophages activity with $\hat{c}_3 = 2\hat{c}$ coupled with an increased production of TGF-β from intimal endothelial cells, with $\hat{c}_2 = 1.5\hat{c}$.

For the different scenarios, results are computed via steps 2–4 in Table 4.2 and by employing values of parameters for the transport-remodeling problems summarized in Table 4.4.

4.4.5.1 *Scenario S1*

Concentration profiles $\tilde{c}_q$ obtained from Eqs. (4.23) are shown in Fig. 4.8. Results show that TGF-β (resp., IL) diffuse in aortic thickness starting from the source at the intima-media layer $\rho = 0$ (resp., at the media-adventitial layer $\rho = 1$). The resulting concentrations profiles $\tilde{c}_2$ and $\tilde{c}_3$ respectively represent the consumption and the source terms for MMPs. In turn, the latter exhibit a symmetric concentration profile (due to the symmetry in the employed values of intimal and adventitia permeability $\hat{\lambda}$) with maximum value at $\rho = 0.5$.

Based on the obtained solution for the transport problem, a null remodeling is predicted from Eqs. (4.25) since $\tilde{c}_1(r) \leq C_R$ and $\tilde{c}_2(r) \leq C_R$. Accordingly, since it results $\mathcal{S}(r,t) = \bar{\mathcal{S}}(r)$ for

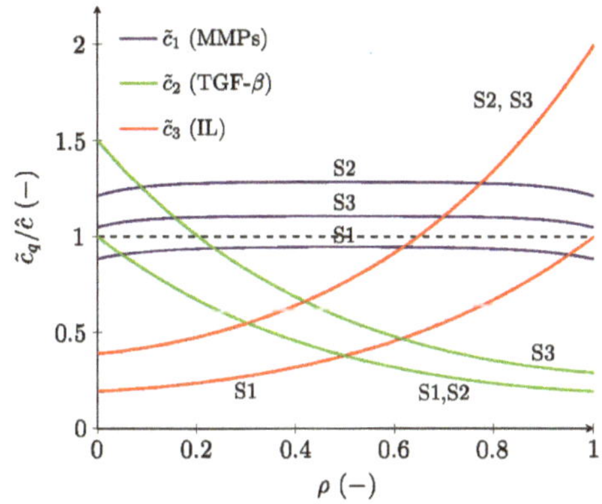

FIGURE 4.8 Normalized concentration profiles $\tilde{c}_q/\hat{c}$ vs. normalized radial coordinate ρ for the scenarios S1, S2 and S3; the dashed line denotes the value $C_R/\hat{c}$ corresponding to the activation of remodeling. Values of parameters in Tables 4.3 and 4.4.

any t, arterial mechanics from Eq. (4.18) results to be constant in time, obtaining the pressure–radius relationship $P_i(r_i, \bar{\mathcal{S}})$, shown in Fig. 4.7, under homeostatic conditions.

4.4.5.2 Scenario S2

The increment of macrophages activity is addressed as a possible pathological mechanism and it induces an higher source of IL from the adventitial layer. As shown in Fig. 4.8, a higher concentration of IL is obtained in the media layer from Eqs. (4.23), while TGF-β are clearly unaffected. Accordingly, the source term for MMPs (namely, the average $\tilde{c}_2$ in the overall thickness) increases and $\tilde{c}_1(r) > C_R$ is obtained.

This outcome is associated with remodeling activation in Eqs. (4.25). In particular, the rearrangement of the orientations of collagen fibers (namely, $\beta(r,t)$) is obtained as shown in Fig. 4.6, which corresponds to a more isotropic-distribution of fibers with respect to the homeostatic state. For instance, the number of fibers along the preferred directions $\hat{\beta} = \pm 25°$ decreases up to 40%. Moreover, Fig. 4.9 shows the evolution in time of elastin and collagen volume fraction, as well as of fiber radius, associated with the remodeling. A significant degradation of constituents, whose values nonlinearly evolve in time, is obtained: the greatest degradation is associated with the initial stage of the investigated pathological mechanism and a final decrease of about 30% is obtained in the steady-state at $t = T_R$. In the present scenario, inter-molecular cross-links density is not affected (namely, $\Lambda_c(r,t) = \bar{\Lambda}_c(r)$, corresponding results are not shown for the sake of compactness) because Eq. (4.25d) does not depend on $\tilde{c}_1$ and $\tilde{c}_2(r) \leq C_R$.

As a result of the obtained remodeling, arterial mechanics is significantly affected. The time-dependent pressure–radius relationship $P_i(r_i, \mathcal{S}(r,t))$ along the remodeling path is reported in Fig. 4.9, showing a steady-state enlargement (at $t = T_R$) of about 10% at $P_i = \bar{P}_i$ with respect to $t = 0$.

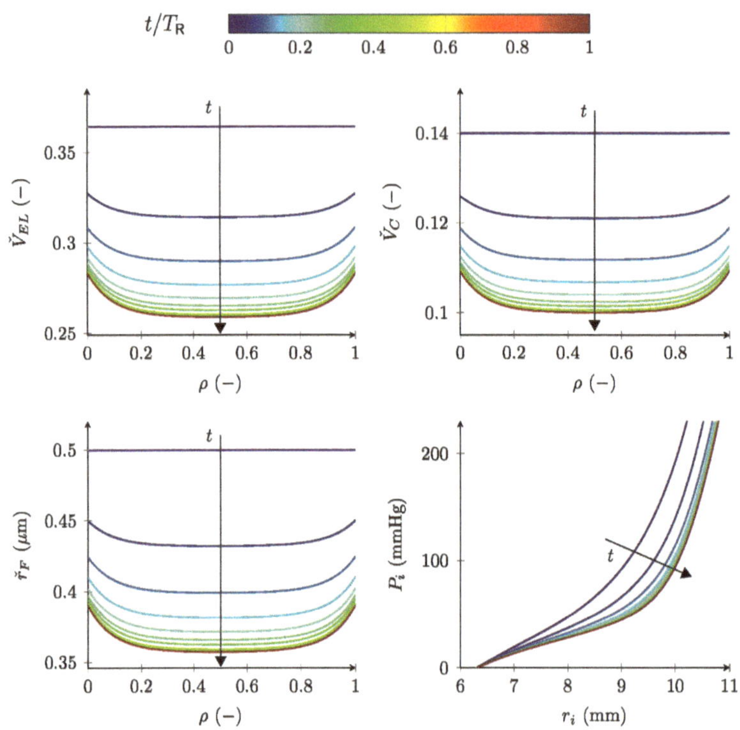

FIGURE 4.9 Scenario S2: time-dependent structural features $s_j(\rho, t)$ and pressure–radius relationship $P_i(r_i, \mathcal{S}(r,t))$ (bottom right); $\check{V}_{EL}$ (top left), average of V_{EL} within media lamellar units (MLUs); $\check{V}_C$ (top right), average of V_C within the interlamellar substance of each MLU; $\check{r}_F$ (bottom left), average of r_F within the interlamellar substance of each MLU. Color map denotes different values of time $t \in [0, T_R]$. Values of parameters in Tables 4.3 and 4.4.

4.4.5.3 *Scenario S3*

The increment in concentrations of both TGF-β and IL with respect to the homeostatic state in scenario S1 (see Fig. 4.8) determines a smaller increase of MMPs concentration with respect to scenario S2.

Nevertheless, since $\tilde{c}_1(r) > C_R$, the degradation of constituents associated with remodeling laws in Eqs. (4.25) is activated. The MMPs-related degradation is anyway mitigated by the depositing properties of TGF-β, since $\tilde{c}_2(r) > C_R$ in about the inner half of arterial thickness. As a matter of fact, the effects of remodeling are shown in Fig. 4.10 where elastin and collagen volume fractions, fiber radius, and inter-molecular cross-links density are shown in the steady-state $t = T_R$, and compared with the ones obtained from different scenarios. Although IL concentration is unaffected with respect to scenario S2, constituents degradation is significantly lower due to the counteracting activity induced by TGF-β. Besides, the value of structural parameters increases in the inner half of arterial thickness, wherein $\tilde{c}_2 > \tilde{c}_1$, and the average variation of the value of structural parameters is less than 10%. In particular, inter-molecular cross-links density increases only because it is not affected by MMPs in the

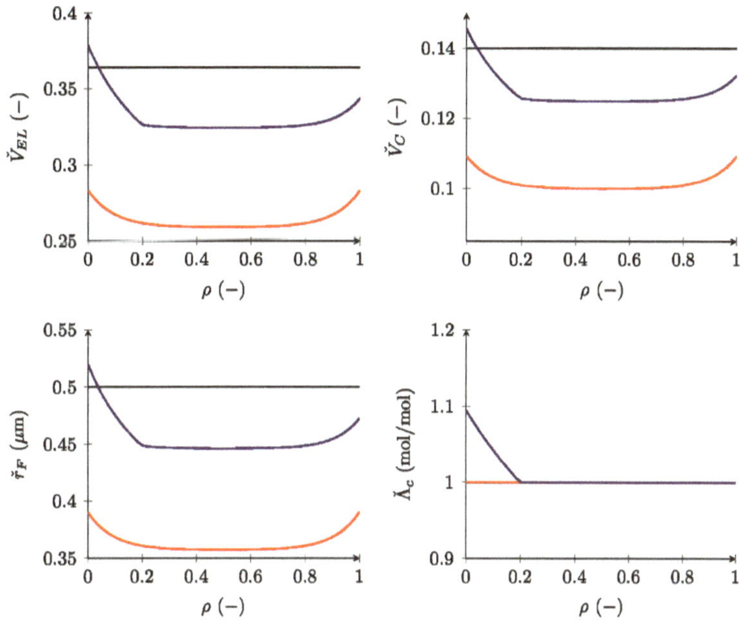

FIGURE 4.10 Structural features $s_j(\rho, T_R)$ at steady-state $t = T_R$ for the scenarios S1 (in black), S2 (in red), and S3 (in blue). $\check{V}_{EL}$ (top left), average of V_{EL} within media lamellar units (MLUs); $\check{V}_C$ (top right), average of V_C within the interlamellar substance of each MLU; $\check{r}_F$ (bottom left), average of r_F within the interlamellar substance of each MLU; $\check{\Lambda}_c$ (bottom right), average of Λ_c within the interlamellar substance of each MLU. Values of parameters in Tables 4.3 and 4.4.

present model. Furthermore, as shown in Fig. 4.6, collagen fiber angles are characterized by a symmetric bimodal distribution after the remodeling, similarly to the homeostatic state.

As a consequence, the pressure–radius relationship $P_i(r_i, \mathcal{S}(r, T_R))$ obtained for the present scenario is more similar to the homeostatic scenario S1 than to the pathological one S2 (see Fig. 4.7). In particular, at $P_i = \bar{P}_i$, an aortic enlargement of about 3% is obtained with respect to $t = 0$.

4.5 CONCLUSIONS

Arterial physiopathological behavior involves multiphysics mechanisms, as the result of the complex interplay between microscale transport phenomena and mechanical equilibrium. Indeed, vascular mechanics highly depends on arterial wall constituents whose structural organization and properties are driven by remodeling mechanisms. In turn, the latter are governed by the proteolytic activity of enzymes diffusing within aortic thickness or by cell–cell signaling pathways mediated by soluble factors.

Arterial behavior has been herein modeled and analyzed through a multiphysics strategy that couples macroscopic mechanical description, molecular transport phenomena, and remodeling laws of tissue micro- and nano-structural features. Numerical results have been obtained by addressing a case study which shows the effects of ECM remodeling induced by intra-arterial-wall transport of MMPs, TGF-β, and IL on the compliance of an axisymmetric arterial segment.

Remarkably, obtained results show that the present approach is able to capture arterial dilation as a consequence of alterations in tissue biochemical environment and/or cellular activity. For instance, addressing an increased activity of macrophages in the production of cytokines, an enlargement of about 10% is predicted, associated with the proteolytic activity of MMPs. On the other hand, aortic dilation reduce to only 3% when MMPs activity is counterbalanced by TGF-β. This agrees well with available evidence that confirms the protective role of TGF-β whose increased production might represent an internal feedback mechanism or be a consequence of a pharmacological treatment (Dai et al., 2011).

Future studies will address some limitations of present work. The active contraction of smooth muscle cells and the unsteady fluid–structure interaction effects characterizing arterial mechanics have not been considered; no mass growth or constituents turn-over has been investigated; no distinction has been made between latent and active concentrations of molecular species; simplifying assumptions have been introduced (e.g., no mechanical feedback in the transport problem) in order to reduce arterial multiphysics to an open-loop coupled system which can be solved via a staggered and semi-analytical approach.

Acknowledgements

Present research study has been supported by the Italian Minister of University and Research, MIUR (Program: "Consolidate the Foundations 2015"; Project: BIOART; Grant number (CUP): E82F16000850005). Moreover, M. Marino gratefully acknowledges a post-doctoral fellowship from the Alexander von Humboldt Foundation.

APPENDIX A ALONG-THE-CHORD COLLAGEN FIBER TANGENT MODULUS

By following the Maceri–Marino–Vairo multiscale rationale (Maceri et al., 2013; Marino and Vairo, 2013, 2014), the mechanics of crimped collagen fibers is described by introducing an analytical expression for their along-the-chord tangent modulus C_F that depends on main geometric features (period L_F, amplitude H_F, and radius r_F) as well as on material nanoscale mechanisms.

Since fibers are collection of fibrils, material features are described in term of fibril mechanics. Fibril stretch λ_f depends on molecular elongation λ_m (namely, $\lambda_f = \lambda_f(\lambda_m)$) which, in turn, is a function of entropy-related λ_m^s and energy-related λ_m^h molecular stretches (namely, $\lambda_m = \lambda_m(\lambda_m^s, \lambda_m^h)$). Accordingly, fibril elastic modulus E_f results in

$$E_f(\lambda_f) = E_f(\lambda_m^s, \lambda_m^h) = \frac{E_m(\lambda_m^s, \lambda_m^h)\, \Lambda_c\, k_c\, \ell_{m,o}}{[\Lambda_c\, k_c\, \ell_{m,o} + A_m\, E_m(\lambda_m^s, \lambda_m^h)]}, \tag{A.1a}$$

with k_c being cross-link stiffness and E_m the elastic modulus of collagen molecules, defined as

$$E_m(\lambda_m) = E_m(\lambda_m^s, \lambda_m^h) = \frac{E_m^s(\lambda_m^s)\, E_m^h(\lambda_m^h)}{E_m^s(\lambda_m^s) + E_m^h(\lambda_m^h)}, \tag{A.1b}$$

where entropy-related E_m^s and energy-related E_m^h collagen moduli are

$$E_m^s(\lambda_m^s) = \frac{k_B\, T\, \ell_{m,o}}{\ell_p\, \ell_c\, A_m} \left(\frac{\ell_c^3}{2[\ell_c - \ell_{m,o}\lambda_m^s]^3} + 1 \right), \tag{A.1c}$$

$$E_m^h(\lambda_m^h) = \frac{\ell_{m,o}}{\ell_c} \left[\frac{\hat{E}}{1 + \exp\{-\eta[\ell_{m,o}(\lambda_m^h - 1)/\ell_c - \varepsilon_o^h]\}} + \hat{E}_o \right], \tag{A.1d}$$

with k_B being the Boltzmann constant and T the absolute temperature. Other symbols are defined in Section 4.3.3.1. The multiscale problem is closed via interscale compatibility relationships between atomistic- and nanoscales:

$$\frac{d\lambda_m^s}{d\lambda_m} = \Phi_{ms}(\lambda_m^s, \lambda_m^h) = \frac{E_m(\lambda_m^s, \lambda_m^h)}{E_m^s(\lambda_m^s)}, \tag{A.2a}$$

$$\frac{d\lambda_m^h}{d\lambda_m} = \Phi_{mh}(\lambda_m^s, \lambda_m^h) = \frac{E_m(\lambda_m^s, \lambda_m^h)}{E_m^h(\lambda_m^h)}, \tag{A.2b}$$

and between nanoscale and mesoscale:

$$\frac{d\lambda_m}{d\lambda_f} = \Phi_{fm}(\lambda_m^s, \lambda_m^h) = \frac{E_f(\lambda_m^s, \lambda_m^h)}{E_m(\lambda_m^s, \lambda_m^h)}. \tag{A.3}$$

A closed-form expression for C_F can be obtained from the application of the Principle of Virtual Works by assuming a piecewise-linear fiber-shape and by coupling geometric and material nonlinearities (Marino and Wriggers, 2016). Geometric nonlinearities are introduced by accounting for the functional dependence of fiber amplitude on along-the-chord fiber elongation λ_F, namely $H_F = H_F(\lambda_F)$. Moreover, by definition, it results in $L_F = L_F(\lambda_F) = \lambda_F L_{F,o}$ where $L_{F,o}$ represents fiber period in the reference configuration. Material nonlinearities are accounted for by means of fibril tangent modulus $E_f = E_f(\lambda_f) = E_f(\lambda_m^s, \lambda_m^h)$ in Eqs. (A.1). In order to couple material nonlinearities with geometric ones, an inter-scale compatibility relationship between fibril stretch λ_f and along-the-chord fiber stretch λ_F is introduced by identifying λ_f as coincident with the stretch of the fiber centerline, resulting in

$$\frac{d\lambda_f}{d\lambda_F} = \Phi_f(\lambda_F, H_F) = \frac{\lambda_F \ell_{F,o}^2 + H_F \frac{dH_F}{d\lambda_F}}{\sqrt{(\lambda_F^2 \ell_{F,o}^2 + H_F^2)(\ell_{F,o}^2 + H_{F,o}^2)}}, \tag{A.4}$$

where $\ell_F = L_F/4$ is the fiber quarter-period and $H_{F,o}$ is the fiber amplitude in the reference configuration.

Therefore, omitting functional dependencies for the sake of compactness, it results in

$$C_F = E_f \frac{\ell_F^2 + H_F^2}{\sqrt{\ell_{F,o}^2 + H_{F,o}^2}} \left[\ell_F + \frac{4H_F^2}{3r_F^2\,\ell_F} \left(\ell_F^2 + H_F^2 \right) \right]^{-1}, \tag{A.5}$$

coupled with the following interscale differential equations:

$$\frac{d\lambda_m^s}{d\lambda_F} = \Phi_{ms}(\lambda_m^s, \lambda_m^h)\Phi_{fm}(\lambda_m^s, \lambda_m^h)\Phi_f(\lambda_F, H_F), \tag{A.6a}$$

$$\frac{d\lambda_m^h}{d\lambda_F} = \Phi_{mh}(\lambda_m^s, \lambda_m^h)\Phi_{fm}(\lambda_m^s, \lambda_m^h)\Phi_f(\lambda_F, H_F), \tag{A.6b}$$

$$\frac{dH_F}{d\lambda_F} = -\frac{\ell_F H_F \left[4(\ell_F^2 + H_F^2) - 3r_F^2\right]}{\lambda_F \left[4H_F^2(\ell_F^2 + H_F^2) + 3\ell_F^2 r_F^2\right]}, \tag{A.6c}$$

where Eqs. (A.2), (A.3), and (A.4) are taken into account. Accordingly, from the solution of the system of differential equations (A.6), functions $\lambda_m^s = \lambda_m^s(\lambda_F)$, $\lambda_m^h = \lambda_m^h(\lambda_F)$, and $H_F = H_F(\lambda_F)$ are obtained, resulting in $C_F = C_F(\lambda_F)$. Moreover, in order to highlight the explicit dependence of predicted fiber mechanics on the set of structural parameters $\mathcal{S}_F$ (defined in Eqs. (4.4)), the functional form $C_F = C_F(\lambda_F, \mathcal{S}_F)$ is conveniently employed.

APPENDIX B MICROSTRUCTURE OF AORTIC MEDIA LAYER

Aortic media layer is modeled as comprising N_M iso-width perfectly-bonded layers representing a single media lamellar unit (MLU). Each of them comprises n_M iso-width perfectly-bonded sub-layers. Accordingly, $s_j \in \mathcal{S}$ are piecewise constant functions in r where, for the sake of notation, $s_j^{k,p}$ denotes the value of s_j in the pth sub-lamella of the kth lamella. Moreover, let subscript k take values in $\{1, \dots, N_M\}$, p in $\{1, \dots, n_M\}$, and c in $\{2, \dots, n_M\}$.

The kth lamella is considered as made up of one elastin layer (namely, $V_C^{k,1} = 0$) and $n_M - 1$ sub-lamellae reinforced by collagen fibers ($V_C^{k,c} > 0$).

For describing the homeostatic state, let us introduce $\bar{V}_{EL}^{el}$, $\bar{V}_{EL}^{is}$, and $\bar{V}_C^{is}$ as constant values. The elastin layer is assumed to be characterized by $\bar{V}_{EL}^{k,1} = \bar{V}_{EL}^{el}$ for any MLU (namely, for any $k = 1, \dots, N_M$). Moreover, we choose $\bar{V}_{EL}^{k,p} = \bar{V}_{EL}^{is}$ and $\bar{V}_C^{k,c} = \bar{V}_C^{is}$, constant within and among MLUs. Furthermore, the homeostatic orientation of collagen fibers in the inter-lamellar substance at the reference configuration is defined by the function $\bar{\beta}_o = \bar{\beta}_o(r)$, with

$$\bar{\beta}_o^{k,c} = \varpi_k(r)\vartheta_{c-1}^{ani}(\hat{\beta}) + (1 - \varpi_k(r))\vartheta_{c-1}^{iso} \tag{B.1a}$$

where

$$\varpi_k(r) = \begin{cases} 2(\rho_k(r) - r_i)/(r_e - r_i) & \text{for } \rho_k(r) < (r_e + r_i)/2, \\ 1 & \text{for } \rho_k(r) \geq (r_e + r_i)/2, \end{cases} \tag{B.1b}$$

$$\rho_k(r) = (r_{e,k} + r_{i,k})/2 \quad \text{for } r_{i,k} \leq r < r_{e,k}\,, \tag{B.1c}$$

$$r_{i,k} = r_i + (k-1)(r_e - r_i)/N_M\,, \tag{B.1d}$$

$$r_{e,k} = r_{i,k} + (r_e - r_i)/N_M\,, \tag{B.1e}$$

and ϑ^{ani}_{c-1} (resp., ϑ^{iso}_{c-1}) is the $(c-1)$th component of vector $\boldsymbol{\vartheta}^{ani}(\hat{\beta}) \in \mathbb{R}^{n_M-1}$ (resp., $\boldsymbol{\vartheta}^{iso} \in \mathbb{R}^{n_M-1}$) representing a fiber angle distribution with preferred direction $\hat{\beta} \in (0, \pi/2)$ (resp., without any preferred direction). In the numerical simulations conducted in the present work, we choose $n_M = 18$ and

$$\boldsymbol{\vartheta}^{iso} = (-\hat{\boldsymbol{\vartheta}}^{iso}, 0, \hat{\boldsymbol{\vartheta}}^{iso})\,, \qquad \boldsymbol{\vartheta}^{ani}(\bar{\beta}^*) = (-\hat{\boldsymbol{\vartheta}}^{ani}(\hat{\beta}), 0, \hat{\boldsymbol{\vartheta}}^{ani}(\hat{\beta}))\,, \tag{B.1f}$$

where

$$\hat{\boldsymbol{\vartheta}}^{iso} = (10, 20, 30, 40, 50, 60, 70, 80)\pi/180\,, \tag{B.1g}$$

$$\hat{\boldsymbol{\vartheta}}^{ani}(\hat{\beta}) = (1/2, 2/3, 2/3, 1, 1, 1, 4/3, 5/3)\,\pi\hat{\beta}/180\,. \tag{B.1h}$$

Finally, for the sake of describing the remodeling of fiber orientation, let β_{iso} in Eq. (4.24e) be a piecewise constant function with values

$$\beta^{k,c}_{iso} = \vartheta^{iso}_{c-1}\,. \tag{B.2}$$

References

Auricchio, F., Da Veiga, L.B., Lovadina, C., Reali, A., Taylor, R.L., Wriggers, P., 2013. Approximation of incompressible large deformation elastic problems: some unresolved issues. Comput. Mech. 52 (5), 1153–1167.

Auricchio, F., Conti, M., Ferrara, A., 2014. How constitutive model complexity can affect the capability to fit experimental data: a focus on human carotid arteries and extension/inflation data. Arch. Comput. Methods Eng. 21 (3), 273–292.

Baek, S., Rajagopal, K.R., Humphrey, J.D., 2006. A theoretical model of enlarging intracranial fusiform aneurysms. J. Biomech. Eng. 128 (1), 142–149.

Bailey, A.J., 2001. Molecular mechanisms of ageing in connective tissues. Mech. Ageing Dev. 122 (7), 735–755.

Behmoaras, J., Osborne-Pellegrin, M., Gauguier, D., Jacob, M.P., 2005. Characteristics of the aortic elastic network and related phenotypes in seven inbred rat strains. Am. J. Physiol., Heart Circ. Physiol. 288 (2), H769–H777.

Brüel, A., Ørtoft, G., Oxlund, H., 1998. Inhibition of cross-links in collagen is associated with reduced stiffness of the aorta in young rats. Atherosclerosis 140 (1), 135–145.

Buehler, M.J., Wong, S.Y., 2007. Entropic elasticity controls nanomechanics of single tropocollagen molecules. Biophys. J. 93 (1), 37–43.

Carmo, M., Colombo, L., Bruno, A., Corsi, F.R.M., Roncoroni, L., Cuttin, M.S., Radice, F., Mussini, E., Settembrini, P.G., 2002. Alteration of elastin, collagen and their cross-links in abdominal aortic aneurysms. Eur. J. Vasc. Endovasc. Surg. 23 (6), 543–549.

Clark, J.M., Glagov, S., 1985. Transmural organization of the arterial media. The lamellar unit revisited. Arterioscler. Thromb. Vasc. Biol. 5 (1), 19–34.
Chen, H., Liu, Y., Slipchenko, M.N., Zhao, X., Cheng, J.X., Kassab, G.S., 2011. The layered structure of coronary adventitia under mechanical load. Biophys. J. 101 (11), 2555–2562.
Dabagh, M., Jalali, P., Tarbell, J.M., 2009. The transport of LDL across the deformable arterial wall: the effect of endothelial cell turnover and intimal deformation under hypertension. Am. J. Physiol., Heart Circ. Physiol. 297, H983–H996.
Dai, J., Michineau, S., Franck, G., Desgranges, P., Becquemin, J.P., Gervais, M., Allaire, E., 2011. Long term stabilization of expanding aortic aneurysms by a short course of cyclosporine a through transforming growth factor-beta induction. PLoS ONE 6, e28903.
Deguchi, J.O., Aikawa, E., Libby, P., Vachon, J.R., Inada, M., Krane, S.M., Whittaker, P., Aikawa, M., 2005. Matrix metalloproteinase-13/collagenase-3 deletion promotes collagen accumulation and organization in mouse atherosclerotic plaques. Circulation 112, 2708–2715.
D'Errico, M., Sammarco, P., Vairo, G., 2015. Analytical modeling of drug dynamics induced by eluting stents in the coronary multi-layered curved domain. Math. Biosci. 267, 79–96.
Ethier, C.R., Simmons, C.A., 2007. Introductory Biomechanics – From Cells to Organisms. Cambridge University Press, Cambridge.
Farlow, S.J., 1982. Partial Differential Equations for Scientists and Engineers. Wiley, New York.
Figueroa, C.A., Baek, S., Taylor, C.A., Humphrey, J.D., 2009. A computational framework for coupled fluid–solid growth modeling in cardiovascular simulations. Comput. Methods Appl. Mech. Eng. 198 (45), 3583–3602.
Fratzl, P., 2008. Collagen: Structure and Mechanics. Springer, New York.
Freedman, B.R., Bade, N.D., Riggin, C.N., Zhang, S., Haines, P.G., Ong, K.L., Janmey, P.A., 2015. The (dys)functional extracellular matrix. Biochim. Biophys. Acta, Mol. Cell Res. 1853 (11), 3153–3164.
Fuster, V., Kelly, B., 2010. Promoting Cardiovascular Health in the Developing World: A Critical Challenge to Achieve Global Health. National Academies Press, Washington DC.
García-Alvarez, J., Ramirez, R., Checa, M., Nuttall, R.K., Sampieri, C.L., Edwards, D.R., Selman, M., Pardo, A., 2006. Tissue inhibitor of metalloproteinase-3 is up-regulated by transforming growth factor-β1 in vitro and expressed in fibroblastic foci in vivo in idiopathic pulmonary fibrosis. Exp. Lung Res. 32, 201–214.
Hallock, P., Benson, I.C., 1937. Studies on the elastic properties of human isolated aorta. J. Clin. Invest. 16 (4), 595–602.
Holzapfel, G.A., Ogden, R.W., 2010. Modelling the layer-specific three-dimensional residual stresses in arteries, with an application to the human aorta. J. R. Soc. Interface 7, 787–799.
Holzapfel, G.A., Gasser, T.C., Ogden, R.W., 2000. A new constitutive framework for arterial wall mechanics and a comparative study of material models. J. Elast. 61 (1–3), 1–48.
Humphrey, J.D., 2002. Cardiovascular Solid Mechanics: Cells, Tissues, and Organs. Springer, New York.
Humphrey, J.D., Rajagopal, K.R., 2002. A constrained mixture model for growth and remodeling of soft tissues. Math. Models Methods Appl. Sci. 12 (3), 407–430.
Humphrey, J.D., 2009. Vascular mechanics, mechanobiology, and remodeling. J. Mech. Med. Biol. 9 (2), 243–257.
Jones, J.A., Spinale, F.G., Ikonomidis, J.S., 2009. Transforming growth factor-β signaling in thoracic aortic aneurysm development: a paradox in pathogenesis. J. Vasc. Res. 46 (2), 119–137.
Kassab, G.S., 2006. Biomechanics of the cardiovascular system: the aorta as an illustratory example. J. R. Soc. Interface 3, 719–740.
Kucich, U., Rosenbloom, J.C., Abrams, W.R., Rosenbloom, J., 2002. Transforming growth factor-β stabilizes elastin mRNA by a pathway requiring active Smads, protein kinase C-δ, and p38. Am. J. Respir. Cell Mol. Biol. 26 (2), 183–188.
Lee, E., Grodzinsky, A.J., Libby, P., Clinton, S.K., Lark, M.W., Lee, R.T., 1995. Human vascular smooth muscle cell–monocyte interactions and metalloproteinase secretion in culture. Arterioscler. Thromb. Vasc. Biol. 15 (12), 2284–2289.
Leemasawatdigul, K., Gappa-Fahlenkamp, H., 2012. Development of a mathematical model to describe the transport of monocyte chemoattractant protein-1 through a three-dimensional collagen matrix. Cardiovasc. Pathol. 21 (3), 219–228.
Lilly, B., 2014. We have contact: endothelial cell–smooth muscle cell interactions. Physiology 29 (4), 234–241.
Maceri, F., Marino, M., Vairo, G., 2012. Elasto-damage modeling of biopolymer molecules response. Comput. Model. Eng. Sci. 87, 461–481.

Maceri, F., Marino, M., Vairo, G., 2013. Age-dependent arterial mechanics via a multiscale elastic approach. Int. J. Comput. Methods Eng. Sci. Mech. 14, 141–151.
Maiellaro, K., Taylor, W.R., 2007. The role of the adventitia in vascular inflammation. Cardiovasc. Res. 75 (4), 640–648.
Marino, M., Vairo, G., 2013. Multiscale elastic models of collagen bio-structures: from cross-linked molecules to soft tissues. In: Gefen, A. (Ed.), Multiscale Computer Modeling in Biomechanics and Biomedical Engineering. Springer, Berlin, Heidelberg.
Marino, M., Vairo, G., 2014. Computational modelling of soft tissues and ligaments. In: Jin, Z. (Ed.), Computational Modelling of Biomechanics and Biotribology in the Musculoskeletal System. Woodhead Publishing, Cambridge.
Marino, M., 2016. Molecular and intermolecular effects in collagen fibril mechanics: a multiscale analytical model compared with atomistic and experimental studies. Biomech. Model. Mechanobiol. 15, 133 154.
Marino, M., Wriggers, P., 2016. Finite strain response of crimped beams under uniaxial traction: an analytical approach applied to collagen. J. Mech. Phys. Solids. http://dx.doi.org/10.1016/j.jmps.2016.05.010. In press.
Merrilees, M.J., Tiang, K.M., Scott, 1987. Changes in collagen fibril diameters across artery walls including a correlation with glycosaminoglycan content. Connect. Tissue Res. 16 (3), 237–257.
Michel, J.B., Thaunat, O., Houard, X., Meilhac, O., Caligiuri, G., Nicoletti, A., 2007. Topological determinants and consequences of adventitial responses to arterial wall injury. Arterioscler. Thromb. Vasc. Biol. 27 (6), 1259–1268.
Nardinocchi, P., Teresi, L., 2007. On the active response of soft living tissues. J. Elast. 88 (1), 27–39.
O'Connell, M.K., Murthy, S., Phan, S., Xu, C., Buchanan, J., Spilker, R., Dalman, R.L., Zarins, C.K., Denk, W., Taylor, C.A., 2008. The three-dimensional micro- and nanostructure of the aortic medial lamellar unit measured using 3D confocal and electron microscopy imaging. Matrix Biology 27, 171–181.
Pontrelli, G., De Monte, F., 2007. Mass diffusion through two-layer porous media: an application to the drug-eluting stent. Int. J. Heat Mass Transf. 50 (17), 3658–3669.
Rachev, A., Stergiopulos, N., Meister, J.J., 1996. Theoretical study of dynamics of arterial wall remodeling in response to changes in blood pressure. J. Biomech. 29, 635–642.
Rastogi, A., Kim, H., Twomey, J.D., Hsieh, A.H., 2013. MMP-2 mediates local degradation and remodeling of collagen by annulus fibrosus cells of the intervertebral disc. Arthritis Res. Ther. 15 (2), 1.
Rezakhaniha, R., Fonck, E., Genoud, C., Stergiopulos, N., 2011. Role of elastin anisotropy in structural strain energy functions of arterial tissue. Biomech. Model. Mechanobiol. 10 (4), 599–611.
Sagi, I., Gaffney, J.P. (Eds.), 2015. Matrix Metalloproteinase Biology. Wiley, Hoboken.
Sasaki, N., Odajima, S., 1996. Elongation mechanism of collagen fibrils and force–strain relations of tendon at each level of structural hierarchy. J. Biomech. 29, 1131–1136.
Schriefl, A.J., Zeindlinger, G., Pierce, D.M., Regitnig, P., Holzapfel, G.A., 2012. Determination of the layer-specific distributed collagen fibre orientations in human thoracic and abdominal aortas and common iliac arteries. J. R. Soc. Interface 9, 1275–1286.
Svensson, R.B., Mulder, H., Kovanen, V., Magnusson, S.P., 2013. Fracture mechanics of collagen fibrils: influence of natural cross-links. Biophys. J. 104, 2476–2484.
Tarbell, J.M., 2003. Mass transport in arteries and the localization of atherosclerosis. Annu. Rev. Biomed. Eng. 5 (1), 79–118.
Taylor, C.A., Humphrey, J.D., 2009. Open problems in computational vascular biomechanics: hemodynamics and arterial wall mechanics. Comput. Methods Appl. Mech. Eng. 198, 3514–3523.
Truskey, G.A., Yuan, F., Katz, D.F., 2010. Transport Phenomena in Biological Systems. Pearson Prentice Hall Bioengineering. Pearson Education, New Jersey, USA.
Vairo, G., Cioffi, M., Cottone, R., Dubini, G., Migliavacca, F., 2010. Drug release from coronary eluting stents: a multidomain approach. J. Biomech. 43, 1580–1589.
Vandermeer, J., 2010. How populations grow: the exponential and logistic equations. Nat. Educ. Knowl. 3 (10), 15.
van der Slot, A.J., Van Dura, E.A., De Wit, E.C., Degroot, J., Huizinga, T.W.J., Bank, R.A., Zuurmond, A.M., 2005. Elevated formation of pyridinoline cross-links by profibrotic cytokines is associated with enhanced lysyl hydroxylase 2b levels. Biochim. Biophys. Acta, Mol. Basis Dis. 1741 (1), 95–102.
Visse, R., Nagase, H., 2003. Matrix metalloproteinases and tissue inhibitors of metalloproteinases structure, function, and biochemistry. Circ. Res. 92 (8), 827–839.
Volokh, K.Y., Vorp, D.A., 2008. A model of growth and rupture of abdominal aortic aneurysm. J. Biomech. 41 (5), 1015–1021.
Wada, S., Karino, T., 1999. Theoretical study on flow-dependent concentration polarization of low density lipoproteins at the luminal surface of a straight artery. Biorheology 36, 207–233.

Watton, P.N., Ventikos, Y., Holzapfel, G.A., 2009. Modelling the growth and stabilization of cerebral aneurysms. Math. Med. Biol. 26 (2), 133–164.

Wolinsky, H., Glagov, S., 1964. Structural basis for the static mechanical properties of the aortic media. Circ. Res. 14 (5), 400–413.

Wolinsky, H., Glagov, S., 1967. A lamellar unit of aortic medial structure and function in mammals. Circ. Res. 20 (1), 99–111.

Wriggers, P., 2008. Nonlinear Finite Element Methods. Springer, Berlin, Heidelberg.

Zhang, W., Herrera, C., Atluri, S.N., Kassab, G.S., 2005. The effect of longitudinal pre-stretch and radial constraint on the stress distribution in the vessel wall: a new hypothesis. Mech. Chem. Biosyst. 2, 41–52.

CHAPTER

5

Modeling Cystic Fibrosis and Mucociliary Clearance

R. Chatelin*, D. Anne-Archard†, M. Murris-Espin‡,§, D. Sanchez¶, M. Thiriet‖, A. Didier§, P. Poncet**,#

*Université de Lyon, ENI Saint Etienne, LTDS, UMR 5513 CNRS, Saint-Etienne, France †Institut de Mécanique des Fluides de Toulouse (IMFT) - Université de Toulouse, CNRS-INPT-UPS, Toulouse, France ‡Adult Cystic Fibrosis Center (CRCM), CHU de Toulouse, Hôpital Larrey, Toulouse, France §Service de pneumologie et allergologie, CHU de Toulouse, Hôpital Larrey, Toulouse, France ¶Toulouse Institute of Mathematics, UMR CNRS 5219, Toulouse, France ‖EPI INRIA–UPMC–CNRS REO, Université Pierre et Marie Curie, Lab. J-L. Lions, Paris, France **University Pau & Pays Adour, Lab. Mathematics and their Applications (LMAP), UMR CNRS 5142, IPRA, Pau, France

5.1 MUCOCILIARY CLEARANCE AND CYSTIC FIBROSIS

This chapter considers the investigation of airway clearance efficiency and dysfunction using rheological measurements, modeling and simulations of mucus flows. The work is mainly dedicated to the respiratory epithelium dysfunction subsequently to the consumption of cilia-inhibiting drugs (e.g. nicotine) or viral and/or bacterial infections, and cystic fibrosis, although many respiratory diseases are associated with an altered mucus transport. The work presented here is a review based, amongst others, on Chatelin et al. (2015, 2016), Chatelin and Poncet (2013, 2014, 2016), Chatelin (2013).

The first part of this chapter focuses on the trachea–bronchial tree configuration, the respiratory epithelium, the dimensioning aspects and the cystic fibrosis pathology. After a factual dimensioning, the causes and effects of the cystic fibrosis are described. A mathematical model is finally built upon the nature of mucus and its surrounding epithelium environment.

The second part is devoted to the Newtonian modeling: efficient numerical methods are described. The algorithm is based on a penalization technique for the moving cilia-shaped

Chapter coordinator.

geometry, with a one way coupling (the cell motion is prescribed). The asymptotic analysis and the well-posedness of the mathematical model are discussed. Specific numerical methods are introduced with their inherent features (convergence, computational time and memory use) as well as their use when mucociliary clearance is impaired and mucus viscosity rises.

The third part deals with non-Newtonian aspects. Measurements of cystic fibrosis patient mucus is described and summarized, leading to its rheological characterization. Mixing models for non-Newtonian rheology are based upon a continuous evolution from a low-viscosity Newtonian behavior, close to the respiratory epithelium wetted surface, to a strongly shear-thinning behavior in the mucus layer. Finally, an extension of the previous numerical method is set up in order to handle the non-Newtonian features, exhibiting non-reversibility of mucus motion and inefficiency of mucus transport at the cell scale for cystic fibrosis patient.

5.1.1 Airway Wall Environment and Mucociliary Clearance

The mucus in the human respiratory tract is a non-Newtonian fluid secreted by the respiratory epithelium. It protects the respiratory mucosa from dehydration and traps inhaled particles (dust, allergens, carcinogens, micro-organisms and inflammatory debris). It cleans airways, flowing from either the trachea–bronchial tree and larynx or nose and pharynx to the oesophagus entry, where it is swallowed (or expectorated), as displayed in Fig. 5.1. An adequate mucus mobility is necessary for a proper airway clearance. Otherwise pathogens can proliferate, and harmful agents can enter the respiratory mucosa.

The airway surface fluid is heterogeneous with a near-wall liquid layer containing mostly water and slightly polymerized mucins and a mucus layer in which completely polymerized mucins cross-link and build a meshwork, without discontinuity at the microscopic scale. In addition, the airway surface fluid moves due to the cyclic displacement of cilia. The moving fluid is thus associated with a multiscale problem both in time and space.

Respiratory ciliated epithelial cells have a collection of cilia on their lumenal (wetted) surface. Over a given small airway segment, these cilia beat uniformly in the periciliary fluid (PCF) at a frequency of 4 to 20 Hz, quite different from that of the respiratory cycle (0.2–0.25 Hz). Mucus motion engendered by beating cilia of respiratory epithelial cells is influenced by its geometrical and rheological properties. The viscosity of the mucus film (thickness 10–100 μm) increases from its outer (close to the respiratory epithelium) to inner part (close to the core air flow).

Later in this section, a description is provided of the model that is used for numerical simulations of the motion of cilia immersed in a heterogeneous fluid with variable viscosity in the direction normal to the wall. The manifestation of cystic fibrosis is characterized by a stagnation of mucus, allowing a proliferation of pathogen agents and thus major infections, whose inflammation effects tend to destroy the epithelium ciliated cells, leading to sharper stagnation. It results that dangerous bacteria, such as *Pseudomonas aeruginosa* releasing pyocyanin, can then colonize the lung. Computing the resulting fluid displacement enables us to study the impact of different physical and biological parameters on the mucus transport at the scale

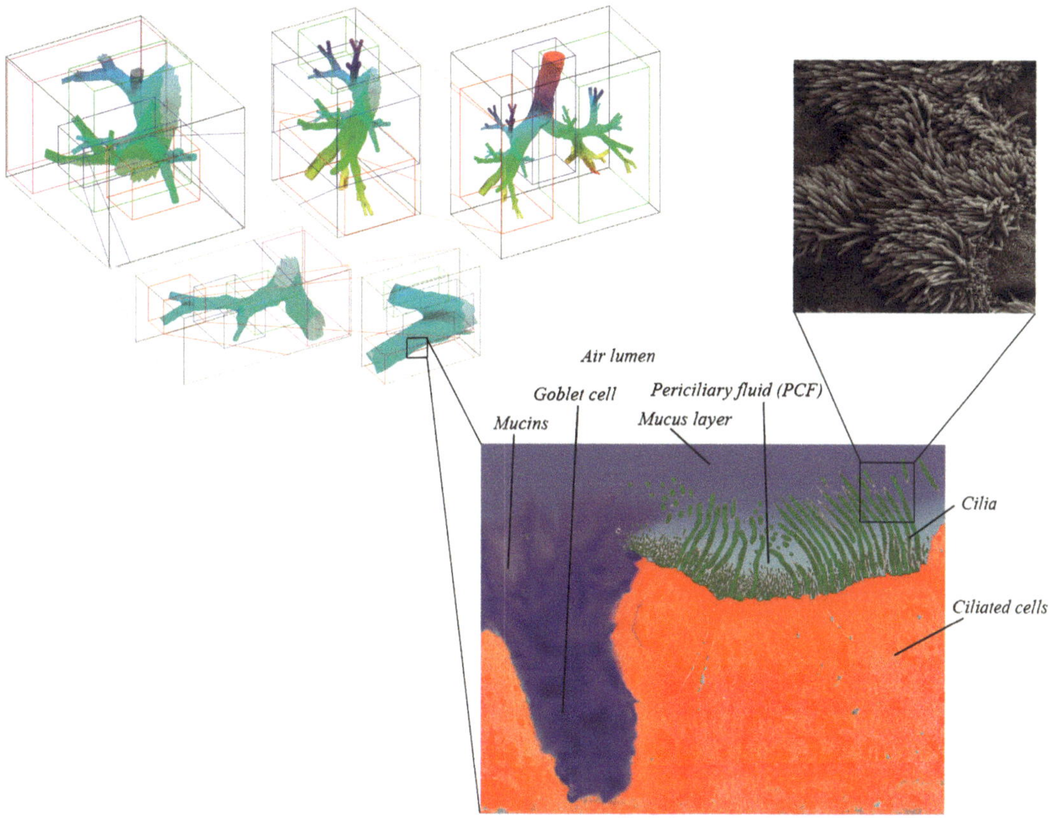

FIGURE 5.1 Multiscale feature of the tracheobronchial tree along its 6 first generations, from the trachea to the ciliated cells covering a large part of the lung epithelium, taken from Chatelin and Poncet (2013).

of the cell, pointing out mechanisms involved in cystic fibrosis, the major pathological target of the present work.

5.1.2 The Cystic Fibrosis Pathology

Cystic fibrosis (CF) is a monogenic autosomal recessive disorder. This is the most frequent life-shortening genetic disease in the Caucasian population caused by mutations in the cystic fibrosis transmembrane conductance regulator (CFTR) gene on chromosome 7. These mutations lead to the production of a dysfunctional CFTR protein.

The prevalence of the disease varies according to location and ethnic background. CF affects more than 70,000 people worldwide: 30,000 in the USA, 6000 in France (Hoffman and Ramsey, 2013). It is observed in 1 infant out of 2500 in Europe and North America. The prevalence of cystic fibrosis in the United States of America is 0.797 per 10,000 inhabitants and 0.737 in the 27 European Union countries, with a highest value of 2.98 in the Republic of Ireland (Farrell, 2008).

More than 2000 mutations have been identified in the CFTR gene. Despite the allelic diversity in this genetic disease, about 85% of Caucasian people[1] with CF pathology carry at least one copy of the CFTR gene mutation ΔF508 ($F508^{del}$ CFTR channel variant, cf. Quon and Rowe, 2016). Approximately 4% of all Whites carry a single cystic fibrosis mutation. Carriers of a single cystic fibrosis mutation do not generally exhibit any phenotypic abnormality. Two mutations are required to be present for clinical manifestations of the disease.

In 1969, CF patients in industrialized countries had a mean survival of 14 years. Because of better understanding of relationships between CFTR gene defects and organ pathophysiology, life expectancy has improved substantially over the last 40 years. By 2010, the median predicted survival rate approached 40 years (Clancy and Jain, 2012).

The CFTR protein belongs to the large family of adenine nucleotide-binding cassette transporters consisting of two transmembrane domains, two nucleotide binding domains (NBDs) and a unique regulatory domain. This cAMP-regulated anion channel[2] is expressed in the airway epithelium, submucosal glands and many other epithelial organs (e.g. biliary and pancreatic ducts) (Mall and Hartl, 2014). It transports anions (mainly chloride and bicarbonate) through the apical membrane of epithelial cells, thereby creating an osmotic gradient for fluid secretion; the active ion transport is coupled with water efflux. The basic defect causes an imbalance between secretion and absorption of the epithelial lining fluid rendering CF airway surfaces vulnerable to dehydration. The induced decrease in the thickness of the airway surface liquid is associated with an increased viscosity and thickening of airway secretions, an obstruction of airway lumens, a decreased ability to clear inhaled pathogens, and a dysregulated inflammation.

Studies in cultured primary airway epithelia from CF patients carrying a broad spectrum of CFTR mutations demonstrated that the airway surface liquid is depleted with a collapse of the periciliary layer and a mucociliary dysfunction (Mall and Hartl, 2014). In the following sections, the very different nature between periciliary fluid and mature mucus is investigated, and our results confirm the drop of the mucociliary clearance in the case of PCL collapsing.

Static airway mucus triggers inflammation of epithelial tissue and forms a nidus for bacterial infection. Defective CFTR-mediated bicarbonate transport leads to an acidic airway surface liquid that results in a decrease in antimicrobial activity (Haq et al., 2016). The epithelial sodium channel (ENaC) activity is upregulated in CF, which leads to sodium hyperabsorption and contributes to airway surface dehydration. ENaC activation occurs via various mechanisms including the cleavage of its subunit by endogenous proteases. Indeed, a damaging vicious circle of mucus retention, inflammation, infection is created (Hoffman and Ramsey, 2013). Many factors mediate the inflammatory response to chronic bacterial infection such as proinflammatory cytokines (IL-1β, IL-6, IL-8, GM-CSF, TNF-α). Sputum of CF contains elevated levels of neutrophils and hence exhibits a high concentration of neutrophil-derived proteases such as neutrophils elastase (Twigg et al., 2015).

Clinically, CF is a complex multi-organ disease that leads to a significant chronic sinopulmonary disease, pancreatic exocrine and endocrine impairments (diabetes), bowel disease,

[1] In the general meaning of people from European extraction, including American white people and intermingling in former Europe.

[2] cAMP stands for cyclic adenosine monophosphate, involved in the cAMP receptor proteins (CRP) binding tightly to a specific DNA.

elevated sweat chloride, and male infertility (Zielenski, 2000). Chronic inflammation and infection may potentiate respiratory tissue injury, thereby altering progressively pulmonary function. The primary cause of death and long-term complications among people with CF is chronic bacterial infection in lungs (Yankaskas et al., 2004). Chronic obstructive lung disease evolves from an early onset of mucus plugging in small airways, chronic neutrophilic airway inflammation and bacterial infection (Zemanick et al., 2010).

Cystic fibrosis patients battle life-long pulmonary infections. Chronic pulmonary infections contributes to a decline in forced expiratory volume in 1 s (FEV1) of approximately 2% per year. They result from a limited number of microbial pathogens, among which the most common species are *Haemophilus influenza*, *Staphylococcus aureus*, *Pseudomonas aeruginosa*, and the *Burkholderia cepacia complex*. CF mucus is a specifically favorable niche for growth of *Pseudomonas aeruginosa* including glycosylation of airway mucins (Jeffries et al., 2016). There is sufficient evidence that eradication of early infection and prevention of chronic *Pseudomonas aeruginosa* infection is associated with clinical benefit (Doring et al., 2012).

Until recently, medical treatments for CF exclusively targeted specific sequelae of the underlying pulmonary disease. For example, intravenous or inhaled antibiotics are administered to limit bacterial infection while macrolides or ibuprofen can be used to reduce inflammation (Doring et al., 2012). Restoring the impaired clearance of airway secretions in order to optimize micro-organisms removal from airways is one of the major goals of therapy in patients with CF (Kurbatova et al., 2015).

Current therapies are varied and include: (i) airway clearance to mobilize bronchopulmonary secretions and interrupt the vicious cycle of inflammation and infection (McShane et al., 2013), (ii) inhaled dornase which, while it does break down DNA and decrease mucus viscosity, has recently been shown to require a peripheral small airway deposit (Kurbatova et al., 2015), (iii) hypertonic saline which may improve the airway surface hydration (Hoffman and Ramsey, 2013), and (iv) mannitol that can enhance the airway mucociliary clearance (Aitken et al., 2012; Bilton et al., 2011).

New pharmacotherapy approaches focus on the development of CFTR modulators. These CFTR-targeted drugs are designed to rescue mutant CFTR function (Mall and Hartl, 2014). Correctors relieve protein misprocessing due to the most common mutation ΔF508, as they facilitate proper maturation and delivery to the plasma membrane. Potentiators are CFTR modulators that increase the time CFTR spends in the open channel configuration and hence chloride transport. They restore airway surface hydration and mucus clearance. Combined therapies (potentiators and correctors) are now available for patients (Quon and Rowe, 2016).

Alternative therapeutic approaches to improve airway surface hydration include stimulation or inhibition of other channels and transporters involved in apical ion transport to obviate the abnormal ion transport, solute transfer, pH and airway surface liquid volume generated by dysfunctional CFTRs. Given hyperactivity of ENaC in CF, inhibitors of the ENaC channel or activators of alternative Cl^- channels (CaCC) have recently been investigated as potential therapies (Haq et al., 2016).

5.1.3 Mathematical Modeling of Lung Wall Environment

Since the 1970s and the paper of Blake (1972), the mucociliary clearance has been frequently studied using mathematical modeling combined with numerical and analytic computations – see Smith et al. (2008) for a review. Since the publication of this work (2008), several researchers published new researches using 2D modeling (Smith et al., 2007b; Lukens et al., 2010; Montenegro-Johnson et al., 2013; Lee et al., 2011; Jayathilake et al., 2015; Sedaghat et al., 2016) or 3D modeling (Jayathilake et al., 2012; Simons et al., 2015; Smith et al., 2007a; Li et al., 2016; Ding et al., 2014; Ding and Kanso, 2015; Guo and Kanso, 2016a, 2016b).

In these papers the airway surface liquid is often considered to have a two-layer structure with constant viscosity in each layer (Lee et al., 2011; Jayathilake et al., 2015; Li et al., 2016). Alternatively, other studies assume a constant viscosity airway surface liquid composed only with one fluid layer (Jayathilake et al., 2012; Simons et al., 2015). The two layer model has also been combined with rheological behaviors in the upper mucus layer, using a shear-thinning modeling (Montenegro-Johnson et al., 2013; Craster and Matar, 2000) or a viscoelastic modeling (Mitran, 2007; Sedaghat et al., 2016; Guo and Kanso, 2016b).

Regarding numerical analysis, the mucociliary clearance has been computed using various techniques. As the flow modeling leads to a Stokes problem, the Stokeslets method (Cortez et al., 2005) has been used by several authors (Smith et al., 2007a, 2007b; Lukens et al., 2010; Montenegro-Johnson et al., 2013; Guo and Kanso, 2016a). The main drawbacks of this method include the computational cost which grows with the number of immersed cilia and becomes prohibitive in 3D, as well as the limitation to constant viscosity flows avoiding the computation of non-linearities such as rheology.

Finite volumes have also been used: a complete discretization coupled to a finite element beam model for cilia has been published (Mitran, 2007), taking into account the fluid–structure interaction. The flow was successfully computed but this technique requires a large amount of computational resources to solve resulting linear systems. The immersed boundary method, introduced by Peskin (2003) was also used in a finite differences context for a parametric study. More recently the immersed boundary method was coupled with Lattice Boltzmann techniques in order to take into account the two-layer structure of the airway surface liquid (Li et al., 2016; Sedaghat et al., 2016; Guo and Kanso, 2016a, 2016b).

Because the characteristic dimensions of the mucus flow are very small (the mucus film is 10–40 μm height and cilia are 5–8 μm long), the inertia terms of the Navier–Stokes equations are negligible, and the conservation of momentum can be represented by the linearized Stokes equations. In fact, at this scale inertia terms vanish in the Navier–Stokes equations. To simulate the mucociliary clearance, it is important to take into account the cilia motion which propels the airway surface liquid by means of the fluid–structure interaction. In the following model description, the fluid–structure interaction is assumed to be "one way": the force exerted by the fluid on the solid is not computed. In fact, this remains an open problem from a numerical point of view, in particular an eventual beating adaptation (when the airway surface liquid becomes more viscous, for example) has never been studied numerically and, to our knowledge, only one experimental work was reported (Hill et al., 2010).

The existence of an interface between the periciliary fluid and mucus is debated (Button et al., 2012; Ehre et al., 2014). In fact, polymerized mucins are released in the periciliary fluid, where they become progressively hydrated during their crossing of the PCL. Mucins accumulate at a given distance normal to the wall, giving rise to a mucin-concentrated phase, the mucus. In this region, mucin polymers interweave and eventually cross-link. The whole set of transformations undergone by mucins during their transport through the ASL is referred to as *maturation* in the present text.

These mucins are primarily responsible for the mucus structure and its highly non-Newtonian behavior. Near the epithilium the mucin concentration is very low and the airway fluid behaves as a Newtonian fluid, while further away from the epithilium, the presence of high mucin concentrations results in non-Newtonian shear thinning behavior (as described on Figure 5.2). In order to represent the airway fluid, and particularity its viscosity, as a continuum instead of a two layer fluid, a variable α is introduced that is a function of position that varies in a smooth sigmoidal manner between 0 and 1. This function is defined in greater detail later in this section.

This means that the viscosity can be seen as an *ad-hoc* function of α: Either $\mu = \Phi(\alpha)$ in the Newtonian case, detailed in Section 5.2, or $\mu = \Phi(u, \alpha)$ in the non-Newtonian case, typically

$$\mu = k(\alpha)(2D : D)^{(n(\alpha)-1)/2}$$

for power-law fluids, or a Carreau law for the fluids whose rheology exhibits thresholds, where D is the strain tensor defined below, detailed in Section 5.3. Then starting from an initial distribution, this variable α which characterizes mucins evolution, is transported and diffused by the mucus flow which depends on the viscosity. Hence α is the solution to a convection–diffusion equation and it is the natural to assume that the viscosity is bounded: $\mu_0 \leq \Phi(\alpha) \leq \mu_1$.

The mathematical problem which has to be solved is the following:

$$\begin{cases} u = \bar{u} & \text{inside cilia,} \\ -\operatorname{div}(2\mu D(u)) = f - \nabla p & \text{inside fluid,} \\ \operatorname{div} u = 0 & \text{inside fluid,} \\ \partial_t \alpha + u \cdot \nabla\alpha - \eta\Delta\alpha = 0 & \text{everywhere in computational domain,} \\ \mu = \Phi(\alpha) \text{ or } \mu = \Phi(\alpha, u) & \text{everywhere in computational domain,} \end{cases} \tag{5.1}$$

where p is the pressure, μ is the fluid viscosity, f are the external forces and $D(u) = (\nabla u + (\nabla u)^T)/2$ is the strain tensor and $\bar{u}$ is the cilia velocity (see computational details in Section 5.2.2.1). With initial and boundary conditions yet to be set.

To take into account the motion of immersed beating cilia, the Penalization method is used (Angot et al., 1999). It consists in adding an extra term to the conservation of momentum to solve a unique equation in all the domain, without any distinction between the cilia region $B(t)$ and the fluid region $\Omega \backslash B(t)$ (assuming that Ω is the computational domain). This extra Penalization term forces the velocity u to be equal to a prescribed solid velocity $\bar{u}$ in the solid region $B(t)$. It transforms the previous problem into the Penalized Stokes problem or the Stokes–Brinkman problem coupled to the same convection–diffusion equation generating a non-linear dynamics:

$$\begin{cases} -\operatorname{div}(2\mu^\varepsilon D(u^\varepsilon)) + \dfrac{\mathbf{1}_{B(t)}}{\varepsilon}(u^\varepsilon - \bar{u}) = f - \nabla p^\varepsilon & \text{in } \Omega, \\ \operatorname{div} u^\varepsilon = 0 & \text{in } \Omega, \\ \partial_t \alpha^\varepsilon + u^\varepsilon \cdot \nabla\alpha^\varepsilon - \eta\Delta\alpha^\varepsilon = 0 & \text{in } \Omega, \\ \mu^\varepsilon = \Phi(\alpha^\varepsilon) \text{ or } \mu^\varepsilon = \Phi(\alpha^\varepsilon, u^\varepsilon) & \text{in } \Omega, \end{cases} \tag{5.2}$$

with $\mathbf{1}_{B(t)}$ the characteristic function of the solid domain and ε the penalty parameter.

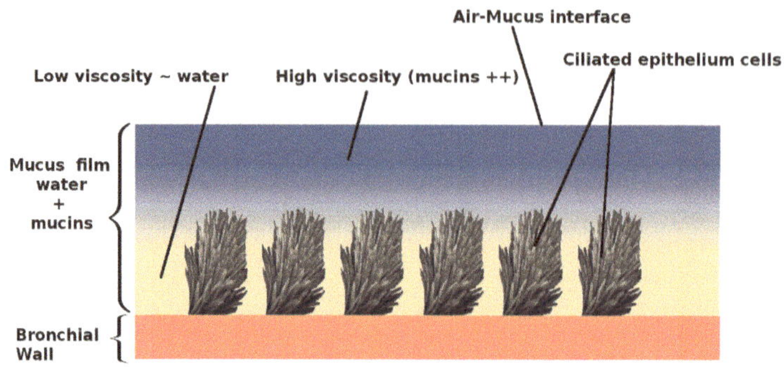

FIGURE 5.2 Sketch of the mucus flow on the bronchial wall: the non-homogeneous mucus, which exhibits a smooth transition between PCL and maturated mucus, is propelled by beating ciliated epithelium cells, from Chatelin and Poncet (2016).

This coupled non-linear problem will be solved using dedicated algorithms in order to get powerful 3D simulations and to allow parametric investigations. As detailed in Section 5.2.2 these algorithms are based on fast solvers on regular Cartesian grids, so that the computational domain Ω is assumed to be a rectangular cuboid.

The Stokes problem is completed with a no slip boundary condition at the epithelium surface (where the periciliary fluid is present), and a free slip boundary condition at the air–mucus interface (which is assumed to be flat due to a strong surface tension effect, Enault et al., 2010; Chatelin et al., 2016) and periodic boundary conditions in the other (proximal and transverse) directions. In the following, let Γ_0 and Γ_L respectively denote the upper and lower boundaries of the rectangular cuboid computational domain. Hence Γ_0 and Γ_L respectively are the boundary regions where the homogeneous Dirichlet and homogeneous Neumann boundary conditions are imposed.

For the convection–diffusion equation, homogeneous Neumann boundary conditions are set at both epithelium–periciliary–fluid-layer and air–mucus interfaces (meaning on $\Gamma_0 \cup \Gamma_L$) and periodic boundary conditions in other directions.

Moreover, an initial condition is provided for the mucins ratio α. A sigmoid profile is used to model the cross-linked mucins ratio which continuously varies from 0 in the deeper periciliary fluid layer to 1 in the upper mucus layer:

$$\alpha(x, y, z, t = 0) = \frac{\arctan(\gamma(z - \delta)) - \arctan(-\gamma\delta)}{\arctan(\gamma(H_z - \delta)) - \arctan(-\gamma\delta)} \tag{5.3}$$

where δ is the ratio between the mucus layer height and the periciliary fluid layer height, and z denotes the coordinate perpendicular to the epithelium designating position in the fluid. A depiction of Eq. (5.3) is presented in Fig. 5.3.

5.2 NEWTONIAN MODELS

A Newtonian fluid behavior means that the viscosity μ only depends on α, and is independent of u. In Eqs. (5.1) and (5.2), the case $\mu = \Phi(\alpha)$ is considered. This section is divided into

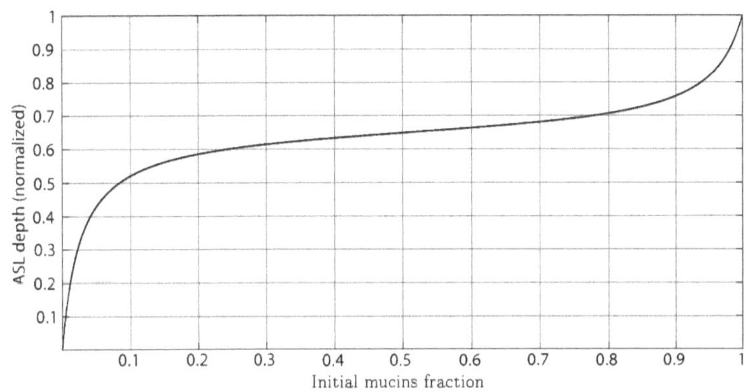

FIGURE 5.3 Initial profile α used for the cross-linked mucins ratio, defined by Eq. (5.3), with respect to the normalized depth $\tilde{z} = z/H$, H being the total mucus height.

three parts: firstly, the well-posedness of Eq. (5.2) is demonstrated; secondly, the numerical algorithms dedicated to the computation of this problem are presented; and finally, several computations are presented as well as phenomenological analysis.

5.2.1 Mathematical Analysis

In order to perform a mathematical analysis of the problem, we consider first a regularization of the domain $B(t)$ described in Fig. 5.4. From a technical point of view, this allows using trace and lifting theorems in these two domains and justifying the pertinence of the Penalization method. For the sake of simplicity, the notation $B(t)$ refers to this regularized domain in the whole Section 5.2.1.

Let $T > 0$ be a final time (not necessarily a cilia period). Let $\Gamma = [0, L_1] \times [0, L_2]$ and $\Omega = \Gamma \times [0, L]$ be the domain of study. We denote by $\Gamma_0 = \Gamma \times \{0\}$ and $\Gamma_L = \Gamma \times \{L\}$ the lower and upper boundaries of Ω.

Since the domain B depends on the time parameter $t \in [0, T]$, we need to track this dependency in the following. We let

$O = [0, T] \times \Omega$ be the whole domain in time and space coordinates,
$O_p = \{(t, x) \in [0, T] \times \Omega,\ \forall t \in [0, T],\ x \in B(t)\}$ be the penalized domain,
$O_p^c = \left\{(t, x) \in [0, T] \times \Omega,\ \forall t \in [0, T],\ x \in \Omega \setminus \bar{B}(t)\right\} = O \setminus \bar{O}_p$, and
$\Sigma = \{(t, x) \in [0, T] \times \Omega,\ \forall t \in [0, T],\ x \in \partial B(t) \setminus \partial\Omega\}$ be the inner boundary of O_p.

Problem (5.1) is rewritten in these spaces as

$$\begin{cases} u = \bar{u} & \text{in } O_p, \\ -\operatorname{div}(2\mu D(u)) = f - \nabla p & \text{in } O_p^c, \\ \operatorname{div} u = 0 & \text{in } O, \\ \partial_t \alpha + u \cdot \nabla \alpha - \eta \Delta \alpha = 0 & \text{in } O, \\ \mu = \Phi(\alpha) & \text{in } O \end{cases} \tag{5.4}$$

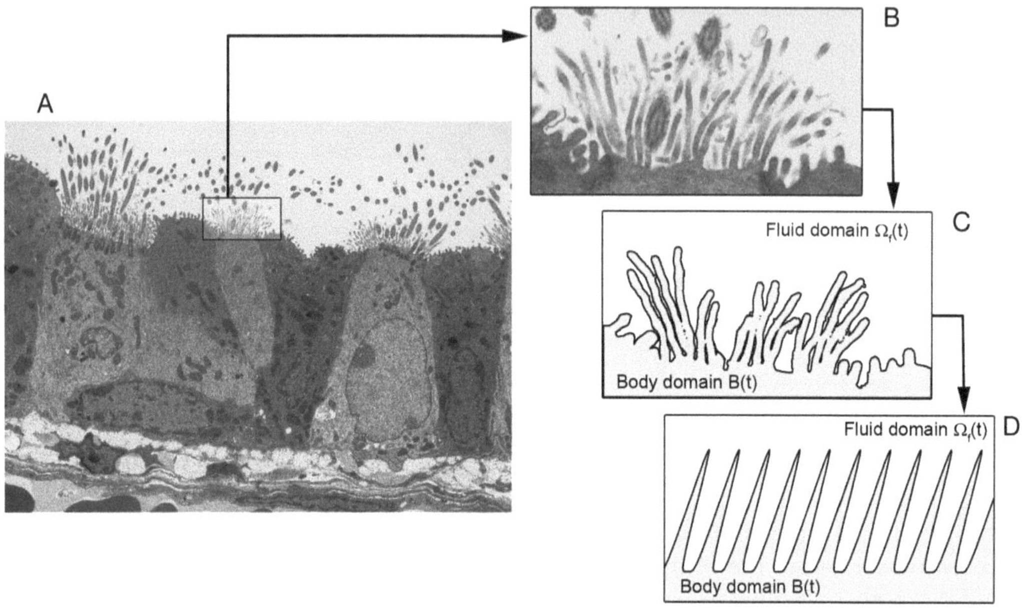

FIGURE 5.4 The moving domain $B(t)$ models the fixed bronchial wall of the lung and the moving cilia. In the following, we consider the model (D) where the boundary $\partial B(t)$ is regular (from Chatelin et al., 2016).

with the initial and boundary conditions described previously:

$$\begin{cases} u = 0 & \text{on } [0,T] \times \Gamma_0, \\ u = \bar{u} & \text{on } \Sigma, \\ u \cdot \nu = \dfrac{\partial u_1}{\partial \nu} = \dfrac{\partial u_2}{\partial \nu} = 0 & \text{on } [0,T] \times \Gamma_L, \\ u \text{ is periodic} & \text{on } [0,T] \times \partial\Gamma \times [0,L], \\ \dfrac{\partial \alpha}{\partial \nu} = 0 & \text{on } [0,T] \times (\Gamma_0 \cup \Gamma_L), \\ \alpha \text{ is periodic} & \text{on } [0,T] \times \partial\Gamma \times [0,L], \\ \alpha(0,\cdot) = \alpha_0 & \text{in } \Omega, \end{cases} \tag{5.5}$$

where $\eta > 0$ is the diffusion coefficient, $\Phi \in W^{3,\infty}(\mathbb{R})$ and there exists a real number $\mu_0 > 0$ such that for all $\alpha \in \mathbb{R}$, $\mu_0 \leq \phi(\alpha)$.

According to the model, the penalized problem is rewritten as

$$\begin{cases} -\operatorname{div}(2\mu^\varepsilon D(u^\varepsilon)) + \dfrac{\mathbf{1}_{B(t)}}{\varepsilon}(u^\varepsilon - \bar{u}) = f - \nabla p^\varepsilon & \text{in } O, \\ \operatorname{div} u^\varepsilon = 0 & \text{in } O, \\ \partial_t \alpha^\varepsilon + u^\varepsilon \cdot \nabla \alpha^\varepsilon - \eta \Delta \alpha^\varepsilon = 0 & \text{in } O, \\ \mu^\varepsilon = \Phi(\alpha^\varepsilon) & \text{in } O, \end{cases} \tag{5.6}$$

with its initial and boundary conditions coming straight from Eqs. (5.5):

$$\begin{cases} u^\varepsilon = 0 & \text{on } [0,T]\times\Gamma_0, \\ u^\varepsilon\cdot\nu = \dfrac{\partial u_1^\varepsilon}{\partial\nu} = \dfrac{\partial u_2^\varepsilon}{\partial\nu} = 0 & \text{on } [0,T]\times\Gamma_L, \\ u^\varepsilon \text{ is periodic} & \text{on } [0,T]\times\partial\Gamma\times[0,L], \\ \dfrac{\partial\alpha^\varepsilon}{\partial\nu} = 0 & \text{on } [0,T]\times(\Gamma_0\cup\Gamma_L), \\ \alpha^\varepsilon \text{ is periodic} & \text{on } [0,T]\times\partial\Gamma\times[0,L], \\ \alpha^\varepsilon(0,\cdot) = \alpha_0 & \text{in } \Omega. \end{cases} \tag{5.7}$$

Finally, the following functional spaces are introduced:

Definition 1. Let $\mathcal{F}(\Omega; X)$ be the set of functions defined on Ω with values in X and

$$\begin{aligned} \mathbb{L}_0^2(\Omega) &= \left\{v\in\mathbb{L}^2(\Omega),\ \textstyle\int_\Omega v\,dx = 0\right\}, \\ V(\Omega) &= \big\{v\in\mathbb{H}^1(\Omega),\ \operatorname{div} v = 0 \text{ on } \Omega, u = 0 \text{ on } \Gamma_0, \\ &\qquad u\cdot\nu = 0 \text{ on } \Gamma_L,\ u \text{ has periodic conditions on } \partial\Gamma\times[0,L]\big\} \end{aligned}$$

with the classical $\mathbb{H}^1$-norm. Thanks to the generalized Poincaré's inequality (see Boyer and Fabrie, 2005, Section III.6), there exist two constants depending only on Ω such that

$$c_1\|\nabla v\|_{\mathbb{L}^2(\Omega)} \le \|v\|_{\mathbb{H}^1(\Omega)} \le c_2\|\nabla v\|_{\mathbb{L}^2(\Omega)}.$$

The following existence and uniqueness results require some regularity assumptions on the obstacle's time evolution (H1)–(H2) and on the velocity $\bar{u}$ inside the obstacle (H3):

Hypothesis 1 (H1).

- Domain B is an open set in Ω having the uniform C^2-regularity property (see Adams and Fournier, 2003, Section 4 for the definition) up to the boundary of Ω in order to obtain a periodic smooth domain,
- $\exists(l,l')\in(\mathbb{R}^+)^2$ such that $0<l<l'<L$ and $\Gamma\times[0,l]\subset B\subset\Gamma\times[0,l']$, *i.e.* we assume that the bronchial wall is the fixed domain $\Gamma\times[0,l]$ to which cilia are attached and that the cilia never reach the top of the domain Ω,
- $\Omega\setminus\bar{B}$ is an open set in Ω having the uniform C^2-regularity property,
- $\Omega\setminus\bar{B}$ is simply connected.

Hypothesis 2 (H2).

- There exists a function $\Psi\in C^0([0,T];C^2(\Omega))$ such that for all $t\in[0,T]$ $\Psi(t)_{|[0,l]\times\Gamma} = \mathrm{Id}_{[0,L]\times\Gamma}$, $\Psi(t)([0,l']\times\Gamma)\subset[0,l']\times\Gamma$ and $\Psi(t)$ has periodic boundary conditions on $[0,L]\times\partial\Gamma$,
- $\forall t\in[0,T]$ $\Psi(t)$ is a C^2-diffeomorphism on Ω,
- $\forall t\in[0,T]$ $B(t)=\Psi(t,B)$.

Remark. Here the assumption has been made that the bronchial wall never moves and that the cilia never reaches the top of the domain. Additionally, $\Omega\setminus\bar{B}(t)=\Psi(t,\Omega\setminus\bar{B})$.

Hypothesis 3 (H3).

- $\bar{u} \in \mathcal{C}^0([0,T];\mathbb{H}^2(B(\cdot)))$,
- $\forall t \in [0,T]$, $\operatorname{div} \bar{u}(t,\cdot) = 0$ in $B(t)$,
- $\forall t \in [0,T]$, $\bar{u}(t,\cdot)$ fulfills

$$\forall t \in [0,T] \;\; \bar{u}(t,\cdot)_{|\Gamma_0} = 0,$$

- $\forall t \in [0,T]$, $\bar{u}(t,\cdot)$ has periodic boundary conditions on $\partial B(t) \cap (\partial\Gamma \times [0,L])$.

Definition 2. $(u^\varepsilon, p^\varepsilon, \alpha^\varepsilon)$ is a weak solution to (5.6)–(5.7) if it fulfills

- $u^\varepsilon \in \mathbb{L}^q(0,T;V(\Omega))$, $p^\varepsilon \in \mathbb{L}^q(0,T;\mathbb{L}^2_0(\Omega))$ with $2 \le q \le +\infty$, $\alpha^\varepsilon \in \mathbb{L}^\infty(0,T;\mathbb{H}^2(\Omega)) \cap L^2(0,T;\mathbb{H}^3(\Omega)) \cap \mathbb{H}^1(0,T;\mathbb{H}^1(\Omega))$,
- For all $v \in \mathcal{C}_c(0,T;V(\Omega))$,

$$\iint_{(0,T)\times\Omega} 2\mu^\varepsilon D(u^\varepsilon) : D(v)\,dx\,dt + \frac{1}{\varepsilon}\iint_{O_p} (u^\varepsilon - \bar{u}) \cdot v\,dx\,dt = \iint_{(0,T)\times\Omega} f \cdot v\,dx\,dt, \tag{5.8}$$

- α^ε is the solution to

$$\begin{cases} \partial_t \alpha^\varepsilon + u^\varepsilon \cdot \nabla\alpha^\varepsilon - \eta\Delta\alpha^\varepsilon = 0 & \text{in } \Omega, \\ \dfrac{\partial\alpha^\varepsilon}{\partial\nu} = 0 & \text{on } \Gamma_0 \cup \Gamma_L, \\ \alpha^\varepsilon \text{ has periodic conditions} & \text{on } \partial\Gamma \times [0,L], \\ \alpha^\varepsilon(0,\cdot) = \alpha_0 & \text{in } \Omega, \end{cases} \tag{5.9}$$

- $\mu^\varepsilon = \Phi(\alpha^\varepsilon)$.

Theorem 1. *Let $0 < \mu_0$ and $\varepsilon > 0$. Let $\alpha_0 \in \mathbb{H}^2(\Omega)$ and $\Phi \in W^{2,\infty}(\mathbb{R})$ such that $\forall x \in \mathbb{R}$, $\mu_0 \le \Phi(x)$. We moreover assume (H1), (H2), (H3) and that for all $T > 0$, $f \in \mathbb{L}^\infty(0,T;\mathbb{L}^2(\Omega))$. There exists then a unique weak solution $(u^\varepsilon, p^\varepsilon, \alpha^\varepsilon)$ to (5.6) such that*

$$\begin{cases} \alpha^\varepsilon \in \mathbb{L}^\infty(0,T;\mathbb{H}^2(\Omega)) \cap \mathbb{L}^2(0,T;\mathbb{H}^3(\Omega)) \cap \mathbb{H}^1(0,T;\mathbb{H}^1(\Omega)), \\ u^\varepsilon \in \mathbb{L}^q(0,T;V(\Omega)) \;\; \textit{for all } 2 \le q \le +\infty, \\ p^\varepsilon \in \mathbb{L}^q(0,T;\mathbb{L}^2_0(\Omega)) \;\; \textit{for all } 2 \le q \le +\infty. \end{cases}$$

Moreover, there exist some constants independent of ε such that u^ε fulfills

$$\begin{aligned} \mu_0\|D(u^\varepsilon)\|^2_{\mathbb{L}^\infty([0,T];\mathbb{L}^2(\Omega))} \quad & + \frac{1}{\varepsilon}\|u^\varepsilon - \bar{u}\|^2_{\mathbb{L}^\infty([0,T];\mathbb{L}^2(B(\cdot)))} \\ & \le C\|f\|_{\mathbb{L}^\infty([0,T];\mathbb{L}^2(\Omega))}\|\bar{u}\|_{\mathbb{L}^\infty([0,T];\mathbb{H}^2(B(\cdot)))} \\ & \quad + C'\left(\|f\|^2_{\mathbb{L}^\infty([0,T];\mathbb{L}^2(\Omega))} + \|\Phi\|^2_{\mathbb{L}^\infty(\mathbb{R})}\|\bar{u}\|^2_{\mathbb{L}^\infty([0,T];\mathbb{H}^2(B(\cdot)))}\right), \\ \mu_0\|D(u^\varepsilon)\|^2_{\mathbb{L}^2([0,T];\mathbb{L}^2(\Omega))} \quad & + \frac{1}{\varepsilon}\|u^\varepsilon - \bar{u}\|^2_{\mathbb{L}^2([0,T];\mathbb{L}^2(B(\cdot)))} \\ & \le C\|f\|_{\mathbb{L}^2([0,T];\mathbb{L}^2(\Omega))}\|\bar{u}\|_{\mathbb{L}^\infty([0,T];\mathbb{H}^2(B(\cdot)))} \\ & \quad + C'\left(\|f\|^2_{\mathbb{L}^2([0,T];\mathbb{L}^2(\Omega))} + \|\Phi\|^2_{\mathbb{L}^\infty(\mathbb{R}))}\|\bar{u}\|^2_{\mathbb{L}^\infty([0,T];\mathbb{H}^2(B(\cdot)))}\right). \end{aligned}$$

Definition 3. (u, p, α) is a weak solution to (5.4) with the conditions (5.5) if it fulfills

- For all $2 \leq q \leq +\infty$, $u \in \mathbb{L}^q(0, T; V(\Omega))$, $p \in \mathbb{L}^q(0, T; \mathbb{L}_0^2(\Omega \setminus \bar{B}(\cdot)))$ and $\alpha \in \mathbb{L}^\infty(0, T; \mathbb{H}^2(\Omega)) \cap \mathbb{L}^2(0, T; \mathbb{H}^3(\Omega)) \cap \mathbb{H}^1(0, T; \mathbb{H}^1(\Omega))$,
- $u = \bar{u}$ in O_p,
- $(u, p) \in \mathbb{L}^q(0, T; \mathbb{H}^1(\Omega \setminus \bar{B}(\cdot))) \times \mathbb{L}^q(0, T; \mathbb{L}_0^2(\Omega \setminus \bar{B}(\cdot)))$, with $2 \leq q \leq +\infty$, is the solution to

$$\begin{cases} -\operatorname{div}(2\mu D(u)) = f - \nabla p \text{ in } \mathbb{L}^2(0, T; \mathbb{H}^{-1}(\Omega \setminus \bar{B}(\cdot))), \\ \operatorname{div} u = 0 \text{ in } O_p^c, \\ u_{|\Sigma} = \bar{u}_{|\Sigma} \text{ a.e.}, \\ u \cdot \nu = \dfrac{\partial u_1}{\partial \nu} = \dfrac{\partial u_2}{\partial \nu} = 0 \quad \text{on } [0, T] \times \Gamma_L, \\ u \text{ has periodic conditions on } [0, T] \times \partial\Gamma \times [0, L], \end{cases}$$

- α is the solution to (5.9),
- $\mu = \Phi(\alpha)$.

Theorem 2. *Under the assumptions of Theorem 1 and with the extra assumption $\Phi \in W^{3,\infty}(\mathbb{R})$, there exists a unique weak solution (u, p, α) to (5.4)–(5.5). Moreover, the weak solution $(u^\varepsilon, p^\varepsilon, \alpha^\varepsilon)$ of (5.6) converges weakly towards (u, p, α) as ε goes to zero.*

We now present the sketches of the proofs of Theorems 1 and 2 and highlight the key points in getting these theorems.

5.2.1.1 Sketch of the Proof of Theorem 1

To prove the existence of a solution to this coupled non-linear problem, introduce the sequence (α^n, u^n, p^n) defined by:

1. $\alpha^0 = \alpha_0$,
2. $\mu^n = \Phi(\alpha^n)$,
3. (u^n, p^n) is the solution to

$$\begin{cases} -\operatorname{div}(2\mu^n D(u^n)) + \dfrac{\mathbf{1}_{B(t)}}{\varepsilon}(u^n - \bar{u}) = f - \nabla p^n & \text{in } \Omega, \\ \operatorname{div} u^n = 0 & \text{in } \Omega, \\ u^n = 0 & \text{on } \Gamma_0, \\ u \cdot \nu = \dfrac{\partial u_1}{\partial \nu} = \dfrac{\partial u_2}{\partial \nu} = 0 & \text{on } \Gamma_L, \\ u^n \text{ has periodic conditions} & \text{on } \partial\Gamma \times [0, L], \end{cases}$$

4. α^{n+1} is the solution to

$$\begin{cases} \partial_t \alpha^{n+1} + u^n \cdot \nabla \alpha^{n+1} - \eta \Delta \alpha^{n+1} = 0 & \text{in } \Omega, \\ \dfrac{\partial \alpha^{n+1}}{\partial \nu} = 0 & \text{on } \Gamma_0 \cup \Gamma_L, \\ \alpha^{n+1} \text{ has periodic conditions} & \text{on } \partial\Gamma \times [0, L], \\ \alpha^{n+1}(0, \cdot) = \alpha_0 & \text{in } \Omega. \end{cases}$$

We let $\alpha^{n+1} = G(\alpha^n)$.

The study of this recurring sequence leads us to the study of the convection–diffusion equation, on the one side, and the penalized Stokes equation, on the other side.

Theorem 3. *Let $\eta > 0$ and $u \in \mathbb{L}^q(0, T; V(\Omega))$ for all $2 \le q \le +\infty$. Let $\alpha_0 \in \mathbb{H}^1(\Omega)$. There exists a unique solution $\alpha \in \mathbb{L}^\infty(0, T; H^1(\Omega)) \cap \mathbb{L}^2(0, T; \mathbb{H}^2(\Omega)) \cap \mathbb{H}^1(0, T; \mathbb{L}^2(\Omega))$ to the problem (5.9).*

Moreover, if $\alpha_0 \in \mathbb{H}^2(\Omega)$ then $\alpha \in \mathbb{L}^\infty(0, T; \mathbb{H}^2(\Omega)) \cap \mathbb{L}^2(0, T; \mathbb{H}^3(\Omega)) \cap \mathbb{H}^1(0, T; \mathbb{H}^1(\Omega))$, and there exist some constants C and C' depending only on Ω and η such that for all $t \in [0, T]$:

$$\begin{aligned}
&\|\alpha(t, \cdot)\|_{\mathbb{H}^2(\Omega)} \\
&\quad \le C\|\alpha_0\|_{\mathbb{H}^2(\Omega)} \exp\left(C' \int_0^t \left(1 + \|u(s, \cdot)\|^2_{\mathbb{H}^1(\Omega)} + \|u(s, \cdot)\|^4_{\mathbb{H}^1(\Omega)} + \|u(s, \cdot)\|^8_{\mathbb{H}^1(\Omega)}\right) ds\right), \\
&\|\alpha\|_{\mathbb{L}^2(0,t;\mathbb{H}^3(\Omega))} \\
&\quad \le C\|\alpha_0\|_{\mathbb{H}^2(\Omega)} \exp\left(C' \int_0^t \left(1 + \|u(s, \cdot)\|^2_{\mathbb{H}^1(\Omega)} + \|u(s, \cdot)\|^4_{\mathbb{H}^1(\Omega)} + \|u(s, \cdot)\|^8_{\mathbb{H}^1(\Omega)}\right) ds\right).
\end{aligned}$$

Proof. A classical Galerkin decomposition is performed, and energy estimates are made in order to obtain this result (see also Dautray and Lions, 2000, Vol. 5, Section XVIII). ■

Theorem 4. *Assume that hypothesis (H1), (H2) and (H3) are satisfied. Let $\mu_0 > 0$, $\varepsilon > 0$ and $T > 0$. Let $\mu \in L^\infty([0, T] \times \Omega)$ such that $\mu_0 \le \mu(t, x)$ a.e. in $[0, T] \times \Omega$. Let $f \in \mathbb{L}^\infty(0, T; \mathbb{L}^2(\Omega))$. There exists then a unique weak solution (u, p) to the problem*

$$\begin{cases}
-\mathrm{div}(2\mu D(u)) + \dfrac{\mathbf{1}_{B(t)}}{\varepsilon}(u - \bar{u}) = f - \nabla p & \text{in } [0, T] \times \Omega, \\
\mathrm{div}\, u = 0 & \text{in } [0, T] \times \Omega, \\
u = 0 & \text{on } [0, T] \times \Gamma_0, \\
u \cdot \nu = \dfrac{\partial u_1}{\partial \nu} = \dfrac{\partial u_2}{\partial \nu} = 0 & \text{on } [0, T] \times \Gamma_L, \\
u \text{ has periodic conditions} & \text{on } [0, T] \times \partial\Gamma \times [0, L],
\end{cases} \tag{5.10}$$

which fulfills

$$(u, p) \in \mathbb{L}^q([0, T], V(\Omega)) \times \mathbb{L}^q\left([0, T]; \mathbb{L}^2_0(\Omega)\right)$$

for all $2 \le q \le +\infty$.

Additionally, the following inequalities hold:

$$\begin{aligned}
\mu_0\|D(u)\|^2_{\mathbb{L}^\infty([0,T];\mathbb{L}^2(\Omega))} &+ \frac{1}{\varepsilon}\|u - \bar{u}\|^2_{\mathbb{L}^\infty([0,T];\mathbb{L}^2(B(\cdot)))} \\
&\le C\|f\|_{\mathbb{L}^\infty([0,T];\mathbb{L}^2(\Omega))}\|\bar{u}\|_{\mathbb{L}^\infty([0,T];\mathbb{H}^2(B(\cdot)))} \\
&\quad + C'\left(\|f\|^2_{\mathbb{L}^\infty([0,T];\mathbb{L}^2(\Omega))} + \|\mu\|^2_{\mathbb{L}^\infty([0,T]\times\Omega)}\|\bar{u}\|^2_{\mathbb{L}^\infty([0,T];\mathbb{H}^2(B(\cdot)))}\right), \\
\mu_0\|D(u)\|^2_{\mathbb{L}^2([0,T];\mathbb{L}^2(\Omega))} &+ \frac{1}{\varepsilon}\|u - \bar{u}\|^2_{\mathbb{L}^2([0,T];\mathbb{L}^2(B(\cdot)))} \\
&\le C\|f\|_{\mathbb{L}^2([0,T];\mathbb{L}^2(\Omega))}\|\bar{u}\|_{\mathbb{L}^\infty([0,T];\mathbb{H}^2(B(\cdot)))} \\
&\quad + C'\left(\|f\|^2_{\mathbb{L}^2([0,T];\mathbb{L}^2(\Omega))} + \|\mu\|^2_{\mathbb{L}^2([0,T];\mathbb{L}^\infty(\Omega))}\|\bar{u}\|^2_{\mathbb{L}^\infty([0,T];\mathbb{H}^2(B(\cdot)))}\right).
\end{aligned}$$

Moreover, if $\mu \in \mathbb{L}^q(0,T;W^{1,\infty}(\Omega))$ *with* $2 \le q \le +\infty$, *then*

$$(u,p) \in \mathbb{L}^q\left([0,T], H^2(\Omega)\right) \times \mathbb{L}^q\left([0,T]; \mathbb{H}^1(\Omega)\right),$$

and there exists a constant C depending on Ω *such that*

$$\|u\|_{\mathbb{L}^q([0,T];\mathbb{H}^2(\Omega))} + \left\|\frac{p}{\mu}\right\|_{\mathbb{L}^q([0,T];\mathbb{H}^1(\Omega))} \le C\|g\|_{\mathbb{L}^q([0,T];\mathbb{L}^2(\Omega))},$$

where $g = \frac{1}{\mu}\left(2D(u)\nabla\mu + f - \frac{\mathbf{1}_{B(t)}}{\varepsilon}(u-\bar{u}) + p\frac{\nabla\mu}{\mu}\right) \in \mathbb{L}^q([0,T];\mathbb{L}^2(\Omega))$.

Proof. The existence and uniqueness may be obtained classically thanks to Lax–Milgram theorem. The regularity is due to classical Stokes regularity results (see, for example, Boyer and Fabrie, 2005) thanks to the following rewriting of the equation:

$$-\operatorname{div} 2D(u) = \frac{1}{\mu}\left(-2D(u)\nabla u - \frac{\mathbf{1}_{B(t)}}{\varepsilon}(u-\bar{u}) + f - \nabla p\right).$$

The ε-independent estimates are obtained by taking $u - \tilde{u}$ as a test function where $\tilde{u}(t,\cdot)$ is a prolongation of $\bar{u}$ in $\Omega \setminus B(t)$ fulfilling the same boundary conditions as u and the constraint $\operatorname{div}\tilde{u} = 0$. To prove the estimates, we need time-independent estimates on this prolongation. These are obtained thanks to standard time-independent trace and lifting theorem applied to $\bar{u}(t,\Psi^{-1}(t,x))$ which is well defined in the non-moving non-deformable domain B thanks to Hypothesis H2. We recover time-independent bounds thanks to the bounds we have on Ψ. ■

Both results prove that the function G is well-defined from $E \cap F$ to itself with

$$E = \mathbb{L}^\infty(0,T;\mathbb{H}^1(\Omega)) \cap L^2(0,T;\mathbb{H}^2(\Omega)) \text{ and } F = \mathbb{L}^\infty(0,T;\mathbb{H}^2(\Omega)) \cap L^2(0,T;\mathbb{H}^3(\Omega)).$$

Moreover, when α and α' are solutions of (5.9) with the respective velocities u and u', u and u' are solutions of (5.10) respectively with the viscosities μ and μ', and $\mu = \Phi(\alpha)$, $\mu' = \Phi(\alpha')$, we obtain

$$\begin{aligned}
\|(\alpha-\alpha')(t,\cdot)\|^2_{\mathbb{H}^1(\Omega)} &\le C\|u-u'\|^2_{\mathbb{L}^2(0,t;H^1(\Omega))},\\
\|(u-u')(t,\cdot)\|_{\mathbb{H}^1(\Omega)} &\le C\|u'(t,\cdot)\|_{\mathbb{H}^2(\Omega)}\|(\mu-\mu')(t,\cdot)\|_{\mathbb{H}^1(\Omega)},\\
\|(\mu-\mu')(t,\cdot)\|_{\mathbb{H}^1(\Omega)} &\le C(1+\|\alpha'\|_{\mathbb{L}^\infty(0,T;\mathbb{H}^2(\Omega))})\|(\alpha-\alpha')(t,\cdot)\|_{\mathbb{H}^1(\Omega)}.
\end{aligned}$$

This leads to

$$\begin{aligned}
&\|(G(\alpha)-G(\alpha'))(t,\cdot)\|_{\mathbb{H}^1(\Omega)}\\
&\quad\le C\|u-u'\|_{\mathbb{L}^2(0,t;H^1(\Omega))}\\
&\quad\le C\left(\int_0^t \|u'(t,\cdot)\|^2_{\mathbb{H}^2(\Omega)}\|(\mu-\mu')(s,\cdot)\|^2_{\mathbb{H}^1(\Omega)}\,ds\right)^{1/2}\\
&\quad\le C\|u'\|_{\mathbb{L}^2(0,t;\mathbb{H}^2(\Omega))}\|\mu-\mu'\|_{\mathbb{L}^\infty(0,t;\mathbb{H}^1(\Omega))}\\
&\quad\le C\|u'\|_{\mathbb{L}^2(0,t;\mathbb{H}^2(\Omega))}\|\alpha-\alpha'\|_{\mathbb{L}^\infty(0,t;\mathbb{H}^1(\Omega))},
\end{aligned}$$

where the constant C only depends on T, Ω, f, $\bar{u}$, α_0, ε and $\|\Phi\|_{\mathbb{L}^\infty(\mathbb{R})}$.

Since

$$\|\nabla\mu'(t,\cdot)\|_{\mathbb{L}^\infty(\Omega)} \leq C\|\Phi'\|_{\mathbb{L}^\infty(\mathbb{R})}\|\alpha(t,\cdot)\|^{1/4}_{\mathbb{H}^2(\Omega)}\|\alpha(t,\cdot)\|^{3/4}_{\mathbb{H}^3(\Omega)},$$

we obtain that $\mu' \in \mathbb{L}^{8/3}(0,T;\mathrm{W}^{1,\infty}(\Omega))$ and then that $u' \in \mathbb{L}^{8/3}(0,T;\mathbb{H}^2(\Omega))$. Finally,

$$\|u'\|_{\mathbb{L}^2(0,t;\mathbb{H}^2(\Omega))} \leq t^{1/8}\|u'\|_{\mathbb{L}^{8/3}(0,T;\mathbb{H}^2(\Omega))}.$$

There exists then a time $T^* \in]0,T[$ such that

$$\|(G(\alpha)-G(\alpha'))(t,\cdot)\|_{\mathbb{L}^\infty(0,T^*;\mathbb{H}^1(\Omega))} \leq \frac{1}{2}\|\alpha-\alpha'\|_{\mathbb{L}^\infty(0,T^*;\mathbb{H}^1(\Omega))}.$$

Moreover, the time T^* only depends on T, Ω, f, $\bar{u}$, α_0, ε and $\|\Phi\|_{\mathbb{L}^\infty(\mathbb{R})}$. Thanks to the estimates in Theorem 3 and to the estimates in Theorem 4 that are independent of α, we obtain the existence of a bound on α valid for all $t \in [0,T]$ and depending only on the data of the problem. Replacing the bounds on α by this new one, we obtain a smaller time T^* valid starting from all $t \in [0,T]$.

Since the function G is a contraction on $[0,T^*]$, we obtain that the recurring sequence (α^n,u^n,p^n) converges to a weak solution on $[0,T^*]$, then on $[T^*,2T^*]$, and finally on $[0,T]$. Moreover, since G is a contraction and thanks to the estimates we have, we prove the uniqueness of the solution.

5.2.1.2 Sketch of the Proof of Theorem 2

The solutions of (5.4)–(5.5) are obtained in the limit as ε tends to zero of the solution of (5.6)–(5.7). According to Theorem 1, we have that

- α^ε is bounded independently from ε in the spaces $\mathbb{L}^\infty(0,T;\mathbb{H}^2(\Omega))$, $\mathbb{L}^2(0,T;\mathbb{H}^3(\Omega))$ and $\mathbb{H}^1(0,T;\mathbb{H}^1(\Omega))$, which also implies that α^ε and $\nabla\alpha^\varepsilon$ are bounded in $\mathbb{H}^1((0,T)\times\Omega)$,
- u^ε is bounded in $\mathbb{L}^q(0,T;V(\Omega))$ for all $2 \leq q \leq +\infty$,
- $\varepsilon^{-1}\|u^\varepsilon-\bar{u}\|^2_{\mathbb{L}^q(0,T;B(\cdot))}$ is bounded for all $2 \leq q \leq +\infty$.

The ε-independent bounds makes it possible to extract subsequences still denoted $(u^\varepsilon,p^\varepsilon,\alpha^\varepsilon)$ that converge weakly or weakly-$\star$ depending on the bounds and such that u^ε converges strongly to $\bar{u}$ in $\mathbb{L}^q(0,T;\mathbb{L}^2(B(\cdot)))$ for all $2 \leq q \leq +\infty$.

One of the keys to passing in the limit for non-linear terms of the equations consists in getting some strong convergence. This is obtained thanks to the bound on α^ε and $\nabla\alpha^\varepsilon$ in $\mathbb{H}^1(O)$ which gives, thanks to Sobolev injection theorem in $\mathbb{R}^4$, the strong convergence of a subsequence of α^ε in $\mathrm{W}^{1,q}(O)$ with $2 \leq q \leq 4$. We can now take the limit for u in $B(t)$ and use Equation (5.8) in the domain $\Omega \setminus \bar{B}(t)$ as well as for α^ε in (5.9). Finally, the uniqueness of the solution can be gained by adapting the estimates that were used to prove that the function G defined in Section 5.2.1.1 is a contraction.

5.2.2 Numerical Analysis

This section presents the numerical algorithms specially developed to compute this mucus flow propelled by the ciliated epithelium cells. The equation set is rewritten as

$$\begin{cases} \partial_t \alpha + u \cdot \nabla \alpha - \eta \Delta \alpha = 0 & \text{in } \Omega, \\ \mu = \Phi(\alpha) & \text{in } \Omega, \\ -\operatorname{div}(2\mu D(u)) + \dfrac{\mathbf{1}_{B(t)}}{\varepsilon}(u - \bar{u}) = f - \nabla p & \text{in } \Omega, \\ \operatorname{div} u = 0 & \text{in } \Omega, \end{cases} \tag{5.11}$$

where all the quantities have the same meaning as in the previous section. The ε indices are removed in this part to improve the algorithm readability.

The first part of this section brings details on the computation of the cilia motion. The second part is dedicated to the computation of the mucins' ratio α (first equation of (5.11)). The last part details the computation of the fluid velocity, namely the Stokes problem (last two equations of (5.11)).

5.2.2.1 *Computation of the Cilia Beating*

The cilia motion is decomposed into two phases: the recovery stroke and the effective stroke. In the recovery stroke, which lasts about $2/3$ of the beating period, cilia are bending and moving backward, close to the bronchial wall and deep in the periciliary fluid layer. During the effective stroke the cilia are more straight and beat forward, penetrating in the mucus layer.

In order to numerically represent a coherent beating, each cilium is modeled by a filament of 0.3 μm diameter. The motion of this filament is governed by a transport equation set on a parametric curve which describes the filament center-line. Eqs. (5.12) are computed to obtain this, which is presented in Fig. 5.5, compared with the state-of-the-art experimental and numerical works.

Eventually, the beating remains planar in this study, and a small decay can be added between two adjacent cilia to create a metachronal synchronization:

$$\begin{aligned} &\frac{\partial L}{\partial t} + v(t)\frac{\partial L}{\partial \theta} = 0, \qquad P(0,t) = 0, \\ &L(0,t) = \partial_\theta P(0,t) = g(t) = (2\cos(2\pi t/T), 0, 1) \end{aligned} \tag{5.12}$$

where θ is the curve parameter, $P(\theta, t)$ is the curve position at time t, $L = \partial_\theta P$ and $v(t) = (1 + 8\cos^2(\pi(t + 0.25T)/T))/T$, with T being the beating period. To obtain a metachronal synchronization, the same shift is added to both g and v.

Eq. (5.12) enables the computation of the filament position at any time t. Then the characteristic function of $B(t)$, the volume occupied by the cilia array, can be deduced from the parametric curves, as well as the cilia velocity field $\bar{u}(t)$. These quantities are used in the following sections.

In all the computations, the counterforce applied by the fluid on the solid is not taken into account in the beating model. To our knowledge, this kind of computation and corresponding beating adaptation was only studied experimentally (Hill et al., 2010), and coherent modeling remains an open problem.

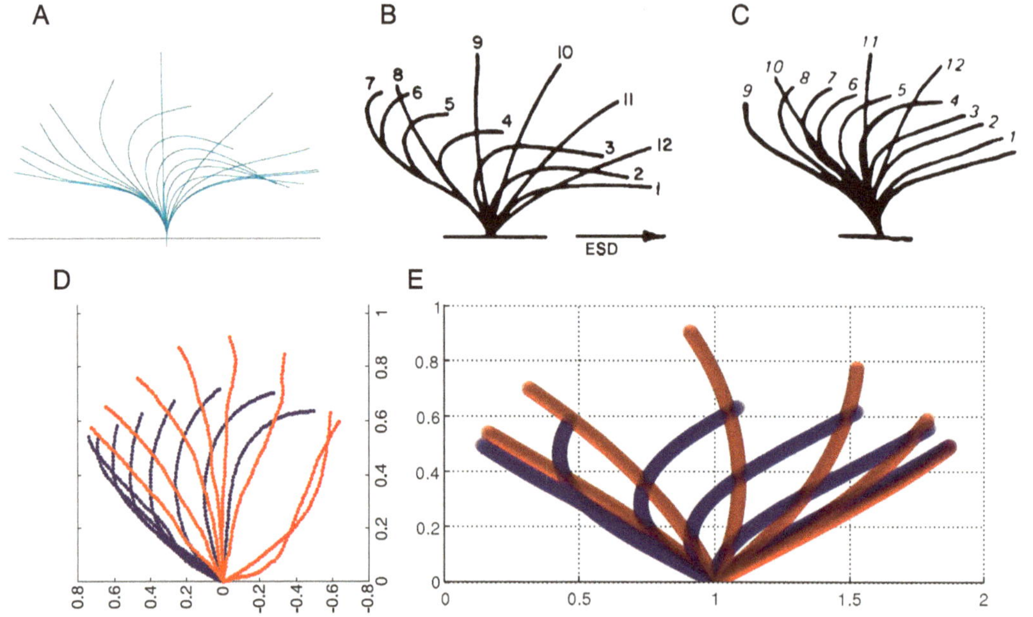

FIGURE 5.5 Trajectory followed by cilia motion for which each panel is representative of a model posed in the literature: (A) Fauci and Dillon (2006), (B) Gheber and Priel (1997), (C) Sanderson and Sleigh (1981) (culture of rabbit cells), and (D) Mitran (2007) (3D elastic beams model) also representative of the model used in this chapter. Different states of the recovery stroke are displayed in blue, and the states of the effective stroke are displayed in red.

5.2.2.2 *Numerical Computation of the Mucins Convection–Diffusion*

The first two equations of (5.11) can be rewritten as an autonomous dynamical system. Indeed, the velocity can be represented a function of the mucins' ratio through the viscosity, $u = F(\alpha)$:

$$\begin{cases} \partial_t \alpha + F(\alpha) \cdot \nabla \alpha - \eta \Delta \alpha = 0 & \text{in } \Omega, \\ \mu = \Phi(\alpha) & \text{in } \Omega. \end{cases} \tag{5.13}$$

The computation of this non-linear convection–diffusion equation can be difficult in a straightforward manner (finite difference discretization, for example) since two CFL stability conditions arise, one for each phenomenon (convection and transport). Hence the time step must usually be very small, imposing a large number of iterations and a large number of linear system assembling and resolution, which is computationally inefficient in 3D.

The viscous splitting (see Beale and Majda, 1981) of Eq. (5.13) enables the separation of the resolution of both operators $\partial_t + F(\alpha) \cdot \nabla$ and $\partial_t - \eta \Delta$, and allows the independent solution of both problems representing each phenomenon: a linear heat equation for diffusion (which is quite common to solve, particularly when η is small enough to allow explicit computations, which is the case in mucus framework) and a non-linear convection equation. The splitting introduced in Chatelin et al. (2015) guarantees a second-order accuracy and permits alternating the resolution of diffusion and convection in order to benefit of diffusion regularizing effects after each sub-step of convection.

In order to compute the convection, a Lagrangian method is used (Cottet and Koumoutsakos, 2000), the mucin's ratio is discretized on a particles cloud:

$$\alpha(t) = \sum_{p=1}^{\mathcal{P}} \alpha_p(t)\delta_{\xi_p(t)} v_p(t) \tag{5.14}$$

where ξ_p and v_p are respectively the position and the volume of particle p which carries a mucins' ratio α_p. Using such a discretization transforms the convection PDE into a set of ODEs:

$$\frac{d\xi_p}{dt} = u(\xi_p(t), t), \qquad \frac{d\alpha_p}{dt} = 0, \qquad \frac{dv_p}{dt} = 0. \tag{5.15}$$

In the last equation, particles' volume does not change since the velocity field is incompressible.

The advantage of solving (5.15) instead of a convection PDE is that there is no CFL restriction, so large time steps can be achieved. Moreover, in (5.15), only the first equation needs to be computed, which is very computationally efficient and is well suited for 3D computations.

An important feature of the present Lagrangian method is that the velocity field is computed on a Cartesian grid, where the viscosity (and thus mucins' ratio) needs to be known. Hence it is necessary to interpolate these quantities between the grid and the particles. To do that, a fourth-order interpolation kernel M_4' (Monaghan, 1985) is used. It guarantees fast computations since the computational cost is also linearly proportional to the number of particles (which is identical to the number of grid points) and accurate results due to the kernel's third order and regularity.

The operator splitting combined with the Lagrangian computation of convection and accurate interpolations with well chosen kernels enables linear computations with respect to the number of grid points. In order to keep the feature, the computations of velocity also needs to be efficient.

5.2.2.3 *An Iterative Projection for the Incompressible Stokes Problem*

The computation of the Stokes problem at any given time t is considered now. The cilia position and velocity are already computed (so $\bar{u}(t)$ and $\mathbf{1}_{B(t)}$ are known) and the convection–diffusion of α was achieved so $\mu(t)$ is also computed. The last two equations of the set in (5.11) remain to be solved:

$$\begin{cases} -\mathrm{div}(2\mu D(u)) + \dfrac{\mathbf{1}_{B(t)}}{\varepsilon}(u - \bar{u}) = f - \nabla p & \text{in } \Omega, \\ \mathrm{div}\, u = 0 & \text{in } \Omega \backslash B(t), \end{cases} \tag{5.16}$$

To compute this problem an iterative projection method has been developed (Chatelin and Poncet, 2013). This class of methods has been widely used for non-stationary Navier–Stokes equations (Guermond et al., 2006). The principle of all these methods is quite similar: the first step consists in solving the momentum equation without the pressure gradient. Hence an intermediate velocity field u^* is introduced that is not required to satisfy the incompressibility condition, $\operatorname{div} u = 0$.

A projection is then computed and the intermediate velocity is corrected in order to obtain a divergence-free velocity. This is performed by means of the projector $\mathbb{P}(u^*) = u^* - \nabla\zeta$ where ζ is solution of $-\Delta\zeta = -\operatorname{div} u^*$ in the whole domain, and satisfying the boundary condition $\partial\zeta/\partial n = u \cdot n$.

The main drawback of these methods is that the projector is usually computed by solving a Poisson problem with an homogeneous Neumann boundary condition. After the correction, the velocity field usually does not satisfy the original boundary condition (Guermond et al., 2006).

That the boundary conditions are not satisfied becomes more pronounced at low Reynolds numbers. This is the reason that projection methods developed for Navier–Stokes flows cannot be applied to the Stokes equations (Chatelin, 2013). To overcome this problem, a new iterative projection method was introduced in Chatelin and Poncet (2013). The algorithm is presented as follows:

1. (Initialization) $u_0^* = u_0$ is provided.
2. (Fixed point iterations, where i denotes the iteration number)
   (a) u_i^* is known, as well as $u_i = \mathbb{P}(u_i^*)$ and $\nabla\zeta_i = u_i^* - \mathbb{P}(u_i^*)$,
   (b) **(Prediction)** u_{i+1}^* is the solution to

$$\begin{cases} -\mu\Delta u_{i+1}^* + \dfrac{\mathbf{1}_{B(t)}}{\varepsilon}(u_{i+1}^* - \nabla\zeta_i - \overline{u}(t)) = f + \left(2D(u_i) + \operatorname{div} u_i^*\right)\nabla\mu & \text{in } \Omega, \\ u_{i+1}^* = g - \nabla\zeta_i & \text{on } \partial\Omega, \end{cases} \tag{5.17}$$

   where the last term of (5.17) comes from the substitution $u^* = u + \nabla\zeta$.
   (c) **(Correction)** Compute ζ_{i+1}, the solution to a Poisson problem

$$\begin{cases} -\Delta\zeta_{i+1} = -\operatorname{div} u_{i+1}^* & \text{in } \Omega, \\ \dfrac{\partial\zeta_{i+1}}{\partial n} = 0 & \text{on } \partial\Omega, \end{cases} \tag{5.18}$$

   and project, namely let $u_{i+1} = \mathbb{P}(u_{i+1}^*) = u_{i+1}^* - \nabla\zeta_{i+1}$.
   (d) Repeat steps 2a through 2c and increment i until convergence is satisfied.
3. The solution to the Penalized Stokes problem is u_{i+1}.

This algorithm only involves the computation of elliptic problems. In particular, the computation of the projector can be achieved using an FFT or multigrid solver (Swarztrauber and Sweet, 1975; Adams, 1989) which are very powerful. Indeed, using a regular Cartesian grid, the structure of the discretized operator is known *a priori* and no matrix assembling or linear system resolution is required. Hence the computational cost is quasi-linear with respect to N, the number of computational grid points, that is to say, with a complexity $\mathcal{O}(N \log N)$. Furthermore, it guarantees an accurate computation of boundary conditions, immersed solid velocity, inviscid velocity and non-homogeneous effects.

In order to conserve this computational efficiency, which is very attractive for 3D computations, the penalized Stokes equation (5.17) and the projection using FFT need to be computed with a similar computational cost. The next section describes a numerical method for the penalized Stokes equation using only FFTs.

5.2.2.4 A Sherman–Morrison–Woodbury Approach to Compute Penalized Poisson Problems

To compute the Penalized Poisson problem of the first step of the previous iterative projection, a special algorithm has been developed (Chatelin and Poncet, 2014), based on the algebraic perturbation relation, the Sherman–Morrison–Woodbury formula (Hager, 1989). This formula enables the computation of the inverse of an invertible perturbed matrix

$$\left(A + E^T C E\right)^{-1} = A^{-1} - A^{-1} E^T \left(C^{-1} + E A^{-1} E^T\right)^{-1} E A^{-1}. \tag{5.19}$$

The problem to solve (first step of the previous iterative projection) can be rewritten as follows (gathering all the terms depending on previous iteration k in the RHS):

$$\begin{cases} -\Delta u + cu = f & \text{in } \Omega, \\ u = 0 & \text{on } \partial\Omega, \end{cases} \tag{5.20}$$

where c represents the Penalization term and iteration subscripts were removed. Hence this problem can be seen as a Poisson problem perturbed in the cilia volume. Then, applying a generic discretization to (5.20), it can be exactly rewritten as an algebraic perturbation problem, and the SMW formula can be applied:

$$\left(A + E^T C E\right) U = F \Longleftrightarrow U = A^{-1} F - A^{-1} E^T \left(C^{-1} + E A^{-1} E^T\right)^{-1} E A^{-1} F. \tag{5.21}$$

In this equation, each product involving A^{-1} can be achieved without matrix assembling and inversion by using the Poisson solver. The key point is to invert the term inside parenthesis without matrix inversion and assembling, which is achieved with an iterative GMRES method. The global algorithm reads in four steps:

1. Solve the Poisson problem $AW = F$.
2. Solve $QZ = EW$ with $Q = C^{-1} + EA^{-1}E^T$. This is performed by means of the GMRES method that requires the product evaluation $Z \to QZ$ as follows:
   (a) Solve the Poisson problem $A\Theta = E^T Z$;
   (b) Set $QZ = C^{-1} Z + E\Theta$.
3. Solve the last Poisson problem $AY = E^T Z$.
4. Update the solution $U = W - Y$.

Steps 2a and 2b are computed at each Krylov iteration so the total number of Poisson problems to solve is $2 + N_{Kry}$ where N_{Kry} is the typical size of the Krylov space. The other steps are negligible from a computational point of view: as C is a diagonal matrix, its inverse is immediately computed, restriction and inclusion matrices E and E^T are deduced from $\mathbf{1}_{B(t)}$ and never assembled.

The overall algorithm, presented in the previous sections has the following convergence features:

- A $\mathcal{O}(\sqrt{\varepsilon})$ convergence order for the Penalization method, which is a standard result (Carbou and Fabrie, 2003).

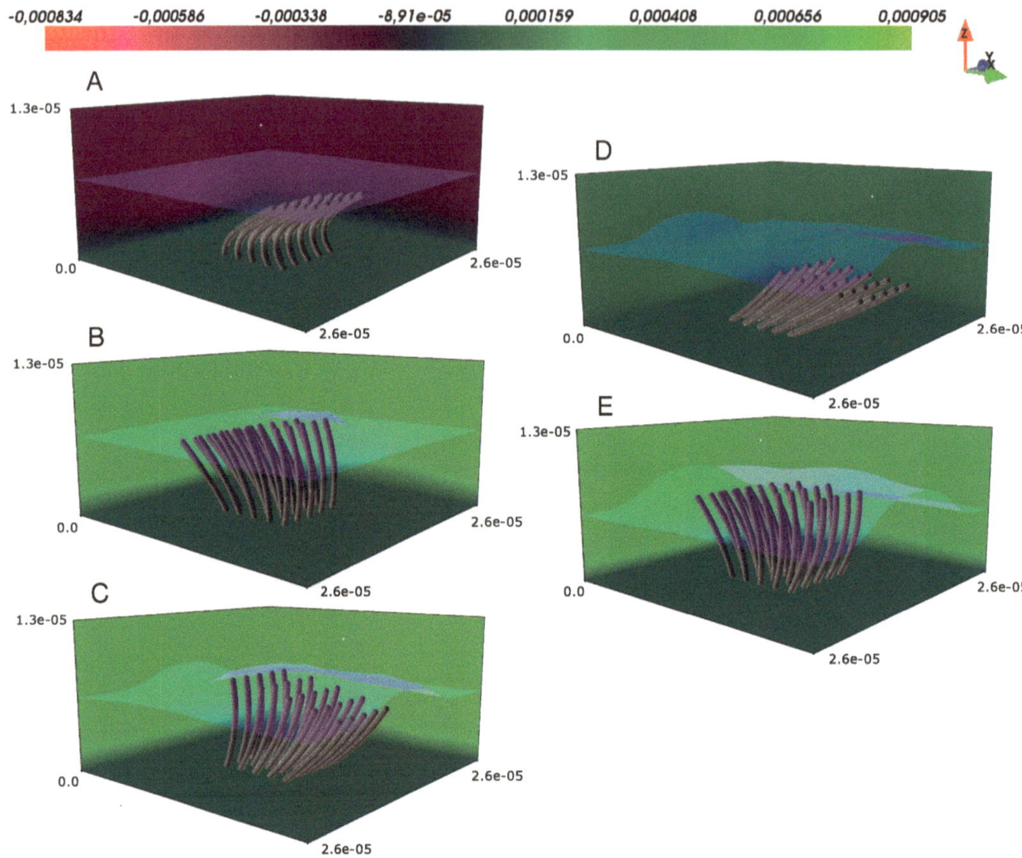

FIGURE 5.6 Snapshots of the mucus propulsion by a carpet of 25 ciliated epithelium cells at five different instants. The colormap shows the magnitude of the first component of the velocity field. At the beginning of the computation the cross-linked mucins ratio α is initialized according to Eq. (5.3). The periciliary fluid viscosity is equal to that of water (10^{-3} Pa s) and the viscosity of the mucus is 0.1 Pa s. An isosurface of $\alpha = 0.5$ is presented in blue. In white, a passive particle clearly shows that the mucus is transported toward trachea. Panels (A) to (E) are snapshots at times $t = 3.33 \times 10^{-2}$, 0.77, 0.28, 0.29 and 0.88 s.

- A second-order temporal accuracy, obtained with the splitting algorithm (Chatelin et al., 2015).
- A second-order spatial accuracy which is given by the fast FFT solver (Swarztrauber and Sweet, 1975).
- A very negligible mass diffusion in particle transport and interpolation processes (Chatelin and Poncet, 2013).

5.2.3 Numerical Computations of the Mucus Propelled by Ciliated Epithelium

The first simulation presented involves a simplified epithelium cell composed of 25 cilia regularly arranged as a 5×5 array. Snapshots are presented in Fig. 5.6. This simulation was

TABLE 5.1 Reference parameters for the numerical simulations (gathered from Fahy and Dickey, 2010; Sanderson and Sleigh, 1981; Smith et al., 2008; Thiriet, 2012)

Name	Value
Cilia length	8 µm
Cilia diameter	0.3 µm
periciliary fluid viscosity (η_{PCF})	0.001 Pa s
Viscosity ratio ($\beta = \eta_{ML}/\eta_{PCF}$)	100
Viscosity transition stiffness (γ)	10
Transition length (δ)	7 µm
airway surface liquid height (H_z)	13 µm
Beating frequency	10 Hz
Mucins diffusion (ν)	10^{-11} m^2 s^{-1}

performed with the parameters gathered of Table 5.1. This figure presents an isosurface of cross-linked mucins ratio, velocity magnitude in the proximal direction and a passive black particle, advected by the flow. On average, this passive tracer travels forward (in the proximal direction to the trachea) which shows that the mucociliary clearance is effective.

The second simulation (see Fig. 5.7) presents the mucociliary transport with a carpet of 300 cilia, still using the parameters of Table 5.1. This computation presents the effect of the metachronal synchronization (when a shift is added in the beating between adjacent cilia, creating the metachronal wave) for a large number of cilia. This figure shows that the metachronal wave travels backward, generating a positive first velocity component wave in the mucus layer whereas the velocity is zero or lightly negative in the periciliary fluid layer.

As in the previous figure, the passive white particle clearly shows that the upper part of the airway surface liquid travels forward, thus the mucociliary clearance is efficient. This simulation shows that a "leading cavity" effect is created when cilia are in recovery stroke, maximizing the mucociliary transport. These results, also reported in Hussong et al. (2011) using simplified simulations, help to validate the present methodology.

5.2.4 Phenomena Analysis

In order to quantify the mucociliary clearance efficiency, the mean velocity of the mucus layer is averaged over several beating periods T:

$$\overline{U} = \frac{1}{3T H_x H_y (H_z - \delta)} \int_0^{3T} \int_\delta^{H_z} \int_0^{H_y} \int_0^{H_x} u_x(x, y, z, t) dx dy dz dt \tag{5.22}$$

where δ is the transition height between PCL and mucus, introduced in Eq. (5.3).

Using different parameter sets, this quantity enables the comparison between different simulations (Chatelin and Poncet, 2016). This can be used in the identification of the meaningful parameters and can help to identify the influence of pathological situations in mucociliary transport. In this study, the following meaningful parameters were identified:

- The beating frequency $2\pi/T$ is an important parameter of the modeling. Indeed, as a consequence of the time-dilatation of the quasi-static Stokes equations, the mucociliary transport

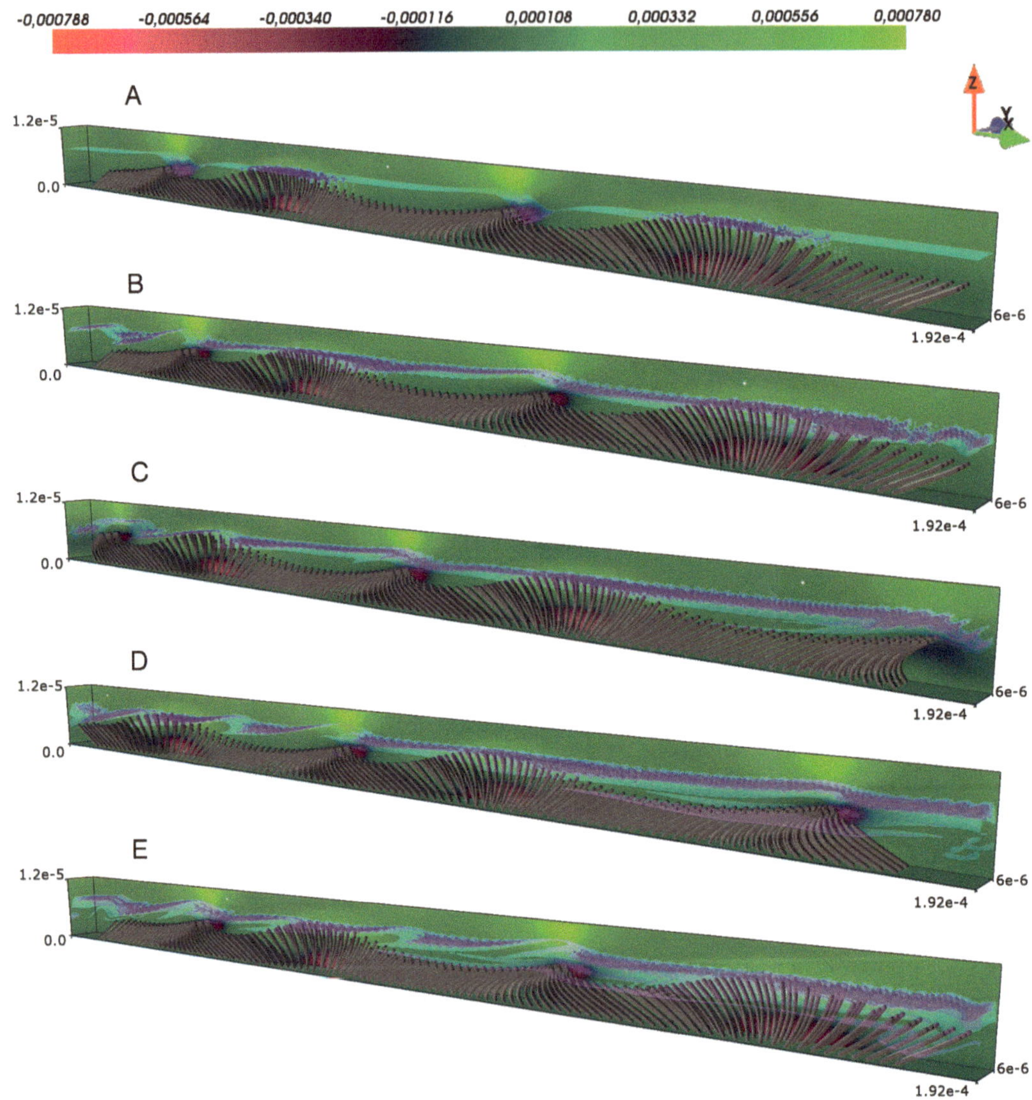

FIGURE 5.7 Snapshots of the mucus propulsion by a carpet of 300 ciliated epithelium cells at five different times. Colormap and fluid features are the same as in Fig. 5.6, except that the isosurface level is $\alpha = 0.65$. Panels (A) to (E) are snapshots at times $t = 6.67 \times 10^{-3}$, 0.2, 0.23, 0.35 and 0.5 s.

varies linearly with respect to this parameter. This result was also reported using 2D computations with a bi-fluid modeling (Lee et al., 2011).

- It was also demonstrated that the cilia length is very important. In fact, when cilia are too small the mucociliary clearance is deteriorated. This paper shows that it can be divided by 10 if cilia are 3 μm long instead of 9. It was demonstrated that the most efficient result is obtained when cilia penetrate the mucus layer during the effective stroke and remain in the periciliary fluid layer during the recovery stroke. This result confirms this assumption formulated in Liron and Rozenson (1983) and contradicts the results of Fulford and Blake (1986).

- As expected, the viscosity ratio between mucus and periciliary fluid has also a great importance: when the mucus becomes highly viscous, the mucociliary clearance is also deteriorated, it can be divided by two when the mucus is 1000 times more viscous than periciliary fluid layer. It is important to notice that the most efficient mucociliary clearance is not obtained for a ratio of 1 (meaning an airway surface liquid only composed of periciliary fluid) but when the mucus is about 10 times more viscous. It means that the mucociliary clearance permits to handle the accumulation of cross-linked mucins in the mucus layer, ensuring a good protection of the airways.
- The transition length δ between mucus and periciliary fluid layer is also very important. It is the equilibrium between both these characteristic lengths which leads to the penetration (or not) of the cilia tips in the mucus during the effective stroke. In particular, when this transition length is too small (this is representative of the case in which nearly all of the airway surface liquid is composed of mucus) the mucociliary clearance can be divided by 2.5.
- Finally, the transition stiffness γ between the periciliary fluid layer and the mucus is also an important parameter. It means that to achieve an efficient mucociliary clearance, two layers have to be clearly identified. This also enables the penetration in the mucus only during the effective stroke.

5.3 RHEOLOGY OF MUCUS AND NON-NEWTONIAN MODELS

In this section, the methods and data available in the literature are presented, followed by our own methods and results. The power law and Carreau law are then considered the best fitting models, and introduced in the mathematical model and in the numerical scheme. A discussion confronting experimental data and results from numerical simulation is finally performed, with a description of the medical outcomes.

5.3.1 Rheometry Data on Lung Mucus from the Literature

Looking at studies concerned with rheological characterizations of pulmonary mucus, two points are noticeable. First, two kinds of equipment are used: as expected, classical rheometers, but also magnetic micro-rheometers. Second, rheological characterizations are often limited to the determination of the dynamic storage modulus (G') and loss modulus (G'') for two frequencies, 1 and 100 rad/s, which are supposed to be characteristic frequencies of mucociliary clearance and cough clearance, respectively (App et al., 1998).

5.3.1.1 The Different Measurement Methods

Considering rheological measurements on lung mucus, there are almost as many studies with magnetic micro-rheometers as with classical rheometers. The magnetic micro-rheometer uses either a single steel bead (App et al., 1998; Redding, 2008) or multiple magnetite particles (Edwards and Yeates, 1992; Ziemann et al., 1994) positioned in a few μL of fluid and oscillated by means of an electromagnetic field to obtain a sinusoidal force. In the first case, the bead diameter ranges from 50 to 100 μm while particles dimension ranges from 2 to 6 μm in the second case. The motion of the single bead or of the particles embedded in the fluid is then quantified, generally by fast image processing techniques; see King and Macklem (1977) and Ziemann et al. (1994) for the detail of rheological parameters extraction. Note that King and Macklem (1977) have incorporated Thomas and Walters (1966) results to take into account elasticity of the fluid. This technique has largely been used for respiratory mucus (Jeanneret-Grosjean et al., 1988; Rubin et al., 1990; App et al., 1998; Zayas et al., 1999; Redding, 2008).

The main advantage of the micro-rheometer lies in the very small volume of fluid that is necessary (a few μL). This is of particular interest for mucus that is generally not available in large quantities except for major pathologies such as cystic fibrosis. It is important to note that in general, mucus type fluids are not homogeneous, and that larger sample volumes would allow a more representative measurement of the global rheological behavior and a better precision. Additional drawbacks are that only oscillatory experiments are possible with such a micro-rheometer and that the flow is not a simple shear flow as it is theoretically required for rheometry (Barnes et al., 1989).

Classical rheometry has predominantly been used in recent years (Sanders et al., 2000; Dawson et al., 2003; Nielsen et al., 2004; Innes et al., 2009; Serisier et al., 2009; Daviskas et al., 2010; Tomaiuolo et al., 2014). It is performed using controlled-stress rheometer equipped with cone and plate geometry or parallel plates geometry. These are generally chosen to minimize the fluid volume: small diameter/small angle for the cone (1° or 2°), or small gap for parallel plates (between 50 and 250 μm when specified).

Note that the term "micro-rheology" may be confusing as it is also used to represent length scales below 0.1–0.3 μm, for instance, to examine virus or nano-particle motion through mucus (Sanders et al., 2000; Dawson et al., 2003; Lai et al., 2009; Schuster et al., 2013). Viscosity experienced by these objects is close to that of water: it is linked to the mesh size of the network structure of mucus which is larger than (or similar to) the object size. Nevertheless, for the mucus clearance process larger length scales are relevant. The viscosity has to be considered macroscopic, which is anticipated to greatly exceed the viscosity of water and is associated to a complex rheological behavior.

5.3.1.2 Mucus Rheology from Literature

A general presentation of mucus, its role, chemical structure and physiological properties can be found in Cone (2005). Reviews of the rheological properties are presented by Yeates et al. (1997) and Picout and Ross-Murphy (2003). Considering human airway mucus, few data are available in the literature for steady state viscosity while data for dynamic viscosity $\eta' = G'/\omega$ or complex viscosity $|\eta^*| = |G^*|/\omega$, from oscillatory measurements, are more numerous. When it is reported, the measurement temperature is 20, 25, or 37°C. This might appear as a large discrepancy but variances caused by temperature differences are small compared to those from the non-Newtonian feature, as it was demonstrated by Tomaiuolo et al. (2014).

When a rheogram is proposed, all authors agree to recognize a strong shear thinning behavior. Some authors (Moores et al., 1992; Dawson et al., 2003; Suk et al., 2009) observed a tendency to a Newtonian plateau at low shear rates with a transition towards power-law behavior between 0.1 s^{-1} and 1 s^{-1}, depending on authors. On the other hand Yeates et al. (1997) did not observe any trend to a plateau although exploring shear rate from 0.2 s^{-1} upwards. The same observation was made by Puchelle et al. (1985, 1987) for shear rate starting from 0.01 and 0.1 s^{-1}, respectively. In the shear thinning region, the behavior can be modeled by a power-law $\tau = k\dot{\gamma}^n$, where τ and $\dot{\gamma}$ are the shear stress and the shear rate respectively, k the fluid consistency and n the power-law index. Values for k and n given in literature, or extracted from literature data, are reported in Table 5.2. Note the low values of the power-law index (between 0.04 and 0.26) which are characteristic of a strong shear-thinning behavior.

Thixotropy is the property of a fluid to show a reversible decreasing viscosity when submitted to a constant shear rate (i.e. observation of a decreasing shear stress as a function of time). Note that thixotropy is sometimes confused with shear-thinning, although rheological definition is unambiguous (Picout and Ross-Murphy, 2003). Thixotropy leads to a hysteresis phenomena that has been observed on viscosity measurements at increasing shear rates followed by measurements at decreasing shear rates (Nielsen et al., 2004; Tomaiuolo et al., 2014). Conversely, note that hysteresis is not necessarily evidence of thixotropy because elasticity may also be responsible for hysteresis, especially when the characteristic time of the fluid is comparable to (or larger than) the measurement time scale. References to a characteristic time for mucus are not numerous in the literature: Besseris and Yeates (2007) gave 42 s for canine tracheal mucus, while for human mucus Moores et al. (1992) obtained 20 s from creep tests, and Nielsen et al. (2004) estimated it to be of the order of thousands of seconds from oscillatory measurements. Academic thixotropy measurements were realized by Puchelle et al. (1985, 1987) who observed a shear stress overshoot after a shear rate step. Such an overshoot is linked to elasticity, but successive experiments with decreasing rest time after previous shearing show a decreasing overshoot. This has allowed Puchelle et al. to identify the presence of thixotropy for reconstituted mucus (Puchelle et al., 1985) and native mucus (Puchelle et al., 1987).

A few authors have identified a viscoplastic behavior with a yield stress varying between 0.05 to 0.26 Pa (Edwards and Yeates, 1992) until 100 Pa (Silberberg, 1983). For instance, this last value was used by Craster and Matar (2000) to numerically analyze the surfactant transport of mucus film in the pulmonary airways. More recently, Broughton-Head et al. (2007) proposed a value of 14 Pa for cystic fibrosis mucus. As far as we know, these are the only publications identifying a viscoplastic behavior and giving yield stress values. This can be connected to specific rheometers used or to the conduction of the rheological characterizations.

On the other hand, elasticity has been clearly identified in the literature. This is due in part to the widespread use of the magnetic micro-rheometer which is restricted to small-amplitude oscillatory shear experiments. The storage modulus $G'(\omega)$ and the loss modulus $G''(\omega)$ are the most common data encountered in the literature at the two frequencies $\omega = 1$ and $\omega =$ 100 rad/s. The moduli G' and G'' are the imaginary and real parts of the complex modulus $G^* = \tau/\gamma$, where γ is the deformation. A phase shift between shear stress and shear rate $\dot{\gamma}$, i.e. a non-zero G' value, demonstrate the elastic feature of the fluid.

Data are expressed in terms of modulus or, in equivalent form, using the complex modulus G^* and the loss tangent $\tan\delta = G''/G'$ (not to be confused with the PCL height in the previous sections), or G' and the so-called dynamic viscosity $\eta' = G''/\omega$. The values of G' and G'' for two specific values of the frequency, 1 and 100 rad/s, are often reported in

TABLE 5.2 Mucus rheological characterization, power-law parameters; CR, Classical rheometer; MμR, Magnetic μ-rheometer; CF, Cystic Fibrosis; Br, Bronchiectasis; COPD, Chronic Obstructive Pulmonary Disease; CB, Chronic Bronchitis; LS, Light Smokers; HS, Heavy Smokers; XS, ExSmokers; C, Cancer; FEV1%, ratio of the Forced Expiratory Volume in 1 s (FEV1) to the Forced Vital Capacity (FVC)

Source	Alias	Rheometer	Pathology	Additional information	n	k	n^*	k^*
Puchelle et al. (1987)		CR	Chronic airway hypersecretion		0.18	18.38		
Jeanneret-Grosjean et al. (1988)	HL6	MμR	Healthy		–	–	0.19	14.13
	HL7		Healthy	Control	–	–	0.20	15.49
			Healthy	After administration of Atropine	–	–	0.20	70.79
			Healthy	After administration of Diazepam	–	–	0.15	20.42
Rubin et al. (1990)		MμR	Healthy	Taken from endotracheal tube at the end of a surgery	–	–	0.20	10.96
			CF		–	–	0.29	23.99
Zayas et al. (1990)	HL8	MμR	Healthy		–	–	0.22	13.18
			LS	Without cancer, without obvious airway infection	–	–	0.17	7.08
			HS	Without cancer, without obvious airway infection	–	–	0.15	18.62
			XS	Without cancer, without obvious airway infection	–	–	0.16	6.92
			C	Without obvious airway infection	–	–	0.15	11.75
Moores et al. (1992)		CR	COPD	Infective exacerbation	0.04	15.00	0.13	1.71
Dawson et al. (2003)	CFL5	CR	CF	–	0.05	18.00	0.17	7.34
Nielsen et al. (2004)		CR	CF	–	0.27	5.40	0.163	12.45
Redding (2008)		MμR	Br	Idiopathic bronchiectasis (children)	–	–	0.20	6.84
			CF	Children	–	–	0.16	33.43
			CB	Adults	–	–	0.14	53.03
Innes et al. (2009)		CR	Healthy	Sputum induction performed with nebulized 3% saline for 20 min	–	–	0.18	3.34
Serisier et al. (2009)	HL9	CR	Healthy	Sputum induction with 4.5% hypertonic saline	–	–	0.48	0.35
	CFL1		CF	Clinically stable	–	–	0.41	0.54
			COPD	Clinically stable	–	–	0.34	1.29
Suk et al. (2009)		CR	CF	–	0.16	2.42		
Daviskas et al. (2010)		CR	Br	Clinically stable	–	–	0.41	36.31
		CR	Br	Clinically stable – After repetitive voluntary coughing –	–	–	0.45	24.76
Schuster et al. (2013)		CR	Healthy	Taken from endotracheal tube at the end of a surgery	–	–	0.16	12.97
Tomaiuolo et al. (2014) according to FEV1	CFL2	CR	CF	FEV1% > 50%	0.23	2.18	0.18	1.40
	CFL3		CF	30% < FEV1% < 50%	0.21	12.81	0.17	2.25
	CFL4		CF	FEV1% < 30%	0.26	6.87	0.16	2.42
Tomaiuolo et al. (2014) according to main colonization		CR	CF	*S. aureus*	0.20	3.04	–	–
			CF	*P. aeruginosa*	0.17	14.10	–	–
			CF	*B. cepacia* and *S. maltophila*	0.14	10.86	–	–

the literature because they were proposed by some authors (Jeanneret-Grosjean et al., 1988; Zayas et al., 1990, 1999; Shah et al., 1996; App et al., 1998; Innes et al., 2009) to define a mucociliary clearability index (conventionally called *MCI*) and a cough clearability index (*CCI*), which respectively quantify the clearability by normalized ciliary function (using values at 1 rad/s) and by cough (using values at 100 rad/s). As far as we know, the mucociliary clearability index was first proposed by Jeanneret-Grosjean et al. (1988). *MCI* and *CCI* are then mentioned in Zayas et al. (1990). Note that these indices were also used by Tomaiuolo et al. (2014) at a frequency of 10 rad/s for *CCI*. These indices are of particular interest in the evaluation and quantification of the influence of pathologies on the mucus clearance for smokers and lung cancer patients (Zayas et al., 1990, 1999), for asthma (Innes et al., 2009), and for the influence of treatments or physiotherapy (Shah et al., 1996; App et al., 1998).

In the study of the flow of mucus, the essential parameter is the viscosity. One can ask whether the Cox–Merz rule is sufficient to evaluate $\eta(\dot{\gamma})$ using oscillatory measurement data. The Cox–Merz rule is considered as very useful when only linear viscoelastic data are available (Bird et al., 1977). It is expressed as:

$$|\eta^*(\omega)| = \sqrt{\eta'(\omega)^2 + \eta''(\omega)^2} = \eta(\dot{\gamma})|_{\dot{\gamma}=\omega}. \tag{5.23}$$

The Cox–Merz rule is well-adapted for polymer solutions with entangled network structure. It fails (with generally $|\eta^*| > \eta$) for cross-linked systems or for highly concentrated entangled systems. Cross-linked polymers are characterized by the non-existence of a Newtonian plateau for $\eta(\dot{\gamma})$ and a weak evolution of elastic and viscous modulus with frequency. Based on such observations, numerous authors think that mucus is of cross-linked type (Puchelle et al., 1985; Innes et al., 2009; Schuster et al., 2013) but Yeates et al. (1997) developed arguments for an entangled "sliplinked" strand system. Looking at Table 5.2, there are only six cases where the power-law parameters (n, k) and (n^*, k^*) (from power-law model applied to $\eta^*(\omega)$ available data) can be estimated and compared. The consistency k is underestimated when assimilated to k^* for five of the six cases, while no real tendency can be observed on (n, n^*) data (see Fig. 5.8). This point will be re-examined later.

Information related to pathology are presented in Fig. 5.9. For this representation, the rheological variables $|G^*|$ and $\tan\delta$ were chosen as they are often considered as relatively independent from one another while G' and G'' are not (Jeanneret-Grosjean et al., 1988; King, 1980). Nevertheless, $|G^*|$ varies with respect to viscosity while $\tan\delta$ compares viscous and elastic response of the fluid (with $\tan\delta$ infinite for purely viscous fluids). Concerning healthy patients, it is noticeable that the two points corresponding to the lower $|G^*|$-value were obtained after nebulization with a saline solution to induce sputum. This probably led to the thinning of the mucus and also helps to explain these lower values as well as the large $\tan\delta$ value.

These comparisons have to be considered cautiously because conditions differ between studies (how mucus is taken and prepared, how rheological characterizations are conducted, for example). It is noteworthy to point out that mucus samples of patients with pathology do not strongly differ from samples of healthy patients (except in some cases of bronchiectasis and COPD). On this graph, the majority of data corresponds to stable patients, or healthy, one case corresponds to exacerbated situation and no information is given for five cases. In a

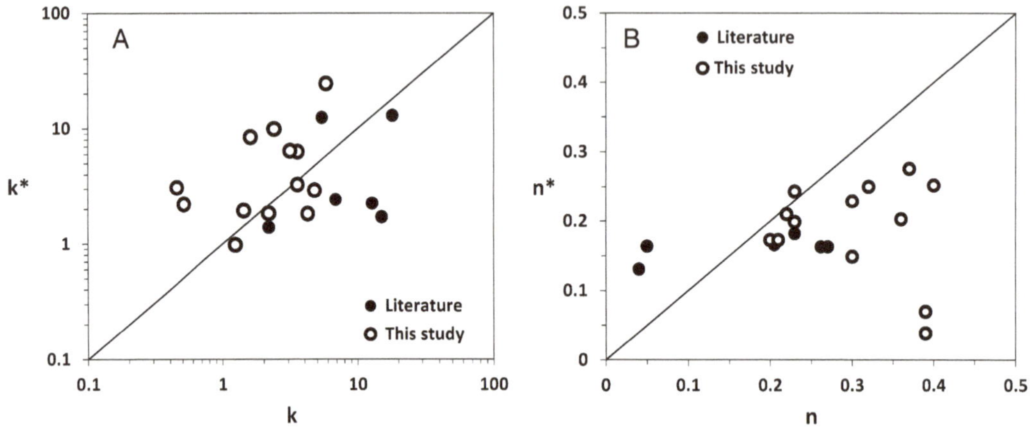

FIGURE 5.8 Comparison between consistency k and power-law index n issued from viscosity curves and estimations (k^*, n^*) issued from the Cox–Merz rule. Note that each point from literature is a mean value while a point marked "this study" corresponds to one of the tested mucus. The solid line denotes the identity.

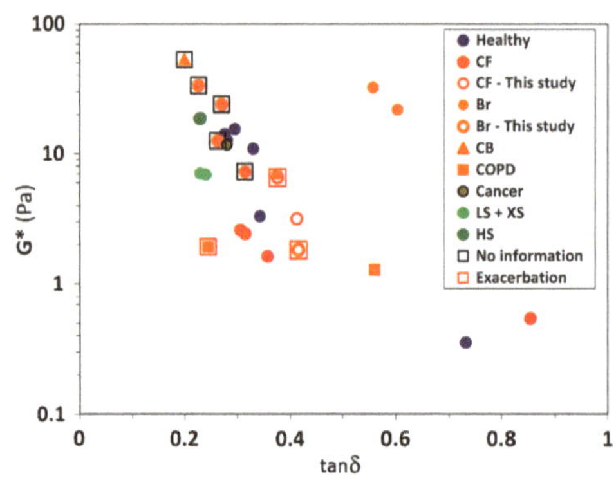

FIGURE 5.9 Rheological parameters G^* and $\tan\delta$ (measured at 1 rad/s) for different pathologies. Symbols are the same as in Table 5.2. All cases with simple symbols concern studies of patients in a stable state. Symbols with a surrounding square indicate a study for patients in exacerbation or a study without information.

study on rheological properties of CF mucus, King (1981) proposed to classify mucus from its visual aspect in three categories (linked to the nature and to the degree of infection): mucoid, muco-purulent and purulent. Comparing G' and η' for three frequencies, he observed that rheological properties for mucoid and muco-purulent human CF mucus are similar to control canine tracheal ones while purulent mucus showed higher values and a diminution of $\tan\delta$ (i.e. higher elasticity). This was also underlined by Deneuville et al. (1997), Lopez-Vidriero and Reid (1978). Although this distinction between mucus seemed no longer considered for several years, more recent studies have confirmed the interest of this observation, using stable/exacerbated distinction rather than visual criteria: Serisier et al. (2009) and Tomaiuolo et al. (2014) for CF sputum, and Tabatabaei et al. (2015) for COPD.

5.3.2 Mucus Sample Analysis and Rheological Results

5.3.2.1 Mucus Samples Collecting

Consecutive adult patients with cystic fibrosis (CF) bronchiectasis (Br) were recruited when admitted in the regional adult CF center of Toulouse. After information and oral consent, sputum samples were collected under supervision of a chest physiotherapist. Patients were classified as either having a stable status or having an exacerbation. The classification of exacerbation represented an increased cough or an increased bronchorrhea or purulence sputum, and/or an increased dyspnea, and/or a tiredness and/or a loss of weight). Sputum samples were frozen at −20°C before rheological analysis.

5.3.2.2 Mucus Sample Analysis

In this study rheological characterizations were evaluated on a Mars III rheometer (Thermo Sc.) equipped with a cone and plate geometry (1° angle, 35 mm diameter) which requires 0.2 mL of fluid. Measurements were conducted at 37°C. The lower plate Peltier system for temperature regulation was completed by a sample hood that was custom made in-house with a specific temperature regulation and wet sponges to saturate the measuring chamber and prevent evaporation of the sample. As previously mentioned, samples were stored frozen at −20°C. The influence of freezing on the rheological properties of mucus has previously been investigated by Rubin et al. (1990), Sanders et al. (2000), Gastaldi et al. (2000) and Serisier et al. (2009) for different conditions (temperature, freezing duration). These authors all conclude that freezing does not affect rheological behavior. This result was confirmed in the current study on three samples that were characterized when fresh and again at 21 or 31 days under frozen conditions. Mucus samples were thawed keeping them in a water–ice mixture just before and during measurements to prevent degradation (Sanders et al., 2000).

Once a sample was loaded in the rheometer, the following protocol was followed: an initial dynamical strain sweep at a fixed frequency of 1 Hz was conducted to determine the linear domain. Then the storage modulus $G'(\omega)$ and the loss modulus $G''(\omega)$ were determined using an oscillatory frequency sweep for a fixed strain in the linear domain (between 1 and 3% according to the sample). Then, the sample was submitted to very large deformations in which the viscosity $\eta(\dot{\gamma})$ was determined by a steady-shear flow test for shear rates in the range from 10^{-3} to 10^{3} s^{-1}. Careful attention was paid to reaching the steady state regime for cases with low shear rate values. Additional measurements were sometimes conducted to examine the stability of a sample. A power-law model $\eta = k\dot{\gamma}^{n-1}$ (k, consistency; n, power-law index) was used to fit the rheological data. When the sampled volume was sufficient, two or three different tests were performed.

5.3.2.3 Rheological Results

Initially a few samples were used in order to determine the measurement strategy. Then 15 mucus samples were analyzed. Two of the samples were rejected because they were visually too heterogeneous and consequently gave rise to a large dispersion in viscosity and modulus values. The remaining 13 mucus samples were then considered for data analysis.

To examine thixotropy, experiments were conducted that were based on the observation of the shear stress overshoot after a shear rate step (similar to those of Puchelle et al., 1987). Because the results of those experiments were found to be inconclusive, another approach was

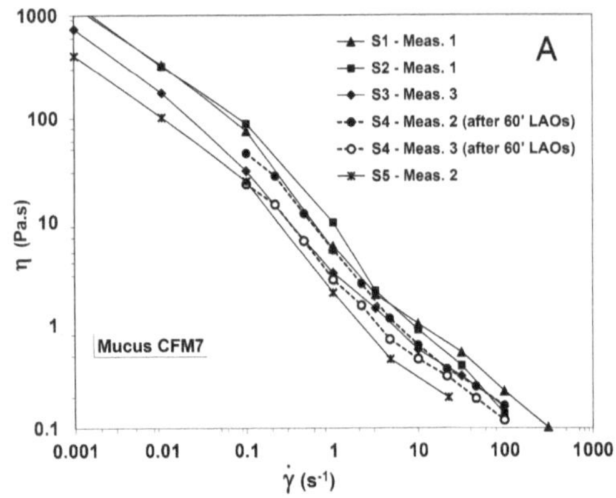

FIGURE 5.10 Viscosity curves, on the left for mucus CFM7: 'S' standing for Sample and 'Meas.' for measurement. Measurements on sample 4 were done after a sequence of large amplitude oscillations (LAOs) at 300% and 1 Hz during 60 min. Right picture shows rheograms for three different mucus.

adopted. On one of the mucus samples (CFM7), viscosity measurements were conducted on four samples while on the fifth one, a large amplitude (300%, i.e. far from the linear domain) oscillatory sequence (LAOs) was first applied during 60 min. After that a viscosity measurement was done. The results are reported in Fig. 5.10A. The viscosities for samples S1, S2, S3 and S5 reveal the variability between samples and between measurements. Fig. 5.10 reports the data indicating the three measurement numbers. For measurements 2 and 3 (Meas. 2 and Meas. 3), only small amplitude oscillatory measurements were conducted before (strain sweep and/or frequency sweep). No specific classification linked to solicitations to which the sample was previously submitted appears in these results, especially after a strong and long solicitation such as 60 min of LAOs (S4, Meas. 2 and Meas. 3). Fig. 5.10B presents steady-state shear viscosity curves established for mucus CFM2, CFM6 and BM12. For each of them, two or three measurements were conducted, thus illustrating both the dispersion observed on the same mucus and the variability observed between mucus. The dispersion observed on all these series are similar, which implies that thixotropy is weak and comparable here to the natural dispersion arising from heterogeneity of samples and from the measurement difficulties.

As previously reported (Chatelin et al., submitted for publication), a strongly shear-thinning behavior is observed all over the shear rate range studied. Mucus samples were characterized on a wide shear rate range starting at 10^{-3} s^{-1} and no trend to a Newtonian plateau was observed, corroborating the results of Puchelle et al. (1985, 1987) and Yeates et al. (1997). Note that the same observation can be made on the viscosity calculated from the Cox–Merz rule, i.e. from data for which the fluid has only been subjected to small deformations (in this case, the equivalent shear rate range covered is [0.01; 30] s^{-1}). This behavior is commonly associated with cross-linked polymers and is sometimes referred to as "yield stress" behavior (leading to the commonly used term "viscoelastic gel"). While data for shear rates as low as 10^{-3} s^{-1} were conducted, the $\tau(\dot{\gamma})$ curves did not show any trend to a yield stress.

TABLE 5.3 Power-law parameters, relaxation time and power-law parameters from Cox–Merz rule from collected samples. (n, k) data are from Chatelin et al. (submitted for publication)

	CFM1	CFM2	CFM3a	CFM3b	CFM4	CFM5	CFM6
n	0.20	0.39	0.32	0.39	0.30	0.30	0.23
k	5.80	0.45	1.24	1.60	3.53	2.37	4.22
$T_C(s)$	> 20	$\gg 20$	$\gg 20$	–	> 20	> 20	≈ 20
n^*	0.17	0.07	0.25	0.04	0.15	0.23	0.24
k^*	24.6	3.08	0.98	8.45	6.32	9.92	1.82

	CFM7	CFM8	CFM9	CFM10	CFM11	BM12
n	0.21	0.23	0.37	0.36	0.22	0.40
k	3.10	3.53	0.51	2.17	4.75	1.43
$T_C(s)$	≈ 100	≈ 100	≈ 50	≈ 30	> 20	≈ 80
n^*	0.17	0.20	0.28	0.20	0.21	0.25
k^*	6.43	3.28	2.20	1.85	2.92	1.95

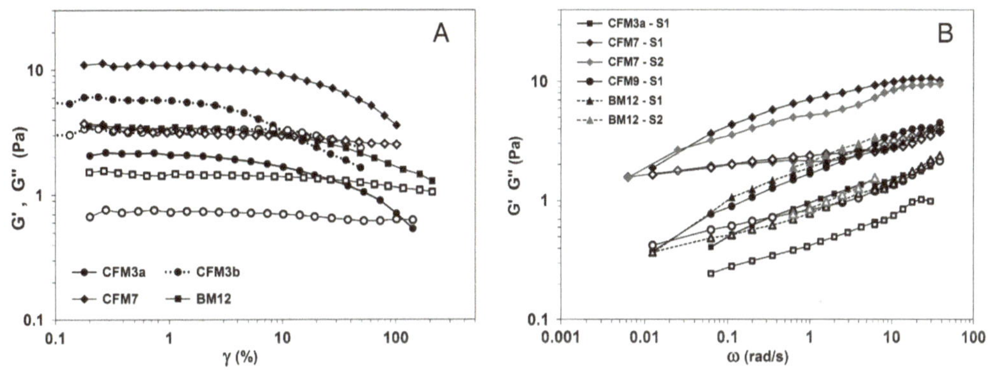

FIGURE 5.11 Dynamic oscillatory shear tests (filled symbols, elastic modulus G'; hollow symbols, viscous modulus G''): (A) strain amplitude sweep, (B) frequency sweep.

A power-law model was fitted on the shear rate range [0.1; 100] s^{-1} which is representative of shear rate involved in the flow. The power-law parameter (index n and consistency k) are reported in Table 5.3 (from Chatelin et al., submitted for publication for (n, k) values). The power-law indexes n are in the range 0.2 to 0.4 and are thus similar to values obtained by Puchelle et al. (1987), Nielsen et al. (2004), Suk et al. (2009) and Tomaiuolo et al. (2014) while Moores et al. (1992) and Dawson et al. (2003) observed lower n-values. Values of consistency k, and then for viscosity, are also similar to the values published in the literature. The large shear rate range explored in this study gives an appropriate description of viscosity for the present study, which means this range is coherent with the shear rates observed in the numerical simulations.

Oscillatory measurements are presented in Fig. 5.11. Strain sweeps are presented for four mucus samples (Fig. 5.11A) and clearly show a large linear domain as it extends until 5 to 10%. Fig. 5.11B presents the storage modulus G' and the loss modulus G'' from a frequency sweep. These curves clearly display the viscoelastic character of mucus: exploring the low frequencies (until 10^{-3} Hz, i.e. more than 1000 s for a point), the $G'(\omega)$ and $G''(\omega)$ curves present the typical viscoelastic frequency-evolution: although slightly varying with frequency, these curves cross each other at a frequency ω_c which allows the determination of a characteristic times $T_c = 1/\omega_c$. This cross frequency was not reached for each mucus sample, but values or estimates are reported in Table 5.3. They range from 20 up to 100 s or more, thus highlighting the precautions to be taken for measurements that must correspond to a steady regime.

As a great number of studies only consider dynamic measurements, it is of interest to examine the validity of steady shear viscosity estimated from these measurements. For polymer solutions and polymer melts, the Cox–Merz rule (Eq. (5.23)) is considered as an acceptable way to evaluate viscosity. Parameters k^* and n^* of the power-laws fitted on $|\eta^*(\omega)|$ data are reported in Table 5.3 and presented in Figs. 5.8A–B for k vs k^* and n vs n^*, respectively.

Results are slightly different from those extracted from the literature: for consistency, which can be regarded as viscosity at 1 s^{-1}, k^* is similar or higher than k. This is coherent with what is known on the validity of the Cox–Merz rule which overestimates the shear viscosity (Picout and Ross-Murphy, 2003). Concerning the power-law index, n^* is similar to n for five cases but significantly lower for others, i.e. indicates a more shear-thinning behavior. Considering both k^* and n^*, this signifies that the difference between dynamic viscosity and shear viscosity becomes increasingly large with decreasing shear rates: considering the viscosity at 1 s^{-1}, a relative differences of more than 100% is observed 7 times out of 13. For a shear rate of 0.01 s^{-1}, this score reaches 9 out of 13.

Nevertheless, a power-law index of 0.2 (resp. 0.4) leads to changes in viscosity in a ratio 6.3 (resp. 4) over an order of magnitude on shear rate. It follows that variations of viscosity in such a mucus flow are most of the time considerably larger than variations emanating from the discrepancy induced by the Cox–Merz rule. In this sense, dynamic measurements for viscosity can be considered when shear viscosity cannot be measured.

5.3.3 Mathematical Modeling of the Rheology and Related Numerical Analysis

As presented in the previous section, mucus exhibits a non-Newtonian behavior with shear-thinning effects. The previous section shows that the power-law is a good representation of the rheological behavior of the mucus. As presented in Table 5.3, the exponent of the shear rate is negative (since n was identified below unity). From a mathematical and a computational point of view, this can be critical because the viscosity can tend toward infinity when the shear rate tends toward zero, in particular when the flow is at rest (at any given time and given point) or when any infinitely small fluid volume undergoes a solid-like displacement (i.e. with no strain).

In order to avoid the singularity that arises at rest, a limitation of the viscosity at low shear rates is introduced. Furthermore, the aqueous polymer mixture cannot be less viscous than its solvent, which means that at high shear rate, the decreasing viscosity has a lower bound at the PCL viscosity (Newtonian with viscosity equal to that of water at 20°C). Among the different constitutive laws, the Carreau model is interesting as it is a power-law with two Newtonian limits:

$$\eta(\dot{\gamma}) = \eta_{min} + (\eta_{max} - \eta_{min}) \left[1 + (\beta\dot{\gamma})^2\right]^{\frac{n-1}{2}}. \tag{5.24}$$

The transition between the plateau with viscosity η_{max}, and the power-law behavior occurs at $\dot{\gamma} = 1/\beta$. So β can be used for smoothing and avoiding the singularity at rest, and therefore can be chosen as large as wanted. The parameter η_{max} is then calculated from the power-law parameters and the β-value. In our simulations, β is set to 4000 s^{-1}, this corresponds to a shear rate transition lower than the lowest shear rates explored in mucus rheometry. This β value was sufficient to ensure that the results do not depend on the regularization process. On the other hand, viscosity is limited to values higher than η_{min} (water viscosity here) for high shear rates.

Furthermore, the previous non-Newtonian results are given for the mucus layer, and the periciliary fluid layer remains a Newtonian fluid whose viscosity is equal to that of water. Hence it is necessary to model viscosity as a function of α: with a transition from the Newtonian periciliary fluid when $\alpha = 0$ to the non-Newtonian Carreau model when $\alpha = 1$.

From a numerical point of view, the only difference is the definition of the viscosity which is now a function of the velocity, through the shear rate $\dot{\gamma}(u) = (2\,D : D)^{1/2}$. The problem to solve is the same as (5.11), except for the second equation which is replaced by an α-dependent rheological constitutive equation (Chatelin et al., submitted for publication):

$$\eta(\dot{\gamma}) = \Phi(u,\alpha) = \eta_{PCF} + (\eta_0(\alpha) - \eta_{PCF})\left[1 + (\beta\dot{\gamma})^2\right]^{\frac{n(\alpha)-1}{2}} \tag{5.25}$$

where

$$\eta_0(\alpha) = \eta_{PCF}\left(\frac{k_{ML}}{\eta_{PCF}}\right)^{\alpha} \beta^{\alpha(1-n_{ML})} \text{ and } n(\alpha) = \alpha \cdot n_{ML} + (1-\alpha). \tag{5.26}$$

As mentioned previously, these equations state that the viscosity of the mucus cannot be smaller than the periciliary fluid viscosity, even when the shear rate tends toward infinity. They are introduced in Eq. (5.2), which reads:

$$-\mathrm{div}(2\eta(\dot{\gamma})D) + \frac{\mathbf{1}_{B(t)}}{\varepsilon}(u - \bar{u}) = f - \nabla p \quad \text{and} \quad \mathrm{div}\, u = 0 \quad \text{in } \Omega \tag{5.27}$$

with the adequate boundary and initial conditions, coupled to the model followed by α, either the transport equation (5.13) or the stratified law displayed on Fig. 5.3.

Hence the computational algorithm presented in Section 5.2.2 is slightly modified. Indeed, the computation of α remains identical whereas a non-linearity appears in the conservation of momentum equation. This non-linearity is treated "explicitly" in the same fixed point as the projection iterations. Hence the only modification is in the step 2b of this algorithm (see Section 5.2.2.3) which becomes a multi-criteria fixed-point algorithm. Denoting $\eta_i = \Phi(u_i, \alpha)$, the new step 2b reads:

$$-\eta_i \Delta u^*_{i+1} + \frac{\mathbf{1}_{B(t)}}{\varepsilon}(u^*_{i+1} - \nabla\zeta_i - \bar{u}(t)) = f + \left(2D(u_i) + \mathrm{div}\, u^*_i\right)\nabla\eta_i. \tag{5.28}$$

The rest of the algorithm presented in Section 5.2.2 remains unchanged. In particular, the SMW approach (see Section 5.2.2.4) is still applicable to compute (5.28) as well as the operator splitting and the Lagrangian computation of the convection of the mucin's ratio.

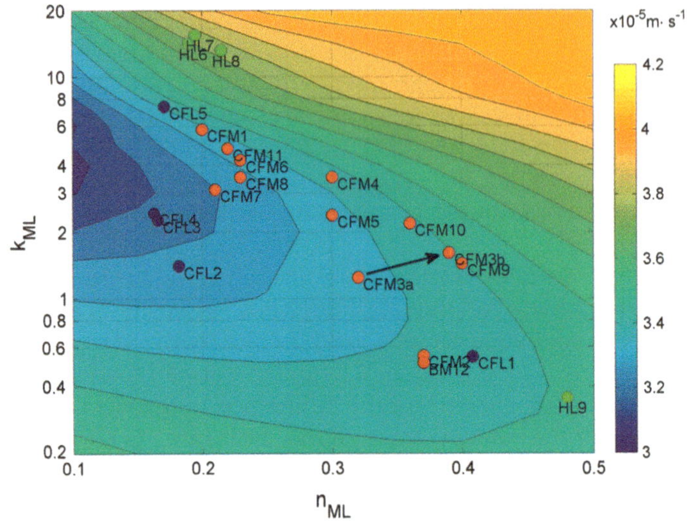

FIGURE 5.12 Aliases from Tables 5.2 and 5.3. Blue and red markers are Cystic Fibrosis samples respectively from the literature and from our study, while green markers denote healthy samples from the literature.

5.3.4 Medical Outcomes

Regarding the Newtonian case, a parametric study was performed to investigate the influence of the rheological parameters on the mucociliary clearance efficiency (Chatelin et al., submitted for publication). The same quantity, defined by Eq. (5.22), is computed for this discussion. In Fig. 5.12, the results are presented for 169 (n_{ML}, k_{ML}) pairs using the α-dependent rheological constitutive equations previously introduced. The mucus samples presented in Section 5.3, in Tables 5.2 and 5.3, are also reported on this phase cartography.

This figure clearly shows that the mucociliary transport does not evolve linearly with respect to the power-law coefficients n and k. This is a direct consequence of the non-linearities of the modeling. In fact, it is not meaningful to discuss viscosity variations for shear-thinning fluids and these power-law coefficients are more significant. It is noteworthy to observe that the CF mucus generally takes lower consistency values k than healthy mucus samples, as well as lower power-law exponent n (except for the HL9).

Fig. 5.12 shows that a drop of 6–15% mobility may be observed for the CF mucus compared to the healthy samples. Even if this decrease is not spectacular, it is sufficient to trigger pathological infections such as *Pseudomonas aeruginosa* or *Burkholderia cepacia* (Eberl and Tümmler, 2004). Moreover, it is important to notice that even if the datum of HL9 (Serisier et al., 2009) is very different from other healthy mucus (Jeanneret-Grosjean et al., 1988; Zayas et al., 1990), the mean mucus velocity remains similar, and we can conclude that the associated mucociliary transports are comparable.

Fig. 5.12 demonstrates that the shear-thinning behavior has a very strong effect: when the power-law exponent decreases to 0.2 (which is the measured value for 5 samples), the mucociliary transport is highly deteriorated. This is counterintuitive, but it is the consequence of a higher velocity during the recovery stroke in the mucus layer, since the "apparent viscosity" decreases.

From a clinical point of view, two types of CF mucus can be distinguished: stabilized and decompensated (with acute secondary infections). The mucus of same patient was sampled in both states: two days after hospitalization (CFM3a, decompensated) and after 17 days of treatment, combining antibiotics and steroids (CFM3b, stabilized). The computations presented in Fig. 5.12 clearly show an increase in the mucociliary transport for this patient, thus we conclude that the treatment was beneficial.

Finally, similarly to the Newtonian case, several computations were performed for different heights of periciliary fluid and still 169 values of (n_{ML}, k_{ML}) pairs – see Chatelin et al. (submitted for publication) for details. It was demonstrated that this periciliary fluid height has a very important effect on the mucociliary transport with up to a 21% decrease in mucus mean velocity. This parameter represents the effect of a variable CFTR deficiency (Thornton and Sheehan, 2004; Ehre et al., 2014) and it also shows the lack of mucociliary clearance in cystic fibrosis, allowing pathological proliferations.

This effect had been summarized in Fig. 5.13, where three groups are displayed:

- A group $k = 1.5, n = 0.35$, corresponding to a center of gravity of samples CFM3–5, 9, 10, which is moderately impacted by PCL rarefaction,
- A group $k = 4$, $n = 0.25$, corresponding to a center of gravity of samples CFM1, 6–8, 11, which is strongly impacted by PCL rarefaction, decreased up to -13% compared to healthy cases, which means a -6% due to lack of mucociliary clearance efficiency, made worse by -7% due to PCL height decrease.
- A location $k = 8$, $n = 0.15$, corresponding to case CFL5 reported in Dawson et al. (2003), corresponding to a situation extremely impacted by PCL rarefaction, with a decrease of -23% compared the healthy configuration.

5.4 CONCLUDING REMARKS

The lung is resistant to damages following exposure to pathogens and more or less toxic chemical pollutants in inhaled air. Airway mucus represents the first line of defense. Inhaled particles (allergens, dust, micro-organisms, and cellular debris) trapped in the mucus gel are removed from the respiratory tract by the mucociliary clearance. Mucus cleans airways, as it flows from either the tracheobronchial tree or upper airways toward the pharynx, where it is swallowed (or expectorated).

Ciliary motions enable mucus propulsion in the human respiratory tract. The beating cycle of respiratory cilia is composed of 3 phases: a rest phase, a recovery stroke, and an effective stroke. During the recovery stroke, cilia (height 1–7 μm according to the location in the respiratory tract and investigators) sweep backward remaining close to the epithelium and immersed in the periciliary fluid without contact with mucus. Cilia of a patch of ciliated cells form a coordinated beating cluster (beating frequency 10–20 Hz). The cilium tips impact mucus during a short phase of its beating cycle. Cilia are more rigid in the forward power stroke and more flexible in recovery stroke in preparation for the next effective stroke.

The present mathematical model of airway surface liquid dynamics relies on a heterogeneous fluid that moves due to beating deformable cilia. The airway surface liquid nature evolves continuously in both composition and rheology from the periciliary fluid to the

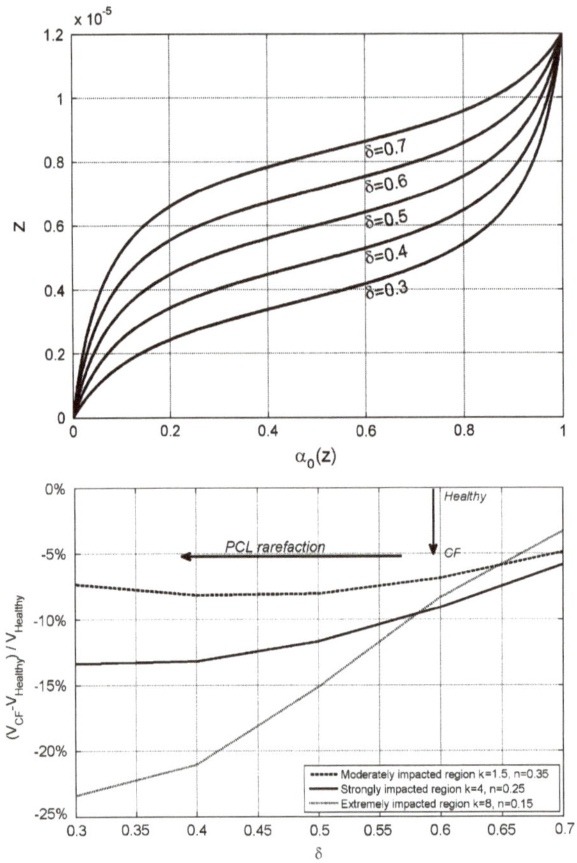

FIGURE 5.13 PCL rarefaction, that is to say, δ decreasing from 0.7 and 0.3 is displayed on the top panel with the corresponding α_0 profile. The impact of the PCL rarefaction on mucus mean velocity in percent of healthy mucus velocity is displayed on the bottom panel. The healthy velocity is the mean velocity of HL6-8 for $\delta = 0.6$. The *Moderately impacted region* corresponds to the samples CFM3, 4, 5, 9, 10, the *Strongly impacted region* corresponds to the samples CFM1, 6, 7, 8, 11, and the *Extremely impacted region* corresponds to the sample CFL5.

mucus. Within the near-wall fluid surrounding the deformable cilia, the mucin polymers released by the epithelium further hydrate. Once they reach a certain distance normal to the wetted surface of the respiratory epithelium, which is slightly lower than the cilium height, mucin polymers accumulate, engendering a mucin-concentrated phase. In this region, mucin polymers interweave and eventually cross-link. This evolving nature of the airway surface liquid affects its rheology. For example, for a airway surface liquid thickness of 12 μm and cilium height of 8 μm, the thickness of the periciliary fluid and mucus are equal to 7 and 5 μm, respectively.

The rheological behavior is associated with two parameters of interest, the fluid consistency index k and the power law index n that describes the difference between the two asymptotic behaviors (μ_0 and μ_∞) and the slope of the shear rate–viscosity curve between these two asymptotic behaviors.

The numerical model is combined with an experimental study of the rheology of sputum from cystic fibrosis patients. The proposed robust numerical method handle variations in shear rate and viscosity of more than 5 orders of magnitude. It leads to a cartography that allows discussing major issues on defective mucociliary clearance observed in cystic fibrosis during periodic check-up of the disease in a steady clinical state (i.e., in the absence of infection exacerbation) for prognosis.

Acknowledgements

This work was partially supported by the ANR Grant BioFiReaDy supervised by Philippe Poncet, under the contract number ANR-2010-JCJC-0113-01. This work was granted access to the HPC resources of P2CHPD (Pôle de Calcul Hautes Performances Dédiés, from Fédération Lyonnaise de Simulation et de Modélisation Numériques), of cluster Hyperion (CALMIP Toulouse, France) under the allocation P0314, and cluster Avakas (MCIA Bordeaux, France). We also acknowledge the use of the rheometer from the cooperative structure of joint research FERMaT (FR3089).

References

Adams, J.C., 1989. mudpack: multigrid portable Fortran software for the efficient solution of linear elliptic partial differential equations. Appl. Math. Comput. 34 (2, Part 2), 113–146.

Adams, R.A., Fournier, J.J.F., 2003. Sobolev Spaces. Academic Press.

Aitken, M.L., Bellon, G., De Boeck, K., Flume, P.A., Fox, H.G., Geller, D.E., Haarman, E.G., Hebestreit, H.U., Lapey, A., Schou, I.M., Zuckerman, J.B., Charlton, B., 2012. Long-term inhaled dry powder mannitol in cystic fibrosis. Am. J. Respir. Crit. Care Med. 185 (6), 645–652.

Angot, P., Bruneau, C.-H., Fabrie, P., 1999. A penalization method to take into account obstacles in incompressible viscous flows. Numer. Math. 81 (4), 497–520.

App, E.M., Kieselmann, R., Reinhardt, D., Lindemann, H., Dasgupta, B., King, M., Brand, P., 1998. Sputum rheology changes in cystic fibrosis lung disease following two different types of physiotherapy: flutter vs autogenic drainage. Chest 114 (1), 171–177.

Barnes, H.A., Hutton, J.F., Walters, K., 1989. An Introduction to Rheology. Elsevier.

Beale, J.T., Majda, A., 1981. Rates of convergence for viscous splitting of the Navier–Stokes equations. Math. Comput. 37 (156), 243–259.

Besseris, G.J., Yeates, D.B., 2007. Rotating magnetic particle microrheometry in biopolymer fluid dynamics: mucus microrheology. J. Chem. Phys. 127 (10), 105106.

Bilton, D., Robinson, P., Cooper, P., Gallagher, C.G., Kolbe, J., Fox, H., Jaques, A., Charlton, B., 2011. Inhaled dry powder mannitol in cystic fibrosis: an efficacy and safety study. Eur. Respir. J. 38 (5), 1071–1080.

Bird, R., Armstrong, R., Hassager, O., 1977. Dynamics of Polymeric Liquids. John Wiley Sons.

Blake, J., 1972. A model for the micro-structure in ciliated organisms. J. Fluid Mech. 55 (01), 1–23.

Boyer, F., Fabrie, P., 2005. Eléments d'analyse pour l'étude de quelques modèles d'écoulements de fluides visqueux incompressibles. Springer.

Broughton-Head, V.J., Shur, J., Carroll, M.P., Smith, J.R., Shute, J.K., 2007. Unfractionated heparin reduces the elasticity of sputum from patients with cystic fibrosis. Am. J. Physiol., Lung Cell. Mol. Physiol. 293 (5), L1240–L1249.

Button, B., Cai, L.-H., Ehre, C., Kesimer, M., Hill, D.B., Sheehan, J.K., Boucher, R.C., Rubinstein, M., 2012. Periciliary brush promotes the lung health by separating the mucus layer from airway epithelia. Science (New York, N.Y.) 337 (6097), 937–941.

Carbou, G., Fabrie, P., 2003. Boundary layer for a penalization method for viscous incompressible flow. Differ. Equ. 8 (12), 1453–1480.

Chatelin, R., 2013. Méthodes numériques pour l'écoulement de Stokes 3D: fluides à viscosité variable en géométrie complexe mobile application aux fluides biologiques. PhD thesis. Université Toulouse 3 Paul Sabatier.

Chatelin, R., Poncet, P., 2013. A hybrid grid-particle method for moving bodies in 3D Stokes flow with variable viscosity. SIAM J. Sci. Comput. 35 (4), B925–B949.

Chatelin, R., Poncet, P., 2014. Hybrid grid–particle methods and penalization: a Sherman–Morrison–Woodbury approach to compute 3D viscous flows using FFT. J. Comput. Phys. 269, 314–328.

Chatelin, R., Poncet, P., 2016. A parametric study of mucociliary transport by numerical simulations of 3D nonhomogeneous mucus. J. Biomech. 49 (9), 1772–1780.
Chatelin, R., Poncet, P., Didier, A., Murris-Espin, M., Anne-Archard, D., Thiriet, M., 2015. Mucus and ciliated cells of human lung: splitting strategies for particle methods and 3d Stokes flows. Proc. IUTAM 18, 114–122.
Chatelin, R., Sanchez, D., Poncet, P., 2016. Analysis of the penalized 3D variable viscosity Stokes equations coupled to diffusion and transport. Modél. Math. Anal. Numér. 50 (2), 565–591.
Chatelin, R., Anne-Archard, D., Murris-Espin, M., Thiriet, M., Poncet, P., submitted for publication. Numerical and experimental investigation of mucociliary clearance breakdown in cystic fibrosis..
Clancy, J.P., Jain, M., 2012. Personalized medicine in cystic fibrosis. Am. J. Respir. Crit. Care Med. 186 (7), 593–597.
Cone, R., 2005. Mucus. In: Mucosal Immunology. Elsevier Academic Press, pp. 49–72. Ch. 4.
Cortez, R., Fauci, L., Medovikov, A., 2005. The method of regularized Stokeslets in three dimensions: analysis, validation, and application to helical swimming. Phys. Fluids 17 (3), 031504.
Cottet, G.-H., Koumoutsakos, P.D., 2000. Vortex Methods: Theory and Practice. Cambridge University Press.
Craster, R.V., Matar, O.K., 2000. Surfactant transport on mucus films. J. Fluid Mech. 425, 235–258.
Dautray, R., Lions, J.L., 2000. Mathematical Analysis and Numerical Methods for Science and Technology: Vols. 1–6. Springer.
Daviskas, E., Anderson, S.D., Young, I.H., 2010. Effect of mannitol and repetitive coughing on the sputum properties in bronchiectasis. Respir. Med. 104 (3), 371–377.
Dawson, M., Wirtz, D., Hanes, J., 2003. Enhanced viscoelasticity of human cystic fibrotic sputum correlates with increasing microheterogeneity in particle transport. J. Biol. Chem. 278 (50), 50393–50401.
Deneuville, E., Perrot-Minot, C., Pennaforte, F., Roussey, M., Zahm, J.-M., Clavel, C., Puchelle, E., Bentzmann, S.D., 1997. Revisited physicochemical and transport properties of respiratory mucus in genotyped cystic fibrosis patients. Am. J. Respir. Crit. Care Med. 156, 166–172.
Ding, Y., Kanso, E., 2015. Selective particle capture by asynchronously beating cilia. Phys. Fluids 27 (12), 121902.
Ding, Y., Nawroth, J.C., McFall-Ngai, M.J., Kanso, E., 2014. Mixing and transport by ciliary carpets: a numerical study. J. Fluid Mech. 743, 124–140.
Doring, G., Flume, P., Heijerman, H., Elborn, J.S., 2012. Treatment of lung infection in patients with cystic fibrosis: current and future strategies. J. Cyst. Fibros. 11 (6), 461–479.
Eberl, L., Tümmler, B., 2004. Pseudomonas aeruginosa and burkholderia cepacia in cystic fibrosis: genome evolution, interactions and adaptation. Int. J. Med. Microbiol. 294 (2–3), 123–131.
Edwards, P.A., Yeates, D.B., 1992. Magnetic rheometry of bronchial mucus. In: Viscoelasticity of Biomaterials. In: ACS Symp. Ser., vol. 489. American Chemical Society, pp. 249–267.
Ehre, C., Ridley, C., Thornton, D.J., 2014. Cystic fibrosis: an inherited disease affecting mucin-producing organs. Int. J. Biochem. Cell Biol. 52, 136–145.
Enault, S., Lombardi, D., Poncet, P., Thiriet, M., 2010. Mucus dynamics subject to air and wall motion. ESAIM Proc. 30, 124–141.
Fahy, J.V., Dickey, B.F., 2010. Airway mucus function and dysfunction. N. Engl. J. Med. 363 (23), 2233–2247. PMID: 21121836.
Farrell, P.M., 2008. The prevalence of cystic fibrosis in the European Union. J. Cyst. Fibros. 7, 450–453.
Fauci, L.J., Dillon, R., 2006. Biofluidmechanics of reproduction. Annu. Rev. Fluid Mech. 38 (1), 371–394.
Fulford, G., Blake, J., 1986. Mucociliary transport in the lung. J. Theor. Biol. 121 (4), 381–402.
Gastaldi, A.C., Jardim, J.R., King, M., 2000. The influence of temperature and length of time of storage of frog mucus samples. Biorheology 37, 203–211.
Gheber, L., Priel, Z., 1997. Extraction of cilium beat parameters by the combined application of photoelectric measurements and computer simulation. Biophys. J. 72 (1), 449–462.
Guermond, J., Minev, P., Shen, J., 2006. An overview of projection methods for incompressible flows. Comput. Methods Appl. Mech. Eng. 195, 6011–6045.
Guo, H., Kanso, E., 2016a. Evaluating efficiency and robustness in cilia design. Phys. Rev. E 93 (3). arXiv:1512.06953.
Guo, H., Kanso, E., 2016b. Mucociliary transport in healthy and diseased environments. arXiv:1605.00688.
Hager, W.W., 1989. Updating the inverse of a matrix. SIAM Rev. 31 (2), 221–239.
Haq, I.J., Gray, M.A., Garnett, J.P., Ward, C., Brodlie, M., 2016. Airway surface liquid homeostasis in cystic fibrosis: pathophysiology and therapeutic targets. Thorax 71 (3), 284–287.
Hill, D.B., Swaminathan, V., Estes, A., Cribb, J., O'Brien, E.T., Davis, C.W., Superfine, R., 2010. Force generation and dynamics of individual cilia under external loading. Biophys. J. 98 (1), 57–66.

Hoffman, L.R., Ramsey, B.W., 2013. Cystic fibrosis therapeutics: the road ahead. Chest 143 (1), 207–213.

Hussong, J., Breugem, W.-P., Westerweel, J., 2011. A continuum model for flow induced by metachronal coordination between beating cilia. J. Fluid Mech. 684, 137–162.

Innes, A.L., Carrington, S.D., Thornton, D.J., Kirkham, S., Rousseau, K., Dougherty, R.H., Raymond, W.W., Caughey, G.H., Muller, S.J., Fahy, J.V., 2009. Ex vivo sputum analysis reveals impairment of protease-dependent mucus degradation by plasma proteins in acute asthma. Am. J. Respir. Crit. Care Med. 180 (3), 203–210.

Jayathilake, P., Le, D., Tan, Z., Lee, H., Khoo, B., 2015. A numerical study of muco-ciliary transport under the condition of diseased cilia. Comput. Methods Biomech. Biomed. Eng. 18 (9), 944–951.

Jayathilake, P., Tan, Z., Le, D., Lee, H., Khoo, B., 2012. Three-dimensional numerical simulations of human pulmonary cilia in the periciliary liquid layer by the immersed boundary method. Comput. Fluids 67, 130–137.

Jeanneret-Grosjean, A., King, M., Michoud, M.C., Liote, H., Amyot, R., 1988. Sampling technique and rheology of human tracheobronchial mucus. Am. Rev. Respir. Dis. 137 (3), 707–710.

Jeffries, J.L., Jia, J., Choi, W., Choe, S., Miao, J., Xu, Y., Powell, R., Lin, J., Kuang, Z., Gaskins, H.R., Lau, G.W., 2016. Pseudomonas aeruginosa pyocyanin modulates mucin glycosylation with sialyl-Lewis(x) to increase binding to airway epithelial cells. Mucosal Immunol. 9 (4), 1039–1050.

King, M., 1980. Rheological requirements for optimal clearance of secretions: ciliary transport versus cough. Eur. J. Respir. Dis. 110, 39–45.

King, M., 1981. Is cystic fibrosis mucus abnormal? Pediatr. Res. 15, 120–122.

King, M., Macklem, P.T., 1977. Rheological properties of microliter quantities of normal mucus. J. Appl. Physiol. 42 (6), 797–802.

Kurbatova, P., Bessonov, N., Volpert, V., Tiddens, H., Cornu, C., Nony, P., Caudri, D., 2015. Model of mucociliary clearance in cystic fibrosis lungs. J. Theor. Biol. 372, 81–88.

Lai, S.K., Wang, Y.-Y., Wirtz, D., Hanes, J., 2009. Micro- and macrorheology of mucus. Adv. Drug Deliv. Rev. 61 (2), 86–100.

Lee, W., Jayathilake, P., Tan, Z., Le, D., Lee, H., Khoo, B., 2011. Muco-ciliary transport: effect of mucus viscosity, cilia beat frequency and cilia density. Comput. Fluids 49 (1), 214–221.

Li, Z., Favier, J., D'Ortona, U., Poncet, S., 2016. An immersed boundary-lattice Boltzmann method for single- and multi-component fluid flows. J. Comput. Phys. 304, 424–440.

Liron, N., Rozenson, M., 1983. Mucociliary transport. J. Submicrosc. Cytol. 15 (1), 317–321.

Lopez-Vidriero, M.T., Reid, L., 1978. Chemical markers of mucous and serum glycoproteins and their relation to viscosity in mucoid and purulent sputum from various hypersecretory diseases. Am. Rev. Respir. Dis. 117, 465–477.

Lukens, S., Yang, X., Fauci, L., 2010. Using Lagrangian coherent structures to analyze fluid mixing by cilia. Chaos 20 (1), 017511.

Mall, M.A., Hartl, D., 2014. CFTR: cystic fibrosis and beyond. Eur. Respir. J. 44 (4), 1042–1054.

McShane, P.J., Naureckas, E.T., Tino, G., Strek, M.E., 2013. Non-cystic fibrosis bronchiectasis. Am. J. Respir. Crit. Care Med. 188 (6), 647–656.

Mitran, S.M., 2007. Metachronal wave formation in a model of pulmonary cilia. Comput. Struct. 85, 763–774.

Monaghan, J., 1985. Extrapolating b splines for interpolation. J. Comput. Phys. 60 (2), 253–262.

Montenegro-Johnson, T.D., Smith, D.J., Loghin, D., 2013. Physics of rheologically enhanced propulsion: different strokes in generalized Stokes. Phys. Fluids 25 (8), 081903.

Moores, C., Hudson, N.E., Davies, A., 1992. The effect of high-frequency ventilation on non-Newtonian properties of bronchial mucus. Respir. Med. 86 (2), 125–130.

Nielsen, H., Hvidt, S., Sheils, C.A., Janmey, P.A., 2004. Elastic contributions dominate the viscoelastic properties of sputum from cystic fibrosis patients. Biophys. Chem. 112 (2–3), 193–200.

Peskin, C.S., 2003. The immersed boundary method. Acta Numer. 11.

Picout, D.R., Ross-Murphy, S.B., 2003. Rheology of biopolymer solutions and gels. Sci. World J. 3, 105–121.

Puchelle, E., Zahm, J.M., Duvivier, C., Didelon, J., Jacquot, J., Quemada, D., 1985. Elasto-thixotropic properties of bronchial mucus and polymer analogs. I – experimental results. Biorheology 22, 415–423.

Puchelle, E., Zahm, J.M., Quemada, D., 1987. Rheological properties controlling mucociliary frequency and respiratory mucus transport. Biorheology 24 (6), 557–563. PMID: 3502756.

Quon, B.S., Rowe, S.M., 2016. New and emerging targeted therapies for cystic fibrosis. BMJ, Br. Med. J. 352, i859.

Redding, G.J., 2008. Physical and transport properties of sputum from children with idiopathic bronchiectasis. Chest 134 (6), 1129.

Rubin, B.K., Ramirez, O., Zayas, J.G., Finegan, B., King, M., 1990. Collection and analysis of respiratory mucus from subjects without lung disease. Am. Rev. Respir. Dis. 141 (4_pt_1), 1040–1043.
Sanders, N.N., Smedt, S.C.D., Rompaey, E.V., Simoens, P., Baets, F.D., Demeester, J., 2000. Cystic fibrosis sputum. A barrier to the transport of nanospheres. Am. J. Respir. Crit. Care Med. 162, 1905–1911.
Sanderson, M.J., Sleigh, M.A., 1981. Ciliary activity of cultured rabbit tracheal epithelium: beat pattern and metachrony. J. Cell Sci. 47 (1), 331–347.
Schuster, B.S., Suk, J.S., Woodworth, G.F., Hanes, J., 2013. Nanoparticle diffusion in respiratory mucus from humans without lung disease. Biomaterials 34 (13), 3439–3446.
Sedaghat, M.H., Shahmardan, M.M., Norouzi, M., Nazari, M., Jayathilake, P.G., 2016. On the effect of mucus rheology on the mucociliary transport. Math. Biosci. 272, 44–53.
Serisier, D.J., Carroll, M.P., Shute, J.K., Young, S.A., 2009. Macrorheology of cystic fibrosis, chronic obstructive pulmonary disease & normal sputum. Respir. Res. 10 (1), 63.
Shah, P.L., Scott, S.F., Knight, R.A., Marriott, C., Ranasinha, C., Hodson, M.E., 1996. In vivo effects of recombinant human DNase I on sputum in patients with cystic fibrosis. Thorax 51 (2), 119–125.
Silberberg, A., 1983. Biorheological matching: mucociliary interaction and epithelial clearance. Biorheology 20, 215–222.
Simons, J., Fauci, L., Cortez, R., 2015. A fully three-dimensional model of the interaction of driven elastic filaments in a Stokes flow with applications to sperm motility. J. Biomech. 48 (9), 1639–1651.
Smith, D., Gaffney, E., Blake, J., 2007a. Discrete cilia modelling with singularity distributions: application to the embryonic node and the airway surface liquid. Bull. Math. Biol. 69 (5), 1477–1510.
Smith, D., Gaffney, E., Blake, J., 2007b. A viscoelastic traction layer model of muco-ciliary transport. Bull. Math. Biol. 69 (1), 289–327.
Smith, D., Gaffney, E., Blake, J., 2008. Modelling mucociliary clearance. Respir. Physiol. Neurobiol. 163, 178–188.
Suk, J.S., Lai, S.K., Wang, Y.-Y., Ensign, L.M., Zeitlin, P.L., Boyle, M.P., Hanes, J., 2009. The penetration of fresh undiluted sputum expectorated by cystic fibrosis patients by non-adhesive polymer nanoparticles. Biomaterials 30 (13), 2591–2597.
Swarztrauber, P., Sweet, R., 1975. Efficient FORTRAN subprograms for the solution of elliptic partial differential equations (abstract). SIGNUM Newsl. 10 (4).
Tabatabaei, S., Jahromi, H., Webster, M., Williams, P., Holder, A., Lewis, K., Davies, G., Griffin, L., Ebden, P., Askill, C., 2015. A CaBER computational–experimental rheological study on human sputum. J. Non-Newton. Fluid Mech. 222, 272–287.
Thiriet, M., 2012. Tissue Functioning and Remodeling in the Circulatory and Ventilatory Systems. Biomathematical and Biomechanical Modeling of the Circulatory and Ventilatory Systems, vol. 5. Springer, Dordrecht.
Thomas, R.H., Walters, K., 1966. The unsteady motion of a sphere in an elastico-viscous liquid. Rheol. Acta 5 (1), 23–27.
Thornton, D.J., Sheehan, J.K., 2004. From mucins to mucus toward a more coherent understanding of this essential barrier. Proc. Am. Thorac. Soc. 1 (1), 54–61.
Tomaiuolo, G., Rusciano, G., Caserta, S., Carciati, A., Carnovale, V., Abete, P., Sasso, A., Guido, S., 2014. A new method to improve the clinical evaluation of cystic fibrosis patients by mucus viscoelastic properties. PLoS ONE 9 (1), e82297.
Twigg, M.S., Brockbank, S., Lowry, P., FitzGerald, S.P., Taggart, C., Weldon, S., 2015. The role of serine proteases and antiproteases in the cystic fibrosis lung the role of serine proteases and antiproteases in the cystic fibrosis lung. Mediat. Inflamm. 2015, e293053.
Yankaskas, J.R., Marshall, B.C., Sufian, B., Simon, R.H., Rodman, D., 2004. Cystic fibrosis adult care: consensus conference report. Chest 125 (1 suppl.), S1–S39.
Yeates, D.B., Besseris, G.J., Wong, L.B., 1997. Physicochemical properties of mucus and its propulsion. In: The Lung: Scientific Foundations. Lippincot-Raven Publishers, Philadelphia, pp. 487–503. Ch. 33.
Zayas, J.G., Man, G.C.W., King, M., 1990. Tracheal mucus rheology in patients undergoing diagnostic bronchoscopy: interrelations with smoking and cancer. Am. Rev. Respir. Dis. 141 (5_pt_1), 1107–1113.
Zayas, J.G., Rubin, B.K., York, E.L., Lien, D.C., King, M., 1999. Bronchial mucus properties in lung cancer: relationship with site of lesion. Can. Respir. J. 6 (3), 246–252.
Zemanick, E.T., Harris, J.K., Conway, S., Konstan, M.W., Marshall, B., Quittner, A.L., Retsch-Bogart, G., Saiman, L., Accurso, F.J., 2010. Measuring and improving respiratory outcomes in cystic fibrosis lung disease: opportunities and challenges to therapy. J. Cyst. Fibros. 9 (1), 1–16.
Zielenski, J., 2000. Genotype and phenotype in cystic fibrosis. Respiration 67 (2), 117–133.
Ziemann, F., Radler, J., Sackmann, E., 1994. Local measurements of viscoelastic moduli of entangled actin networks using an oscillating magnetic bead micro-rheometer. Biophys. J. 66 (6), 2210–2216.

CHAPTER

6

Intracellular Microfluid Transportation in Fast Growing Pollen Tubes

S. Liu*,†, H. Liu†,‡, M. Lin†,‡,#, F. Xu†,‡, T.J. Lu*,†

*State Key Laboratory for Strength and Vibration of Mechanical Structures, School of Aerospace, Xi'an Jiaotong University, Xi'an, PR China †Bioinspired Engineering and Biomechanics Center (BEBC), Xi'an Jiaotong University, Xi'an, PR China ‡MOE Key Laboratory of Biomedical Information Engineering, School of Life Science and Technology, Xi'an Jiaotong University, Xi'an, PR China

6.1 INTRODUCTION

Cytoplasmic streaming, known as cyclosis, is a movement of cytoplasm in various organisms including bacteria, higher plants, and animals (Williamson and Ashley, 1982; Theurkauf, 1994). Cytoplasmic streaming assists in the delivery of nutrients, metabolites, organelles, and other materials to all parts of the cell (Chebli et al., 2013; Van de Meent et al., 2008; Verchot-Lubicz and Goldstein, 2010). It has been demonstrated that cytoplasmic streaming plays key roles in various biological processes, such as fast tip-growth of root hairs and pollen tubes (Justus et al., 2004; Shimmen et al., 1995), movement of chloroplasts in plants (Kadota et al., 2009), asymmetric position of meiotic spindle in mammalian embryos (Munro et al., 2004), and developmental potential of zygote (Ganguly et al., 2012; Hecht et al., 2009). Various flow patterns have been found to exist in plant cells. For example, "fountain streaming" (characterized by a flow near the central axis of the cell that is in the opposite direction to that at the periphery) has been identified in the *lily* pollen tube (Van de Meent et al., 2008). Other flow pattern such as spiral "rotational" streaming in *Chara* (Van de Meent et al., 2008; Verchot-Lubicz and Goldstein, 2010) has also been observed. Among these patterns, substantial attentions have been paid to "fountain streaming" due to its critical effect on longitudinal transportation along plant cells. "Fountain streaming" could be especially prominent in cells with high aspect ratio such as pollen tubes, root hairs, and fungus hypha (Justus et al., 2004; Shimmen et al., 1995; McKerracher and Heath, 1987). Despite its important roles in cell activities, the implications of "fountain streaming" for the fast tip-growth of pollen

Chapter coordinator.

tubes remain elusive (Lazzaro et al., 2013). In this chapter, taking inverse fountain flow as an example, the roles of fountain streaming for the fast tip-growth of pollen tubes are investigated.

"Fountain streaming" is commonly found in pollen tubes, which is the cellular protuberance formed by the male gametophyte in angiosperm species delivering sperm to the ovary of the receptive flower to perform double fertilization. Pollen tubes are associated with high aspect ratios with the length of the cell (~2 mm) measuring up to hundred times as compared with its diameter (~20 μm). Similar to cytoplasmic streaming in other plant cells, "fountain streaming" in pollen tubes, may contribute to three important activities within cells: the positioning of organelles, the transport of nutrients and metabolites, and the generation of convection for enhanced chemical and enzymatic reactions (Chebli et al., 2013). Of significant importance is cyclosis in pollen tubes. Cyclosis provides a driving force, thus enabling the transport of cell wall materials over huge distances in a highly efficient and targeted manner. The growth rate at up to 0.6 mm/h of pollen tube (the fastest growing plant cell types known) is rather impressive, given that growth speed is a direct competitive selection factor for fertilization success (Williams, 2008). It is known that the turgor pressure and the distribution of cell wall materials (e.g., ions, vesicles) in the pollen tube both contribute to its growth. This is because the turgor generates the stress which leads to the expansion of cell wall (Kroeger et al., 2011). The distribution of cell wall materials controls the deposition of new material into the synthetic wall for pollen tube elongation, navigation and self-incompatibility response (Holdaway-Clarke and Hepler, 2003). For example, the well-known Lockhart equation (Kroeger et al., 2011; Lockhart, 1965) states that the rate of the cell wall expansion is proportional to the difference between the turgor pressure and a critical threshold value. For example, the gradient of Ca^{2+} concentration in pollen tube drops off sharply from around 1–5 μM at the apex of a growing tube to basal values of 150–300 nM within 20 μm and the disruption of this gradient invariably causes cessation of pollen tube growth (Holdaway-Clarke and Hepler, 2003). Although these studies show that both the turgor pressure and concentration gradient contribute to pollen tube growth, it remains a mystery how the hydrodynamic property and intracellular transportation processes affect the fast tip-growth of pollen tubes.

Recently, substantial theoretical analysis has been devoted to explore the mechanism underlying the functions of cytoplasmic streaming in plant and animal cells. For instance, Goldstein et al. (Van de Meent et al., 2008; Goldstein et al., 2008) find strongly enhanced lateral transport and longitudinal homogenization by solving the dynamics of fluid flow and diffusion appropriate to archetypal "rotational streaming" (in *Chara* and *Nitella*). In migrating cells, cytoplasmic streaming enables transport of actin monomers in the direction of cell movement (Keren et al., 2009; Prass et al., 2006). For instance, at the one-cell stage of the Caenorhabditis elegans embryo, cortical flow transports the proteins essential for cell polarity (e.g., PAR-3, PAR-6, and PKC-3) toward the anterior. Thus cortical flow contributes to the establishment of the anterior–posterior polarity of the embryo (Munro et al., 2004; Schonegg et al., 2007; Niwayama et al., 2011). Regarding early drosophila embryos, cytoplasmic streaming has been shown to improve the scale of bicoid morphogen gradient and has been shown to affect the position-dependent differentiation (Hecht et al., 2009). In spite of all these efforts, further work is needed to explore how fountain streaming affects the generation of turgor pressure gradients in fast tip-growing pollen tubes. With this in mind, the question must be answered,

FIGURE 6.1 Collection, germination and observations of pollen grains (Liu et al., 2016). (A) A blooming Asiatic lily *Lilium Casablanca* from Yunnan, China, with stigma (1), style (2), anther (3), filament (4), petal (5), ovule (6) and pedicel (7); (B) Scanning electron microscopy image of a pollen grain; (C) Microscope image of a fast tip-growing pollen tube, I ($t = 0$), II ($t = 30$ min) and III ($t = 60$ min). The scale bars are (A) 1 cm, (B) 10 μm, (C) 20 μm.

"How does the streaming regulate the distribution of wall materials that contribute to the tip-growth of pollen tube and plant sexual reproduction?"

Here, motivated by the phenomenology of Asiatic lily *Lilium Casablanca* (Fig. 6.1A), this chapter proposes a mathematical model to provide insights into the previously asked question by examining the most basic aspects of fountain streaming flows. The aim of this chapter is to illustrate how fountain streaming contributes to the fast tip-growth of pollen tubes in terms of driving force and materials supply. To this end, mathematical models of hydrodynamic and advection–diffusion dynamics for fountain streaming flow have been previously developed. The developed models demonstrate the enhancement of turgor pressure and concentration of cell wall synthesis materials at the apical region, both of which are correlated directly with the fountain geometry of flow. The roles of fountain streaming in fast tip-growing pollen tubes are examined by disordering the hydrodynamics through applying the osmotic stress *in vitro*.

6.2 MODELING FLUID FLOW OF FOUNTAIN STREAMING IN POLLEN TUBES

6.2.1 Pollen Collection and Germination

In the experiment, the mature anther of Asiatic lily *Lilium Casablanca* (Fig. 6.1A) were picked and placed in the Petri dish. The scattered pollen grains on the bottom of Petri dish were collected and stored in a centrifugal tube at −70°C. Silicone balls were applied in the Petri

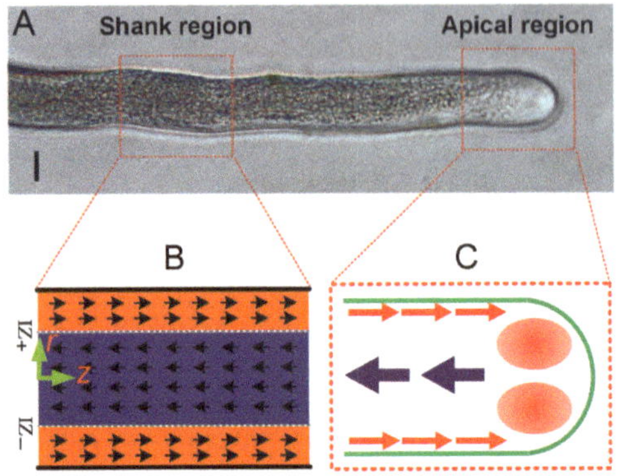

FIGURE 6.2 Observations and simplified model of fountain streaming in pollen tubes (Liu et al., 2016). Fountain streaming in a pollen tube (A); in the shank region with indifferent zones $IZ_{\pm}$ in cylindrical coordinates (B), and an idealized streaming pattern in the apical region of pollen tubes (C).

dish and centrifugal tube for drying. To observe the cytoplasmic streaming in pollen tube, the pollen grains were first thawed in a moist chamber for ~10 hours at 4°C and 2 hours at room temperature, and then cultured in the medium (1.27 mM $CaCl_2$, 0.162 mM H_3BO_3, 0.99 mM KNO_3, 290 mM sucrose, pH 5.2) for germination for 3 hours (Chang et al., 2009).

6.2.2 Living Image and Observation

Several different microscopes and camera are used for the characterization of pollen grains, pollen tubes and their cytoplasmic streaming. In the scanning electron microscope, the pollen grain appears as a cage (Fig. 6.1B). After being cultured for a period of time, the pollen grain begins germinating, and a tube protrudes from the pollen grain and elongation is accompanied with fast tip-growth (Fig. 6.1C). The obvious fountain streaming in the fast tip-growing pollen tube is observed (Fig. 6.2A) on the research level inverted microscope, the Olympus IX81. Fig. 6.1A has been obtained with a camera Nikon D90, and Fig. 6.1B has been obtained with an ultra-high resolution field emission scanning electron microscope (SEM), ZEISS MERLIN Compact.

6.2.3 Hydrodynamics of Fountain Streaming

In order to simplify the theoretical description of the streaming in the pollen tube, biological complexities (e.g., geometry defects, tonoplast, distinction between cytoplasm and vacuole) can be stripped away. When this is done, the structure is simplified to the idealized geometry represented by a cylinder of long distance shuttling region (Fig. 6.2B) and an apical tip of short distance targeting region (Fig. 6.2C). In this approach, the inner surface has the equivalent velocity boundary conditions ($u_z = \overline{u}_z$) although the cytoplasmic streaming emerges naturally by the microfilament self-organization when the myosin motors move on the actin bundle tracks at the periphery. The terms involving the local inertia force, the inertia force of convection, and the body force can be neglected due to the small Reynolds number of fluid flow pollen tubes (Re $= 2.9 \times 10^{-6}$) (Kroeger et al., 2009). For an incompressible Newtonian fluid with velocity $\boldsymbol{u}$ ($\nabla \cdot \boldsymbol{u} = 0$) and pressure p, the viscous Stokes equation is given by

$$\eta \nabla^2 \boldsymbol{u} = \nabla p \tag{6.1}$$

where η is the dynamic viscosity (Pa s). Considering the growth rate of pollen tube V_{Pole}, the flux through its cross-section shows

$$\int_0^R 2\pi r u_z dr = V_{Pole} \pi R^2. \tag{6.2}$$

Following the approach of earlier studies, the hydrodynamics of fountain streaming of the shunting region in cylindrical coordinates can be represented using an analytical method. For the region of long distance shuttling, the possible effects at both ends can be discarded because the streaming rates become independent of cell length (L) for $L > 5R$ (Pickard, 1974). Thus the velocity components are $u_z = u_z(r)$, $u_r = u_\theta(r) = 0$ in a cylindrical system, and the pressure distribution is $p = p(z)$. The velocity components automatically satisfy the incompressible condition of the continuity equation. Then the momentum equation (6.1) becomes:

$$\frac{1}{r}\frac{d}{dr}\left(r\frac{du}{dr}\right) = \frac{1}{\eta}\frac{dp}{dz}. \tag{6.3}$$

Integrating Eq. (6.3), the velocity along the length of pollen tube:

$$u_z = \frac{1}{\eta}\frac{dp}{dz}\frac{r^2}{4} + A \ln r + B. \tag{6.4}$$

Because of $u_z(r=0) \neq \infty \Rightarrow A = 0$ and $u_z(r=R) = \overline{u}_z \Rightarrow B = \overline{u}_z - \frac{1}{\eta}\frac{dp}{dz}\frac{r^2}{4}$, Eq. (6.4) becomes

$$u_z = \overline{u}_z - \frac{1}{4\eta}\frac{dp}{dz}(R^2 - r^2). \tag{6.5}$$

Substituting Eq. (6.5) into Eq. (6.3), the gradient of turgor pressure is given by

$$\frac{dp}{dz} = \frac{8\eta}{R^2}(\overline{u}_z - V_{Pole}). \tag{6.6}$$

Substituting Eq. (6.6) into Eq. (6.5), the normalized velocity is obtained as

$$\frac{u_z}{\overline{u}_z} = 1 - 2\left(1 - \frac{V_{Pole}}{\overline{u}_z}\right)\left(1 - \frac{r^2}{R^2}\right). \tag{6.7}$$

Integrating Eq. (6.6), we get

$$\frac{p}{p_o} = 1 + \frac{8\eta(\overline{u}_z - V_{Pole})}{R^2 p_o} z. \tag{6.8}$$

Ultimately, the normalized velocity distribution at radial direction is simplified as

$$\frac{u_z}{\overline{u}_z} = 1 - 2(1-\xi)\left(1 - \frac{r^2}{R^2}\right), \tag{6.9}$$

and the normalized turgor pressure along the length direction is simplified as

$$\frac{p}{p_o} = 1 + \frac{8(1-\xi)}{\psi}\frac{z}{L} \tag{6.10}$$

where $\xi = V_{Pole}/\overline{u}_z$ and $\psi = R^2 p_o/\eta L\overline{u}_z$. Each pollen tube inherently has a pair of dimensionless ratios ξ and ψ, which determine the hydrodynamic property of fluid flow in the pollen tube. By setting Eq. (6.9) equal to zero, the indifferent zone is obtained at $|r/R|_{IZ} = 1/\sqrt{2(1-\xi)}$, where a vesicle-free line separates apically and basipetally directed cytoplasmic streams.

6.3 MODELING INTRACELLULAR MICROFLUID TRANSPORTATION IN POLLEN TUBES

6.3.1 Analysis for the Properties of Microfluid Transportation in Pollen Tubes

Another critical factor for the fast tip-growth of a pollen tube is the material supply for cell wall synthesis, which mostly depends on the intracellular transportation of fountain streaming in pollen tubes. While the movement of species of cell wall materials in pollen tubes is a fascinating field of research, there are currently only a few theoretical studies that offer the underlying details (Verchot-Lubicz and Goldstein, 2010). An early study of the interplay between calcium gradients and cytoplasmic streaming (Lazzaro et al., 2005) provides an important step towards a deeper understanding of the competition between advective and diffusional transportation of ions and microscopic particles in the tip-growth of plant cell. In that study, the advection–diffusion dynamics of viscous Stokes flow of fountain streaming is investigated in the apical region of pollen tube. To find out whether advection or diffusion acts as the key transport mode for inclusive microscopic particles (e.g., small molecules,vesicles or organelles and ions) in fountain streaming, the Péclet number (Pe) of the fluid flow in pollen tube is analyzed. The Pe is defined as the ratio of the time it takes for organelles to diffuse versus the time it takes for the organelles to be adverted across a certain distance ($\mathrm{Pe} = Lv/D$, where L is a characteristic distance, v is the velocity of particles and D is the diffusion coefficient of the particles). Here, the medium sized inclusions – vesicles – were taken as an example with the characteristic distance such as the tip radius of $L = 10$ μm, the vesicles velocity (v) of the order of 0.3–0.6 μm/s (Chebli et al., 2013), and the diffusion coefficient of the vesicles $D = 0.6$ μm^2/s (Kroeger et al., 2009). The Péclet number Pe can be estimated to be in the range of 5–10, indicating that both diffusion and advection have significant contribution to vesicles movement. For smaller species (e.g., ions and small molecules) across smaller length scales (short-distance targeting region), the Péclet number Pe is very small and hence diffusion will become dominant. In contrast, for larger species (e.g., large proteins, vesicles, and organelles) across larger length scales (long-distance shutting region), the Péclet number Pe is large so that advection will become dominant.

6.3.2 Simulation for the Coupling Advection–Diffusion of Fountain Streaming

The advection–diffusion dynamics of the moving species in pollen tube can be described by the concentration $\boldsymbol{C}$. Due to diffusion and the cytoplasmic flow of fountain streaming, the time-dependent concentration of a species can be generally described as:

$$\frac{\partial C}{\partial t} = -\boldsymbol{u} \cdot \nabla C + D\nabla^2 C + R \tag{6.11}$$

where D is the diffusion coefficient and $\boldsymbol{u}$ is the flow velocity. The term on the left hand side of this equation is the change of species concentration for mass transfer. On the right hand side of this equation, the 1st term is the mass advection; the 2nd term is the mass diffusion; and the last term represents any phenomena that act as mass "sources" or "sinks". Considering a constant growth rate at normal conditions, the time-independent case can be assumed under equilibrium state. Neglecting any sources of ion channels localized at the tip of pollen tubes, the steady-state version is

$$D\nabla^2 C = \boldsymbol{u} \cdot \nabla C. \tag{6.12}$$

To highlight the important role of fountain streaming, the simulation of advection–diffusion dynamics in the shank region of the pollen tube can be performed by using an axisymmetric geometry with boundary conditions of the sliding velocity and flux (Fig. 6.7), where the position of sliding velocity at cell wall is based on the distribution of microfilament in the pollen tube (Kroeger et al., 2009; Bove et al., 2008).

Due to the coupling of hydrodynamics and advection–diffusion, the concentration distribution of cell wall materials at the apical region of pollen tube can be investigated through numerical simulation performed by using software of COMSOL Multiphysics 4.4. The parameters used in the simulation are shown in Table 6.1.

6.4 RESULTS AND DISCUSSION

6.4.1 Hydrodynamics of Fountain Streaming in Pollen Tubes

In this section, a numerical simulation is performed to simulate the fluid flow in the shank region of pollen tube by building a cylinder with the radius of 10 μm and length of 200 μm (a cylinder with aspect ratio of 20 as a representative model of pollen tube). This can be seen in Fig. 6.3A. Although the aspect ratio of mature pollen tube reaches values of up to 100, it is reasonable to perform the numerical simulation in a cylinder with an aspect ratio of 20 because when the aspect ratio is more than 5 ($L > 5R$), the streaming rates become independent of pollen tube length (L) (Pickard, 1974). The sliding velocity $\overline{u} \sim 10$ μm/s at $r = R$ and the slip boundary at either end are also applied. The fluid flow is simplified as

TABLE 6.1 Parameters of pollen tube and properties of hydrodynamics and mass transfer of fountain streaming

Physical quality	Value	References
Density (ρ)	10^3 kg/m^3 (water)	(Kroeger et al., 2009; Keren et al., 2009)
Dynamic viscosity (η)	10^{-3} kg/m/s (Pa s) (water)	(Kroeger et al., 2009; Keren et al., 2009)
Diffusion constant (D)	10^{-9} m^2/s (smallest molecules)	(Goldstein et al., 2008; Kroeger et al., 2009)
	10^{-13}–10^{-9} m^2/s (vesicle)	(Kroeger et al., 2009)
	6.25×10^{-13} m^2/s (vesicle)	
Speed of cyclosis (U)	10^{-4} m/s (chara corallina)	(Goldstein et al., 2008)
	4.5×10^{-7} m/s (pollen tube)	(Kroeger et al., 2009)
	2–9 $\times 10^{-7}$ m/s (pollen tube)	(Lovy-Wheeler et al., 2007)
Radius (R)	5×10^{-4} m (chara corallina)	(Goldstein et al., 2008)
	6.5×10^{-6} m (*Lilium longiflorum* pollen tubes)	(Kroeger et al., 2009)
	10^{-5} m (pollen tube of lily, measuring)	(Kroeger et al., 2009)
	7.5×10^{-8} m (vesicle)	
Péclet number ($Pe = \frac{UR}{D}$)	50 (smallest molecules)	(Goldstein et al., 2008; Van de Meent et al., 2008)
	500–1000 (larger proteins)	(Goldstein et al., 2008; Van de Meent et al., 2008)
	10^2–10^3	(Goldstein et al., 2008; Van de Meent et al., 2008)
	0.054	(Kroeger et al., 2009)
Reynolds number ($Re = \frac{UR}{v}$)	2.9×10^{-6} (*Lilium longiflorum* pollen tubes)	(Kroeger et al., 2009)
Turgor pressure (p)	0.1–0.4 MPa (pollen tube of lily)	(Pietruszka, 2013; Prass et al., 2006; Munro et al., 2004)
Growth rates (V_{Pole})	≤ 20 μm/h (in gymnosperms)	(Vidali et al., 2001)
	80 to 600 μm/h (in ancient angiosperm lineages)	(Williams, 2008)
	0.1–0.35 μm/s	(Vidali et al., 2001)

a viscous Stokes flow described by Eq. (6.1). The 3D view of velocity field of the fountain streaming is presented in Fig. 6.3B. An inverse fountain is present: flow forward from the end to the tip in the inner region and backward from the tip to the end in the outer region.

To obtain velocity distribution along the diameter, a section view of the velocity field at the middle position is presented that neglects the end effects (Fig. 6.4A). A parabolic velocity distribution along the diameter of pollen tube shows good agreement with the analytical solution of Eq. (6.9) (Fig. 6.4B).

The streaming in the algae (e.g., cross-sectional velocity distribution) has been characterized in detail through both experimental measurement and numerical simulation (Goldstein et al., 2008; Mustacich and Ware, 1977). There is little experimental data for streaming velocity in pollen tube. To ensure the reliability of the simulated results, the velocity of cytoplasmic streaming in *characean algae* was firstly reconstructed with a similar method reported in the literature (Goldstein et al., 2008; Mustacich and Ware, 1977). The case of $V_{Pole} = 0$ are inves-

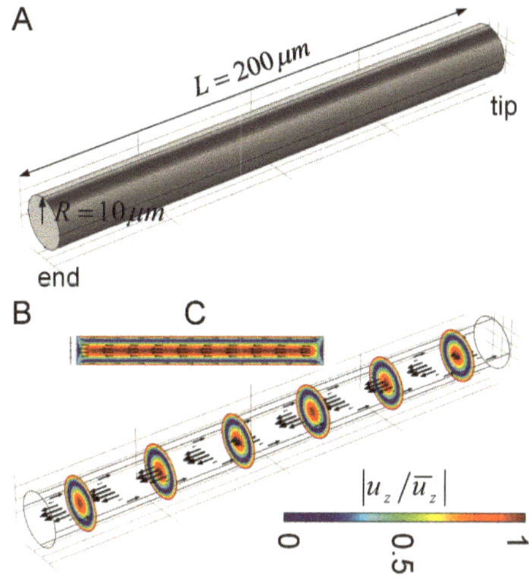

FIGURE 6.3 Hydrodynamics of fountain streaming in the shank region of pollen tube (Liu et al., 2016). (A) Simplified geometry for the shank region of pollen tube; (B) 3D section slides of velocity field; (C) Slide view of the velocity field.

tigated by setting the velocity boundary conditions to those used in Goldstein et al. (2008), except for the helicity of the flow. The 3D section slides of the velocity field (Fig. 6.5A) as well as its section view (Fig. 6.5B) are asymmetric. And the cross-sectional velocity distribution well reconstructed the sigmoidal velocity distribution experimentally observed *in vivo* (Fig. 6.5C, green dots) (Mustacich and Ware, 1977). This suggests that the numerical method developed here is capable of simulating fountain streaming in pollen tubes.

In the case of the lily pollen tube, the growth rate of the pollen tube is $V_{Pole} \sim 10^{-3}$–10^{-1} μm/s and the maximum velocity of fluid flow is $\bar{u} \sim 10^{-1}$–1 μm/s (Table 6.1), so the ratio ξ is in the range of 10^{-2}–10^{-1}. In the case of $\xi = 0$, the resulting dimensionless velocity distribution $u_z/\bar{u}_z \approx 1 - 2(1 - r^2/R^2)$ and the resulting indifferent zone $|r/R|_{IZ} = \sqrt{2}/2$ together imply a stable geometry of fluid flow in the pollen tube. As for the case when the pollen tube radius is on the order of $R \sim 1$–10 μm, the pollen tube has a length of in the order of $R \sim 1$–10^3 μm, the initial pressure of pollen tube is on the order of $p_o \sim 10^{-1}$ MPa, the dynamic viscosity is on the order of $\eta \sim 10^{-3}$–10^{-2} Pa s, and the ratio ψ falls in the range of 10^4–10^{11}. Therefore, the dimensionless gradient of turgor pressure $8(1 - \xi)/\psi \sim 10^{-10}$–$10^{-3}$ is small and negligible, making $p/p_o \approx 1$. However, for the abnormal cases (e.g., under osmotic stress), the gradient of turgor pressure will increase when the initial pressure p_o decreases because ψ becomes smaller with decreasing p_o when the pollen tubes are subjected to osmotic stress.

To illustrate the effects of fountain streaming on regulating the gradients of turgor pressure, the turgor pressure distribution can be calculated along the length of pollen tube. In normal conditions (initial pressure $p_o \sim 1$ MPa), turgor pressure increases linearly with a small gradient along the length of pollen tube (Fig. 6.6A), which is opposite to that of an ordinary pipe with natural pressure drop due to the energy loss according to the Darcy–Weisbach

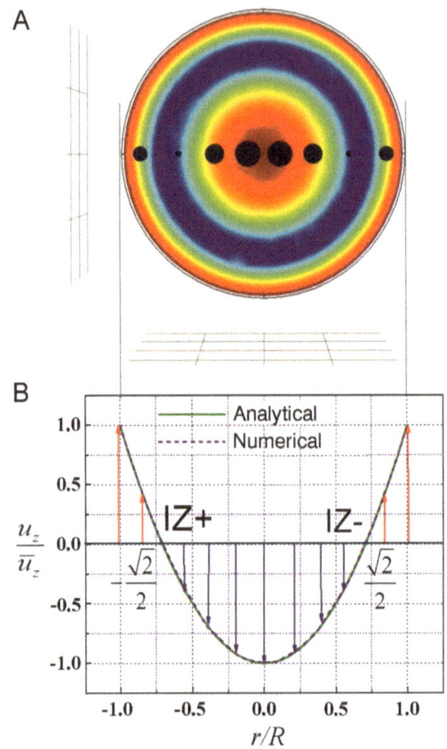

FIGURE 6.4 Hydrodynamics of fountain streaming in the shank region of pollen tube. (Liu et al., 2016) (A) Section view of the velocity field; (B) Comparison with the numerical and analytical velocities along a diameter of pollen tube.

equation (Tan and Takeuchi, 2007). In other words, the peculiar hydrodynamic properties of fountain streaming in pollen tube continuously provide an undamped driving force via balancing the natural pressure drop. In abnormal conditions (initial pressure $p_0 < 1$ MPa), the gradient of turgor pressure increases as the initial pressure decreases (Fig. 6.6B).

By considering the theoretical predictions above, the mechanisms of turgor pressure regulation can be qualitatively interpreted. Although the longitudinal array of actin microfilaments is deemed to drive the polar elongation in a pollen tube (Vidali et al., 2001), the turgor pressure is generally assumed to act as the most important driving force for its fast tip-growth (Benkert et al., 1997; Winship et al., 2011). According to the Lockhart theory (Kroeger et al., 2011; Lockhart, 1965), the rate of cell wall expansion is proportional to the difference between the actual turgor pressure and a threshold pressure. Thus a normal pollen tube, on the one hand, is of relatively high tip-growth rate driven by relatively high turgor pressure; and, on the other hand, maintains its elongation through the fountain streaming. This results in a balancing of the natural pressure drop and maintains an undamped driving force (without significant decrease).

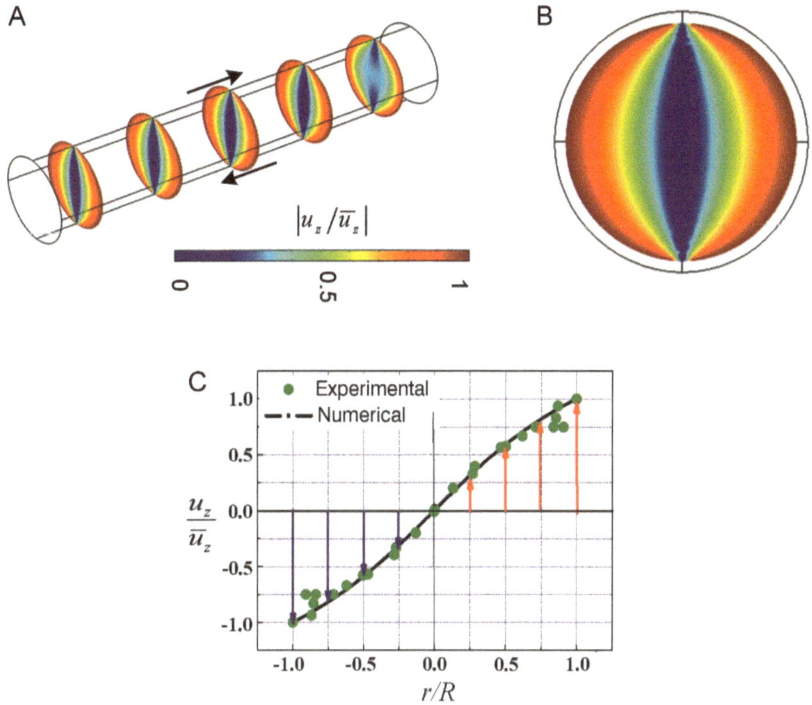

FIGURE 6.5 Comparison of numerical and experimental hydrodynamics in *characean algae* cell (Liu et al., 2016). (A) 3D section slides of velocity field; (B) Section view of the velocity field; (C) Comparison of the numerical velocities along the diameter with the experimental results from Mustacich and Ware (1977).

6.4.2 Intracellular Transportation of Fountain Streaming in Pollen Tubes

In the simulation, the apical region of pollen tube was modeled as a cylinder with an apical cap where the boundary conditions of the sliding velocity and flux. The normalized inlet concentration ($\overline{C} = 1$) and boundary flux ($\gamma = 0.1$) were used (Fig. 6.7A). The section slides of the normalized flow velocity shows the pattern representing an inverse fountain (forward motion near the central axis of the cell and backward motion near the periphery) near the tip (Fig. 6.7B). This is consistent with the experimental observations. To reveal the roles of fountain streaming in regulating the material supply for fast tip-growth, the influence of Péclet numbers on the distribution of concentration of cell wall materials is investigated (Fig. 6.7C). Firstly, sharp patterns of concentration distribution are observed at high Péclet numbers ($\text{Pe} = 10^2 \sim 10^3$), which indicates that the fountain streaming accumulates the materials for cell wall synthesis near the tip of pollen tube. However, at medium Péclet numbers ($\text{Pe} = 4 \sim 10$, around the previously estimated range), the pattern becomes obtuse and reconstructs the spatial configuration of the inversely apical cone in experimental observations (Kroeger et al., 2009; Zonia and Munnik, 2009). At low Péclet numbers ($\text{Pe} = 1 \sim 2$), the concentration distribution tends to be uniform in the apical pollen tube. These results indicate

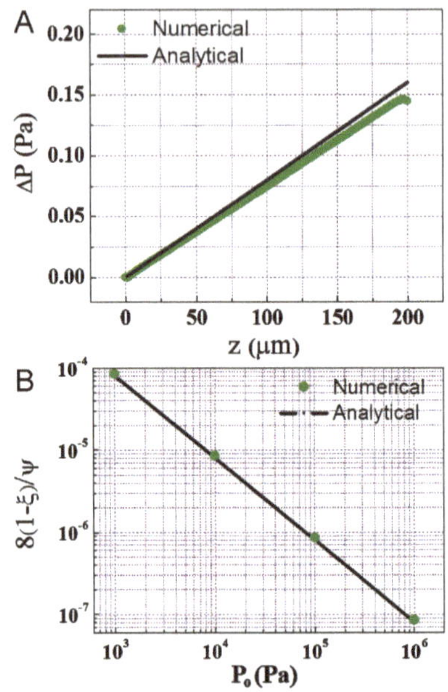

FIGURE 6.6 Turgor pressure and its gradients in pollen tube and its effect on the shape of growing tip of pollen tube (Liu et al., 2016). (A) The analytical and numerical turgor pressure change linearly along the length of pollen tube; (B) The analytical and numerical gradient of the turgor pressure along the length of pollen tube changes with the initial pressure.

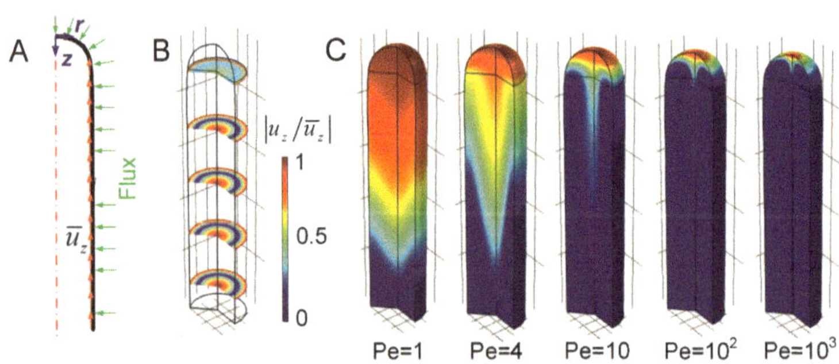

FIGURE 6.7 Concentration in the apical region of pollen tube (Liu et al., 2016). (A) The axisymmetric geometry and boundary conditions of the sliding velocity and flux; (B) The section slides of the normalized fluid velocity; (C) The patterns of normalized concentration with different Péclet numbers.

that fountain streaming plays crucial roles in the distribution of cell wall materials, especially for the small species with medium and higher Péclet numbers.

To quantify the effects of fountain streaming on materials accumulation, the normalized concentration of cell wall materials along the symmetry axis of the pollen tube was investigated for selected Péclet numbers (Fig. 6.8A). Interestingly, the styles of concentration distribution are very different at different Péclet numbers. At low Péclet numbers ($Pe = 1 \sim 2$), the

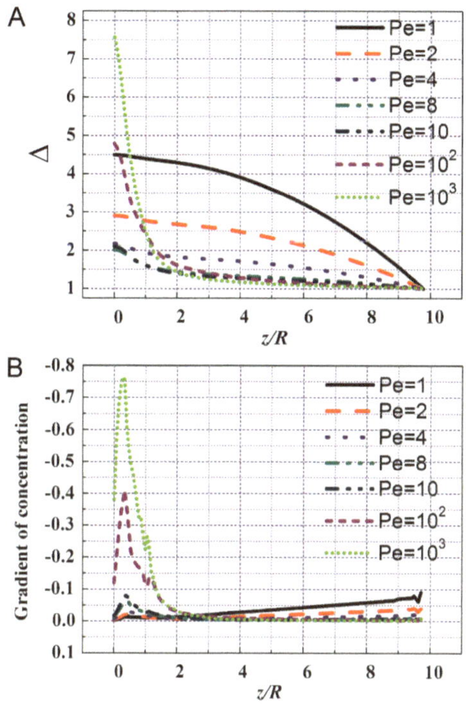

FIGURE 6.8 Concentration in the apical region of pollen tube (Liu et al., 2016). (A) The normalized concentrations along the axis of symmetry of pollen tube with different Péclet numbers; (B) The gradient of concentrations along the axis of symmetry of pollen tube with different Péclet numbers.

concentration increases slowly from the end to the tip. At high Péclet numbers ($Pe = 10^2 \sim 10^3$), the concentration distribution increases exponentially which agrees with the previously mentioned Ca^{2+} concentration distribution (Holdaway-Clarke and Hepler, 2003). Additionally, the gradients of the concentration distribution along the symmetry axis of pollen tube increase when the Péclet numbers become larger (Fig. 6.8B), demonstrating that material accumulation is more efficient at higher Péclet numbers.

Although ion channels localized at the tip of the pollen tube can generate the internal gradient of ions, the fountain streaming might be another mechanism contribute to the transportation and accumulation of cell wall materials (i.e., ions, molecules and vesicles) for the fast tip-growth of pollen tube. As it is known, steep concentration gradients of cell wall materials are essential for polar cell growth (Hepler et al., 2001). In the apical pollen tube, both the calcium concentration of cytoplasmic and the pH value in pollen tube change dramatically over short distances (Holdaway-Clarke and Hepler, 2003; Wilsen et al., 2006). This is because these ion gradients can provide pollen tube with a regulatory spatial framework for its polarity, and are also involved in the temporal regulation of pollen tube elongation. In contrast, the disruption of the gradient will invariably cause cessation of pollen tube growth (Holdaway-Clarke and Hepler, 2003). Given that the concentration gradients are necessary for fast tip-growth of pollen tube, the fountain streaming contributes to the fast tip-growth by regulating the supply of cell wall material.

6.5 CONCLUSIONS

For flowering plants, the pollen tube is necessary for the transport of sperm cells from the pollen grain to the ovary of the receptive flower to perform double fertilization. This process is of critical importance for fast and polar tip-growth of the pollen tube. Both the analytical methods and the numerical methods introduced in this chapter are used to provide understanding of the effects of hydrodynamics and advection–diffusion of fountain streaming in fast-tip growing pollen tube. These findings demonstrate that there are positive gradients of turgor pressure and concentration along the length of pollen tube. Fountain streaming with unique properties of hydrodynamics and transport, generates an undamped driving force for the elongation of pollen tube and also provides cell wall materials over huge distances in a highly efficient and targeted manner. These may be helpful for further understanding the sexual reproduction and infertility of flowering plants under normal conditions and drought stress.

References

Benkert, R., Obermeyer, G., Bentrup, F.W., 1997. The turgor pressure of growing lily pollen tubes. Protoplasma 198, 1–8.

Bove, J., et al., 2008. Magnitude and direction of vesicle dynamics in growing pollen tubes using spatiotemporal image correlation spectroscopy and fluorescence recovery after photobleaching. Plant Physiol. 147, 1646–1658.

Chang, L.C., et al., 2009. Pollen-specific SKP1-like proteins are components of functional SCF complexes and essential for lily pollen tube elongation. Plant Cell Physiol. 50, 1558–1572.

Chebli, Y., Kroeger, J., Geitmann, A., 2013. Transport logistics in pollen tubes. Mol. Plant 6, 1037–1052.

Ganguly, S., Williams, L.S., Palacios, I.M., Goldstein, R.E., 2012. Cytoplasmic streaming in Drosophila oocytes varies with kinesin activity and correlates with the microtubule cytoskeleton architecture. Proc. Natl. Acad. Sci. USA 109, 15109–15114.

Goldstein, R.E., Tuval, I., Van de Meent, J.W., 2008. Microfluidics of cytoplasmic streaming and its implications for intracellular transport. Proc. Natl. Acad. Sci. USA 105, 3663–3667.

Hecht, I., Rappel, W.J., Levine, H., 2009. Determining the scale of the Bicoid morphogen gradient. Proc. Natl. Acad. Sci. USA 106, 1710–1715.

Hepler, P.K., Vidali, L., Cheung, A.Y., 2001. Polarized cell growth in higher plants. Annu. Rev. Cell Dev. Biol. 17, 159–187.

Holdaway-Clarke, T.L., Hepler, P.K., 2003. Control of pollen tube growth: role of ion gradients and fluxes. New Phytol. 159, 539–563.

Justus, C.D., Anderhag, P., Goins, J.L., Lazzaro, M.D., 2004. Microtubules and microfilaments coordinate to direct a fountain streaming pattern in elongating conifer pollen tube tips. Planta 219, 103–109.

Kadota, A., et al., 2009. Short actin-based mechanism for light-directed chloroplast movement in *Arabidopsis*. Proc. Natl. Acad. Sci. USA 106, 13106–13111.

Keren, K., et al., 2009. Intracellular fluid flow in rapidly moving cells. Nat. Cell Biol. 11, 1219–1224.

Kroeger, J.H., Zerzour, R., Geitmann, A., 2011. Regulator or driving force? The role of turgor pressure in oscillatory plant cell growth. PLoS ONE 6, e18549.

Kroeger, J.H., Daher, F.B., Grant, M., Geitmann, A., 2009. Microfilament orientation constrains vesicle flow and spatial distribution in growing pollen tubes. Biophys. J. 97, 1822–1831.

Lazzaro, M.D., Marom, E.Y., Reddy, A.S., 2013. Polarized cell growth, organelle motility, and cytoskeletal organization in conifer pollen tube tips are regulated by KCBP, the calmodulin-binding kinesin. Planta 238, 587–597.

Lazzaro, M.D., et al., 2005. Calcium gradients in conifer pollen tubes; dynamic properties differ from those seen in angiosperms. J. Exp. Bot. 56, 2619–2628.

Liu, S.B., et al., submitted for publication. Fountain streaming contributes to fast tip-growth through regulating the gradients of turgor pressure and concentration in pollen tubes. Under review.

Lockhart, J.A., 1965. An analysis of irreversible plant cell elongation. J. Theor. Biol. 8, 264–275.

Lovy-Wheeler, Cárdenas, A.L., Kunkel, J.G., Hepler, P.K., 2007. Differential organelle movement on the actin cytoskeleton in lily pollen tubes. Cell Motil. Cytoskelet. 64, 217–232.

McKerracher, L.J., Heath, I.B., 1987. Cytoplasmic migration and intracellular organelle movements during tip growth of fungal hyphae. Exp. Mycol. 11, 79–100.

Mustacich, R.V., Ware, B.R., 1977. Velocity distributions of the streaming protoplasm in Nitella flexilis. Biophys. J. 17, 229.

Munro, E., Nance, J., Priess, J.R., 2004. Cortical flows powered by asymmetrical contraction transport PAR proteins to establish and maintain anterior–posterior polarity in the early *C. elegans* embryo. Dev. Cell 7, 413–424.

Niwayama, R., Shinohara, K., Kimura, A., 2011. Hydrodynamic property of the cytoplasm is sufficient to mediate cytoplasmic streaming in the *Caenorhabditis elegans* embryo. Proc. Natl. Acad. Sci. USA 108, 11900–11905.

Pickard, W.F., 1974. Hydrodynamic aspects of protoplasmic streaming in *Chara braunii*. Protoplasma 82, 321–339.

Pietruszka, M., 2013. Pressure–induced cell wall instability and growth oscillations in pollen tubes. PLoS ONE 8, e75803.

Prass, M., Jacobson, K., Mogilner, A., Radmacher, M., 2006. Direct measurement of the lamellipodial protrusive force in a migrating cell. J. Cell Biol. 174, 767–772.

Schonegg, S., Constantinescu, A.T., Hoege, C., Hyman, A.A., 2007. The Rho GTPase-activating proteins RGA-3 and RGA-4 are required to set the initial size of PAR domains in *Caenorhabditis elegans* one-cell embryos. Proc. Natl. Acad. Sci. USA 104, 14976–14981.

Shimmen, T., et al., 1995. Roles of actin filaments in cytoplasmic streaming and organization of transvacuolar strands in root hair cells of *Hydrocharis*. Protoplasma 185, 188–193.

Tan, W.H., Takeuchi, S., 2007. A trap-and-release integrated microfluidic system for dynamic microarray applications. Proc. Natl. Acad. Sci. USA 104, 1146–1151.

Theurkauf, W.E., 1994. Premature microtubule-dependent cytoplasmic streaming in cappuccino and spire mutant oocytes. Science 265, 2093–2096.

Van de Meent, J.W., Tuval, I., Goldstein, R.E., 2008. Nature's microfluidic transporter: rotational cytoplasmic streaming at high Péclet numbers. Phys. Rev. Lett. 101, 178102.

Verchot-Lubicz, J., Goldstein, R.E., 2010. Cytoplasmic streaming enables the distribution of molecules and vesicles in large plant cells. Protoplasma 240, 99–107.

Vidali, L., Mckenna, S.T., Hepler, P.K., 2001. Actin polymerization is essential for pollen tube growth. Mol. Biol. Cell 12, 2534–2545.

Williamson, R.E., Ashley, C.C., 1982. Free Ca^{2+} and cytoplasmic streaming in the alga Chara. Nature 296, 647–651.

Williams, J.H., 2008. Novelties of the flowering plant pollen tube underlie diversification of a key life history stage. Proc. Natl. Acad. Sci. USA 105, 11259–11263.

Wilsen, K.L., et al., 2006. Imaging the actin cytoskeleton in growing pollen tubes. Sex. Plant Reprod. 19, 51–62.

Winship, L.J., Obermeyer, G., Geitmann, A., Hepler, P.K., 2011. Pollen tubes and the physical world. Trends Plant Sci. 16, 353–355.

Zonia, L., Munnik, T., 2009. Uncovering hidden treasures in pollen tube growth mechanics. Trends Plant Sci. 14, 318–327.

CHAPTER

7

Microorganisms and Their Response to Stimuli

R.J. Clarke[#]

Department of Engineering Science, University of Auckland, Auckland, New Zealand

7.1 INTRODUCTION

The first microorganisms are believed to be the prokaryotes, single celled organisms with no distinct internal structures (organelles). Some of the earliest microfossil evidence suggests that microorganisms could have been binding grains together to create superstructures known as stromatolites up to 3.5 billion years ago, in the Archean Period (Awramik, 1992; Allwood et al., 2006). There is even some evidence, albeit less direct, to suggest that prokaryotes emerged even earlier. This evidence comes from examining the abundance of certain carbon isotopes in rocks, since carbon produced through biological processes leaves a distinctive ratio of carbon-12 to carbon-13. Using this method, Bell et al. (2015) suggest that microorganisms could have been around as early as 4.1 billion years ago. This would mean that Life was already underway by the end of the Hadean Epoch, just a few million years after the formation of the Earth about 4.5 billion years ago. During this time the Earth was regularly bombarded by asteroids (referred to as the Late Heavy Bombardment), and has not traditionally been thought of as a period conducive to Life.

Later, about 2 billion years ago in the Proterozoic era, it is believed came the more complex unicellular eukaryotes (Knoll et al., 2006). Eukaryotic cells display a greater degree of internal organization, and form the basis of the multicellular Life that begin to appear in abundance following the Cambrian Explosion 540 million years ago. The early unicellular eukaryotes likely evolved from the prokaryotes, although there is some debate about the exact route that their evolutionary path took (Margulis, 1970). However, it does seem likely that structures such as their mitochondria came about by assimilating prokaryotes that had already mastered the biochemistry behind aerobic respiration (using oxygen to generate energy).

However, at the time unicellular eukaryotes emerged, aerobic respiration was a relatively new innovation. Since the Hadean Period oxygen levels had been less than 1%. Instead the

[#] Chapter coordinator.

atmosphere primarily consisted of nitrogen and carbon dioxide, and it is believed that some early prokaryotes used photosynthesis to fuel their metabolism. In photosynthesis electrons are stripped from a suitable reducing molecule, typically water, and hydrogen is released. The hydrogen is used to produced adrenosine triphosphate (ATP) and nicotinamide adenine dinucleotide phosphate (NADPH), effectively capturing the energy within the photons in a form that can be utilized within the cell. Specifically, these molecules provide the necessary energy for the freed electrons to convert carbon dioxide into a carbohydrate (stored energy) and a waste product. When the cell needs this energy, the process is effectively reversed (cellular respiration).

However, it is believed that the form of photosynthesis used by the earliest prokaryotes did not use water as a source of electrons, but rather hydrogen and hydrogen sulphide gas (Olson, 2006). This process does not produce oxygen as a waste produce, and as such is known as anoxygenic photosynthesis. (Note that most photosynthesizing bacteria still around today, such as purple bacteria that can be found in hot springs, still do not produce oxygen. Some produce sulphur or a sulfide as a waste product, and are hence known as purple sulphur bacteria.) Eventually, however, one highly-successful prokaryote, cyanobacteria, did develop oxygenic photosynthesis. Here water is a used instead of hydrogen or hydrogen sulphide, and oxygen is produced as a waste product. As a result, about 2.5 billion years ago the oxygen levels in the atmosphere began to rise, in what has become known as the Great Oxygenation Event. It may well have been that the cyanobacteria were around and producing oxygen well before 2.5 billion years ago, but that oxygen levels in the atmosphere would only have begun to rise once the potential for the oxygen to react with minerals such as iron had been exhausted. When it did eventually occur, this influx of oxygen into the atmosphere was not an immediate positive for life (Cowie, 2012). As well as practically wiping out the oxygen intolerant anaerobic microorganisms that dominated at the time, by drastically reducing the greenhouse gas methane from the atmosphere through oxidation it is also thought to have lead to perhaps the first, and longest, ice age in history (the Huronion glaciation, also referred to as Snowball Earth I, which lasted 300 million years).

Eventually it is believed that volcanic activity pumped enough carbon dioxide back into the atmosphere to thaw the planet, and halt disaster (Kirschvink, 1992). As the planet rewarmed aerobic life flourished, and multicellular eukaryotic life began to emerge. However, these organisms began processing the carbon dioxide that was keeping the Earth warm into biological carbonates, and so began a second cooling of the Earth about 700 million years ago. This lead to two further ice ages, both much colder than the first, and covering a span of 100 million years in a period. This has become known as the Cryogenian (Snowball Earth II). Once again, volcanic activity is thought to have reintroduced significant amounts of carbon dioxide back into the atmosphere over the course of several million years, causing the planet to rewarm, and saving it from remaining a largely lifeless frozen world (clinging, perhaps, to hydrothermal vent ecosystems in the depths of the ocean).

Today many will know that photosynthesis continues to play a vital role in regulating atmospheric levels of oxygen and carbon dioxide, and will typically associate this process with plants. However, what is perhaps less well known is that even now plants account for only about one quarter of the oxygen produced. The majority is still produced by photosynthesizing microorganisms, specifically the phytoplankton that occupy the Earth's oceans, seas, lakes and rivers. The phytoplankton consist not only of the cyanobacteria, but also a huge variety of other eukaryotic algae (algae are a broad class of organisms that includes not only the microscopic phytoplankton, but also multicellular organisms, like giant kelp that can grow to tens of meters). Remarkably, the microscopic phytoplankton account for at least half of the photosynthesis on the planet.

However, it is not believed that the algae and plants rediscovered photosynthesis independently. Rather, in a similar manner to the way in which eukaryotes are thought to have acquired their mitochondria from the more ancient prokaryotes, their photosynthesizing organelle, the chloroplast, is also believed to have been acquired from an earlier prokaryote (Gray and Doolittle, 1982). Gene analysis suggests that at some point in the distant past a cyanobacterium was engulfed by a unicellular ancestor, which consequently assimilated its ability to photosynthesize (Bhattacharya and Melkonian, 1995; McFadden and Evolution, 2004). This is an example of primary endosymbiosis, and can be seen in phytoplankton such as green algae. These eukaryotic beneficiaries could themselves become engulfed by other eukaryotes, which as a result also obtained the ability to photosynthesize. This is known as secondary endosymbiosis, and can be detected by the fact that the chloroplasts within these cells are surrounded by three or four membranes (chloroplasts formed by primary endosymbiosis are enclosed by a double-membrane, which originates from the cell wall of the engulfed cyanobacterium. From these, other double-membraned organelles – plastids – have evolved, some of which undertake functions not associated with photosynthesis). Since it is now believed that the land plants evolved from green algae (Keeling, 2004), it appears that photosynthesizing organisms both large and small owe their energy-producing abilities to the cyanobacteria.

It was therefore the microorganisms that terraformed our planet from a toxic barren world, into one that now supports such a rich diversity of complex multicellular life. At the same time as oxygenating the atmosphere, they also fixed the inorganic carbon in the atmosphere into organic forms that provide the basis of the food chain. For this they needed (and still need) to be able to locate the necessary dissolved molecules within the vast bodies of water that they inhabit, as well as the light which often powers their biochemistry. In their strategies for doing so we can witness rudimentary forms of the types of sensing methods used by much larger multicellular organisms. In this chapter we will endeavor to give an overview of these strategies, as well as how mathematical models can help us quantify why they are so effective.

7.2 SWIMMING DYNAMICS

For many microorganisms, moving equates to swimming. However, the manner in which microorganisms swim is very different to their larger multicellular counterparts, such as fish and cephalopods, which rely on flow inertia to produce their forward motion (e.g. vortex shedding). Flow inertia is almost non-existent on the scale of a microorganism, and so they instead rely upon viscous resistances to generate thrust. Many microorganisms are endowed with one or more tails or, more properly in this context, *flagella*.

For prokaryotes, these flagella are relatively stiff, usually coiled, and driven by molecular motors embedded in the cell membrane. By rotating one of their flagella in the opposite direction to the others, some prokaryotes are able to stop and turn very quickly. This enables them to undergo run-and-tumble swimming, where periods of swimming are punctuated by sudden changes in direction, giving them the appearance of undergoing Brownian-like motion (Berg, 2004).

Eukaryotes can also use flagella to propel themselves, although the structure of their flagella are very different from those used by prokaryotes. Rather than being passive filaments rotated by molecular motors at their base, eukaryotic flagella actively bend. This is possible because the eukaryotic flagella contain microtubules that are connected together by the protein Nexin, and powered by Dynein motors. This structure is known as the 9+2 axoneme, and the coordinated sliding of its microtubules causes the eukaryotic flagella to beat. The form of this beat can vary from a simple planar wave in some eukaryotes, to more complex helical beating patterns in others (e.g. certain spermatazoa). There is a specific form of the eukaryotic flagellum, the cilium, that is relatively short (5–10 μm in length) and covers the surface of some microorganisms. These cilia beat with a power stroke, followed by a recovery stroke that is more tangential to the cell surface. Since the power stroke generates greater hydrodynamic forces than the recovery stroke, there is a net propulsive force. By harnessing this net thrust across all of the cilia coating its surface in a coordinated fashion, the cell can propel itself with appreciable velocity. Eukaryotes that exploit this method of propulsion are known as ciliates.

There has been a great deal of past work dedicated to understanding the hydrodynamics of microorganism motility, although we shall not concern ourselves too greatly with the specifics here. (For a more complete discussion, the interested reader is referred to Clarke, 2015). In what follows we will simply assume that whatever its swimming mechanism, a microorganism is able to propel itself with swimming speed U_0, although we shall pay attention to whether the microorganism is propelled from the back of its cell body (a *pusher*), or the front (a *puller*). We are able to overlook the specific hydrodynamics in any more detail because in what follows we concern ourselves only with microorganism populations that are sufficiently dilute that cell–cell hydrodynamic interactions are expected to be of secondary importance. As soon as this ceases to be the case, then the exact nature of the flowfield generated by a microorganism can become very important (see, for example, Ishikawa and Pedley, 2008).

When modeling the response of swimming microorganisms to external stimuli (*taxes*), the swimming direction of the cell, $\boldsymbol{p}$, is of course of crucial importance. If the cell body is assumed to be well represented as a prolate spheroid, the dynamic equation for the rate of change of orientation, $\dot{\boldsymbol{p}}$, comes from a classical deterministic fluid-dynamical torque balance (which must be struck since there is effectively no inertia at these microscopic scales) (Pedley and Kessler, 1990)

$$\dot{\boldsymbol{p}} = \frac{1}{2}\boldsymbol{\omega} \times \boldsymbol{p} + \Gamma \boldsymbol{p} \cdot \dot{\boldsymbol{\gamma}} \cdot (\boldsymbol{I} - \boldsymbol{p} \otimes \boldsymbol{p}) + \boldsymbol{L} \tag{7.1}$$

(dots denote derivatives with respect to time). The background flow exerts hydrodynamic torques on the body, and these are given on the right-hand side. In particular, the body experiences torques due to the local vorticity, $\boldsymbol{\omega} = \nabla \times \boldsymbol{u}$, and the local rate of strain, $\dot{\boldsymbol{\gamma}} = (\nabla \boldsymbol{u} + (\nabla \boldsymbol{u})^T)/2$, where $\boldsymbol{u}$ is the flow velocity (and superscript T denotes transpose). Here $\boldsymbol{I}$ is the identity matrix, $\Gamma = (\alpha^2 - 1)/(\alpha^2 + 1)$ is a shape parameter with α being the ratio of major and minor spheroid axes, $\boldsymbol{p} \otimes \boldsymbol{p}$ is the outer product of the orientation vector, and has components $(p_i p_j)$. Since the body must be torque-free, these applied torques cause the body to rotate such that there is an equal-and-opposite viscous torque applied by the body onto the flow (left-hand side). As we shall see, cell response to external stimuli can also give rise to an additional applied torque, $\boldsymbol{L}$, which is also included on the right-hand side of this expression.

When the cell body is more rod-shaped than spheroidal it is possible to replace the dynamic equation (7.1) with one obtained using slender-body theory (Saintillan and Shelley, 2013)

$$\dot{\boldsymbol{p}} = (\boldsymbol{I} - \boldsymbol{p} \otimes \boldsymbol{p}) \cdot \nabla \boldsymbol{u}(\boldsymbol{x}_c) \cdot \boldsymbol{p} + \boldsymbol{L}. \tag{7.2}$$

Here $\boldsymbol{x}_c$ is the center of mass, which moves according to

$$\dot{\boldsymbol{x}}_c = V_0 \boldsymbol{p} + \boldsymbol{u}(\boldsymbol{x}_c), \tag{7.3}$$

where V_0 is the translational velocity of a single body, and depends upon the precise details of its shape. This kinetic equation predicts that rod-like bodies will align in a shear flow, unlike prolate spheroids which rotate (the well known Jeffery orbits). It is worth emphasizing that both (7.1) and (7.2) are derived in the absence of any nearby surfaces.

For a motile body, the swimming velocity of an individual cell is then given by $\boldsymbol{U} = U_0 \boldsymbol{p}$ (assuming that swimming speed, U_0, is independent of swimming direction). Of course, within a suspension of microorganisms there is biological variability between cells that is not captured by a model like (7.2), that treats each body as identical. This can be accommodated for by assuming that $\boldsymbol{p}$ is a statistical quantity, with probability distribution $f(\boldsymbol{p})$, which is taken to satisfy the steady Fokker–Planck equation (Pedley and Kessler, 1992)

$$\nabla_p \cdot (\dot{\boldsymbol{p}} f) = D_R \nabla_p^2 f. \tag{7.4}$$

Here D_R is a coefficient of rotational diffusivity (which takes account of biological variations between individuals in swimming mechanism), and ∇_p and ∇_p^2 are the gradient operator and the Laplacian on the sphere, respectively. (It is worth mentioning that although the steady Fokker–Planck equation has traditionally been assumed, there has been a recent reexamination of its appropriateness (Pedley, 2010), as we shall discuss more in Section 7.5.) Once f has been found, it then defines an average swimming velocity for a cell

$$\langle \boldsymbol{U} \rangle = U_0 \int \boldsymbol{p} f(\boldsymbol{p}) \mathrm{d}\boldsymbol{p}. \tag{7.5}$$

7.3 RESPONSE TO STIMULI

As discussed in the Introduction, microorganisms have certain metabolic requirements, and will alter their swimming motions in pursuit of these survival necessities.

Phototrophs are organisms that obtain their energy from sunlight, typically through photosynthesis, which is a reduction reaction. Land plants, of course, fall into this category, although here we will focus on the microscopic phytoplankton. In order to photosynthesize, phytoplankton need to position themselves near to the surface of the body of water that they inhabit, where a sufficient number of photons are available. They can be assisted in this by the Earth's gravity (gyrotaxis), as well as by directly sensing light levels and adjusting their swimming accordingly (phototaxis).

By contrast chemotrophs obtain their energy not from light, but by oxidization. The oxidizing agent can be oxygen (aerobic respiration) or substances such as sulphur compounds (anaerobic respiration). However, the molecule being oxidized is often an organic molecule, the origins of which can be traced back to sun-powered photosynthesis. However, in places where there is virtually no sunlight, such as in the deep oceans, microorganisms can still thrive. Here they can obtain their energy by oxidizing iron or manganese brought up to the ocean floor by underwater volcanoes.

Whether phototrophic or chemotrophic, many microorganisms are chemotactic, meaning that they have the ability to detect gradients in desirable dissolved molecules (chemoattractants) and swim towards the source. Chemotaxis takes a different form between eukaryotes and prokaryotes. Chemotactic eukaryotes are endowed with a distribution of receptors, which can detect spatial variations of the chemoattractant across the length of their cell body. By contrast, the small size of prokaryotes means that it is unlikely that any chemoattractant gradients would be large enough to be detectable over their body length. As a result, they move throughout their local environment, scanning the chemoattractant concentration at each location, to build up a spatial map of the concentration distribution (Wadhams and Armitage, 2004).

7.3.1 Gyrotactic Phototrophs

Since microorganisms are typically denser than the fluid they inhabit, it perhaps appears paradoxical to suggest that gravity could assist them in staying close to the well-lit water surface. Were they non-motile, they would, of course, slowly sediment into the dark depths. However, when combined with their ability to self-propel, gravity can provide microorganisms with a means to track towards the light, even if they cannot sense the illumination directly. By having a center of mass that is below their center of buoyancy, they will on average swim upwards due to an additional external gravitational torque. This is known as gyrotaxis. (If the center of mass is above their center of buoyancy then they will on average swim downwards, which is sometimes referred to as gravitaxis.) This gyrotaxis can be incorporated into the swimming kinetics by including a gravitational torque

$$\boldsymbol{L} = \frac{1}{2B}\left(\hat{z} - (\hat{z}\cdot\boldsymbol{p})\boldsymbol{p}\right) \tag{7.6}$$

($\hat{z}$ is the unit vertical) in the kinematic equation (7.1) (Pedley and Kessler, 1990), i.e.

$$\dot{\boldsymbol{p}} = \frac{1}{2}\boldsymbol{\omega}\times\boldsymbol{p} + \Gamma\boldsymbol{p}\cdot\dot{\boldsymbol{\gamma}}\cdot(\boldsymbol{I} - \boldsymbol{p}\otimes\boldsymbol{p}) + \frac{1}{2B}\left(\hat{z} - (\hat{z}\cdot\boldsymbol{p})\boldsymbol{p}\right) \tag{7.7}$$

(assuming that gravity is acting in the direction of the unit vertical, $\hat{z}$). Here

$$B = \mu\nu\alpha_{\perp}/2hmg \tag{7.8}$$

is known as a gyrotactic orientation timescale, where μ is the fluid's dynamic viscosity, v is the cell's volume, m is its mass, h is the radial offset of the cell's center-of-mass from its line of symmetry, g is gravitational acceleration and $\alpha_\perp$ gives the hydrodynamic torque that arises due to the spheroid rotating perpendicularly to its axis of symmetry. Non-dimensionalizing $\dot{\boldsymbol{p}}$ on B^{-1} and $\boldsymbol{u}$ on $U_0^2\tau/H$ exposes the importance of two non-dimensional parameters in the Fokker–Planck equation, which now takes the explicit form (Pedley and Kessler, 1992)

$$\begin{aligned}\hat{\boldsymbol{k}} \cdot \nabla_p f - 2\hat{\boldsymbol{k}} \cdot \boldsymbol{p} f + \eta \big[\boldsymbol{\omega} \cdot (\boldsymbol{p} \times \nabla_p f) \\ + 2\Gamma \left(\boldsymbol{p} \cdot \dot{\gamma} \cdot \nabla_p f - 3\boldsymbol{p} \cdot \dot{\boldsymbol{\gamma}} \cdot \boldsymbol{p} f\right)\big] = \lambda^{-1} \nabla_p^2 f,\end{aligned} \tag{7.9}$$

where

$$\lambda = 1/(2BD_R) \tag{7.10}$$

and

$$\eta = BU_0^2\tau/H^2, \tag{7.11}$$

respectively, measure the relative importance of rotational diffusivity and rotations due to the background flow in relation to cell bottom-heaviness. Here H is the depth of the suspension, and τ is the correlation time that a cell spends traveling in a given direction.

7.3.2 Photosensitive Phototrophs

Phenomenological Model: Light Absorption Only

The risks of UV radiation, photoinhibition, photobleaching can mean that some phytoplankton avoid light intensities that are too great, and consequently being as close as possible to the water's surface may not be optimal. Such microorganisms cannot therefore rely on gyrotaxis alone to position them in the ideal light conditions. Thus many phytoplankton also contain photoreceptors, such as the protein Channelrhodopsin-2, that enables them to detect light directly. However, unlike larger multicellular organisms, photosensitive microorganisms have a number of different photoreceptors at their disposal (Sgarbossa and Lenci, 2002). (It is interesting to note that there are even current research efforts aimed at exploring therapeutic use of these algal proteins to restore sight lost through conditions such as *Retinitis pigmentosa*, Doroudchi et al., 2011.)

Some microorganisms are also photokinetic, meaning that their swimming speeds change as a function of the level of illumination. In prokaryotes, photoreceptors can be embedded either in the cell wall, or within the cell cytoplasm. However, it has recently been shown that certain spherical cyanobacterium can use their cell body as a lens to focus the incoming light (much in the same manner as an eyeball), to better detect the direction of the light source (Schuergers et al., 2016). In eukaryotes the photoreceptors can be more complex still.

It is possible to incorporate such photoresponsiveness into the swimming kinetics. The simplest approach involves treating the influence of light on cell swimming velocity in a phenomenological manner, rather than through consideration of a torque balance (7.1). In this vein, Vincent and Hill (1996) took the cell swimming velocity to be proportional to light intensity, I,

$$\boldsymbol{U} = U_0 T(I)\hat{\boldsymbol{e}}, \tag{7.12}$$

where the unit vector $\hat{\boldsymbol{e}}$ points in the direction of the incoming light, and T is a prescribed taxis function. Since the light can be absorbed by the microorganisms, its intensity is a function of cell spatial concentration. This shading effect was incorporated using the Lambert–Beer Law

$$I(\boldsymbol{x}) = I_s \exp\left(-\alpha \int_{\boldsymbol{r}} n(\boldsymbol{x}')\mathrm{d}\boldsymbol{x}'\right), \tag{7.13}$$

where I_s is the intensity of the light source, α quantifies the strength of light absorption by the cells (and supposes that light absorption by the fluid itself is minimal), and the path of integration is along the ray $\boldsymbol{r}$ that connects the light source to a point $\boldsymbol{x}$ in the suspension. The taxis function, $T(I)$, is chosen so that below some critical light intensity, I_c, the microorganisms swim towards the source of the light, and above I_c they swim away from it. The exact form of $T(I)$ and value of I_c will depend upon the specific microorganism. However, if the source light intensity is close to this critical level, the following linearized form can be assumed:

$$T(I) = -\Xi(I - I_c), \tag{7.14}$$

where Ξ is some empirical constant. Hence one of the main characteristics of phototaxis that distinguishes it from gyrotaxis is that its direction can switch at a location in the bulk fluid. As a result, the highest density layer of cells does not necessarily occur at the water surface, as is the case with gyrotactic microorganisms, but instead can occur at some depth where the light conditions are optimal.

Phenomenological Model: Light Absorption and Scattering

Whilst the Lambert–Beer Law (7.13) takes into account reduction in illumination due to light absorption, it does not take into account the fact that light can be scattered by cells. This effect was modeled by Ghorai et al. (2010) by defining the light intensity, $J(\boldsymbol{x}, \theta, \phi)$, at a spatial position $\boldsymbol{x}$, on a ray $\boldsymbol{r}$ pointing in a direction $(\cos\theta\cos\phi, \sin\theta\sin\phi, \cos\theta)$, where θ and ϕ are spherical angles. Assuming that the light source is positioned vertically above, this light intensity is then governed by the following integro-differential equation:

$$\frac{dJ}{ds} + \kappa_0 n J = \frac{\sigma n}{4\pi} \int J(\boldsymbol{x}, \theta', \phi') \sin\theta' \, \mathrm{d}\theta' \, \mathrm{d}\phi', \tag{7.15}$$

here s is an arclength parameter along the ray $\boldsymbol{r}$, κ_0 is an absorption coefficient, and σ a scattering coefficient. It proves convenient to define the scattering albedo

$$\omega = \sigma/\kappa_0. \tag{7.16}$$

When $\omega = 0$ there is only absorption of light (i.e. no scattering), and when $\omega = 1$ there is no absorption, only scattering. This assumes that a light ray is equally likely to be scattered in any direction, and that both absorption and scattering are linearly proportional to cell density, n. In the presence of scattering, the relationship between mean swimming direction and illumination (7.12) needs to be modified to Ghorai et al. (2010)

$$\boldsymbol{U} = -U_0 T(I)\boldsymbol{q}/|\boldsymbol{q}|, \tag{7.17}$$

where

$$I(\boldsymbol{x}) = \int J(\boldsymbol{x}, \theta, \phi)\sin\theta \, d\theta \, d\phi, \quad \boldsymbol{q} = \int \boldsymbol{r} J(\boldsymbol{x}, \theta, \phi)\sin\theta \, d\theta \, d\phi, \tag{7.18}$$

with $\boldsymbol{q}$ known as the intensity flux.

Dynamic Model: Light Absorption Only

The phototactic laws specified in (7.12) and (7.17) are phenomenological in nature, in the sense that they do not take into account the torque balance considered in (7.1), and are also deterministic. In acknowledgment of these shortcoming, Williams and Bees (2011) proposed three alternatives for modeling the response of microorganism to light (assuming that scattering is weak) in a manner which allows a torque balance to be met.

Their first approach models the swimming velocity in the following manner

$$\boldsymbol{U} = -\xi(I - I_c)\boldsymbol{p}. \tag{7.19}$$

Here the swimming direction, $\boldsymbol{p}$, is unaffected by the illumination, and is governed by torque-free requirements (7.1). The sensitivity to illumination comes through its influence upon swimming speed alone, and is therefore referred to as a photokinesis model.

Their second model introduces a gyrotactic torque of the same form as given in (7.7), but where the effective center-of-buoyancy, h, is a function of illumination

$$h = -\xi'(I - I_c), \tag{7.20}$$

where ξ' is some constant of proportionality. This changes the value of B in (7.7), and hence the preferential swimming direction through (7.4).

Their third model directly introduces into (7.7) a photo-reactive torque of the form

$$\boldsymbol{L}(\boldsymbol{p}) = t(I)\boldsymbol{p} \times \left(\beta_1 \hat{\boldsymbol{z}} + \beta_2 \nabla I\right) \tag{7.21}$$

assuming illumination from above ($\hat{z}$ is again the unit vertical). Also the function

$$t(I) = F_0 I(I - I_c), \tag{7.22}$$

ensures that there is no torque in the absence of light, and when the light levels are at the critical value I_c where the microorganisms are not compelled to swim either towards, or away from, the light source. Here F_0, β_1 and β_2 are empirical constants, and light intensity is modeling using the Lambert–Beer Law (7.13), i.e. scattering effects are not included.

7.3.3 Chemotactic Microorganisms

Phenomenological Model

The propensity of cells to swim up gradients in chemical concentration, known as chemotaxis, can be modeled using the Keller–Segel model (Keller and Segel, 1971). Here the relationship between swimming velocity, $\boldsymbol{U}$, and the concentration of the chemoattractant, C, is given by

$$\boldsymbol{U} = aU_0\nabla C, \tag{7.23}$$

where a is an empirically-determined constant specific to a given microorganism. A non-dimensional oxygen concentration is defined as

$$\theta = \frac{C - C_m}{C_0 - C_m}, \tag{7.24}$$

where C_m is the critical oxygen concentration, below which the microorganisms become inactive, and C_0 is the concentration at source.

The transport of the dissolved chemical, and its consumption by microorganisms, can be modeled using an advection–diffusion–reaction equation (Hillesdon et al., 1995)

$$\frac{\partial C}{\partial t} + \nabla \cdot (C\boldsymbol{u} - D_c\nabla C) - Kn = 0, \tag{7.25}$$

where C and D_c are the concentration and diffusivity of the chemical, respectively, and K is the rate of chemical uptake by the cells. The boundary conditions are ones of zero cell flux, and prescribed chemical concentration, C_0, at the surface.

Microorganisms can also become more, or less, active depending upon the chemical concentration (especially in the case of oxygen consuming microorganisms). This was also modeled by Hillesdon et al. (1995) by allowing the swimming speed (and consequently diffusivity), and oxygen uptake rate to be functions of (non-dimensional) chemical concentration

$$U = U_0W(\theta), \quad K = K_0W(\theta). \tag{7.26}$$

The saturating function $W(\theta)$ was modeled either as a Heaviside step function, $W = H(\theta)$, or a smoothed step function, $W = (1 - \exp(-\theta/\theta_1))H(\theta)$.

It is worth noting that a modification of the Keller–Segel model which allows for a cutoff oxygen level, ϵ_0, below which no oxytaxis occurs

$$\boldsymbol{U} = k\xi H(\xi)\hat{\boldsymbol{e}}, \tag{7.27}$$

where $\hat{\boldsymbol{e}}$ is the direction of the oxygen gradient, $\xi = |\nabla\theta| - \epsilon_0$, although this was shown by Hillesdon et al. (1995) to have minimal impact upon the cell and oxygen gradients.

A difference between the impact of oxytaxis (chemotaxis with the consumed chemical being oxygen, which typically has its greatest concentration at the water–air interface) upon the cell concentrations, as opposed to that due to gyrotaxis, is that the oxygen depletes with depth. This means that the cells sufficiently far below the surface are dormant, as there is insufficient oxygen for them to respire, and few cues to direct them towards the more oxygen-rich upper regions. Hence it seems reasonable to expect that the cell profiles might take on a different form in a suspension deep enough for the chemical to become depleted, than in a shallow suspension (and we shall see this to be the case later).

It should further be noted that both (7.23) and (7.27) neglect the influence of the background flow and torque balance (7.1) upon the swimming direction. In that sense they are phenomenological in nature, in similar fashion to the models (7.12) and (7.17) of phototaxis, and preclude study of any simultaneous effects of both chemotaxis and gyrotaxis.

Dynamic Model

One way that some models have addressed this shortcoming for chemotactic run-and-tumble bacteria is by linking the chemical concentration to the length of time a cell spends traveling in a given direction. Run-and-tumble bacteria change their swimming direction randomly (tumble) at intermittent times. If they find themselves swimming along favorable chemical gradients, then the length of time until the next tumble is longer. Equivalently, their tumble rate, Λ, is lower. For weak chemotaxis this can be model as (Bearon and Pedley, 2000)

$$\Lambda = \Lambda_0 (1 - \alpha \nabla C \cdot \boldsymbol{p}) \tag{7.28}$$

where α is some constant of proportionality and Λ_0 is the tumble rate in the absence of any stimulus. More recently, Ezhilan et al. (2012) have proposed the alternative

$$\Lambda = \Lambda_0 \exp\left(-\alpha \frac{DC}{Dt}\right), \quad \frac{DC}{Dt} = \frac{\partial C}{\partial t} + (U_0 \boldsymbol{p} + \boldsymbol{u}) \cdot \nabla C. \tag{7.29}$$

7.4 NON-FLOWING SUSPENSIONS

In Section 7.3 we observed how the swimming motion of individual cells can be affected by external stimuli. We now wish to see how these swimming responses, or *taxes*, affect the bulk dynamics of a suspension containing large numbers of cells. For the moment we shall assume that the fluid in which the cells are immersed is not flowing (we shall consider the case of flowing suspensions in the next section).

A widely-used framework for modeling such suspensions was proposed by Pedley and Kessler (1990), who invoked a cell conservation argument to obtain an advection–diffusion equation description for cell density, $n(\boldsymbol{x}, t)$,

$$\frac{\partial n}{\partial t} + \nabla \cdot \boldsymbol{J} = 0, \quad \boldsymbol{J} = n\boldsymbol{U} - \boldsymbol{D} \cdot \nabla n, \tag{7.30}$$

where $\boldsymbol{U}$ is the swimming velocity (and may take the form of the average swimming velocity, $\langle \boldsymbol{U} \rangle$, in situations where the swimming direction has been modeled as a stochastic quantity, in accordance with (7.5)). It is then possible to examine solutions to (7.30) for the different forms for $\boldsymbol{U}$ considered in Section 7.3. The cell conservation equation (7.30) is typically solved subject to no flux of cells through boundaries or free surfaces, i.e. $\boldsymbol{J} \cdot \boldsymbol{n} = 0$, where $\boldsymbol{n}$ is a unit normal to the boundary.

The translational diffusivity of cells, as characterized by the diffusivity tensor $\boldsymbol{D}$, is due not so much to molecular transport but rather larger magnitude effects such as biological variability in swimming actions. It can be defined in terms of the statistical variance of the orientation vector (Pedley and Kessler, 1990)

$$\boldsymbol{D} = U_0^2 \tau \operatorname{var}(\boldsymbol{p}), \tag{7.31}$$

where τ is a direction correlation time for the microorganism. This model of the diffusivity is acceptable, provided that the rate of strain of the background flow is not too high. In that case, a more sophisticated approach can be taken, which uses Generalized Taylor Dispersion theorem (Hill and Bees, 2002; Manela and Frankel, 2003). An alternative dispersion theorem which is less comprehensive, but perhaps easier to implement, has also been put forward by Bees and Croze (2010).

Whereas the cell conservation equation provides the total density of cells at a given location, sometimes it is necessary to also know their orientations. The Smoluchowski equation considers the density of particles at a location with a particular direction, $\Psi(\boldsymbol{x}, \boldsymbol{p})$. In a sense, the Smoluchowski equation captures the information contained in both the cell conservation equation and Fokker–Planck equation (7.4) (but also contains the influence of translation velocity and diffusivity upon the swimming direction, which the latter does not). It takes the form

$$\frac{\partial \Psi}{\partial t} + \nabla \cdot \boldsymbol{J}_T + \nabla_p \cdot \boldsymbol{J}_R + g(\Psi) = 0, \tag{7.32a}$$

with

$$\boldsymbol{J}_T = U_0 \boldsymbol{p} \Psi - \boldsymbol{D} \cdot \nabla \Psi, \quad \boldsymbol{J}_R = \dot{\boldsymbol{p}} \Psi - D_R \nabla_p \Psi. \tag{7.32b}$$

Here the function $g(\Psi)$ captures any influence of external stimuli, and will prove to be useful when considering chemotaxis of run-and-tumble bacteria (see Section 7.4.3). The Generalized Taylor Dispersion for such run-and-tumble motion on the dispersion, and consequently the tensor $\boldsymbol{D}$, has been considered by Bearon (2003).

7.4.1 Gyrotactic Phototrophs

In the absence of any flow ($\dot{\boldsymbol{\gamma}} = \boldsymbol{\omega} = \boldsymbol{0}$), the solution to the Fokker–Planck equation (7.4) when there is a gyrotactic torque (7.7) gives the following form for the swimming distribution function (Brenner and Weissman, 1972)

$$f = \frac{\lambda \exp(\lambda \boldsymbol{p} \cdot \hat{\boldsymbol{z}})}{4\pi \sinh \lambda}, \tag{7.33}$$

recalling that $\lambda = 1/2BD_R$ where D_R is the rotational diffusivity in the Fokker–Planck equation (7.4). This leads to an average swimming orientation given by Pedley and Kessler (1992)

$$\langle \boldsymbol{p} \rangle = K_1 \hat{\boldsymbol{z}}, \tag{7.34}$$

where

$$K_1 = (\coth \lambda - 1/\lambda), \tag{7.35}$$

and a diffusivity tensor that is axisymmetric about $\hat{z}$

$$D = D_0 \begin{pmatrix} K_1/\lambda & 0 & 0 \\ 0 & K_1/\lambda & 0 \\ 0 & 0 & K_2 \end{pmatrix}, \tag{7.36}$$

where $D_0 = U_0^2 \tau$ and

$$K_2 = 1 - \coth^2 \lambda + 1/\lambda^2. \tag{7.37}$$

Assuming that the surface of the water lies at $z = 0$, the (non-dimensional) cell concentration profile is given by

$$n(z) = \exp(\kappa z), \tag{7.38}$$

where

$$\kappa^{-1} = K_2 U_0 \tau / K_1, \tag{7.39}$$

is referred to as the sublayer depth, and is the characteristic depth for the cell-rich layer at the free surface due to upwards swimming.

7.4.2 Photosensitive Phototrophs

Without Scattering

In the absence of any background flow, we can consider how the distribution of cells in a profile are affected under the dynamic models of phototaxis considered in Section 7.3.2.

Assuming the phenomenological relationship between illumination levels and swimming direction (7.19), and assuming absorption only (7.13) (i.e. no scattering), the swimming distribution function f remains as given by (7.33), with the value of B still given by (7.8) (since this phototaxis model affects swimming speed, but not swimming direction). In a shallow suspension the steady cell density profile has been shown to satisfy (Williams and Bees, 2011)

$$\frac{d^2 m}{dz^2} - d\,(1 - \chi \exp(\kappa m))\,\frac{dm}{dz} = 0, \tag{7.40}$$

where $m = -\int_z^H n(z) \mathrm{d}z$ and

$$\chi = I_s / I_c, \quad d = \frac{\bar{K}_1 H U(0)}{\bar{K}_2 D_0}, \tag{7.41}$$

where $U(0)$ is the swimming speed in the absence of any illumination, K_1 and K_2 are respectively given by (7.35) and (7.37), and $\bar{K}_1$, $\bar{K}_2$ refer to the values of K_1 and K_2 evaluated at a specific value of λ (in their results Williams and Bees, 2011, use $\lambda = 2.2$, which are appropriate to the algae *C. nivalis*). The parameter d is a ratio of the sublayer depth for pure gyrotaxis (7.39) and the overall depth of the suspension, H.

When phototaxis is instead accounted for through an effective gyrotactic torque, (7.20), the swimming orientation function and diffusivity tensor still take the same form given in (7.33) and (7.36), respectively, with λ now a function of cell density

$$\lambda = \frac{1}{2B(I)D_R}(1 - \chi \exp(\alpha m(z))), \tag{7.42}$$

where the parameter B now depends upon illumination level I through the cell's center-of-mass (7.20). Under this phototaxis model, the cell concentration satisfies the following equation (Williams and Bees, 2011)

$$\frac{d^2m}{dz^2} - \frac{H\bar{K}_2}{\bar{K}_1}\frac{K_1(\lambda)}{K_2(\lambda)}\frac{dm}{dz} = 0, \tag{7.43}$$

where $K_1(\lambda)$ and $K_2(\lambda)$ are given by (7.35) and (7.37), but with λ now given by (7.42).

Finally, when phototaxis is instead modeled using a photo-reactive torque (7.21), the Fokker–Planck equation governing direction probability is more difficult to solve. For certain specific values of β_1 and β_2 the swimming direction probability function once again takes the form (7.33) but with (Williams and Bees, 2011)

$$\lambda = \frac{1}{2BD_R}\left[1 - \xi\chi \exp(\alpha m(z))(\chi \exp(\alpha m(z)) - 1)\right], \tag{7.44}$$

($\beta_1 = 1$ and $\beta_2 = 0$) and

$$\lambda = \frac{1}{2BD_R}\left[1 - \zeta\chi\frac{dm}{dz}\exp(2\alpha m(z))(\chi \exp(\alpha m(z)) - 1)\right], \tag{7.45}$$

($\beta_1 = 0$ and $\beta_2 = 1$), where

$$\zeta = 4t_m/h(0)mg \tag{7.46}$$

with t_m the maximum value of t (the strength of the photo-reactive torque (7.22)). The cell density is once again governed by (7.43), but with λ now given by (7.44) or (7.45) (for these special values of β_1 and β_2).

Fig. 7.1 shows the predictions made by the different models of Williams and Bees (2011). Plots A–D show cell densities from the photokinesis model (7.40) as a function of depth, at different illumination levels $\chi = I_s/I_c$ (recalling that I_c denotes the optimum light level, at which the cells have zero swimming speed). The cells are taken to be bottom heavy, with $\kappa = 1.2$ and $d = 20$. As χ increases from 0 (no light) to 1 (optimal light), cell swimming speeds decrease, and it is evident that the high-density surface layer of cells (which exists due to their bottom-heaviness) becomes less localized to the surface. Once the light levels become greater than the optimal level ($\chi > 1$) it can be seen that the competition between upwards swimming due to bottom-heaviness, and downwards swimming due to repulsion from the light at the surface, leads to the highest concentration of cells being located at some intermediate depth in the suspension. As χ is increased still further, the high-density continues to

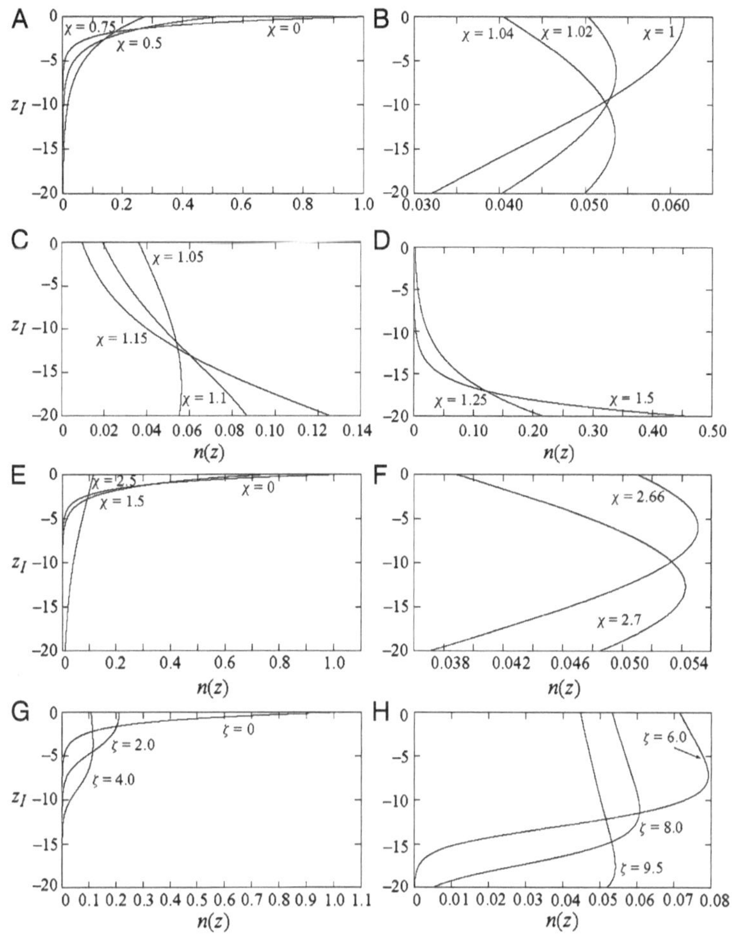

FIGURE 7.1 Phototaxis: Cell densities as a function of depth. Predictions made using: (A)–(D) the photokinesis model (7.40) for different values of $\chi = I_s/I_c$ ($d = 20$, $\kappa = 1.2$); (E)–(H) a photo-reactive torque (7.43) with λ given by (E)–(F) (7.44) and χ varying ($d = 20$, $\kappa = 1$, $\zeta = 0.236$) or (G)–(H) (7.45) and ζ varying ($d = 20$, $\kappa = 1$, $\chi = 2$ and $\eta = 0.01$) (reproduced from Williams and Bees, 2011 with permission).

move downwards, until the bottom of the suspension is reached. Although not shown, similar quantitative behavior has been predicted by the models with a light-dependent center of mass (7.20) (Williams and Bees, 2011).

Figs. 7.1E–F show the cell concentrations predicted using the photo-reactive torque model (7.43), when $\beta_1 = 1$ and $\beta_2 = 0$ (7.44), again as a function of scaled illumination level χ. Once again the cells are also bottom heavy as well as being phototactic ($\kappa = 1$ and $d = 20$), and the non-dimensional strength of the photo-reactive torque here is $\zeta = 0.236$. The trends are similar to those predicted by the photokinesis model, however, the illumination level needs to significantly exceed the optimal light level before the cell-rich region moves below the surface of the water.

Finally, the influence of the strength of the photo-reactive torque, ζ, is explored in plots G–H, for the case where $\beta_1 = 0$ and $\beta_2 = 1$, i.e. (7.45). Here the level of illumination is fixed at $\chi = 2$, double the optimal value, and the cells are bottom heavy with parameters $\kappa = 1.2$ and $d = 20$. It can be see that as the strength of the photosensitivity increases, and the cells are more readily able to move away from the undesirably high levels of light at the surface, the region of highest cell concentration shifts to some level below the water surface.

With Scattering

Using the phenomenological model described in Section 7.3.2, which neglects any balance of torque considerations but does describe cell scattering, Ghorai et al. (2010) have considered how isotropic scattering (i.e. light is scattered equally in all directions) affects the cell distribution through the layer. Using the phototaxis model that incorporates light scattering by cells (7.17), the cell concentration equation takes the form

$$\frac{dn}{dz} = U_0 T(I) n, \tag{7.47}$$

where

$$I = \exp\left(-\kappa_0 \int_z^H n(z')\,\mathrm{d}z'\right) + \int J'(z, \theta, \phi) \sin\theta \,\mathrm{d}\theta \,\mathrm{d}\phi, \tag{7.48}$$

with J' satisfying the integro-differential equation

$$z' \frac{dJ'}{dz} + \kappa_0 n J' = \frac{\omega \kappa_0 n}{4\pi} I, \tag{7.49}$$

where $z' = dz/ds$, subject to the boundary conditions

$$J'(0, \theta, \phi) = 0, \quad 0 \le \theta \le \pi/2, \quad J'(H, \theta, \phi) = 0, \quad \pi/2 \le \theta \le \pi, \tag{7.50}$$

where the surface of the water lies at $z = H$. Note that when $\omega = 0$ (i.e. no scattering), we have $J' = 0$ as a solution, and so we recover the Lambert–Beer Law for light intensity. Ghorai et al. (2010) have numerically solved (7.47)–(7.50) for different values of the absorption and scattering coefficients κ_0 and σ, using a taxis function that takes its critical value at $I = I_c$.

Fig. 7.2A shows that in the presence of scattering ($\omega \neq 0$), there are greater levels of light at any given depth. In the absence of scattering the cell concentrations are greatest at the surface, where the light intensity is also highest (see Fig. 7.2C). However, as the level of scattering is increased, the maximum cell concentration and peak in light intensity occur at some (non-coincident) level below the water surface. This is reminiscent of what was observed in the predictions of the photo-reactive torque, although the mechanism here is different (there is no bottom-heaviness here as in the photo-reactive torque model). As the scattering is pushed to still greater levels, so that it completely dominates over absorption, Ghorai et al. (2010) show that accumulation of cells can in fact occur at two depths. The first accumulation takes place at the water surface, with a second accumulation occurring at a finite depth below the surface (see Fig. 7.2E).

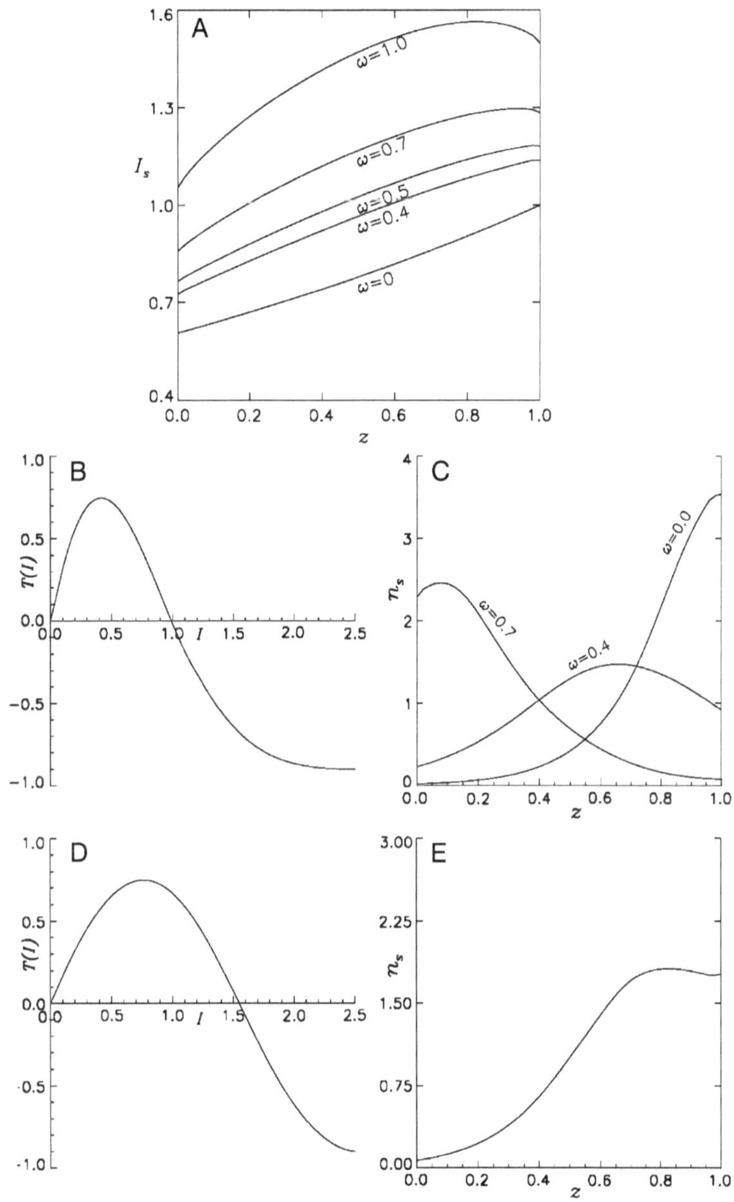

FIGURE 7.2 (A) Light intensity at different depths with absorption $\kappa = 0.5$, and different levels of scattering, ω. (C) Cell concentration as a function of depth for the taxis function T shown in (B), for different levels of scattering ω. Here the critical illumination is given by $I_c = 1$, the absorption is $\kappa = 0.5$. (E) Cell concentration as a function of depth for the taxis function T shown in (D) when scattering is increased to $\omega = 1$ (here $I_c = 1.55$, $\kappa = 0.5$) (reproduced from Ghorai et al., 2010 with permission).

One further refinement to the scattering model has been added by Ghorai and Panda (2013), who consider anisotropic scattering. This is in acknowledgment of the fact that algae tend to scatter more light forwards, than backwards. Under anisotropy the scattering law (7.49) is modified to

$$z'\frac{dJ'}{dz} + \kappa_0 n J' = \frac{\omega\kappa_0 n}{4\pi}\int I(\boldsymbol{x}, \theta', \phi')\Phi(\theta', \phi', \theta, \phi)\sin\theta' d\theta' d\phi', \tag{7.51}$$

where Φ is a scattering phase function, which Ghorai and Panda (2013) take to have the simple form

$$\Phi = 1 + A\cos\theta\cos\theta' \tag{7.52}$$

(θ is the polar angle, measured from the vertical). The effect of anisotropy on the light intensity within the suspension is to increase it in the upper parts of the suspension, and decrease it in the lower parts, as might be expected. There can also be situations where increasing the degree of anisotropy takes the suspension from a state where there is one dense cell layer in the absence of anisotropy, to a configuration where there are two dense layers.

7.4.3 Chemotaxis

Using the stimulus models described in Section 7.3.3, we can also consider the influence of chemical gradients upon cell densities within suspensions. In the absence of any background flow, by solving (7.30) alongside (7.23) and (7.25) (assuming $W(\theta) = H(\theta)$, i.e. a Heaviside step profile), Hillesdon et al. (1995) found the following concentration profiles for chemical and cell concentrations, respectively:

$$C(z) = 1 - \frac{2}{\gamma}\ln\left(\frac{\cos(A_1(1-z)/2)}{\cos(A_1/2)}\right), \tag{7.53}$$

$$n(z) = \frac{A_1^2}{2\gamma\beta}\sec^2(A_1(1-z)/2), \tag{7.54}$$

and A_1 solves $\tan(A_1/2) = \gamma\beta/A_1$, where

$$\gamma = aU_0/D_0, \quad \beta = K_0 N_0 H^2/D_C(C_0 - C_m) \tag{7.55}$$

(recognizing that (7.26) and (7.36) imply a diffusivity of the form $D = D_0 W(\theta)^2$). Here N_0 is the initial (uniform) cell concentration. (Note that we have used here the notation of Hillesdon and Pedley, 1996; in Hillesdon et al., 1995, $A_1 \mapsto \gamma A_1$ and $\beta \mapsto \beta^2$.) Here γ can be thought of as a measure of transport due to swimming in response to oxygen gradients, as compared with that due to diffusivity. β is a ratio of oxygen consumption by the cell, to oxygen diffusion.

By contrast, in a deep suspension, there is a critical depth, z_c, below which cells become inactive due to lack of oxygen. Above this critical depth, the cell and oxygen concentrations

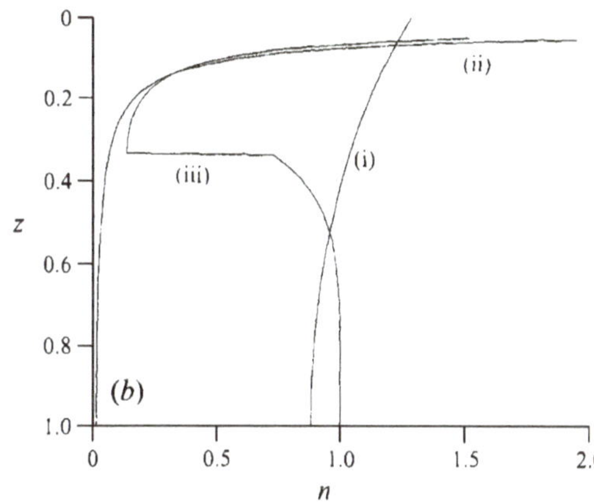

FIGURE 7.3 Chemotaxis: Density profiles computed for oxygen-sensitive (oxytactic) cells in both a shallow suspension (curves (i) and (ii)), and a deep suspension (iii). (Note that both deep and shallow suspensions z has been scaled on the maximum depth.) When $\gamma\beta = 300$ we see a high-density cell region near the upper surface (curves (ii) and (iii)), which is not there for lower values of $\gamma\beta$ (curve (i), $\gamma = 0.8$) (reproduced from Hillesdon and Pedley, 1996 with permission).

assume the following profiles

$$C(z) = \frac{2}{\gamma} \ln\left(\sec(A_2(z_c - z)/2)\right), \tag{7.56}$$

$$n(z) = \frac{A_2^2}{2\gamma\beta} \sec^2\left(A_2(z_c - z)/2\right) \tag{7.57}$$

where $A_2 = \gamma\beta \int_0^{z_c} n dz / \sqrt{\exp(\gamma) - 1}$. However, the value of the critical depth, z_c, must be found by solving the full initial value problem.

Steady cell concentration profiles are shown in Fig. 7.3, which have been computed by Hillesdon et al. (1995) from a well-mixed initial state ($n = 1$). Curve (ii) corresponds to a shallow suspension, whereas curve (iii) corresponds to a deep suspension. In both cases $\gamma\beta = 300$, and vertical distances are scaled on their respective suspension depths (and so non-dimensional depths in the figures are both normalized to unity). In both cases it can be observed that a high concentration layer of cells form near the water surface. In the case of the shallow suspension, the cell density continues to drop with depth, as there is a sufficient oxygen gradient throughout the fluid layer to draw the cells up towards the surface. By contrast, in the deep suspension we see that below a certain depth the cell level remains the same as in the original mixed state ($n = 1$), since there is insufficient oxygen at these depths to stimulate upwards swimming. At lower depths there is a depletion layer, where the cells begin to sense the oxygen gradients and swim up towards the surface. The form of this depletion layer is seen to take the form of a gradual decrease in cell concentration, followed by an abrupt drop in cell concentration. Above this depth the density profile then follows quite closely that predicted for the shallow suspension. As either diffusivity in either cell motion or oxygen consumption increases relative to other effects, which corresponds to a decrease in either γ or β, the high-density cell region near the water surface in a shallow suspension disappears, and the cell profile curve gradually decreases with depth (as shown by curve (i) in Fig. 7.3, computed for $\gamma\beta = 0.8$).

The above predictions are made using the fairly generic and phenomenological relationship between swimming velocity and concentration gradients (7.23). For the case of bacteria which follow gradients using run-and-tumble strategies, Bearon and Pedley (2000) proposed a model based upon the Smoluchowski equation which better captures the details of the chemotaxis. In particular, in (7.32) they posed the following form for the stimulus function g:

$$g(\Psi, \boldsymbol{p}) = -\Lambda(\boldsymbol{p})\Psi + \int T(\boldsymbol{p}, \boldsymbol{p}')\Lambda(\boldsymbol{p}')\Psi(\boldsymbol{p}')\mathrm{d}\boldsymbol{p}'. \tag{7.58}$$

Here $T(\boldsymbol{p}, \boldsymbol{p}')$ is a transition function which encodes the likelihood of a bacterium tumbling from orientation $\boldsymbol{p}'$ into orientation $\boldsymbol{p}$. For simplicity this is taken by Bearon and Pedley (2000) to be isotropic, i.e. $T(\boldsymbol{p}, \boldsymbol{p}') = 1/4\pi$, and the tumbling rate Λ depends upon the chemical concentration gradient, as specified in (7.28).

7.5 FLOWING SUSPENSIONS

The effect of different stimuli on cell populations considered in the previous section assume that the fluid in which the cells are suspended does not flow, and that the only transport of cells comes about due to their inherent self-motility. However in practice, even when not externally driven (by pressure gradients, for example) the immersing fluid can still flow. This is because some of the cell configurations previously described are hydrodynamically unstable, and when they destabilize they can produce visible patterns on a scale many hundreds or thousands of times larger than the size of individual cells.

In these situations the fluxes in the cell conservation equation (7.30), and the Smoluchowski equation (7.32), respectively, extend to the forms

$$\boldsymbol{J} = n\,(\boldsymbol{u} + \langle \boldsymbol{U} \rangle) - \boldsymbol{D} \cdot \nabla n, \tag{7.59}$$

$$\boldsymbol{J}_T = \Psi\,(\boldsymbol{u} + U_0\boldsymbol{p}) - \boldsymbol{D} \cdot \nabla\Psi, \quad \boldsymbol{J}_R = \dot{\boldsymbol{p}}\Psi - D_R\nabla_p\Psi, \tag{7.60}$$

where $\boldsymbol{u}$ is the velocity of the immersing fluid. However, microorganisms are not just passively transported by the background flow, but rather their presence can modify the flow itself. This can occur because the cells have a different density and rheology to the immersing fluid, i.e. they do not deform in the same manner to the equivalent volume of fluid that they have displaced. Their motility can also generate additional sources of stress in the fluid, which we shall come to shortly.

In order to model these effects, it is necessary to supplement the continuum model discussed in Section 7.4 with suitable flow equations. In their most general form, the governing flow equations are

$$\rho\left(\frac{\partial \boldsymbol{u}}{\partial t} + \boldsymbol{u} \cdot \nabla\boldsymbol{u}\right) = -\nabla p + \mu\nabla \cdot \boldsymbol{\tau} + \boldsymbol{F} = \boldsymbol{0}, \quad \nabla \cdot \boldsymbol{u}, \tag{7.61}$$

where $\boldsymbol{u}$ is the flow velocity, p is flow pressure, ρ is the density of the fluid, $\boldsymbol{F}$ represents any body forces on the fluid (and can depend upon the cell concentration, n), and $\boldsymbol{\tau}$ is the deviatoric stress tensor.

The stress tensor is in general the sum of four contributions, i.e. $\boldsymbol{\tau} = \boldsymbol{\tau}_1 + \boldsymbol{\tau}_2 + \boldsymbol{\tau}_3 + \boldsymbol{\tau}_4$, where $\boldsymbol{\tau}_1$ is due to the fluid phase, $\boldsymbol{\tau}_2$ is due to the presence of the cells, $\boldsymbol{\tau}_3$ is due to the fact that the cells are motile and $\boldsymbol{\tau}_4$ is due to the presence of any external torques. In the case where the fluid phase is Newtonian, the stress is linearly proportional to the flow's rate of strain $\dot{\boldsymbol{\gamma}}$, i.e.

$$\boldsymbol{\tau}_1 = 2\mu\dot{\boldsymbol{\gamma}}, \tag{7.62}$$

where μ is the fluid's dynamic viscosity, and

$$\dot{\boldsymbol{\gamma}} = \frac{1}{2}\left(\nabla\boldsymbol{u} + (\nabla\boldsymbol{u})^T\right) \tag{7.63}$$

(although it is worth noting that there is increasing interest in microorganism suspensions in non-Newtonian fluids, Lauga, 2006, 2011). The general form of additional stress due to the presence of the rigid particles (in this case cells) is given by Batchelor (1970) as

$$\begin{aligned}\boldsymbol{\tau}_2 = &\alpha_1\left(\boldsymbol{p}\otimes\boldsymbol{p} - \frac{1}{3}\boldsymbol{I}\right) + \alpha_2\left(\boldsymbol{p}\cdot\dot{\boldsymbol{\gamma}}\cdot\boldsymbol{p}\right)\boldsymbol{I} + \alpha_3\dot{\boldsymbol{\gamma}} \\ &+ \alpha_4\left(\boldsymbol{p}\cdot\dot{\boldsymbol{\gamma}}\cdot\boldsymbol{p}\right)\boldsymbol{p}\otimes\boldsymbol{p} + \alpha_5\left(\boldsymbol{p}\otimes(\dot{\boldsymbol{\gamma}}\cdot\boldsymbol{p}) + (\dot{\boldsymbol{\gamma}}\cdot\boldsymbol{p})\otimes\boldsymbol{p}\right).\end{aligned} \tag{7.64}$$

The coefficients α_2, α_3, α_4 and α_5 are functions of the cell geometry, and are given in Pedley and Kessler (1990). The coefficient α_1 multiples the force dipole contribution to stress (i.e. the stresslet), and also consists of a (less important) component due to rotary diffusion. The third contribution in $\boldsymbol{\tau}_2$ takes the same form as the Newtonian stress tensor (7.62), and so has the effect of increasing the dynamic viscosity by an amount $\alpha_3/2$. The remaining coefficients α_2, α_4 and α_5 are all zero for cells that are spherical, and so are often neglected in models of nearly spherical microorganisms such as the algae *C. Nivalis*, or protozoa such as *Opalina*. However, for many bacteria which have elongated cells, these coefficients are non-zero. Indeed, in models of fibre suspensions they are known to be important (Eriksen, 1960).

For motile microorganism there is another component of stress, that also takes the form of a stresslet

$$\boldsymbol{\tau}_3 = nS\left(\boldsymbol{p}\otimes\boldsymbol{p} - \frac{1}{3}\boldsymbol{I}\right) \tag{7.65}$$

where S is a measure of swimming strength, and which is positive for *pullers* and negative for *pushers*. The terms *pushers* and *pullers* describe the way in which microorganisms use their swimming apparatus to move through the fluid. An example of such a *puller* is the alga *Chlamydomonas reinhardtii*, which has two forward pointing flagella, which it uses to breaststroke through the fluid. In contrast *pushers* tend to generate thrust at their rear, with flagellated bacteria being a good example. For gyrotactic microorganisms such as algae it has traditionally been thought that the active stress given by $\boldsymbol{\tau}_3$ does not play a significant role (Pedley and Kessler, 1990; Pedley, 2010) (although can be important for non-gyrotactic microorganisms; see the section on bacterial chemotaxis below).

If there is an external torque acting upon the cells, as is the case for bottom-heavy microorganisms, then formally it is also necessary to include the gravitational torque into the particle stress as well (Batchelor, 1970)

$$\tau_4 = \frac{1}{2}n\left(\langle \epsilon \cdot \langle \boldsymbol{L}' \rangle + \frac{1}{2}\alpha_0 \left((\boldsymbol{p} \times \boldsymbol{L}')\, \boldsymbol{p} + \boldsymbol{p}\, (\boldsymbol{p} \times \boldsymbol{L}') \right) \right), \tag{7.66}$$

where

$$\boldsymbol{L}' = -\rho v g h \boldsymbol{p} \times \hat{\boldsymbol{z}}. \tag{7.67}$$

Here $\boldsymbol{\epsilon}$ is the Levi-Civita tensor and $\alpha_0 = (a^2 - b^2)/(a^2 + b^2)$ with a and b being the minor and major axes. Once again, this is not thought to play a major role in the bulk dynamics of gyrotactic suspensions.

So, in summary, (7.61)–(7.67) together with (7.30) and (7.59) (or sometimes (7.32) and (7.60)) model the coupled dynamics between the suspension of cells, and the fluid in which they are immersed. It should be noted that no cell–cell interactions are included here, which limits the applicability of the model to semi-dilute suspensions. The term semi-dilute refers to situations where the concentration of cells is sufficiently high that the suspension can be treated as a continuum, but not so high that cell–cell hydrodynamic interactions are expected to play a major role.

7.5.1 Gyrotactic Focusing

One of the most straightforward examples of how a background flow can affect the spatial distribution of cells occurs when a suspension of gyrotactic microorganisms is placed in vertical pipe or channel flow. If there were no flow, the gravitational torque would result in all of the microorganisms swimming (on average) upwards, and so we would not expect to see any redistribution of the cells in the radial direction. However, the no-slip condition on the pipe walls sets up a radial velocity gradient across the pipe, which exerts an additional viscous torque upon the cells. The balance between these two torques endows the microorganisms with a mean swimming direction that is away from the vertical. When the background flow is upwards, the viscous torque acts in a clockwise direction (see Fig. 7.4A), the gravitational torque attempts to restore the swimming direction to the vertical, and consequently the mean swimming direction is radial outwards. Hence the cells tend to aggregate on the pipe wall (which could have important implications for the formation of biofilms). On the other hand, if the pipe flow is downwards, the viscous torque acts in an anti-clockwise direction (see Fig. 7.4B), the gravitational torque again acts to restore the swimmer to the vertical, and the cells on average swim inwards. This leads to an accumulation, or focusing, of the cells along the pipe's centerline (see Fig. 7.4C). Cell focusing of this type can be predicted from the cell conservation models. For Poiseuille flow through a pipe of radius R, and with centerline ve-

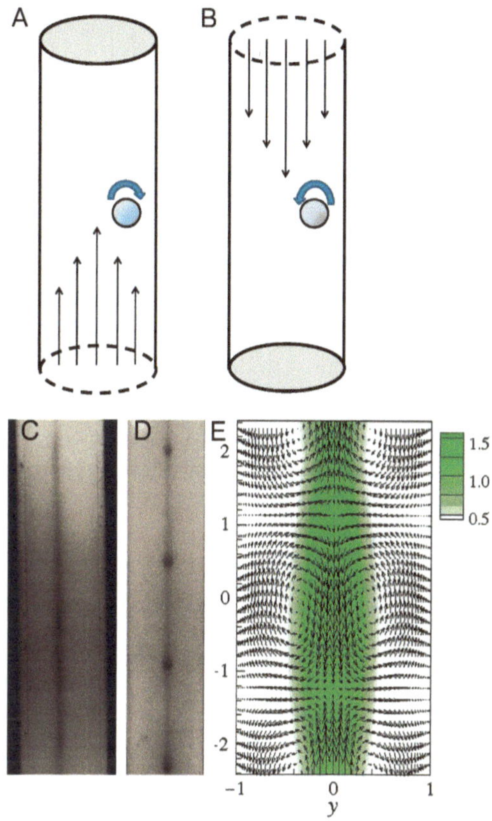

FIGURE 7.4 Gyrotactic focusing: (A) pipe flow that is directed upwards exerts a clockwise viscous torque upon bottom-heavy cells, which is countered by a gravitational torque which attempts to restore an upwards swimming direction, leading to an outwards average swimming; (B) pipe flow that is directed downwards exerts an anti-clockwise viscous torque upon the cells, which leads to an inwards mean swimming. Experimental using the swimming algae *C. nivalis* suspended in vertical downflow in a pipe (reproduced from Kessler, 1986, with permission). (C) Cells initially focus into a relatively uniform beam at the pipe's center, (D) at later times this beam can develop *blips*, (E) computed flowfield generated by the (varicose) blip instability in a 2D channel, with the color representing cell concentrations (reproduced from Hwang and Pedley, 2014 with permission).

locity W_0, the equilibrium swimming direction for a spherical swimming cell in cylindrical coordinates can be found from (7.7) to be (Pedley and Kessler, 1992)

$$\mathbf{p} = \left(-2BW_0 r/R^2, 0, (1 - 4B^2 W_0^2 r^2/R^4)^{1/2}\right) \tag{7.68}$$

provided that $2BW_0 r/R^2 < 1$ for all $r < R$. If this is used directly in the cell conservation equation (7.30) (i.e. assuming no stochasticity in swimming direction), and prescribing a uniform cell distribution n_0 at $z = z_0$ (the pipe inlet), then the following implicit relationship for cell

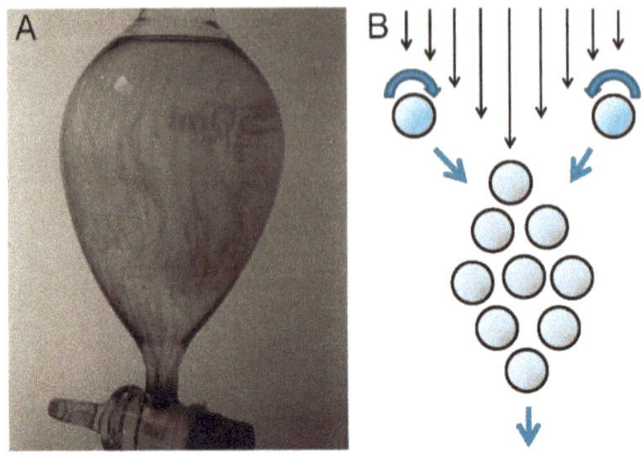

FIGURE 7.5 Gyrotactic plume formation: (A) experimental image showing the formation of plumes within a suspension of the swimming algae *Chlamydomonas nivalis*. The image is taken approximately 5 minutes after the suspension is stirred into a well-mixed state (reproduced from Kessler, 1986, with permission); (B) illustration of the basic mechanisms behind plume formation. A sedimenting bottom-heavy region of cells sets up a trailing flowfield, which causes the neighboring bottom-heavy cells to be turned and recruited into the high-density region.

concentration is obtained (Kessler, 1985)

$$\frac{4BU_0}{R^2}(z_0 - z) = \ln\left(\frac{n}{n_0}\right) + \frac{r^2}{R^2}\left(1 - \frac{n}{n_0}\right) \tag{7.69}$$

(assuming that cell swimming speeds are much less than W_0).

This beam of cells can subsequently destabilize, leading to the appearance of *blips* spaced along the beam (see Fig. 7.4D). These blips travel faster than the background flow, and develop into toroidal vortices downstream (Kessler, 1986). Hwang and Pedley (2014) have recently performed a stability analysis of the hydrodynamically-focused state, and elucidated the mechanism of blip formation. The flowfield predicted by this analysis for a varicose instability is shown in Fig. 7.4E, which shows a structure reminiscent of the blip. There are a number of candidates for the physical mechanism behind the instability, including inflexional points in the flow, as well as gravitational effects. However, it was shown by Hwang and Pedley (2014) that physically the instability is dominated by spatial gradients in $\langle \boldsymbol{p} \rangle$ (although this is not to say that the other instability mechanisms do not play a role at all; mechanisms related to gravity are also found to be important).

7.5.2 Gyrotactic Plumes

Gyrotaxis can also lead to the spontaneous formation of high-density striations of cells within the bulk suspension. The experiments of Kessler (1986) showed how dense plumes of cells could appear out of an initially uniform suspension of swimming algae (see Fig. 7.5A). This occurs because even for a suspension in which the spatial distribution of cells is initially uniform, local high-density regions will inevitably form. This local agglomeration of cells will sediment slightly faster than the surrounding cells, which setups a trailing downwards flow. Cells in the vicinity will then be turned towards this sinking agglomeration due to their

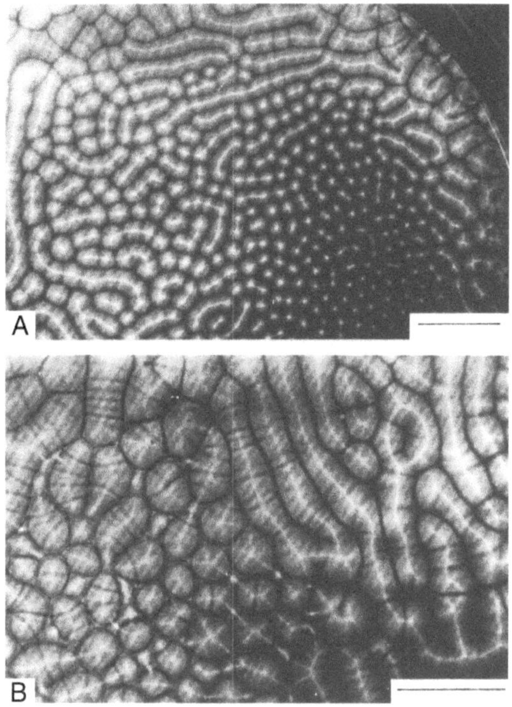

FIGURE 7.6 Bioconvection: Experiments showing cell pattern formation due to bioconvection in a shallow layer of fluid. Plan view of a circular Petri dish, depth 2.5 mm, containing a suspension of the bacterium *Bacillus subtilis* (initial concentration $n \approx 10^6$ cells mm^3). The photographs were taken with dark-field illumination; the white regions correspond to relatively large concentrations of cells and regions low in cell concentration are black. The scale marks indicate 1 cm: (A) near the wall of the Petri dish, and (B) near the center of the Petri dish, where the fluid was slightly deeper than in (A) (reproduced from Hillesdon et al., 1995 with permission).

bottom-heaviness, and be recruited into it (as illustrated in Fig. 7.5B). This process of entrainment continues, ultimately leading to observable plumes of cells.

7.5.3 Bioconvection

One of the best-studied examples of the rich dynamics that can emerge from coupling between motile microorganisms and their immersing fluid arises because microorganisms are typically slightly more dense than the fluid (see Fig. 7.6). This leads to a buoyancy term in the flow equations (7.61), namely

$$\boldsymbol{F} = -n\nu\Delta\rho g\hat{z}, \tag{7.70}$$

where $\Delta\rho$ is the excess density of the cell over that of water and ν is the cell volume.

When considering bioconvection the actual swimming action of the microorganisms could potentially have some relevance. Specifically here, the importance of the active stress term (7.65) seems to depend upon whether the microorganisms is a *pusher* or *puller*. For a bottom-heavy *puller*, such as many alga, it is generally considered to be of secondary importance to the effects of gyrotaxis.

Non-denationalizing distances upon the depth of the suspension, H, cell concentration on some reference value, n_0, times on the timescale for cell diffusion, $H^2/U_0^2\tau$, and flow velocities on $U_0^2\tau/H$, the flow is governed by the non-dimensional Boussinesq equation (assuming that the immersing fluid is considered to be Newtonian)

$$\mathrm{Sc}^{-1}\left(\frac{\partial \boldsymbol{u}}{\partial t} + \boldsymbol{u}\cdot\nabla\boldsymbol{u}\right) = -\nabla p + \nabla^2\boldsymbol{u} - \mathrm{Ra}\,n\hat{z} = \boldsymbol{0}, \quad \nabla\cdot\boldsymbol{u} \tag{7.71}$$

(formally $\boldsymbol{\tau}_2$, $\boldsymbol{\tau}_3$ and $\boldsymbol{\tau}_4$ should be included, however, as discussed above, these play a minimal role in this context, Pedley and Kessler, 1992). Here

$$\mathrm{Sc} \equiv \frac{\mu}{\rho U_0^2\tau}, \quad \mathrm{Ra} \equiv \frac{\nu\Delta\rho n_0 g h^3}{\mu U_0^2\tau} \tag{7.72}$$

are the Schmidt and Rayleigh numbers, respectively.

It is well known that flows governed by the Boussinesq equation are unstable in situations where the fluid becomes stratified in such a way that a denser layer of fluid sits above a less dense layer. When this density stratification occurs due to thermal heating of the fluid from below (since hot fluid is usually less dense than cold fluid), this can lead to the well-known Rayleigh–Benard instability. Here convective cells can spontaneously appear above some critical Rayleigh number, which transports less dense fluid up to the surface, and the denser fluid down (Drazin and Reid, 2004).

For the case of microorganisms, we have already seen examples in Section 7.3 and Section 7.3.3 of how cells response to stimuli can also produce unstable density stratifications (see Figs. 7.1, 7.2 and 7.3). The instability that ensues in this context is known as *bioconvection*, a term first used by Platt (1961). As might be expected, the details of the instability, and the pattern that forms, depends upon the nature of the stimulus, or stimuli, causing the density stratification.

The earliest studies of bioconvection were concerned with bottom-heavy microorganisms. In addition to the effects of buoyancy, the cells are also subject to the gyrotactic effects discussed in Section 7.4.1. Bees and Hill (1998) performed the first linear stability analysis on the equilibrium profile of a bottom-heavy suspension given by (7.38). Assuming normal modes of the form

$$n(x, y, z, t) = \Phi(z)\exp(\mathrm{i}lx + \mathrm{i}my + \sigma t), \tag{7.73}$$

$$u_3(x, y, z, t) = W(z)\exp(\mathrm{i}lx + \mathrm{i}my + \sigma t) \tag{7.74}$$

(as with the classical thermal convection problem, it is only necessary to consider the vertical component of velocity, u_3; the remaining components come from Conservation of Mass). Assuming no flow (i.e. solid upper boundary, rather than a free surface) and no cell fluxes at the top and bottom surfaces, they obtained the following eigenvalue relations

$$\left(\sigma \mathrm{Sc}^{-1} + k^2 - \frac{d^2}{dz^2}\right)\left(k^2 - \frac{d^2}{dz^2}\right) W = -\mathrm{Ra}\, k^2 \Phi,$$
$$\left(K_2 \frac{d^2}{dz^2} - K_2 d \frac{d}{dz} - \frac{K_1 k^2}{\lambda} - \sigma\right)\Phi = d e^{dz}\left(1 - \eta\left[P_5 \frac{d^2}{dz^2} + P_6 d \frac{d}{dz} - P_7 k^2\right]\right) W, \qquad (7.75)$$

where $k = |(l, m)|$ and $d = H\kappa$ measures the relative depth of the sublayer to the overall depth of the suspension. (Note that their definition of the Rayleigh number is slightly different to that given above.) Here K_1 and K_2 are given by (7.35) and (7.37), and P_5, P_6 and P_7 are involved functions of λ.

Solving (7.75) numerically (as well as asymptotically in certain limiting cases) Bees and Hill (1998) were able to show that in the absence of any background flow ($\eta = 0$), the most unstable mode has infinite wavelength. As the importance of the background flow increases with respect to gyrotaxis, as measured by η (7.11), in both deep and shallow suspensions the wavelength of the most unstable mode decreases. In a shallow suspension, the critical Rayleigh number continues to decrease as the relative magnitude background flow is increased (at least up to the point where the critical Rayleigh number is small, i.e. less than 1). By contrast, in deep suspensions the critical Rayleigh number initially increases, then decreases, with η.

Since these initial studies examining bioconvection due to gyrotaxis there have been many subsequent investigations into bioconvection driven by different stimuli, and combinations of stimuli (such as thermol-gyrotaxis Kuznetsov, 2005; Karimi and Ardekani, 2013, and thermo-oxytaxis, Avramenko and Kuznetsov, 2010). It is also worth noting that models of bioconvections have traditionally assumed that the distribution function governing the swimming direction can reliably be modeled as being steady (7.4). However, it has recently been demonstrated that if this requirement is relaxed, then there is a new class of instabilities applicable to top-heavy swimmers, i.e. where the center-of-buoyancy lies above the center of mass (Pedley, 2010). Hwang and Pedley (2014a) have also recently examined the role of background shear on the onset of bioconvection, showing that it can have a destabilizing influence. The full list of theoretical studies into bioconvection is too extensive to discuss in any detail here, however, we provide a (non-exhaustive) summary in Table 7.1. The review article by Hill and Pedley (2005) also provides further details of the bioconvection literature up to 2005.

Interesting buoyancy effects can also lead to novel dynamics at contact lines. Tuval et al. (2005) have demonstrated that when the free surface is tilted, such as is the case inside a droplet, oxytaxis can lead to an effect that is comparable to the Boycott effect observed with pure sedimentation. The denser layer of cells that have accumulated adjacent to the tilted free surface have negative buoyancy, and slide down this interface. This motion of the high-density layer, in turn, stirs up the fluid in the bulk of the droplet, leading to the formation of a vortex near the droplet's contact line with the solid underlying surface, thereby improving overall oxygen levels throughout the droplet.

Bioconvection Plumes

Whilst a linear stability analysis can provide useful insights into the onset of bioconvection, the subsequent dynamics have been observed to be highly nonlinear and complex. The experimental study of Bees and Hill (1997) describe how the two-dimensional bioconvective rolls

TABLE 7.1 Some of the theoretical studies into bioconvection, grouped by source of stimulus: Gravity (G), Chemical Gradient (C), Light but no Scattering (LnS), Light with Scattering (LS), Thermal Effects (T). Models are also categorized as modeling orientation in a phenomenological way (P), deterministic dynamical way (that takes into account a torque balance, DD) or a stochastically dynamical way (SD). Types of analysis include: Linear Stability (LS), Weakly Nonlinear Stability (WNS), Full Numerical Computation (C). Media considered include clear fluid (F) and a porous medium (P)

Study	Stimulus	Orientation	Analysis	Medium
Pedley et al. (1988)	G	DD	LS	F
Hill et al. (1989)	G	DD	LS	F
Bees and Hill (1998)	G	SD	LS	F
Bees and Hill (1999)	G	SD	WNS	F
Ghorai and Hill (1999)	G	DD	C	F
Ghorai and Hill (2000b)	G	DD	C	F
Ghorai and Hill (2007)	G	DD	C	F
Hwang and Pedley (2014)	G	SD	LS	F
Kuznetsov (2005)	G/T	DD	LS	F
Karimi and Ardekani (2013)	G/T	DD	LS	F
Hillesdon and Pedley (1996)	C	P	LS	F
Metcalfe and Pedley (1998)	C	P	WNS	F
Avramenko and Kuznetsov (2010)	C/T	P	LS	F
Vincent and Hill (1996)	LnS	P	LS	F
Ghorai and Hill (2005)	LnS	P	LS	F
Ghorai et al. (2010)	LS	P	LS	F
Williams and Bees (2011)	LS	SD	LS	F
Panda and Ghorai (2013)	LS	P	C	F
Panda and Ghorai (2016)	LnS	P	C	F
Kuznetsov and Avramenko (2002), Kuznetsov and Jiang (2003)	G	DD	LS	P

predicted by the linear stability analysis can subsequently succumb to a secondary three-dimensional instability, breaking down into plumes that descend down into the suspension. Although similar in appearance, the origin of these plumes is different from those described by Kessler (1986) that occur in the bulk suspension. The latter arise from spontaneous nucleation (localized increase in cell density), with the resulting plumes often disconnected from any of the container surfaces. By contrast, the plumes generated by the overturning instability fall from the free surface (although gyrotactic effects are still present in the subsequent development of the plumes).

Although these plumes are fully nonlinear in their dynamics, in the case of gyrotactic bioconvection, Bees and Hill (1999) demonstrated that the spacing between (infinitely long) plumes was well-predicted by the critical wavelengths in the linear stability analysis. Moreover, they were able to demonstrate that the plumes can drift horizontally over time, and also support the kinds of vertical traveling pulses seen in the experiments of Kessler (1985). When the suspension is sufficiently deep, the plumes themselves can succumb to further instabilities, as demonstrated by full numerical treatment of the governing equations (Ghorai and Hill, 1999, 2000a, 2002). Initially axisymmetric plumes are susceptible to both varicose (which is seen as *pulses* traveling down the plume, see Fig. 7.7) and sinusoidal (or meandering) in-

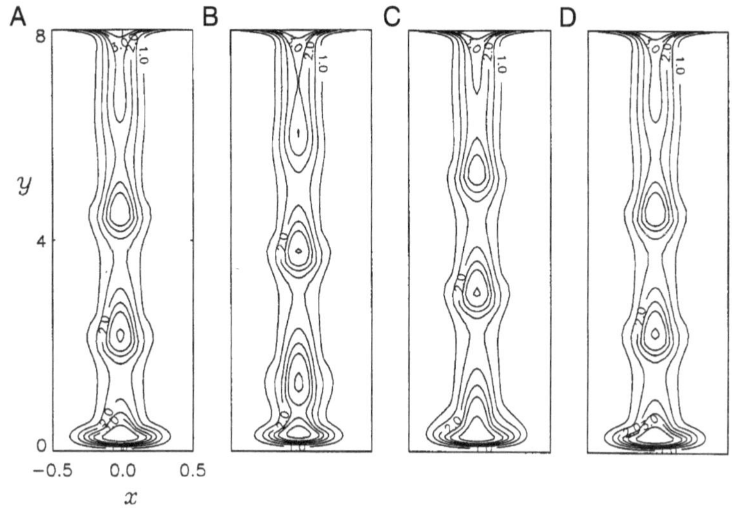

FIGURE 7.7 Bioconvection: Numerical simulation showing snapshots of the time evolution of a two-dimensional gyrotactic bioconvective plume that forms part of a periodic array. Lines show contours of cell concentation, with a (non-dimensional) time interval of 0.15 between each snapshot (the left image being the earliest). Pulses of cells are seen to travel down the length of the plume, as a result of a varicose instability (reproduced from Ghorai and Hill, 2000b with permission).

stabilities. The long-time dynamics of the are shown to be complex, with the configuration of the plumes constantly changing in time (Ghorai and Hill, 2000b, 2007). The numerical studies also noted that as the gyrotactic number B is increased, these pulses are seen to become smaller, until disappearing altogether. Also, that if the swimming speed was decreased, the depth to which the plumes reached decreased, with the plumes also achieving a steady state (i.e. no pulses are observed). Both trends are also seen to be affected by the value of the Rayleigh number.

Bioconvective plumes are also reported in experiments where the microorganisms create an unstable high-density layer near the surface due to oxytaxis, rather than gyrotaxis (Kessler, 1986; Hillesdon et al., 1995). An important feature of the plumes in this context is that they can transport oxygen down to the lower depths, and revive cells that are dormant in those locations due to lack of oxygen (Fig. 7.8). These plumes have been analyzed in some depth by Metcalfe and Pedley (2001) for a deep suspension, who were able to demonstrate that the radius of a steady plume scales like the square-root of its depth (although the model did not include effects such as replenishment of the cell-rich layer near the surface with upwards swimming cells in the bulk). Kuznetsov and Avramenko (2004) have performed a similar analysis for plumes of oxytactic microorganisms in a porous medium, which was subsequently compared against full numerical computation by Becker et al. (2004) (where the flow within the plume is driven by a unit axial filtration velocity prescribed within a certain radius at the top of the suspension, and the parabolic equations are spatially marched from this inlet condition).

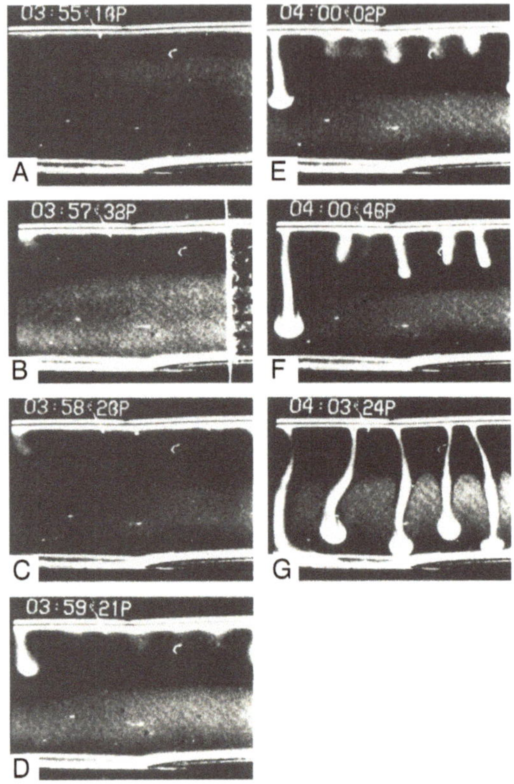

FIGURE 7.8 Side view of a 7–8 mm deep suspension of oxytactic bacteria *B. subtilis*. Dense layer of cells near the open upper surface (where oxygen levels are high) seen to destabilize, resulting in falling plumes of cells. Time of snapshots shown in top left corner (reproduced from Hillesdon et al., 1995 with permission).

7.5.4 Bacterial Chemotaxis

As mentioned earlier, bacteria can follow chemical gradients by undergoing run-and-tumble, and their cell distributions can be modeled by the Smoluchowski equation (7.32) with a transition function given by (7.58). Bottom-heaviness (and hence buoyancy) is generally not considered important with bacterial suspensions, however, the fact that they are pushers does mean that the stresslet (7.65) can play an important role in determining the flows generated (Simha and Ramaswarmy, 2002). It has been found that an orientationally-isotropic suspension can destabilize once cell concentrations exceed some critical density. This results in collective swimming, where the swimming direction of cells becomes spatially correlated (more details can be found in Koch and Subramanian, 2011; Saintillan and Shelley, 2013; Clarke, 2015). To examine the influence of chemical gradients in suspensions of run-and-tumble bacteria, Subramanian et al. (2011) performed a linear stability on an initially homogeneous (e.g. well-mixed) suspension. They found that the presence of a chemical gradient reduced the critical density for the onset of collective swimming. More recently Ezhilan et al. (2012) have performed full numerical simulations of the governing equations for bacterial

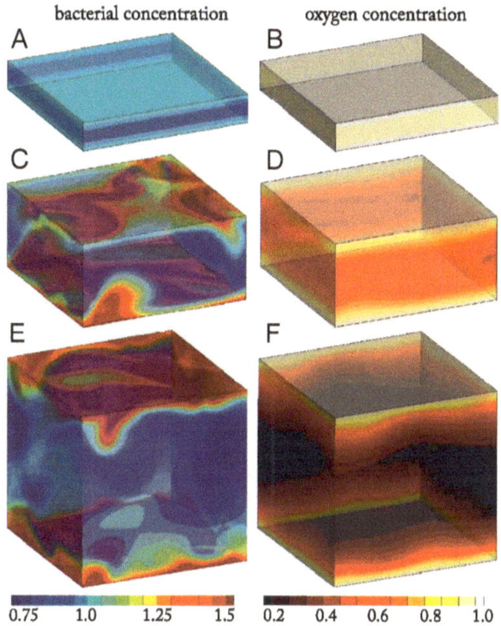

FIGURE 7.9 Steady state (A, C, E) Bacterial cell concentrations and (B, D, F) oxygen concentrations computed in films of thicknesses (A, B) 100 μm, (C, D) 300 μm and (E, F) 600 μm. The geometry is doubly-periodic in the horizontal directions, and top and bottom surfaces are stress free (reproduced from Ezhilan et al., 2012 with permission).

suspensions within fluid films. They demonstrated that not only does the complexity in the cell distributions increase significantly when the film is thicker (see Fig. 7.9), but also that the stresslets play an important role in enhancing oxygen (and cell) transport through the suspension.

Convection also been observed in bacterial suspensions. However, the driving mechanism here is believed not to be buoyancy effects, but rather the fact that an alignment of cells in the direction of the chemical gradient causes local fluid pumping in a manner that reinforces any spatially-periodic distributions of cells (Kasyap and Koch, 2012, 2014).

In the presence of shear, for example as experienced by a bacterial suspension in channel flow, experiments have shown that cells can become localized to regions of high shear (Rusconi et al., 2014), a phenomenon that has been shown by Bearon and Hazel (2015) to be captured using the run-and-tumble Smoluchowski equation (7.32).

7.5.5 Porous Media

Whilst much of the suspension modeling work has considered unhindered immersing fluids, Kutnzetsov and coworkers have considered the impact of porous media upon suspension dynamics. In these situations the Boussinesq equations are replaced by the D'Arcy equations including buoyancy due to cell gyrotaxis (Kuznetsov and Avramenko, 2002; Kuznetsov and

Jiang, 2003)

$$\rho \boldsymbol{u} = -\kappa' \nabla p - n\rho g \hat{z} \tag{7.76}$$

where now $\boldsymbol{u}$ is the (pore averaged) filtration velocity, and κ' is the permeability. There has also been some modeling work examining bioconvection in porous media due to oxytaxis (Kuznetsov and Avramenko, 2003), as well as mixtures of nanoparticles and microorganisms (Kuznetsov, 2012a, 2012b). Some of this work has looked at the effect of pores being clogged by the microorganism cells both in dilute (Kuznetsov and Jiang, 2003) and dense (Kuznetsov and Avramenko, 2015) suspensions.

7.6 CONCLUSIONS

In this chapter we have endeavored to give an overview of the different ways in which microorganisms respond to stimuli within their environment, and how this has been modeled mathematically. As we have seen, these responses can take a variety of different forms. Some microorganisms that obtain their energy from photosynthesis, such as algae, are bottom heavy. This means that they naturally swim (on average) upwards towards the light. Others possess photoreceptors in their cell bodies which allow them to sense the levels of light directly, and position themselves in optimal levels of illumination. The depth at which these optimal levels occur can depend upon both absorption and scattering of light by the microorganisms themselves.

Both photosynthesizing and non-photosynthesizing microorganisms also need to be able to hone in on the dissolved chemicals required for their metabolic processes. As such, many microorganisms are chemotactic, meaning that they can sense chemical gradients. Eukaryotes can be large enough that chemical gradients are appreciable across the length of their cell body, and so can be detected by a distribution of receptors within the cell. By contrast prokaryotes are typically too small for chemical gradients to be measurable over the length of their cell body, and so may undergo run-and-tumble. This entails them undergoing a biased random walk through their environment, sampling the chemical concentrations along the way, and spending longer traveling in directions with a favorable chemical gradient.

Another feature which distinguishes prokaryotes from many eukaryotes like algae is the fact that they are *pushers*, rather than *pullers*. This means that they are propelled from behind, rather than the front, and this generates actives stresses in the fluid which can lead to correlated swimming motions across many individual cells within the suspension. This, in turn, can enhance transport of cells and nutrients within the suspension. For bottom *pullers*, which characterizes many algae, bulk motions that they generate within their immersing fluid are believed to derive more from gyrotactic effects, than active stresses (Pedley and Kessler, 1992). However, we have seen that they can still generate substantial bulk motions in the fluid through buoyancy-driven instabilities. This occurs because microorganisms are typically more dense than the fluid in which they inhabit. When they aggregate in a region in response to a stimulus, this creates a hydrodynamically-unstable density stratification. The best-known example of which is probably bioconvection, where the instability can lead to dense plumes of falling cells.

The cell conservation model of Pedley and Kessler (Pedley et al., 1988; Pedley and Kessler, 1990) has laid the foundations for much of the theoretical work that has been conducted in this area (although for microorganisms such as run-and-tumble bacteria, and strong background flows, an approach that uses the Smoluchowski equation is sometimes preferable (Bearon and Pedley, 2000; Subramanian et al., 2011)). These models are appropriate for semi-dilute suspensions, however, the issue of how to treat cell–cell hydrodynamic interactions at the continuum level remains a continuing challenge. Even under the semi-dilute approximation, there remain some gaps in the models. For example, Williams and Bees (2011) model phototaxis in a way that takes into account torque balances, but does not include scattering effects. Conversely, Ghorai et al. (2010) and Ghorai and Panda (2013) incorporate scattering, but in a manner that does not enforce any torque balance. Similarly, whilst models of chemotaxis involving run-and-tumble microorganisms incorporate a torque balance, models of chemotaxis by non run-and-tumble cells (for example, eukaryotes) are also phenomenological and do not include any torque balances. In relation to the porous models, there are also some potential issues in assuming that a torque balance takes the form given in (7.1) which is derived in the absence of any nearby boundaries.

One area that we have not covered in great detail here is the role of chemotaxis in reproduction. Spermatazoa are believed to use chemical cues to navigate their way to the ovum, and can exhibit behaviors such as hyperactivation (a modification to the beating pattern of their flagella) in response to chemical stimulus. For more information on this area readers are referred to the review articles of Ooi et al. (2014) and Fauci and Dillon (2006). Furthermore, the immersing fluids in these scenarios tend to be non-Newtonian, an area that has not seen a huge amount of theoretical work in terms of collective dynamics modeling.

We began this chapter by discussing the influence of microorganisms upon the development of the Earth's biosphere, a scale that is far removed from the reduced models and experiments that have been conducted to-date. However, it is hoped that by integrating these reduced models of microorganisms into larger scale ecological and climate models it may begin to become possible to quantitatively assess the role of unicellular life upon our planet's ecosystem.

References

Allwood, A.C., Walter, M.R., Kamber, B.S., Marshall, C.P., Burch, I.W., 2006. Stromatolite reef from the Early Archaean era of Australia. Nature 441, 714–718.

Avramenko, A., Kuznetsov, A., 2010. Bio-thermal convection caused by combined effects of swimming of oxytactic bacteria and inclined temperature gradient in a shallow fluid layer. Int. J. Numer. Methods Heat Fluid Flow 20 (2), 157–173.

Awramik, S., 1992. The oldest records of photosynthesis. Photosynth. Res. 33, 75–89.

Batchelor, G., 1970. The stress system in a suspension of force-free particles. J. Fluid Mech. 41, 545–570.

Bearon, R., 2003. An extension of generalized Taylor dispersion in unbounded homogeneous shear flows to run-and-tumble chemotactic bacteria. Phys. Fluids 15 (6), 1552–1563.

Bearon, R., Hazel, A., 2015. The trapping in high-shear regions of slender bacteria undergoing chemotaxis in a channel. J. Fluid Mech. 771, 1–13.

Bearon, R., Pedley, T., 2000. Modelling run-and-tumble chemotaxis in a shear flow. Bull. Math. Biol. 62, 775–791.

Becker, S., Kuznetsov, A., Avramenko, A., 2004. Numerical modeling of a falling bioconvection plume in a porous medium. Fluid Dyn. Res. 33, 323–339.

Bees, M., Croze, O., 2010. Dispersion of biased swimming micro-organisms in a fluid flowing through a tube. Proc. R. Soc. A 466, 2057–2077.

Bees, M.A., Hill, N.A., 1997. Wavelength of bioconvection patterns. J. Exp. Biol. 200, 1515–1526.
Bees, M.A., Hill, N.A., 1998. Linear bioconvection in a suspension of randomly swimming, gyrotactic microorganisms. Phys. Fluids 10, 1864–1881.
Bees, M.A., Hill, N.A., 1999. Non-linear bioconvection in a deep suspension of gyrotactic swimming microorganisms. J. Math. Biol. 38, 135–168.
Bell, E.A., Boehnkea, P., Harrison, T.M., Mao, W.L., 2015. Potentially biogenic carbon preserved in a 4.1 billion-year-old zircon. Proc. Natl. Acad. Sci. USA 112, 14518–14521.
Berg, H., 2004. E. Coli in Motion. Springer.
Bhattacharya, D., Melkonian, M., 1995. The phylogeny of plastids: a review based on comparisons of small-subunit ribosomal RNA coding regions. J. Phycol. 31 (4), 489–498. http://dx.doi.org/10.1111/j.1529-8817.1995.tb02542.x.
Brenner, H., Weissman, M.H., 1972. Rheology of a dilute suspension of dipolar spherical particles in an external field. II. Effect of rotary Brownian motion. J. Colloid Interface Sci. 41, 499–531.
Clarke, R., 2015. Transport in the microbiome. In: Heat Transfer and Fluid Flow in Biological Processes. Ch. 6.
Cowie, J., 2012. Climatic Change: Biological and Human Aspects, second edition. Cambridge University Press.
Doroudchi, M., Greenberg, K.P., Liu, J., Silka, K.A., Boyden, E.S., Lockridge, J.A., Arman, A.C., Janani, R., Boye, S.E., Boye, S.L., Gordon, G., Matteo, B.C., Sampath, A.P., Hauswirth, W.W., Horsager, A., 2011. Virally delivered channelrhodopsin-2 safely and effectively restores visual function in multiple mouse models of blindness. Molec. Ther. 19 (7), 1220–1229.
Drazin, P., Reid, W., 2004. Hydrodynamic Stability, second edition. Cambridge University Press.
Eriksen, J.L., 1960. Transversely isotropic fluids. Colloid Polym. Sci. 173 (2), 117–122.
Ezhilan, B., Pahlavan, A., Saintillan, D., 2012. Chaotic dynamics and oxygen transport in thin films of aerotactic bacteria. Phys. Fluids 24, 091701.
Fauci, L., Dillon, R., 2006. Biofluidmechanics of reproduction. Annu. Rev. Fluid Mech. 38, 371–394.
Ghorai, S., Hill, N.A., 2000a. Wavelengths of gyrotactic plumes in bioconvection. Bull. Math. Biol. 62, 429–450.
Ghorai, S., Hill, N.A., 2000b. Periodic arrays of gyrotactic plumes in bioconvection. Phys. Fluids 12, 5–22.
Ghorai, S., Hill, N.A., 2002. Axisymmetric bioconvection in a cylinder. J. Theor. Biol. 219, 137–152.
Ghorai, S., Hill, N.A., 1999. Development and stability of gyrotactic plumes in bioconvection. J. Fluid Mech. 400, 1–31.
Ghorai, S., Hill, N.A., 2005. Penetrative phototactic bioconvection. Phys. Fluids 17, 074101.
Ghorai, S., Hill, N.A., 2007. Gyrotactic bioconvection in three dimensions. Phys. Fluids 19, 054107.
Ghorai, S., Panda, M., 2013. Bioconvection in an anistropic scattering suspension of phototactic algae. Eur. J. Mech. B, Fluids 41, 81–93.
Ghorai, S., Panda, M.K., Hill, N.A., 2010. Bioconvection in a suspension of isotropically scattering phototactic algae. Phys. Fluids 22, 071901.
Gray, M.W., Doolittle, W.F., 1982. Has the endosymbiont hypothesis been proven? Microbiol. Rev. 46, 1–42.
Hill, N.A., Pedley, T., 2005. Bioconvection. Fluid Dyn. Res. 37, 1–20.
Hill, N.A., Pedley, T., Kessler, J., 1989. The growth of bioconvection patterns in a suspension of gyrotactic microorganisms in a layer of finite depth. J. Fluid Mech. 208, 509–543.
Hill, N.A., Bees, M.A., 2002. Taylor dispersion of gyrotactic swimming micro-organisms in a linear flow. Phys. Fluids 14, 2598–2605.
Hillesdon, A., Pedley, T., 1996. Bioconvection in suspensions of oxytactic bacteria: linear theory. J. Fluid Mech. 324, 223–259.
Hillesdon, A., Pedley, T., Kessler, J., 1995. The development of concentration gradients in a suspension of chemotactic bacteria. Bull. Math. Biol. 57, 298–344.
Hwang, Y., Pedley, T., 2014a. Bioconvection under uniform shear: linear stability analysis. J. Fluid Mech. 738, 522–562.
Hwang, Y., Pedley, T.J., 2014. Stability of downflowing gyrotactic microorganism suspensions in a two-dimensional vertical channel. J. Fluid Mech. 749, 750–777.
Ishikawa, T., Pedley, T., 2008. Coherent structures in monolayers of swimming particles. Phys. Rev. Lett. 100, 088103.
Karimi, A., Ardekani, A., 2013. Gyrotactic bioconvection at pycnoclines. J. Fluid Mech. 733, 245–267.
Kasyap, T., Koch, D., 2012. Chemotaxis driven instability of a confined bacterial suspension. Phys. Rev. Lett. 108, 038101.
Kasyap, T., Koch, D., 2014. Instability of an inhomogeneous bacterial suspension subjected to a chemo-attractant gradient. J. Fluid Mech. 741, 619–657.
Keeling, P.J., 2004. Diversity and evolutionary history of plastids and their hosts. Am. J. Bot. 91.

Keller, E.F., Segel, L., 1971. Model for chemotaxis. J. Theor. Biol. 30, 235–248.

Kessler, J.O., 1985. Co-operative and concentrative phenomena of swimming microorganisms. Contemp. Phys. 26 (2), 147–166. http://dx.doi.org/10.1080/00107518508210745.

Kessler, J.O., 1986. Individual and collective fluid dynamics of swimming cells. J. Fluid Mech. 173, 191–205.

Kirschvink, J., 1992. Late proterozoic low-latitude global glaciation: the snowball earth. In: The Proterozoic Biosphere: A Multidisciplinary Study. Cambridge University Press.

Knoll, A.H., Javaux, E.J., Hewitt, D., Cohen, P., 2006. Eukaryotic organisms in proterozoic oceans. Philos. Trans. R. Soc. B 361, 1023–1038.

Koch, D., Subramanian, G., 2011. Collective hydrodynamics of swimming microorganisms: living fluids. Annu. Rev. Fluid Mech. 43, 637–659.

Kuznetsov, A., 2005. The onset of bioconvection in a suspension of gyrotactic microorganisms in a fluid layer of finite depth heated from below. Int. Commun. Heat Mass Transf. 32, 574–582.

Kuznetsov, A., 2012a. Nanofluid bioconvection in a horizontal fluid-saturated porous layer. J. Porous Media 15, 11–27.

Kuznetsov, A., 2012b. Nanofluid bioconvection in a porous layer: oxytactic microorganisms. J. Porous Media 15, 233–248.

Kuznetsov, A., Avramenko, A., 2002. A 2D analysis of stability of bioconvection in a fluid saturated porous medium-estimation of the critical permeability value. Int. Commun. Heat Mass Transf. 29, 175–184.

Kuznetsov, A., Avramenko, A., 2003. Analysis of stability of bioconvection of motile oxytactic bacteria in a horizontal fluid saturated porous layer. Int. Commun. Heat Mass Transf. 53, 95–104.

Kuznetsov, A., Avramenko, A., 2004. Analytical investigation of a falling plume caused by bioconvection of oxytactic bacteria in a fluid saturated porous medium. Int. J. Eng. Sci. 42, 557–569.

Kuznetsov, A., Avramenko, A., 2015. Effect of fouling on stability of bioconvection of gyrotactic microorganisms in a porous medium. J. Porous Media 8 (1), 45–53.

Kuznetsov, A., Jiang, N., 2003. Bioconvection of negatively geotactic microorganisms in a porous medium: the effect of cell deposition and declogging. Int. J. Numer. Methods Heat Fluid Flow 13.

Lauga, E., 2006. Life at high Deborah number. Europhys. Lett. 86, 64001.

Lauga, E., 2011. Propulsion in a viscoelastic fluid. Phys. Fluids 19, 083104.

Manela, A., Frankel, L., 2003. Generalized Taylor dispersion in suspensions of gyrotactic swimming micro-organisms. J. Fluid Mech. 490, 99–127.

Margulis, L., 1970. Origin of Eukaryotic Cells. Yale University Press.

McFadden, V.D.G., Evolution, G.I., 2004. Red algal genome affirms a common origin of all plastids. Curr. Biol. 14, 514–516.

Metcalfe, A.M., Pedley, T.J., 1998. Bacterial bioconvection: weakly nonlinear theory for pattern selection. J. Fluid Mech. 370, 249.

Metcalfe, A.M., Pedley, T.J., 2001. Falling plumes in bacterial bioconvection. J. Fluid Mech. 445, 121.

Olson, J.M., 2006. Photosynthesis in the Archean era. Photosynth. Res. 88 (2), 109–117.

Ooi, E., Smith, D., Gadelha, H., Gaffney, E., Kirkman-Brown, J., 2014. The mechanics of hyperactivation in adhered human sperm. R. Soc. Open Sci. 1, 140230.

Panda, M., Ghorai, S., 2016. Penetrative phototactic bioconvection in a two-dimensional non-scattering suspension. Phys. Fluids 28, 054105.

Panda, M.K., Ghorai, S., 2013. Penetrative phototactic bioconvection in an isotropic scattering suspension. Phys. Fluids 25, 071902.

Pedley, T., 2010. Instability of uniform micro-organism suspensions revisited. J. Fluid Mech. 647, 335–359.

Pedley, T., Kessler, J., 1990. A new continuum model for suspensions of gyrotactic micro-organisms. J. Fluid Mech. 212, 155–182.

Pedley, T., Hill, N.A., Kessler, J., 1988. The growth of bioconvection patterns in a uniform suspension of gyrotactic microorganisms. J. Fluid Mech. 195, 223–237.

Pedley, T.J., Kessler, J.O., 1992. Hydrodynamic phenomena in suspensions of swimming microorganisms. Annu. Rev. Fluid Mech. 24, 313–358.

Platt, J.R., 1961. 'Bioconvection patterns' in cultures of free-swimming organisms. Science 133, 1766–1767.

Rusconi, R., Guasto, J.S., Stocker, R., 2014. Bacterial transport suppressed by fluid shear. Nat. Phys. 10 (3), 212–217.

Saintillan, D., Shelley, M.J., 2013. Active suspensions and their nonlinear models. C. R. Phys. 14, 497–517.

Schuergers, N., Lenn, T., Kampmann, R., Meissner, M.V., Esteves, T., Temerinac-Ott, M., Korvink, J.G., Lowe, A.R., Mullineaux, C.W., Wilde, A., 2016. Cyanobacteria use micro-optics to sense light direction. eLife 5, e12620.

Sgarbossa, G., Lenci, F., 2002. Photoreception and photomovements of microorganisms. Photochem. Photobiol. Sci. 1.

Simha, R., Ramaswarmy, S., 2002. Hydrodynamic fluctuations and instabilities in ordered suspensions of self-propelled particles. Phys. Rev. Lett. 89, 058101.

Subramanian, G., Koch, D., Fitzgibbon, S., 2011. The stability of a homogeneous suspension of chemotactic bacteria. Phys. Fluids 23, 2011.

Tuval, I., Cisneros, L., Dombrowski, C., Wolgemuth, C.W., Kessler, J.O., Goldstein, R.E., 2005. Bacterial swimming and oxygen transport near contact lines. Proc. Natl. Acad. Sci. USA 102, 2277–2282.

Vincent, R.V., Hill, N.A., 1996. Bioconvection in a suspension of phototactic algae. J. Fluid Mech. 327, 343–371.

Wadhams, G., Armitage, J., 2004. Making sense of it all: bacterial chemotaxis. Nat. Rev. Mol. Cell Biol. 5 (12), 1024–1037.

Williams, C.R., Bees, M.A., 2011. Photo-gyrotactic bioconvection. J. Fluid Mech. 678, 41–86.

CHAPTER

8

Nano-Swimmers in Lipid-Bilayer Membranes

*M.-J. Huang**, *A. Mikhailov*[†,‡,#]

*Chemical Physics Theory Group, Department of Chemistry, University of Toronto, Toronto, ON, Canada [†]Abteilung Physikalische Chemie, Fritz-Haber-Institut der Max-Planck-Gesellschaft, Berlin, Germany [‡]Department of Mathematical and Life Sciences, Hiroshima University, Hiroshima, Japan

8.1 INTRODUCTION

In his classical study (Purcell, 1977), P.M. Purcell has addressed hydrodynamics in the world at micrometer scales where motions of biological micro-organisms take place under the conditions of low Reynolds numbers. In order to swim in a fluid at low Reynolds numbers, micro-organisms must cyclically change their conformations in a nonreciprocal way so that their reverse motion in a cycle does not represent a simple reciprocal of the forward motion. For example, bacteria *E. coli* can propel themselves by rotating their corkscrew flagella and sperm cells are propelled by a snake-like wiggling of their flagella. The swimming behavior in a low Reynolds number regime at micrometer scales has been extensively theoretically studied, and various examples of micro-swimmers have been proposed and investigated.

While experiments have shown chemotaxis-like motions for enzyme molecules (Muddana et al., 2010; Sengupta et al., 2013; Dey et al., 2014; Riedel et al., 2015), specific mechanism resulting of hydrodynamic propulsion at nanoscales are still under debate (Alonso and Mikhailov, 2009; Sakaue et al., 2010; Golestanian and Najafi, 2010; Golestanian, 2010; Astumian, 2014). The operation of molecular machines, such as kinesin and myosin motors that walk along bio-filaments, is known to be based on conformational changes powered by adenosine triphosphate (ATP) hydrolysis (Alberts et al., 2002). Their turnover cycles are continued as long as the system is kept away from equilibrium by supplying chemical fuel molecules and removing the waste. Cyclic conformational motions could be directly observed in single-molecule experiments by using fast atomic force microscopy (Kodera et al., 2010; Uchinashi et al., 2011).

[#] Chapter coordinator.

Coarse-grained simulations for hepatitis C virus (HCV) helicase could reproduce relevant conformation changes in a detailed way (Flechsig and Mikhailov, 2010; Echeverria et al., 2011). It is important to note that such cyclic conformational motions are generally not reciprocal. This is because the reverse motions take place without the ligands and, hence, in a system whose composition and dynamical properties are different. In principle, any protein machine may behave as a swimmer on nanoscales (Sakaue et al., 2010).

Biological membranes formed by lipid bilayers play a fundamental role in a living cell; they build up the wall separating the cell from its environment and make up the vesicles used in intracellular transport. Usually, cellular membranes contain a large number of proteins, and many protein inclusions, such as ion pumps, are active. As a result of protein activity, non-thermal membrane fluctuations can be induced (Prost and Bruinsma, 1996; Sankararaman et al., 2002; Chen, 2004). Furthermore, Turing-like patterns and traveling waves can spontaneously arise in such systems (Chen and Mikhailov, 2010).

Under physiological conditions, biological membranes are in the fluid phase. At submicrometer scales, hydrodynamic coupling to the solvent can be neglected, and biological membranes can be approximated as two-dimensional (2D) fluids (Saffman and Delbruck, 1975; Diamant, 2009). Generally, reduction from three to two dimensions is known to have a profound effect on fluid mechanics (see, e.g., Marchioro and Pulvirenti, 1982). While hydrodynamic interactions in three dimensions are fading away as the inverse of the distance at large separations between the particles, such interactions in 2D fluids have a logarithmic dependence on the distance between the particles and, therefore, do not disappear even at long separations. Since hydrodynamic interactions play a principal role in propulsion phenomena, the behavior of swimmers in 2D systems should be significantly different.

Although structurally resolved coarse-grained numerical modeling of active proteins in lipid bilayers is possible (see Huang et al., 2013), it is computationally intensive. Instead, we have considered a simple model of an active membrane inclusion which represents a chain of three transmembrane domains linked by elastic springs. Our 2D membrane swimmer is similar to that considered by A. Hajafi and R. Golestanian in three dimensions (Najafi and Golestanian, 2004; Golestanian and Najafi, 2010). While being simple, this model molecular swimmer incorporates some essential operation aspects of real protein machines. For example, enzyme adenylate kinase also consists of three domains and distances between them are changing in dependence on the enzyme ligand state (Echeverria et al., 2011).

Below we begin with a description of a coarse-grained mesoscopic model for the lipid bilayers combined with an efficient multiparticle collision dynamics for the surrounding solvent. A simple three-domain membrane swimmer is then introduced and its self-propulsive motion in a lipid bilayer induced by conformational motions is demonstrated (Huang et al., 2012a).

8.2 METHODS

8.2.1 Hybrid MD-MPCD Simulations for Fluid Lipid Bilayers in a Solvent

All-atom molecular dynamics (MD) simulations of lipid bilayers have been performed (Tieleman et al., 1997; Feller, 2000; de Groot and Grubmeller, 2001; Saiz and Klein, 2002; Dickey and Faller, 2008), but they are computationally demanding and therefore limited to

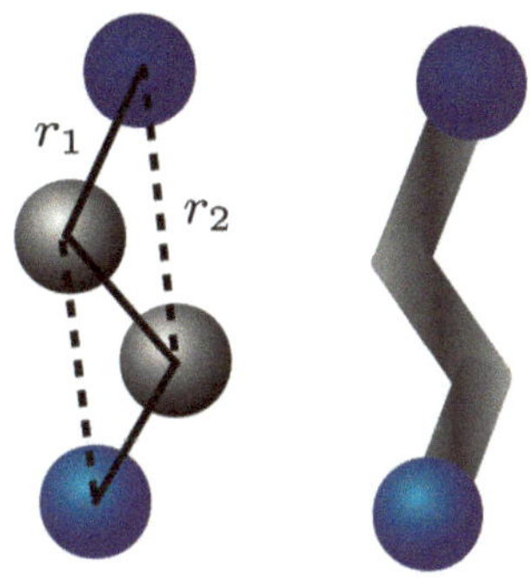

FIGURE 8.1 The lipid chain (left) and its schematic representation as a rod (right). The lipid consists of four beads linked by elastic FENE bonds (solid lines) and straightened by elastic bonds (dashed lines). The first bead (dark blue) is hydrophilic. Three other beads are hydrophobic, the terminal hydrophobic bead is shown as light blue. From Huang et al. (2012b).

relatively small systems (~10 nm) and short time scales (~10 ns). However, there are many important biochemical processes which occur at longer time scales. For instance, it is known that the characteristic mechanochemical motions in proteins that are essential for their enzymatic or motor functions, usually require milliseconds or more for their completion (Alberts et al., 2002). Therefore, microscopic investigations of biomembranes with protein inclusions are beyond the capacity of all-atom MD simulations, especially when the surrounding solvent molecules are explicitly considered. A major portion of the computational time in explicit solvent models is spent simulating the dynamics of a great number of solvent molecules in the system. It is desirable to construct a coarse-grain dynamical scheme that treats the dynamics of the solvent efficiently. Such a scheme is provided by multiparticle collision dynamics (MPCD) (Malevanets and Kapral, 1999b, 1999a; Kapral, 2008).

In this approach, solvent particles, representing coarse-grained real molecules, undergo effective multiparticle collisions at discrete time intervals τ and stream freely between consecutive collisions. When multiparticle collisions in the solvent take place, solvent particles are sorted into the cells of a simple cubic lattice with linear size a_0 and particles in the same cell exchange momentum with each other while the total momentum in the cell is conserved.

Specifically, for the constant temperature version of MPCD (Gompper et al., 2009), if the mean velocity of the solvent particles in the cell ξ is $\mathbf{V}_\xi$, the collision event of particle i inside this cell is modeled by updating its velocity, $\mathbf{v}_i$, so that the new velocity, $\mathbf{v}'_i$, is given by

$$\mathbf{v}'_i = \mathbf{V}_\xi + \mathbf{v}_i^{ran} - \sum_{j \in cell\ \xi} \mathbf{v}_j^{ran}/N_\xi, \tag{8.1}$$

where the components of the random velocity $\mathbf{v}_i^{ran}$ are chosen as Gaussian random numbers with zero mean and variance $k_B T/m$, N_ξ is the number of solvent particles in the cell ξ and the summation is performed over all solvent particles in this cell. For the system with small mean free path of the solvent particles, in order to improve the mixing properties of the dynamics random grid-shifting algorithm is employed (Ihle and Kroll, 2001, 2003).

In our study, lipid molecules are also described at the mesoscopic level, with each lipid molecule modeled as a chain of four beads. The hydrophobic head (h) corresponds to a single bead and the hydrophobic tail (t) consists of three beads (Fig. 8.1). The interactions between the beads in a lipid and between the lipids have the same form as in (Cooke and Deserno, 2005; Cooke et al., 2005; Reynwar et al., 2007; Deserno, 2009). The interaction between two lipid beads is described by the repulsive Lennard-Jones potential,

$$V_{rep} = 4\epsilon_{\alpha\alpha'}\left[\left(\frac{\sigma}{r_{ij}}\right)^{12} - \left(\frac{\sigma}{r_{ij}}\right)^{6} + \frac{1}{4}\right]\Theta(r_c - r_{ij}), \tag{8.2}$$

where $\sigma = 1$ is the particle diameter, $\Theta(r)$ is the Heaviside function, $r_{ij} = |\mathbf{r}_i - \mathbf{r}_j|$ is the distance between the beads i and j, and $r_c = 2^{1/6}\sigma$ is the cutoff length over which the interaction vanishes. The interaction energy between beads i and j takes the value $\epsilon_{\alpha\alpha'}$, where $\alpha, \alpha' \in \{h, t\}$ if bead i is of type α and bead j is of type α'.

Two neighboring beads in a lipid chain are linked by a finite extensible nonlinear elastic (FENE) bond (Kremer and Grest, 2004),

$$V_{bond} = -\frac{1}{2}k_{bond}r_\infty^2 \ln[1 - (r_1/r_\infty)^2], \tag{8.3}$$

where r_1 is the distance between the beads and $r_\infty = 1.5\,\sigma$ is the maximum distance allowed by the FENE bond, and the spring constant is $k_{bond} = 20\,\epsilon/\sigma^2$.

The bending rigidity of a lipid chain is modeled by additional springs connecting next-nearest neighbor beads and is characterized by the potential

$$V_{bend} = \frac{1}{2}k_{bend}(r_2 - 4\sigma)^2, \tag{8.4}$$

where r_2 is the distance between such two beads. The spring constant $k_{bend} = 2.5\,\epsilon/\sigma^2$ and the natural length is 4σ.

The hydrophobic effects are taken into account by adding an attractive interaction potential between the beads that belong to different lipid tails. The effective interaction potential is

$$V_{att}(r_{ij}) = \begin{cases} 4\epsilon_{\alpha\alpha'}\left[\left(\frac{\sigma}{r_{ij}}\right)^{12} - \left(\frac{\sigma}{r_{ij}}\right)^{6}\right], & r_{ij} < r_c, \\ -\epsilon_{\alpha\alpha'}\cos^2\frac{\pi(r_{ij}-r_c)}{2w_{\alpha\alpha'}}, & r_c \le r_{ij} \le r_c + w_{\alpha\alpha'}, \\ 0, & r_{ij} > r_c + w_{\alpha\alpha'}, \end{cases} \tag{8.5}$$

where $w_{\alpha\alpha'} = w_{tt}$ and $\epsilon_{\alpha\alpha'} = \epsilon_{tt}$.

Interactions between solvent particles and lipid beads are described by the intermolecular potential given by Eq. (8.2) with the same cutoff length, but with a different interaction strength $\epsilon_{\alpha\alpha'} = \epsilon_{st}$. Hydrophilic effects between a solvent particle and a lipid head bead are taken into account by introducing an attractive interaction given by Eq. (8.5) with $w_{\alpha\alpha'} = w_{sh}$ and $\epsilon_{\alpha\alpha'} = \epsilon_{sh}$.

The time evolution of the entire system is implemented through a hybrid MD-MPCD scheme (Huang et al., 2012b). This scheme combines molecular dynamics segments for the lipid–lipid, lipid–solvent interactions and multiparticle collisions for solvent particles at discrete time intervals τ. In the MD segments, Newton's equations of motion are used to evolve all particles for a time interval τ. At time τ multiparticle collisions among solvent particles take place. This sequence of MD and MPCD steps is repeated to evolve the entire system. For the MD segments, Newton's equation of motion are integrated using the velocity–Verlet algorithm with a time step of $\delta t = 0.005\, t_0$, where $t_0 = \sqrt{m\sigma^2/\epsilon}$, and the MPC time step was $\tau = 0.2 = 40\, \delta t$.

The energies of the interaction potentials are chosen to be $\epsilon_{ht} = 1\, \epsilon$, $\epsilon_{hh} = \epsilon_{tt} = 0.5\, \epsilon$, $\epsilon_{sh} = 0.05\, \epsilon$, and $\epsilon_{st} = 2.0\, \epsilon$. The interaction ranges for tail–tail and solvent–head interactions are chosen such that $r_c + w_{tt} = 2.6\, \sigma$ and $r_c + w_{sh} = 1.65\, \sigma$. All particles and beads have equal mass m. The lateral size of an MPC cell is $a_0 = \sigma$, and the solvent density is equal to five. Below, the results are presented in the units of length σ, the units of energy ϵ, and the units of time δt. Note that, by comparing lipid diffusion constant obtained from the simulations and from experiments, these units can be mapped onto the SI units as discussed below.

Self-assembled lipid membranes are known to possess a rich phase behavior. At higher temperatures, the lipids in the membrane are in the liquid phase, so that their motion is diffusive and their in-plane orientations are not ordered. As temperature decreases, the membrane undergoes a transition to a gel phase where lipids show a nematic order. Using the simulation method described above, self-assembly of lipid bilayer and various membrane structures could be studied.

Fig. 8.2 shows the time evolution of a self-assembled lipid bilayer membrane starting from a random mixture of lipid molecules and solvent particles. An example of different lipid phases at three temperatures is shown in Fig. 8.3. Below we focus on the liquid-like lipid phase at $k_B T/\epsilon = 1$ (Fig. 8.3B) since it is most relevant for real biological membranes. At this temperature, the membrane thickness determined by the average distance between lipid head beads in two leaflets is close to $5.5\, \sigma$. Comparing this to the thickness of a typical biomembrane ($\sim$ 4.5 nm), the unit length in the simulations is identified to be $\sigma \simeq 0.8$ nm. The time unit δt can be estimated by comparing the typical lipid lateral diffusion constant in the experiments ($D \simeq 4\ \mu\text{m}^2/\text{s}$) with the simulation result ($D = 7.86 \times 10^{-5} \sigma^2/\delta t$). We find $\delta t \simeq 10$ ps. The elastic properties of the model lipid bilayer determined in the simulations are comparable to those of a typical real lipid bilayer. By varying the membrane area, the membrane surface tension is determined and the lateral stretching modulus are $k_A = 53\, k_B T/\text{nm}^2$, close to the experimental values $k_A = 50\text{–}70\, k_B T/\text{nm}^2$ (Phillips et al., 2009). The membrane bending modulus can be derived from the power spectrum of membrane mid-plane hight fluctuations. The value of the bending modulus is $\kappa = 28\, k_B T$, similar to typical value for biomembranes of $\kappa = 10 - 20\, k_B T$ (Phillips et al., 2009).

8.2.2 Simple Model for Membrane Swimmers

In our simulations, a model 2D swimmer consisting of three cylindric domains ($i = 1, 2, 3$) was employed, with their instantaneous centers of mass position given by two-dimensional vectors $\mathbf{R}_i$. As shown in Fig. 8.4, neighboring domains are connected by elastic springs. To prevent bending of the machine in the membrane plane, we have defined θ as the angle between vectors $\mathbf{R}_{12} = \mathbf{R}_1 - \mathbf{R}_2$ and $\mathbf{R}_{23} = \mathbf{R}_2 - \mathbf{R}_3$ and introduced the angular potential $U(\theta) = \beta(1 - \cos\theta)$, where $\cos\theta = \mathbf{R}_{12} \cdot \mathbf{R}_{23}/(|\mathbf{R}_{12}||\mathbf{R}_{23}|)$ with a high stiffness constant β to

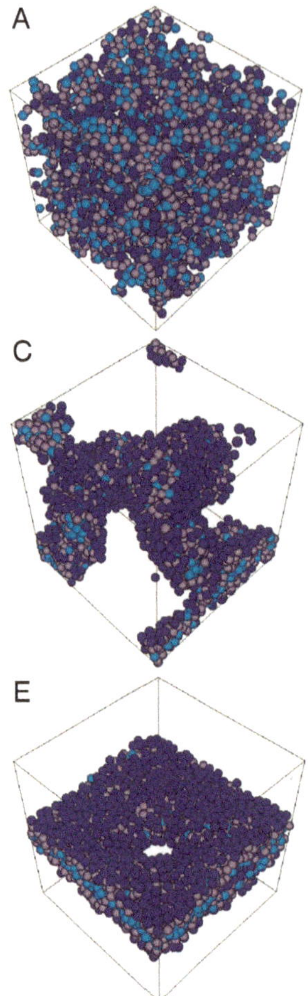

FIGURE 8.2 Self-assembly of a lipid bilayer. Six configurations at time moments: (A) 0 δt, (B) 6000 δt, (C) 18,000 δt, (D) 160,000 δt, (E) 246,400 δt, and (F) 300,000 δt are shown. The initial state (A) corresponds to the uniform mixture of lipids and the solvent. From Huang et al. (2012b).

keep the swimmer straight. The elastic energy of the swimmer is

$$E = \frac{1}{2}k\left[\left(|\mathbf{R}_{12}| - l_{12}(s)\right)^2 + \left(|\mathbf{R}_{23}| - l_{23}(s)\right)^2\right], \tag{8.6}$$

where k is the stiffness constant, and $l_{12}(s)$ and $l_{23}(s)$ are the natural lengths of the links between two neighboring domains that depend on the discrete state variable s. There are three states of the swimmer with different natural lengths of the links. We have

$$l_{12}(s=1) = l_0, \quad l_{12}(s=2) = l_{12}(s=3) = l_0(1-\epsilon'), \tag{8.7}$$

$$l_{23}(s=1) = l_{23}(s=2) = l_0, \quad l_{23}(s=3) = l_0(1-\epsilon'). \tag{8.8}$$

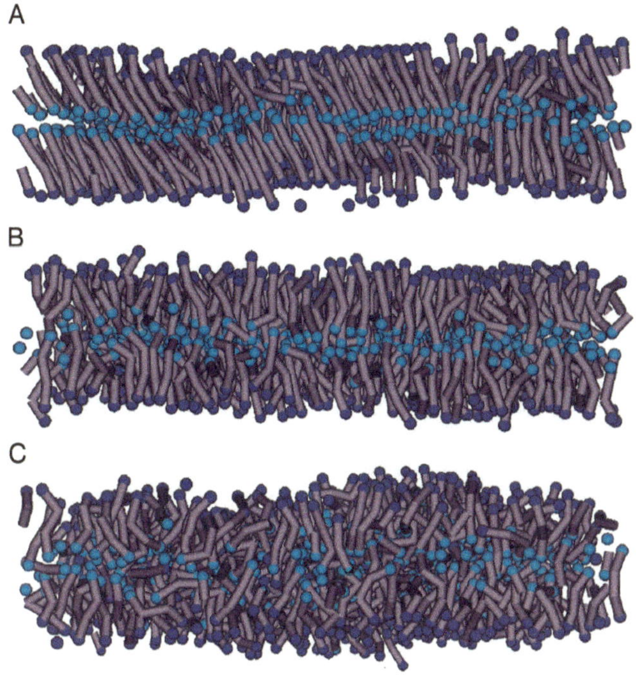

FIGURE 8.3 Membrane structures at three different temperatures: (A) $k_BT/\epsilon = 0.4$, (B) $k_BT/\epsilon = 1.0$, and (C) $k_BT/\epsilon = 2.0$. The rod representation is used to display lipids. The solvent particles are not shown. From Huang et al. (2012b).

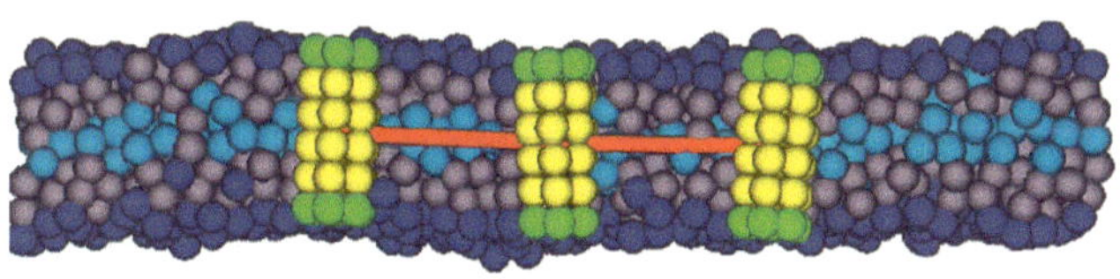

FIGURE 8.4 The microscopic membrane swimmer. Three cylindric domains are immersed into the lipid bilayer. The top and the bottom parts of each domain are hydrophilic, whereas the middle parts are hydrophobic. The domains are linked through elastic interactions applied to their mass centers. The natural lengths of these links are cyclically switched. From Huang et al. (2012a).

Thus, both links are long in the first state. In the second state, the link l_{12} is short, while the link l_{23} is long. In the third state, both links are short. Here ϵ' is a dimensionless parameter that determines the amount of change of the links.

To make this object active, switching between the states in a cyclic manner, i.e., $\cdots \to 1 \to 2 \to 3 \to 1 \cdots$, is introduced. When the equilibrium configuration corresponding to a certain dynamical state s is approached, the state variable s is switched to the next state in the sequence and relaxation to this state begins. The switching conditions between consecutive states are defined in terms of the instantaneous distances between the domains. That is, if $|\mathbf{R}_{12} - l_{12}(s = 1)| < \rho$ and $|\mathbf{R}_{23} - l_{23}(s = 1)| < \rho$, a transition to the state $s = 2$ takes place. If

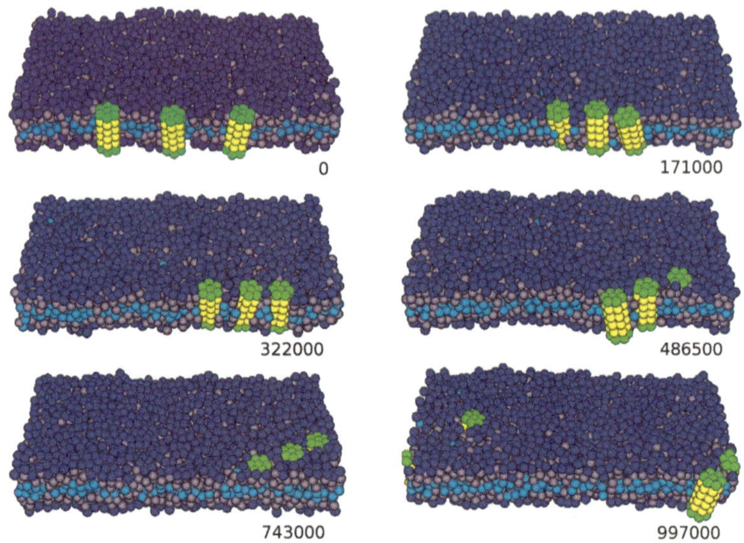

FIGURE 8.5 A series of six consecutive snapshots from a simulation of the membrane swimmer. Time moments (in MD steps) are indicated for each snapshot. The solvent is not displayed and, for clarity, only half of the membrane, with the cross-section along the initial swimmer position, is shown. The elastic links connecting the cylindric domains are not visualized here. Periodic boundary conditions are employed and, therefore, in the last snapshot the swimmer is seen reentering the membrane from the opposite direction. From Huang et al. (2012a).

$|\mathbf{R}_{12} - l_{12}(s=2)| < \rho$ and $|\mathbf{R}_{23} - l_{23}(s=2)| < \rho$, a transition to the state $s = 3$ occurs. Finally, if $|\mathbf{R}_{12} - l_{12}(s=3)| < \rho$ and $|\mathbf{R}_{23} - l_{23}(s=3)| < \rho$, a transition to the state $s = 1$ is implemented. In the simulations, natural lengths of the links, connecting three domains, were equal to $l_0 = 8\,\sigma$ when the swimmer was in state $s = 1$; they were shortened by 50% in every cycle, i.e., we had $\epsilon' = 0.5$. The switching parameter was chosen to be $\rho = 0.2\,\sigma$.

8.3 RESULTS

The initial orientation of the swimmer was in the direction along the x-axis, and the center of mass of the swimmer was in the middle of the membrane. After the equilibration during which the swimmer was kept passive, the swimmers were made active and performing cyclic self-propelled motions. Fig. 8.5 displays several snapshots from a membrane-resolved simulation of the swimmer. In the simulation, about 100 cycles have taken place and the swimmer has moved over a distance comparable to its total length. Fig. 8.6 shows 23 consequent cycles monitoring the total length of the swimmer (i.e., of the distance between centers of mass of the domains 1 and 3), each peak corresponds to the state $s = 1$ of the swimmer. As revealed by this figure, the durations of the cycles are strongly fluctuating and sometimes the swimmer stays for a while in one of its states before the motion can continue. This behavior is due to the fact that the dynamics of the swimmer takes place inside a densely packed medium and, when one of its domains tries to move, it needs to push aside the surrounding lipids.

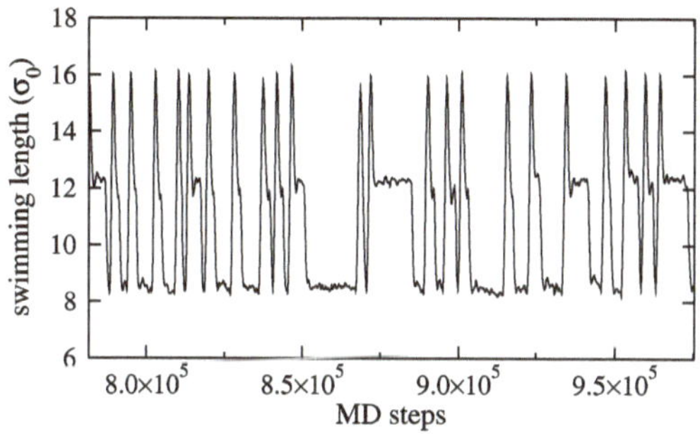

FIGURE 8.6 Total length of the swimmer as a function of time for a relatively short time interval including 23 swimmer cycles. From Huang et al. (2012a).

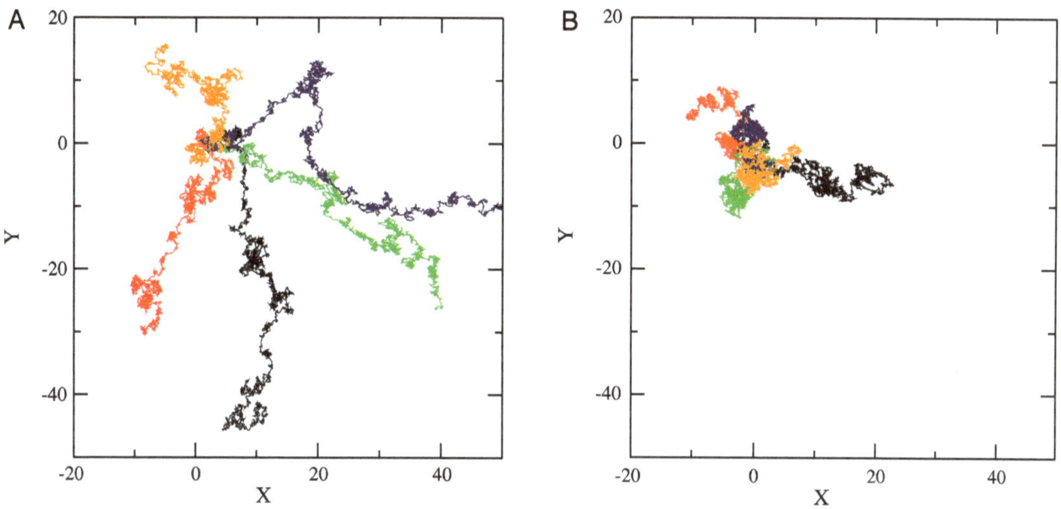

FIGURE 8.7 Five trajectories of the active (A) and passive (B) swimmers. Each trajectory corresponds to 5×10^6 MD steps. In a passive swimmer, active transitions between the conformational states are prohibited, so that it is always in the equilibrium state $s = 1$. From Huang et al. (2012a).

In the presence of thermal fluctuations an important question is whether the effects of propulsion prevail for an active swimmer or its dynamics is still dominated by thermal diffusion. To analyze it, a series of long simulations for the active swimmer and also, for comparison, for the passive swimmer has been performed. For the passive swimmer, active switching between the states has been excluded, so that it was always found in the equilibrium conformation ($s = 1$).

In Fig. 8.7A, five different trajectories of the active swimmer obtained in independent simulations are shown. Each trajectory corresponds to $5 \times 10^6\ \delta t$. A single cycle of the swimmer takes, on average, about 7288 δt, so that these trajectories cover about 675 to 700 swimmer cycles. For comparison, we show in Fig. 8.7B five trajectories of the passive swimmer under the same conditions. Examining the trajectories, one can already notice that the trajectories of an active swimmer are more extended than those of its passive analog.

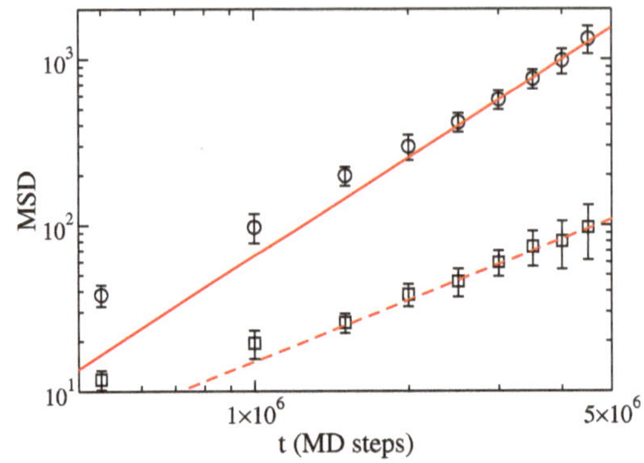

FIGURE 8.8 Log–log plots of the mean squared displacement of the swimmer mass-center showing ballistic motion for active swimmer (circles), MSD(t) $\sim t^2$ (solid line), and diffusive motion for passive swimmer (squares), MSD(t) $\sim t$ (dashed line). From Huang et al. (2012a).

Statistical analysis of the simulation data for active and passive swimmers has been undertaken. Fig. 8.8 shows the dependencies of the mean squared displacement (MSD) on time for the two objects. For the active swimmer (circles), the observed dependence is well approximated by the quadratic law, whereas the classical diffusion law with the MSD linearly proportional to time is found for the passive swimmer under the same conditions (squares).

From the observations described above, the motion of the active swimmer is essentially ballistic on the considered time scales, whereas the passive swimmer performs the classical diffusive motion on the same time scales. From Fig. 8.8, the mean propulsion velocity V of the active swimmer can be estimated, yielding $V = 1.2 \times 10^{-5}\,\sigma/\delta t$. The diffusion constant of the passive swimmer is estimated as $D = 6.3 \times 10^{-6}\,\sigma^2/\delta t$. This implies that for the system under consideration the propulsion effects dominate over the effects of thermal noise.

In addition to the translational motion, characterized by the position of the center of mass of the swimmer, we have also analyzed orientational motions of the active swimmer. The orientation was characterized by the instantaneous direction of the unit vector $\hat{\mathbf{R}}_{31}(t)$, which is pointing from domain 1 to domain 3, and the orientational correlation was computed as the average, $C(t) = \langle \hat{\mathbf{R}}_{31}(t) \cdot \hat{\mathbf{R}}_{31}(0) \rangle$. The computed orientational correlation function for the active swimmer is displayed in Fig. 8.9. As we see, it is well approximated by an exponential law with $C(t)$ proportional to $\exp(-t/\tau_c)$ with the correlation time $\tau_c \simeq 4 \times 10^7\,\delta t$. Note that this correlation time is about ten times longer than the simulation time. This explains why only the ballistic motion of the active swimmer was observed at long times. It should be expected that, on the time scale much longer than the orientational correlation time τ_c, the ballistic motion would be replaced by diffusive motion with an enhanced diffusion constant (Mikhailov and Meinköhn, 1997; Howse et al., 2007). The simulations presented in this chapter were not yet long enough to resolve this asymptotic behavior.

The Reynolds numbers of the swimmer with respect to the solvent and the membrane can be estimated. Using the linear size of a swimmer $L = 16\,\sigma$, the mean velocity $V = 1.2 \times 10^{-5}\,\sigma/\delta t$, and the solvent kinematic viscosity $\nu_s = 0.75\,m(\sigma\tau)^{-1}$ according to (Kapral, 2008; Gompper et al., 2009), the Reynolds number for swimming in the solvent is estimated as Re =

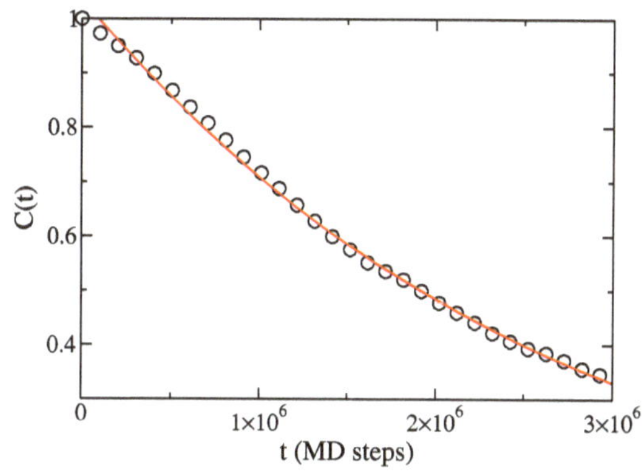

FIGURE 8.9 The plot of the orientational correlation function, C(t), for the active swimmer, where the solid line is the exponential fit. From Huang et al. (2012a).

$VL/\nu_s \simeq 0.00026$. Because the viscosity of the membrane is 1000 times larger than the solvent (Diamant, 2009), the Reynolds number for the swimmers with respect to the membrane is smaller by a factor of 10^3. This indicates that the considered swimmer indeed operated well in the low Reynolds number regime.

In Section 8.1, the length unit in the simulations is identified as $\sigma = 0.8$ nm, and the MD time step size is $\delta t = 10$ ps. By using these values, physical properties of the considered model swimmer can be accessed. For the passive swimmer, we find the diffusion constant $D = 6.3 \times 10^{-6}\sigma^2/\delta t \simeq 0.4\ \mu m^2/s$, which is comparable to the typical diffusion constants of membrane proteins (Gambin et al., 2006). For the active swimmer, it is found that the propulsion velocity is $V = 1.2 \times 10^{-5}\sigma/\delta t = 960\ \mu m/s$. The mean cycle time of the swimmer is $7300\,\delta t \simeq 0.1\ \mu s$. The turnover times of real enzymes vary from a microsecond to milliseconds and even seconds. Thus, the active swimmer presented here would correspond to rapid enzymes with very short turnover cycles. It is possible to include stochastic waiting times for the transitions between the internal swimmer states, and also to consider swimmers with slower internal conformational motions. When the swimmer is cycling less frequently, its velocity would be reduced. Note that, even then, propulsive motions may be visible on the background of thermal diffusion. If the cycle time of the swimmer is 1000 times longer than in the present simulations, the propulsion velocity would be about $V \simeq 1\ \mu m/s$. A swimmer with such a velocity would move ballistically over a micrometer in a second. Since the diffusion constant of the swimmer $D \simeq 0.4\,\mu m^2/s$, it would also move diffusively, on the average, over a distance about a micrometer within the time of a second. Hence, propulsion effects would still be significant for the slower membrane swimmers on such time scales.

8.4 CONCLUSIONS

Our numerical investigations were performed for a model three-domain swimmer that can be viewed as an idealization of an active protein with three domains relative distances between which change within each machine cycle. By using the parameter values characteristic for real biological membranes and actual active proteins, we could demonstrate that the propulsion effect can be strong enough to dominate over the random diffusive motion of proteins inside a membrane. Based on our results, the physical picture of cellular membranes may need to be modified. The proteins inside them are not only subject to thermal Brownian motion, but can also actively travel within the membrane provided that ATP or other chemical "fuel" is supplied. It would be interesting to experimentally verify our theoretical predictions. The potential functional importance of active intramembrane transport in biological cells has to be further discussed.

References

Alberts, B., Johnson, A., Lewis, J., Raff, M., Roberts, K., Walter, P., 2002. Molecular Biology of the Cell, 4th edn. Garland, New York.

Alonso, S., Mikhailov, A.S., 2009. Phys. Rev. E 79, 061906.

Astumian, R.D., 2014. ACS Nano 8, 11917.

Chen, H.-Y., 2004. Phys. Rev. Lett. 92, 168101.

Chen, H.-Y., Mikhailov, A.S., 2010. Phys. Rev. E 81, 031901.

Cooke, I.R., Deserno, M., 2005. J. Chem. Phys. 123, 224710.

Cooke, I.R., Kremer, K., Deserno, M., 2005. Phys. Rev. E 72, 011506.

de Groot, B.L., Grubmeller, H., 2001. Science 294, 2353.

Deserno, M., 2009. Macromol. Rapid Commun. 30, 752.

Dey, K.K., Das, S., Poyton, M.F., Sengupta, S., Butler, P.J., Cremer, P.S., Sen, A., 2014. ACS Nano 8, 11941.

Diamant, H., 2009. J. Phys. Soc. Jpn. 78, 041002.

Dickey, A.N., Faller, R., 2008. Biophys. J. 95, 2636.

Echeverria, C., Togashi, Y., Mikhailov, A.S., Kapral, R., 2011. Phys. Chem. Chem. Phys. 13, 10527.

Feller, S.E., 2000. Curr. Opin. Colloid Interface Sci. 5, 217.

Flechsig, H., Mikhailov, A.S., 2010. Proc. Natl. Acad. Sci. USA 107, 20875.

Gambin, Y., Lopez-Esparza, R., Reffay, M., Sierecki, E., Gov, N.S., Genest, M., Hodges, R.S., Urbach, W., 2006. Proc. Natl. Acad. Sci. USA 103, 2098.

Golestanian, R., 2010. Phys. Rev. Lett. 105, 018103.

Golestanian, R., Najafi, A., 2010. Phys. Rev. Lett. 100, 038101.

Gompper, G., Ihle, T., Kroll, D.M., Winkler, R.G., 2009. Adv. Polym. Sci. 221, 1.

Howse, J.R., Jones, R.A.L., Ryan, A.J., Gough, T., Vahabakhsh, R., Golestanian, R., 2007. Phys. Rev. Lett. 99, 048102.

Huang, M.-J., Chen, H.-Y., Mikhailov, A., 2012a. Eur. Phys. J. E 35, 119.

Huang, M.-J., Kapral, R., Mikhailov, A.S., Chen, H.-Y., 2012b. J. Chem. Phys. 137, 055101.

Huang, M.-J., Kapral, R., Mikhailov, A.S., Chen, H.-Y., 2013. J. Chem. Phys. 138, 195101.

Ihle, T., Kroll, D.M., 2001. Phys. Rev. E 63, 020201.

Ihle, T., Kroll, D.M., 2003. Phys. Rev. E 67, 066795.

Kapral, R., 2008. Adv. Chem. Phys. 140, 89.

Kodera, N., Yamanoto, D., Ishikawa, R., Ando, T., 2010. Nature 468, 72.

Kremer, K., Grest, G.S., 2004. J. Chem. Phys. 92, 5057.

Malevanets, N., Kapral, R., 1999a. J. Chem. Phys. 112, 7260.

Malevanets, N., Kapral, R., 1999b. J. Chem. Phys. 110, 8605.

Marchioro, G., Pulvirenti, M., 1982. Commun. Math. Phys. 84, 483.

Mikhailov, A.S., Meinköhn, D., 1997. Self-motion in physico-chemical systems far from thermal equilibrium. In: Schimansky-Geier, L., Pöschel, Th. (Eds.), Stochastic Dynamics, vol. 484. Springer, Berlin.

Muddana, H.S., Sengupta, S., Mallouk, T.E., Sen, A., Butler, P.J., 2010. J. Am. Chem. Soc. 132, 2110.

Najafi, A., Golestanian, R., 2004. Phys. Rev. E 69, 062901.

Phillips, R., Kondev, J., Theriot, J., 2009. Physical Biology of The Cell. Garland Science.

Prost, J., Bruinsma, R., 1996. Europhys. Lett. 33, 321.

Purcell, E.M., 1977. Am. J. Phys. 45, 3.

Reynwar, B.J., Illya, G., Harmandaris, V.A., Muller, M.M., Kremer, K., Deserno, M., 2007. Nature (London) 447, 461.

Riedel, C., Gabizon, R., Wilson, C.A.M., Hamadani, K., Tsekouras, K., Marqusee, S., Pressé, S., Bustamante, C., 2015. Nature 517, 227.
Saffman, P.G., Delbruck, M., 1975. Proc. Natl. Acad. Sci. USA 72, 3111.
Saiz, L., Klein, M.L., 2002. Acc. Chem. Res. 35, 482.
Sakaue, T., Kapral, R., Mikhailov, A.S., 2010. Eur. Phys. J. B 75, 381.
Sankararaman, S., Menon, G.I., Kumer, P.B.S., 2002. Phys. Rev. E 66, 031914.
Sengupta, S., Dey, K.K., Muddana, H.S., Tabouillot, T., Ibele, M.E., Butler, P.J., Sen, A., 2013. J. Am. Chem. Soc. 135, 1406.
Tieleman, D.P., Marrink, S.J., Berendsen, H.J., 1997. Biochim. Biophys. Acta 1331, 235.
Uchinashi, T., Iino, R., Ando, T., Noji, H., 2011. Science 333, 755.

CHAPTER

9

Phase Field Modeling of Inhomogeneous Biomembranes in Flow

S. Aland[#]

Institute of Scientific Computing, TU Dresden, Dresden, Germany

9.1 MOTIVATION

Membranes are essential components of cells. They not only envelop the cell as a whole but form the basic structural units for compartmentalizing its interior (Bassereau et al., 2014). Biomembranes are complex structures whose fundamental components include a bilayer of amphiphilic molecules (lipids), proteins, and cholesterol. They often exhibit inhomogeneities, for example, from inserted proteins, adsorbed molecules, or different compositions of lipids and cholesterol. The morphology and structure of biomembranes play an important role in their biological function (Lowengrub et al., 2009).

Vesicles are closed biomembranes that are missing any proteins and typically consist of only lipids and cholesterol. One could think of vesicles as fluid filled sacs with sizes ranging from few tens of nanometers (small unilamellar vesicle) to tens of microns (giant unilamellar vesicles). Smaller vesicles appear within cells where they typically split up from a larger membrane in a process called endocytosis. They are used to transport cargo material in nature as well as in technical applications, for example, as drug delivery systems (Chonn and Cullis, 1995; Allen and Cullis, 2004).

Even in equilibrium vesicles can exhibit an amazing variety of shapes, which can be predicted by minimizing a generalized version of the Helfrich energy (Helfrich, 1973). But to understand the rich dynamics of vesicles and membranes in nature, the surrounding fluids have to be taken into account. Due to their simplicity, giant unilamellar vesicles serve as a model system for anucleate cells such as red blood cells (RBCs) to study membrane mechanical properties and their interaction with surrounding viscous fluids. As a result, the behavior of vesicles and membranes in external flows has been of great interest in the last years, experimentally, theoretically and numerically.

[#] Chapter coordinator.

Numerical methods for membranes and vesicles have to deal with the complex interplay between membrane elasticity and hydrodynamic forces acting at microscopic length scales. Starting with single-component vesicles (Seifert et al., 1991), over the years, more and more complex and fundamental components of cell membranes have been numerically taken into account. That includes lipids, proteins and cholesterol and their interplay on the morphology, structure and dynamics of membranes (Allain and Amar, 2004; Lowengrub et al., 2009; Wang and Du, 2007). Adhesive effects to substrates have been investigated in Das and Du (2008).

While there are many different numerical approaches to deal with membrane inhomogeneities, we will focus here on the phase field (or diffuse interface) method, which offers a unique simplicity for incorporation of different physical effects. For example, protrusion and contraction forces have been considered to study cell morphodynamics (Shao et al., 2010, 2012), cell motility was modeled by a signaling network (Marth and Voigt, 2013), collective motion of cells was investigated (Löber et al., 2015) and formation of a vesicle during endocytosis was simulated (Lowengrub et al., 2016).

In this chapter, we give an overview on phase field models for inhomogeneous membranes. Starting from the energy of the system we present the governing sharp interface equations of a membrane in flow and address extensions to account for membrane inhomogeneities. In particular we discuss the definition of a common bending stiffness and spontaneous curvature for membranes with mixed components. We then discuss numerical approaches and review the phase field model to approximate the sharp interface equations where particular focus is put on the realization of the inextensibility constraint arising from membrane stretch elasticity. Finally, we present recent results of phase field models on multi-component membranes and endocytosis.

9.2 ENERGY OF THE SYSTEM

The lipid bilayer can be regarded as a thin fluid material of about 5 nm thickness where each lipid covers an area of approximately 0.7 nm^2 on the bilayer surface. The scale difference of system size, usually in the μm range, and membrane thickness can be exploited to describe the membrane as an elastic surface surrounded by fluids. In the following, we consider closed membranes. Hence, the membrane surface Γ separates two fluid phases Ω_0 and Ω_1 as illustrated in Fig. 9.1. The computational domain comprises all the involved phases $\Omega = \Omega_0 \cup \Omega_1 \cup \Gamma$. Two principal curvatures C_1 and C_2 of the surface can be defined as inverse of the two orthogonal curvature radii, see Fig. 9.2 for an illustration. In this notation, the mean curvature $H = C_1 + C_2$ is equivalent to the total curvature, by contrast some authors define H as $(C_1 + C_2)/2$. Equivalently, H can be defined using the surface normal $\mathbf{n}$ such that $H = \nabla \cdot \mathbf{n}$. Thus, H is positive when $\mathbf{n}$ points out of a convex region, as in Fig. 9.1, and negative when $\mathbf{n}$ points out of a concave region. Further, the Gaussian curvature is $K = C_1 \times C_2$.

To better understand the variety of shape changes that cells experience during their life, Helfrich was one of the first to derive a continuum energy for lipid bilayers (Helfrich, 1973). He assumed Hookean responses of the membrane to bending and stretching that essentially lead to three elastic moduli: stretching modulus k_S, bending modulus k_B, and Gaussian bending modulus k_G. Neglecting the influence of lipid tilt, the Helfrich energy for bending,

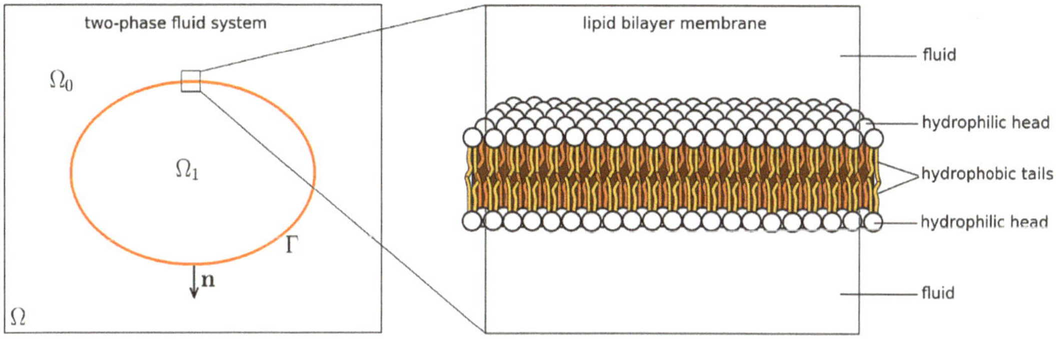

FIGURE 9.1 Illustration of a closed membrane. (Left) Mathematical notation, the membrane surface Γ separates two fluid phases Ω_0 and Ω_1. (Right) A close-up shows the physical form of the membrane as a lipid bilayer.

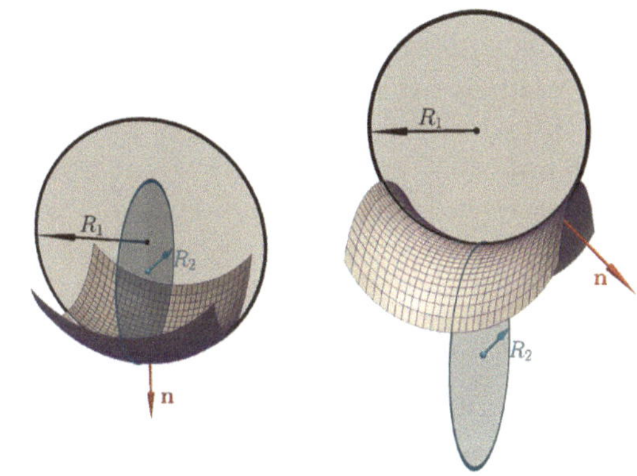

FIGURE 9.2 Illustration of membrane curvature. The two principal curvatures $C_1 = 1/R_1$ and $C_2 = 1/R_2$ are used to calculate mean curvature $H = C_1 + C_2$ and Gaussian curvature $K = C_1 \times C_2$. (Left) Convex membrane ($K > 0$). (Right) Saddle shaped membrane ($K < 0$).

Gaussian bending, and stretching is given by

$$E = E_B + E_G + E_S, \tag{9.1}$$

$$E_B = \int_\Gamma \frac{1}{2} k_B (H - H_0)^2 \,\mathrm{d}A, \tag{9.2}$$

$$E_G = \int_\Gamma k_G K \,\mathrm{d}A, \tag{9.3}$$

$$E_S = \int_\Gamma \frac{1}{2} k_S (J - 1)^2 \,\mathrm{d}A, \tag{9.4}$$

where J is the local area stretch of the membrane and H_0 is a spontaneous curvature, i.e., the preferred curvature of the membrane in absence of stresses. Nonzero H_0 arises, for example, from asymmetries between the to layers of lipids. The local area stretch J is defined as the current area of a surface element divided by the initial area of the same material part of the surface: $J = A/A_0$, hence, $J = 1$ corresponds to the reference state.

The complete energy (9.1) can be seen as an approximation of thin shell theory for vanishing shear modulus due to membrane fluidity. For constant k_G and as long as the membrane topology does not change, the Gaussian curvature term can be neglected, since it remains constant as can be shown by the Gauss–Bonnet formula.

The local area stretch can be computed from an interfacial advection equation

$$d_t J + \mathbf{v} \cdot \nabla J = J \nabla_\Gamma \cdot \mathbf{v} \quad \text{on } \Gamma, \tag{9.5}$$

where $\nabla_\Gamma \cdot$ is the surface divergence operator. The initial condition

$$J(t = 0) = 1 \quad \text{on } \Gamma \tag{9.6}$$

sets an initially unstretched equilibrium state of the membrane. Note that the above equation is very similar to evolution equations for interfacial species concentrations, like surfactants. The difference is the missing diffusion term and the sign of the right hand side.

The areal stiffness k_S of a lipid bilayer is typically very large such that the membrane surface area hardly changes over time. Rupture of a lipid bilayer occurs already at a stretching of a few percent (Hallett et al., 1993). Hence, in many cases it is appropriate to consider a constraint $J = 1$. In this case, the stretching energy E_S can be replaced by

$$E_{\tilde{S}} = \int_\Gamma \gamma (J - 1)\, \mathrm{d}A, \tag{9.7}$$

with a local Lagrange multiplier γ that needs to be determined to satisfy the constraint. Evolution equations for the interaction of the membrane with the surrounding fluids can be derived from variation of this energy.

9.3 HYDRODYNAMIC MODELS

Simulations of non-hydrodynamic models that can be used to describe an equilibrium shape of vesicles have been performed in Du et al. (2004, 2006). However, to understand the complex motions and shape changes of membranes in form of vesicles and red blood cells, the coupling of the dynamics with the surrounding fluid has to be taken into account. Within a low Reynolds number regime, the Stokes limit is valid. Various numerical approaches have been considered in this limit, for example, to analyze the tank-treading and tumbling motion of vesicles and red blood cells (Beaucourt et al., 2004; Kraus et al., 1996; Kim and Lai, 2010; Sohn et al., 2010; Zhao and Shaqfeh, 2011).

In regimes with the Reynolds number being of order unity or higher, the Stokes limit might not be appropriate, for example, when simulating vesicles traversing an artery with clots inducing shear rates of up to 425,000 s^{-1} (Bark and Ku, 2010), resulting in a Reynolds number of the order of 2 (Salac and Miksis, 2012). Here, we consider the general case of arbitrary

Reynolds numbers. Hence, the Navier–Stokes equations govern the evolution of the fluid velocity $\mathbf{v}$ in Ω_i, $i = 1, 2$,

$$\rho_i \left(\frac{\partial \mathbf{v}}{\partial t} + \mathbf{v} \cdot \nabla \mathbf{v} \right) + \nabla p - \nabla \cdot \left(\eta_i (\nabla \mathbf{v} + (\nabla \mathbf{v})^T) \right) = \mathbf{F} \qquad \text{in } \Omega_i, \tag{9.8}$$

$$\nabla \cdot \mathbf{v} = 0 \qquad \text{in } \Omega_i, \tag{9.9}$$

where ρ_i, η_i are the density and viscosity in the respective fluid phases, p is the pressure and $\mathbf{F}$ an external force.

Assuming no tangential slip leads to the interfacial jump condition

$$[\mathbf{v}]_\Gamma = \mathbf{0}, \tag{9.10}$$

where $[f]_\Gamma = f_{|\Omega_1} - f_{|\Omega_0}$ denotes the jump across the interface. Hydrodynamic and membrane forces are balanced by the jump condition

$$\left[-p\mathbf{I} + \eta_i (\nabla \mathbf{v} + (\nabla \mathbf{v})^T) \right]_\Gamma \cdot \mathbf{n} = -\frac{\delta E_B}{\delta \Gamma} - \frac{\delta E_S}{\delta \Gamma}, \tag{9.11}$$

where $\mathbf{I}$ is the identity matrix. The right hand side of Eq. (9.11) contains bending and stretching forces from the lipid bilayer which can be derived by variation of the corresponding energies with respect to changes in the interface position. The first variation $\frac{\delta E}{\delta \Gamma}$ can be defined implicitly using the energy time derivative as

$$d_t E = \int_\Gamma \frac{\delta E}{\delta \Gamma} \cdot \mathbf{v}_\Gamma \, \mathrm{d}A, \tag{9.12}$$

where $\mathbf{v}_\Gamma$ is the interface velocity. To obtain $\frac{\delta E_S}{\delta \Gamma}$, we may compute

$$d_t E_S = \int_\Gamma \frac{1}{2} k_S (J-1)^2 \nabla_\Gamma \cdot \mathbf{v} + k_S (J-1)(d_t + \mathbf{v} \cdot \nabla) J \, \mathrm{d}A, \tag{9.13}$$

where $\nabla_\Gamma \cdot \mathbf{v}$ accounts for changes in the surface area. Using Eq. (9.5) leads to

$$d_t E_S = \int_\Gamma \frac{1}{2} k_S (J-1)^2 \nabla_\Gamma \cdot \mathbf{v} + k_S (J-1) J \nabla_\Gamma \cdot \mathbf{v} \, \mathrm{d}A \tag{9.14}$$

$$= \int_\Gamma \frac{1}{2} k_S (3J^2 - 2J + 1) \nabla_\Gamma \cdot \mathbf{v} \, \mathrm{d}A \tag{9.15}$$

$$= \int_\Gamma \frac{1}{2} k_S (3J^2 - 2J + 1)(\nabla_\Gamma \cdot (P\mathbf{v}) - H\mathbf{n} \cdot \mathbf{v}) \, \mathrm{d}A, \tag{9.16}$$

where $P = \mathbf{I} - \mathbf{n} \otimes \mathbf{n}$ is the surface projection. Integration by parts finally gives

$$d_t E_S = \int_\Gamma \frac{1}{2} k_S \mathbf{v} \cdot (H\mathbf{n} - \nabla_\Gamma)(3J^2 - 2J + 1) \, \mathrm{d}A, \tag{9.17}$$

where ∇_Γ is the surface gradient. By definition (9.12) and a first order Taylor expansion, one obtains

$$\frac{\delta E_S}{\delta \Gamma} \approx k_S(H\mathbf{n} - \nabla_\Gamma)(J - 1), \tag{9.18}$$

which shows that the resulting stretching force effectively is a surface tension of magnitude $k_S(J-1)$. Note, that this surface tension can be negative if the membrane is locally compressed ($J < 1$).

The bending force $\frac{\delta E_B}{\delta \Gamma}$ can be similarly derived by energy variation, see Lowengrub et al. (2016). The result is

$$\frac{\delta E_B}{\delta \Gamma} = - k_B \left(\Delta_\Gamma (H - H_0) + (H - H_0)\|\nabla_\Gamma \mathbf{n}\|_F^2 - \frac{1}{2} H (H - H_0)^2 \right) \mathbf{n}, \tag{9.19}$$

where Δ_Γ denotes the Laplace–Beltrami operator and $\|\cdot\|_F$ the Frobenius norm.

For the case of large stretching modulus k_S, the energy $E_{\tilde{S}}$ may be used to replace E_S, which avoids the use of J and the possibly stiff coupling between surface tension force and membrane stretch evolution. It holds

$$\frac{\delta E_{\tilde{S}}}{\delta \Gamma} = \gamma H \mathbf{n} - \nabla_\Gamma \gamma. \tag{9.20}$$

In this case γ acts as a Lagrange multiplier and can be determined from requiring $J = 1$. In practice, this constraint is usually enforced by the inextensibility condition

$$\nabla_\Gamma \cdot \mathbf{v} = 0 \qquad \text{on } \Gamma \tag{9.21}$$

that needs to be coupled to the Navier–Stokes equation.

The bilayer membrane is impermeable to many molecules. Water can in principle permeate the membrane, but under zero net osmotic pressure any influx of water is relaxed by an outflux of the same magnitude, making the membrane practically impermeable on time scales of minutes. The volume enclosed by the membrane can thus be considered as constant. Mathematically, this amounts to the condition of zero mass transfer, stating that the membrane velocity equals the fluid velocity,

$$\mathbf{v}_\Gamma = \mathbf{v} \qquad \text{on } \Gamma. \tag{9.22}$$

The relatively large membrane viscosity of about 10^{-9} to 10^{-8} Pa s m has recently given rise to computational models that take into account membrane viscosity and membrane inertia. In this case, additional surface Navier–Stokes equations replace the stress balance in Eq. (9.11). We refer to the recent paper of Barrett, Garcke and Nürnberg (Barrett et al., 2015).

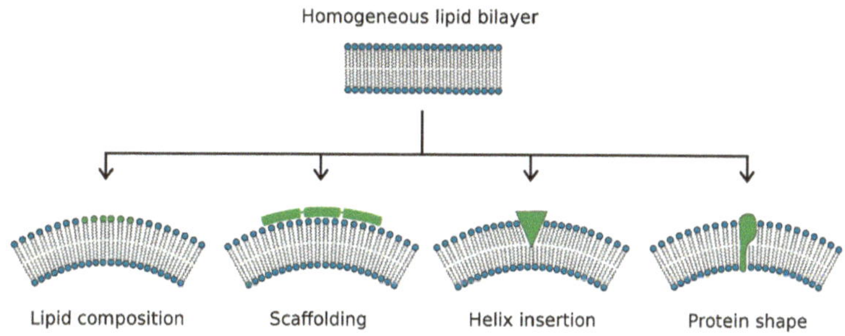

FIGURE 9.3 Basic mechanisms to induce membrane curvature by inhomogeneous lipid composition or adsorption of curvature-inducing molecules.

9.4 INHOMOGENEOUS MEMBRANES

So far, we have considered homogeneous lipid bilayer membranes. In general, however, biomembranes can be much more diverse, including inhomogeneities from inserted proteins, adsorbed molecules, or different compositions of lipids and cholesterol. Typically, such changes in the lipid bilayer change the mechanical membrane properties and induce interface curvature, see Fig. 9.3. We will term additional molecules that lead to membrane curvature as curvature-inducing molecules (CIM).

When several types of lipid and cholesterol are present, phase transformations may occur on the membrane leading to the formation of domains, so-called rafts, with distinct physical properties (Lowengrub et al., 2009). From a biological point of view, the presence of rafts can affect signaling and trafficking (Simons et al., 1997; van Meer and Sprong, 2004) since particular proteins may concentrate at or near rafts (Hess et al., 2005). Rafts may also play a role in other membrane processes such as membrane adhesion and fusion (Takeda et al., 2003) as well as budding and fission during endocytosis and exocytosis (Rauch and Farge, 2000). Along a membrane, the phase separation and the corresponding formation of rafts is strongly coupled to the shape dynamics. Hence, curvature effects can dynamically influence the phase separation and vice versa, leading to a rich dynamical behavior.

Mathematical models have been developed for membranes with variable biophysical properties using discrete and continuum approaches, see, for example, the review (Elson et al., 2010). Discrete modeling approaches, such as dissipative particle dynamics, molecular dynamics, or Monte Carlo methods, have been used to simulate the dynamics of phase-separation and domain formation, vesicle fission and fusion (e.g., see Kumar et al., 2001; Gao et al., 2008; Smith and Uspal, 2007; Grafmüller et al., 2007; Markvoort et al., 2007; Laradji and Kumar, 2005; Huang et al., 2006; Shillcock and Lipowsky, 2005; Laradji and Kumar, 2004).

Continuum methods are an alternative modeling approach to reach larger length and time scales. In continuum models, the effect of inhomogeneities is in general considered by introducing variable bending stiffness and spontaneous curvature. One approach to distinguish membrane regions with different physical properties is to introduce a field c along the membrane. This field can be a purely auxiliary field or physically represent the concentration of a certain type of lipids or of additional membrane molecules. In any case, bending stiffness and spontaneous curvature depend on c: $k_B = k_B(c)$, $H_0 = H_0(c)$, which alters the bending force from Eq. (9.19) such that variations in c are taken into account (Lowengrub et al., 2016),

$$\begin{aligned}\frac{\delta E_B}{\delta \Gamma} =& -\Delta_\Gamma \left(k_B(c)(H - H_0(c))\right)\mathbf{n} \\ &- k_B(c)(H - H_0(c))\|\nabla_\Gamma \mathbf{n}\|^2 \mathbf{n} \\ &+ \frac{k_B(c)}{2}(H - H_0(c))^2 H\mathbf{n} - \frac{\delta E_B}{\delta c}\nabla_\Gamma c,\end{aligned} \tag{9.23}$$

$$\frac{\delta E_B}{\delta c} = \frac{1}{2}\frac{\mathrm{d}k_B}{\mathrm{d}c}(H - H_0(c))^2 - k_B(c)(H - H_0(c))\frac{\mathrm{d}H_0}{\mathrm{d}c}. \tag{9.24}$$

9.4.1 Separated Membrane Components

Consider a membrane with some inhomogeneously distributed curvature inducing molecules (CIM) changing bending stiffness and spontaneous curvature in the membrane region they cover. If the size of these molecules is of the order of the system size, a good modeling approach will consider each molecule and the field c describing their presence will be discontinuous, e.g., $c = 1$ where the molecules are present and $c = 0$ in regions of a clean membrane. The same holds if the membrane is composed of different physical components with different physical properties. Also these components can be identified by the field c and if the size of the each components is comparable to the system size, c would be discontinuous as well.

On the other hand, if the size of the CIM is very small compared to the system size, it might not be possible to take every single CIM into account. Equivalently, if the membrane components are mixed on a small length scale, one might not want to resolve every single lipid or CIM on the microscopic scale. Consequently, a coarse-graining will be necessary and c will represent concentration of mixing. For example, c may account for the normalized amount of CIM per surface area, e.g., $0 \leq c \leq 1$, with $c = 1$ denoting full coverage with CIMs.

In this case, we will speak of mixed membrane components while we speak of separated membrane components when resolving every single of change components along the membrane. Separated components also occur if the curvature-inducing molecules tend to bind to other molecules of their kind or if phase separation occurs, for example, in case of multicomponent membranes. From a modeling perspective, one has to distinguish between separated and mixed components since they involve different energies.

If separated components are considered, typically a line energy has to be defined along the phase boundary between them. Such an energy reads

$$E_L = \int_{\text{Curve}} \tau \, \mathrm{d}s \tag{9.25}$$

where Curve denotes a curve that separates the two phases along the membrane and τ is the line tension. The line tension can be measured in experiments (Tian et al., 2007) and has been shown to have a significant effect on the lateral organization of lipid membrane components (Garcia-Saez et al., 2007). Further, the Gaussian bending energy can no longer be neglected if it differs in the two phases (Haußer et al., 2013). For different constant values of b_G in the different phases, the Gaussian bending energy reduces to

$$E_G = \int_C [b_G] H_G \mathrm{d}s \tag{9.26}$$

where $[b_G]$ denotes the jump of b_G across C and H_G is the geodesic curvature of C. Through a comparison of experimental vesicle geometries and numerical results, the importance of the Gaussian bending energy in multicomponent vesicles has recently been verified in Baumgart et al. (2005). Still, most mathematical methods of multicomponent membranes neglect Gaussian bending, due to numerical difficulties.

9.4.2 Mixed Membrane Components

Now let us address the case of a mixed membrane components, i.e., when c is a coarse grained concentration field. Let us assume c to be normalized such that $c = 1$ and $c = 0$ correspond to the two distinct membrane regions. Typically, the bending stiffness and spontaneous curvatures are known for these two cases, while the bending stiffness of mixed regions, i.e., $0 < c < 1$, is unknown. Let k_B^0, k_B^1, and H_0^0, H_0^1 denote the known bending stiffness and spontaneous curvature for $c = 0$, $c = 1$, respectively.

To state the energy of the two-component membrane with mixed regions (i.e., $0 < c < 1$), we need explicit formulations for k_B and H_0 depending on c. The linear combination $k_B(c) = k_B^0(1 - c) + k_B^1 c$ was used in Nowak and Chou (2008) and is also typically used for diffuse interface models of multicomponent membranes, e.g., in Sohn et al. (2010), Haußer et al. (2013), Sohn et al. (2012). While this linear approach is valid for separated membrane components, we will show in the following that this is not appropriate for mixed components. We will present a derivation where the whole bending energy is coarse-grained to account for different bending of covered and non-covered parts of the membrane.

Let us consider a small control area on the membrane $A \subset \Gamma$. A should be small enough such that the observed curvature H can be assumed as constant within A but still big enough to contain a mixture of both membrane components. We denote the region of the two components within A by A^0 and A^1, with $A^0 \cup A^1 = A$. Under the given premises, the membrane curvature will assume a constant value in A^0 and (a different) constant value in A^1, see Fig. 9.4 for an illustration.

Next, we can introduce the indicator function c as the relative area of component one, $c = |A^1|/|A|$. Since the control volume is very small, the small scale differences in the curvature will not be seen by the observer. The observed coarse-grained curvature H will be related to the microscopic curvatures of the two components by

$$H = cH^1 + (1 - c)H^0. \tag{9.27}$$

Next, consider the bending energy of the control volume, which is

$$E_A = \int_{A^1} \frac{1}{2} k_B^1 (H^1 - H_0^1)^2 \mathrm{d}A + \int_{A^0} \frac{1}{2} k_B^0 (H^0 - H_0^0)^2 \mathrm{d}A \tag{9.28}$$

$$= \int_A \frac{1}{2} k_B^1 (H^1 - H_0^1)^2 c + \frac{1}{2} k_B^0 (H^0 - H_0^0)^2 (1 - c) \mathrm{d}A. \tag{9.29}$$

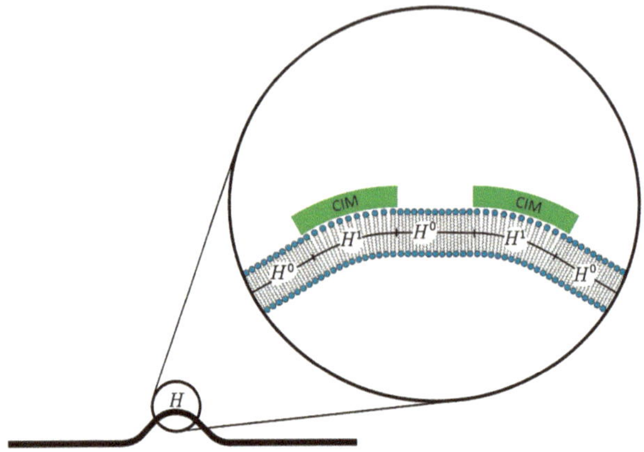

FIGURE 9.4 Schematic of a partially covered membrane. While the observed membrane curvature is H, a close-up shows that the membrane is curved differently in covered and uncovered regions (curvature H^1 and H^0, resp.). The different curvatures are related by Eq. (9.27).

Assume that this energy is minimal, since A is small enough so that the system is in local equilibrium. Subject to Eq. (9.27) and keeping H fixed, E_A is minimized for

$$H^0 = \frac{k_B^1(H - cH_0^1) + k_B^0 c H_0^0}{(1-c)k_B^1 + k_B^0 c}, \tag{9.30}$$

$$H^1 = \frac{k_B^0(H - (1-c)H_0^0) + k_B^1(1-c)H_0^1}{(1-c)k_B^1 + k_B^0 c}, \tag{9.31}$$

which gives the equilibrium energy

$$E_A = \int_A \frac{1}{2} \frac{k_B^0 k_B^1}{(1-c)k_B^1 + ck_B^0} \left(H - H_0^0(1-c) - H_0^1 c\right)^2 \mathrm{d}A. \tag{9.32}$$

Since the above holds for arbitrary A, we may conclude the bending energy of the whole membrane is

$$E = \int_\Gamma \frac{1}{2} k_B(c)\,(H - H_0(c))^2\,\mathrm{d}A, \tag{9.33}$$

with the c-dependent parameters

$$k_B(c) = \frac{k_B^0 k_B^1}{(1-c)k_B^1 + ck_B^0}, \qquad H_0(c) = H_0^0(1-c) + H_0^1 c. \tag{9.34}$$

Hence, the spontaneous curvature of a mixed two-component membrane depends linearly on c, while the bending stiffness is not a simple linear function of c. Intuitively, the expression

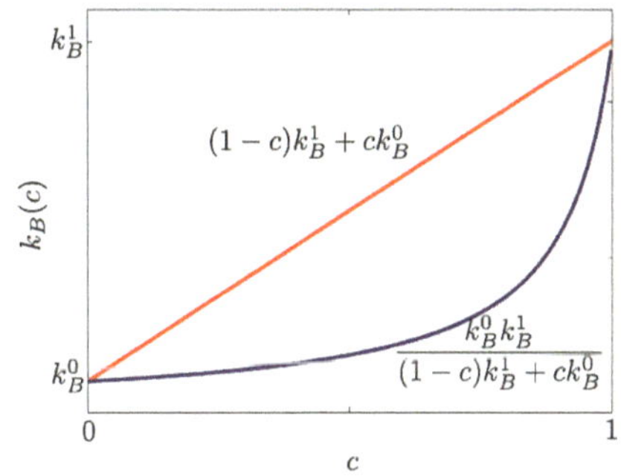

FIGURE 9.5 A comparison of linear and nonlinear laws for the bending stiffness of a membrane of mixed components.

for $k_B(c)$ makes a lot of sense. For example, if one of the two components has zero bending stiffness, all the curvature would be accommodated in the component without bending stiffness, while the other component would just assume its spontaneous curvature. This is reflected in the above definition of the bending stiffness, since $k_B(c) = 0$ in this case. Obviously, in the limits $c = 0, c = 1$ we have $k_B(0) = k_B^0, k_B(1) = k_B^1$, respectively. A comparison of the linear bending stiffness used in Nowak and Chou (2008) and the nonlinear bending stiffness from Eq. (9.34) is shown in Fig. 9.5.

9.5 NUMERICAL METHODS

Many different numerical methods have been proposed to solve the coupled system of Navier–Stokes equations (9.8)–(9.11) and interface advection (9.22). Most of these approaches consider the membrane as an inextensible surface, i.e., E_S is replaced by $E_{\tilde{S}}$ and the Navier–Stokes system is coupled to the inextensibility constraint (9.21).

The boundary-integral method (Pozrikidis, 1992; Kraus et al., 1996; Zhao and Shaqfeh, 2011) uses Green's tensor formalism to calculate velocity fields by integration of the membrane forces over the vesicle surface. The calculated velocity field obeys the Stokes equations per definition and is thus limited to viscous flow. Inertial effects cannot be included.

Particle-based methods define effective particles that interact in such a way that the actual equations of motion are recovered (Abreu et al., 2014). The multi-particle collision dynamics method (Noguchi and Gompper, 2005) takes hydrodynamic interactions, membrane properties, and thermal fluctuations into account. Although partly phenomenological, this numerical scheme is very flexible and allows to simulate complex flow geometries and many-body problems (Abreu et al., 2014).

Immersed boundary methods (Peskin, 2002; Kim and Lai, 2010) simulate elastic membranes immersed in incompressible fluids. Due to their simplicity, these methods are increasingly popular for fluid–structure interaction. Large scale simulations can be done when these methods are coupled to approximate solvers of the Navier–Stokes equations, like Lattice-Boltzmann (Krüger, 2016; Kaoui et al., 2012). However, membrane fluidity still poses a challenge on most IBM methods, since most approaches are based on elasticity of a membrane represented by a grid.

The phase field (Biben et al., 2005) or advected field (Beaucourt et al., 2004; Biben and Misbah, 2003) approach represents the membrane by a continuous field that assumes distinct values in the bulk phases and varies along the interface. The field is advected by a Willmore-flow or special Allen–Cahn equation. The inextensibility constraint can be realized by a penalty approach or by regularizing the inextensibility equation (Aland et al., 2014). In the next sections we will further discuss phase field methods. Other continuum-based numerical schemes include level-set (Salac and Miksis, 2012; Laadhari et al., 2014; Doyeux et al., 2013) and front-tracking (Yazdani and Bagchi, 2013; Luo et al., 2013) methods. A comparison between a level-set and phase field methods is given in Maitre et al. (2012) and between level-set and parametric methods in Benes et al. (2009).

Recently, new methods have been proposed to account for the surface fluidity by additional surface Navier–Stokes equations, which include membrane viscosity and membrane inertia (Barrett et al., 2015; Arroyo and DeSimone, 2009).

9.6 THE PHASE FIELD METHOD

A phase field is an auxiliary function (or field) to model interfacial problems. It takes distinct values in each of the fluid phases (e.g., -1 and 1) with a smooth transition in between, around the interface. Hence, the interface is diffuse with a finite width and an intermediate level set of the phase field (e.g., 0) may be used to get a discrete interface location. Phase field methods were originally developed to describe solid–liquid phase transitions and have since seen an enormous growth in use, see Emmerich (2008) for a review.

Fig. 9.6 shows an example of a phase field. A fine computational grid is needed to numerically resolve the transition layer, which makes phase field computations usually more expensive than other interface capturing methods. On the other hand, phase field methods offer many advantages in the mathematical modeling. One distinct advantage is their simplicity. The equation to advect the phase field is straightforward to discretize and does not require any numerical stabilization techniques. As opposed to Level-Set methods, mass is conserved and reinitialization is not necessary. Phase field methods naturally admit physical energy laws which makes it possible to develop energy stable discretization schemes. Also topological changes, such as during vesicle fission and fusion, are naturally captured and additional physical processes can be easily included within the phase field formulation.

Early phase field methods for biomembranes were used without the coupling to hydrodynamics of the surrounding fluids. These models have been used to describe equilibrium shapes of vesicles in Du et al. (2004, 2006). Analytical studies that discuss existence and convergence can be found in Du and Zhu (2006), Du et al. (2005a). The spontaneous curvature has been taken into account in Du et al. (2005b). Topological considerations including the calculation of the Euler number are addressed in Du et al. (2006), Esedoglu et al. (2013). A non-hydrodynamic diffuse interface model has been coupled to further equations in order to describe cell motility (Shao et al., 2010), creating a system, which is permanently out of equilibrium.

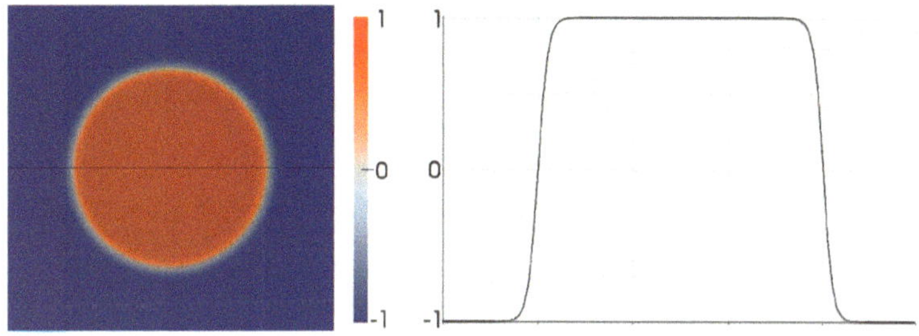

FIGURE 9.6 (Left) Schematic of a phase field. (Right) A cut through the phase field along the black line (left) shows the smooth but rapid transitions across the interface in form of a tangent hyperbolic profile.

Du et al. were the first to develop a thermodynamically consistent hydrodynamic phase field model for lipid bilayers (Du et al., 2007). It has since been used for a variety of applications, for example, to investigate the rheology of red and white blood cells (Lázaro et al., 2014; Marth et al., 2016), to describe cell motility (Marth and Voigt, 2013) or endocytosis (Lowengrub et al., 2016). A slightly different hydrodynamic diffuse interface model has been introduced in Biben and Misbah (2003), Biben et al. (2005) with a different phase field formulation and without an explicit formulation of the diffuse Helfrich energy. More recently, in Farshbaf-Shaker and Garcke (2011) a very general approach was given, that covers both diffuse interface formulations. Reviews of existing phase field models for the minimization of the Helfrich energy can be found in Lázaro et al. (2015), Du (2011).

9.6.1 Phase Field Equations

Given a phase field that represents the membrane position, curvature as well as the interface delta function can be expressed in terms of ϕ. The phase field variant of the bending energy including spontaneous curvature can be formulated as

$$E_{B,\epsilon} = \int_{\Omega} \frac{3k_B}{8\sqrt{2}\epsilon} \left(\epsilon \Delta\phi - \frac{1}{\epsilon} W_0'(\phi) \right)^2 \mathrm{d}V, \tag{9.35}$$

where $W_0'(\phi) = (\phi^2 - 1)(\phi + \sqrt{2}\epsilon H_0)$ is the first derivative of a modified double well potential (Du et al., 2005a). In Du et al. (2005b) and more rigorously in Wang (2008), convergence of $E_{B,\epsilon}$ toward the sharp interface bending energy was shown as $\epsilon \to 0$.

By requiring that the sum of bending and kinetic energy be non-increasing in time, one can derive the Navier–Stokes–Helfrich equations (Du et al., 2007). The Navier–Stokes equations, which are now extended to whole Ω, become

$$\rho \left(\frac{\partial \mathbf{v}}{\partial t} + \mathbf{v} \cdot \nabla \mathbf{v} \right) + \nabla p - \nabla \cdot \left(\eta (\nabla \mathbf{v} + (\nabla \mathbf{v})^T) \right) = \nabla \cdot (\mathbf{P} |\nabla\phi| \gamma) + \frac{\delta E_{B,\epsilon}}{\delta \phi} \nabla \phi, \tag{9.36}$$

$$\nabla \cdot \mathbf{v} = 0, \tag{9.37}$$

where $\rho(\phi), \eta(\phi)$ are interpolations of density and viscosity, e.g.,

$$\rho(\phi) = \rho_1(1+\phi)/2 + \rho_2(1-\phi)/2, \qquad \eta(\phi) = \eta_1(1+\phi)/2 + \eta_2(1-\phi)/2, \tag{9.38}$$

and the surface projection operator $\mathbf{P}$ is defined by

$$\mathbf{P} = \mathbf{I} - \mathbf{n} \otimes \mathbf{n} = \mathbf{I} - \frac{\nabla\phi \otimes \nabla\phi}{|\nabla\phi|^2}. \tag{9.39}$$

Additionally, the phase field needs to be advected with the flow which is usually realized by an advected Willmore flow equation

$$d_t\phi + \mathbf{v} \cdot \nabla\phi = -M\frac{\delta E_{B,\epsilon}}{\delta\phi}. \tag{9.40}$$

Other approaches to advect the phase field are conserved Willmore flow (Lowengrub et al., 2016) and the advected field approach (Biben and Misbah, 2003; Beaucourt et al., 2004)

For the given system of Eqs. (9.36)–(9.40), existence of global weak solutions and uniqueness under extra regularity were proven (Du et al., 2007). Local time existence and uniqueness of strong solutions are shown in Liu et al. (2012).

9.6.2 Inextensibility or Surface Elasticity

As discussed in Section 9.3, there are, in principle, two ways to account for the elasticity of the bilayer surface. The necessary surface tension γ can be calculated either directly from the local area stretch J or as a Lagrange multiplier for the inextensibility constraint $\nabla_\Gamma \cdot \mathbf{v} = 0$ on Γ.

To keep track of the area stretch J, it is convenient to introduce its inverse $\hat{J} = 1/J$. From the evolution equation for J, Eq. (9.5), we obtain

$$d_t\hat{J}^{-1} + \mathbf{v} \cdot \nabla\hat{J}^{-1} = \hat{J}^{-1}\nabla_\Gamma \cdot \mathbf{v} \qquad \text{on } \Gamma, \tag{9.41}$$

or equivalently,

$$d_t\hat{J} + \mathbf{v} \cdot \nabla\hat{J} = -\hat{J}\nabla_\Gamma \cdot \mathbf{v} \qquad \text{on } \Gamma. \tag{9.42}$$

This equation is identical to a surface evolution equation for the surface concentration of any physical interfacial species without diffusion. In particular, $\hat{J}$ is a conserved quantity along the interface. Many numerical methods exist to deal with conserved interfacial quantities. In the context of phase field methods, the diffuse interface approach (Ratz and Voigt, 2006) is the standard technique which we follow here.

Consider the weak formulation of Eq. (9.42) with a test function ψ

$$\int_\Gamma \psi d_t\hat{J} + \psi\nabla_\Gamma \cdot (\mathbf{v}\hat{J})\, \mathrm{d}A = 0. \tag{9.43}$$

From the Leibniz rule we get

$$d_t \int_\Gamma \psi \hat{J} \, dA - \int_\Gamma \hat{J}(d_t\psi + \mathbf{v} \cdot \nabla\psi) \, dA = 0. \tag{9.44}$$

Next, we extend the corresponding surface integrals over whole Ω by multiplication of the integrand with the approximate surface delta function $|\nabla\phi|/2$ and get

$$d_t \int_\Omega \psi \hat{J}|\nabla\phi| \, dA - \int_\Gamma |\nabla\phi|\hat{J}(d_t\psi + \mathbf{v} \cdot \nabla\psi) \, dV = 0. \tag{9.45}$$

Now use that $d_t\psi = 0$ and integration by parts in the last term to obtain

$$\int_\Omega d_t(\hat{J}|\nabla\phi|)\psi + \nabla \cdot (|\nabla\phi|\hat{J}\mathbf{v})\psi \, dV = 0. \tag{9.46}$$

The strong form of this formulation is

$$d_t(\hat{J}|\nabla\phi|) + \nabla \cdot (|\nabla\phi|\hat{J}\mathbf{v}) = 0 \qquad \text{in } \Omega, \tag{9.47}$$

which is a simple evolution equation on the full computational domain. This equation can be solved for $\hat{J}$ with initial condition $\hat{J} = 1$ and provides the surface tension $\gamma = k_S(1/\hat{J} - 1)$ for the membrane stretching according to Eq. (9.18).

Alternatively, instead of considering membrane stretching, inextensibility of the membrane can be enforced. To translate the inextensibility constraint into a diffuse interface context, Aland et al. (2014) have introduced a regularized extension over the whole domain, namely

$$\xi\epsilon^2\nabla \cdot \left(\phi^2\nabla\gamma\right) + |\nabla\phi|\nabla_{\Gamma,\epsilon} \cdot \mathbf{v} = 0, \tag{9.48}$$

where the diffuse inextensibility constraint is defined as

$$\nabla_{\Gamma,\epsilon} \cdot \mathbf{v} = \mathbf{P} : \mathbf{v}. \tag{9.49}$$

The parameter ξ is a regularization parameter independent of ϵ. Eq. (9.48) becomes $\nabla_{\Gamma,\epsilon} \cdot \mathbf{v} = 0$ at the interface where $\phi \approx 0$, and reduces to $\Delta\gamma = 0$ away from Γ since $\phi^2 \approx 1$ and $|\nabla\phi| \approx 0$. This provides a harmonic extension of γ off Γ, while maintaining the local inextensibility constraint near Γ. The formulation of diffuse inextensibility is validated by asymptotic analysis and numerical tests in Aland et al. (2014).

9.7 PHASE FIELD MODELS FOR INHOMOGENEOUS MEMBRANES

In Lowengrub et al. (2016), a phase field model was derived for the hydrodynamic simulation of inextensible membranes with locally varying bending stiffness and spontaneous curvature.

Since k_B and H_0 are now functions of some concentration field c, the variation of the bending energy with respect to ϕ is

$$\frac{\delta E_{B,\epsilon}}{\delta\phi} = \frac{3}{4\sqrt{2}}\left[\Delta(k_B(c)\mu) - \frac{1}{\epsilon^2}k_B(c)\mu\frac{\mathrm{d}W_0'(\phi,c)}{\mathrm{d}\phi}\right], \tag{9.50}$$

$$\mu = \epsilon\Delta\phi - \frac{1}{\epsilon}W_0'(\phi,c). \tag{9.51}$$

The membrane-induced bending forces are derived by energy variation and from the diffuse bending energy (9.35)

$$\mathbf{F} = \frac{\delta E_{B,\epsilon}}{\delta\phi}\nabla\phi + \frac{3}{4\sqrt{2}\epsilon}\left[\frac{1}{2}\frac{\mathrm{d}k_B}{\mathrm{d}c}\mu^2 - \frac{1}{\epsilon}k_B(c)\mu\frac{\mathrm{d}W_0'(\phi,c)}{\mathrm{d}c}\right]\nabla c. \tag{9.52}$$

While the first term on the right hand side was also present for homogeneous membranes (see Eq. (9.36)), the last term is only present if k_B or H_0 vary along the membrane.

9.7.1 Multicomponent Vesicles

Phase separation is a well-studied process in bulk phases. Similarly to bulk phase separation, spinodal decomposition takes place along multicomponent membranes and creates distinct surface phases (e.g., liquid-ordered, liquid-disordered) known as rafts (or domains). These rafts may play an important role in regulating protein activity since certain proteins may be attracted to some rafts (Hess et al., 2005).

However, the dynamics of phase coarsening on a membrane is much more complex than in a bulk material since the coarsening process is strongly coupled to the shape dynamics and curvature effects can dynamically influence the phase separation, and vice versa. Experiments on giant unilamellar vesicles demonstrate that there exists a rich variety of behavior of multicomponent vesicles. Spinodal decomposition, raft coarsening, and viscous fingering can all be accompanied by changes in membrane morphology, vesicle budding, fission, and fusion. See, for example, Veatch and Keller (2003), Baumgart et al. (2003), Beattie et al. (2005), Semrau et al. (2008).

In Lowengrub et al. (2009), a study of the dynamics of two-dimensional vesicles containing two surface phases was performed. It was found that differences between the spontaneous curvatures and the bending rigidities of the surface phases lead to the formation of buds and asymmetric vesicle shapes. Also, an example was given where asymmetric spontaneous curvature induced vesicle fission producing two asymmetric daughter vesicles with each daughter vesicle being rich in one of the surface phases. Most interestingly, it was confirmed that not only spinodal decomposition and coarsening of the surface phases affect the vesicle shape but also that the vesicle shape affects spinodal decomposition.

In a subsequent paper (Sohn et al., 2010), the multicomponent vesicle model was coupled to hydrodynamics and the effect of different surface phases on the morphology of the vesicle and its dynamics was investigated. It was shown that the different phases result in a dramatic evolution in shear flow as the vesicle bends in regions where the bending stiffness and spontaneous curvature are small, see Fig. 9.7.

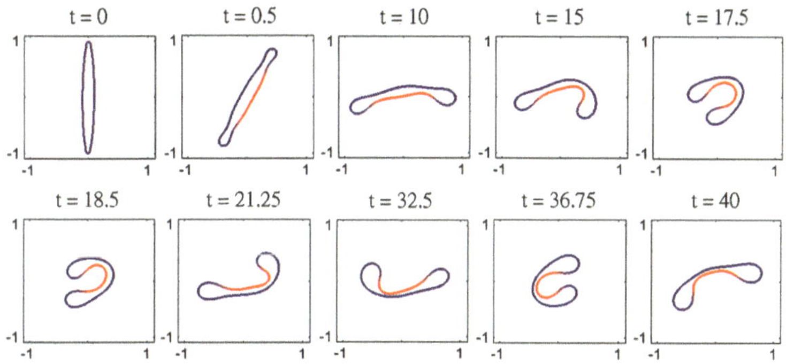

FIGURE 9.7 Multicomponent vesicle evolution under an applied shear with a 30–70 mixture of lipid phases with different bending stiffness and spontaneous curvatures. Figure adopted from Sohn et al. (2010) with permission.

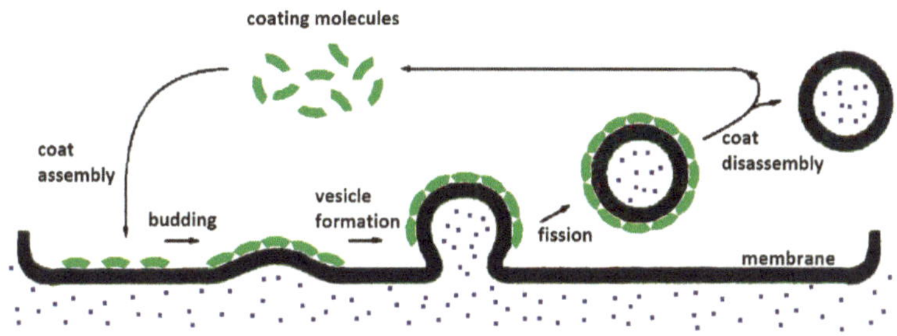

FIGURE 9.8 Schematic of endocytosis mediated by curvature-inducing molecules. Molecules are recruited to the membrane which leads to budding and fission. The coat of the vesicle then disassembles to release the coating molecules into the cytosol. The formed vesicle transports cargo material (blue dots) from the cell exterior into the cell.

9.7.2 Endocytosis

The biological membranes that surround cells and organelles often undergo shape changes as part of cellular processes. In some of these processes, such as endocytosis, inter-organelle transport and virus entry (Lalonde and Sundquist, 2012), a small vesicle is formed from a larger membrane. As discussed earlier, such events are typically orchestrated by curvature-inducing molecules (CIM) that attach to the membrane. Typical CIMs are bar-domain proteins and clathrin, which is a key component in endocytosis. While for a long time it was believed that clathrin assembly alone can force a membrane into a curved state, new findings suggest that an interplay with additional proteins is necessary to effectively curve membranes in vivo.

In the classical scenario, coating molecules are recruited to the membrane where they induce curvature. Once the curvature is large enough, the membrane forms a vesicle which can finally pinch off as triggered by additional molecules as dynamin. The coat of the vesicle then disassembles to release the coating molecules into the cytosol, see Fig. 9.8 for an illustration.

Classical clathrin-mediated endocytosis, which occurs on a timescale of seconds, has been extensively studied, including contributions from mathematical models (Berro et al., 2010;

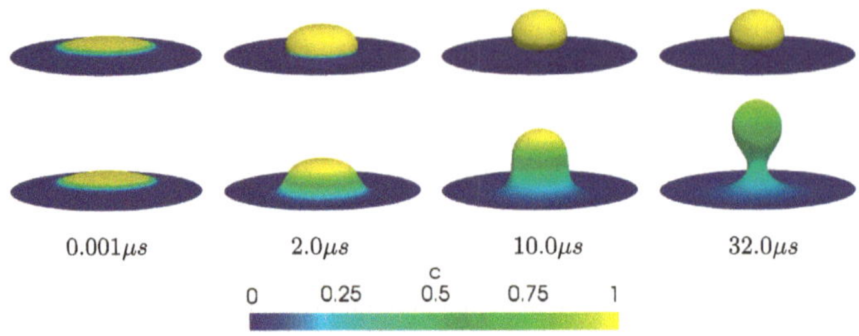

FIGURE 9.9 Comparison of membrane evolution and CIM concentrations c when the curvature-inducing molecules are transported without (top) or with surface diffusion (bottom). In the absence of diffusion, the distinct states values of 1 (covered) and 0 (uncovered) are maintained and a steady-state configuration of a compact vesicle connected to the membrane by a small neck. Surface diffusion decreases the gradients of the CIMs that transiently produces longer necks and more deformed vesicles.

Liu et al., 2009; Nowak and Chou, 2008). However, recent reports demonstrate that in some circumstances vesicles can form de novo in a few milliseconds (Watanabe et al., 2013). For example, ultrafast endocytosis at the neurological synapse has been reported as a two-step process, where within milliseconds excess membrane creates large endocytic vesicles, that then fuse to endosomes, which then are resolved into synaptic vesicles via a clathrin-dependent process (Watanabe et al., 2014).

Typical endocytic vesicles assume diameters as small as 50–100 nm, which makes it difficult for imaging techniques to provide data for a detailed understanding of the process and the functionality of the involved molecules. Numerical simulations provide a means to shed light on the complex interplay of membrane curvature, cytoplasmic flow and arrival of CIMs. A phase field model for the dynamics of inextensible vesicles in a viscous fluid with stiff, curvature-inducing molecules was derived in Lowengrub et al. (2016). The model couples the Willmore flow for the membrane advection (9.40) with the Navier–Stokes equations (9.36)–(9.37) and an improved version of the inextensibility constraint (9.48) that prevents accumulation of stretching/compression errors. The membrane-induced bending forces are as given in Eqs. (9.50)–(9.52). The bending stiffness and spontaneous curvature are assumed to depend linearly on CIM concentration,

$$k_B(\phi) = (1 + 11c) \times 10^{-19}\ \mathrm{kg\,m^2/s^2}, \tag{9.53}$$

$$H_0(\phi) = c \cdot (14.1\ \mathrm{nm})^{-1}. \tag{9.54}$$

The Gaussian curvature energy is neglected.

A circular clathrin coat in the center of a flat axisymmetric membrane section was assumed as initial condition. Two forms of transport of the CIM concentration were considered: Fickian surface diffusion and pure advection. The results show that CIM induced bending forces can enable a very rapid formation of a small vesicle attached to the membrane by a narrow neck, see Fig. 9.9. When the Fickian model is used, this is a transient state with the steady state being a flat membrane with a uniformly distributed molecule concentration due to diffusion. When pure advection of the CIMs is used, molecule concentration gradients are sustained, and the system evolves to a steady-state with a small vesicle attached to the membrane.

The steady-state vesicle morphology was found to depend on the initial coverage of CIMs on the membrane. For increasing CIM coverages a transition from no neck, to a spherical neck, to a mushroom-shaped neck was found. More numerical studies of endocytosis including the hydrodynamics including adsorption of CIM from the bulk fluid to the membrane are still missing.

References

Abreu, D., Levant, M., Steinberg, V., Seifert, U., 2014. Fluid vesicles in flow. In: Advances in Colloid and Interface Science.

Aland, S., Egerer, S., Lowengrub, J., Voigt, A., 2014. Diffuse interface models of locally inextensible vesicles in a viscous fluid. J. Comput. Phys. 277, 32–47.

Allain, J.-M., Amar, M.B., 2004. Biphasic vesicle: instability induced by adsorption of proteins. Physica A 337 (3–4), 531–545.

Allen, T.M., Cullis, P.R., 2004. Drug delivery systems: entering the mainstream. Science 303 (5665), 1818–1822.

Arroyo, M., DeSimone, A., 2009. Relaxation dynamics of fluid membranes. Phys. Rev. E 79 (3), 031915.

Bark, D.L., Ku, D.N., 2010. Wall shear over high degree stenoses pertinent to atherothrombosis. J. Biomech. 43 (15), 2970–2977.

Barrett, J.W., Garcke, H., Nuürnberg, R., 2015. Numerical computations of the dynamics of fluidic membranes and vesicles. Phys. Rev. E 92 (5), 052704.

Bassereau, P., Sorre, B., Lévy, A., 2014. Bending lipid membranes: experiments after W. Helfrich's model. Adv. Colloid Interface Sci. 208, 47–57.

Baumgart, T., Das, S., Webb, W., Jenkins, J., 2005. Membrane elasticity in giant vesicles with fluid phase coexistence. Biophys. J. 89 (2), 1067–1080.

Baumgart, T., Hess, S.T., Webb, W.W., 2003. Imaging coexisting fluid domains in biomembrane models coupling curvature and line tension. Nature 425 (6960), 821–824.

Beattie, M.E., Veatch, S.L., Stottrup, B.L., Keller, S.L., 2005. Sterol structure determines miscibility versus melting transitions in lipid vesicles. Biophys. J. 89 (3), 1760–1768.

Beaucourt, J., Rioual, F., Séon, T., Biben, T., Misbah, C., 2004. Steady to unsteady dynamics of a vesicle in a flow. Phys. Rev. E 69 (1), 011906.

Beneš, M., Mikula, K., Oberhuber, T., Ševčovič, D., 2009. Comparison study for Level set and Direct Lagrangian methods for computing Willmore flow of closed planar curves. Comput. Vis. Sci. 12 (6), 307–317.

Berro, J., Sirotkin, V., Pollard, T.D., 2010. Mathematical modeling of endocytic actin patch kinetics in fission yeast: disassembly requires release of actin filament fragments. Mol. Biol. Cell 21 (16), 2905–2915.

Biben, T., Misbah, C., 2003. Tumbling of vesicles under shear flow within an advected-field approach. Phys. Rev. E 67 (3), 31908.

Biben, T., Kassner, K., Misbah, C., 2005. Phase-field approach to three-dimensional vesicle dynamics. Phys. Rev. E 72 (4), 041921.

Chonn, A., Cullis, P.R., 1995. Recent advances in liposomal drug-delivery systems. Curr. Opin. Biotechnol. 6 (6), 698–708.

Das, S., Du, Q., 2008. Adhesion of vesicles to curved substrates. Phys. Rev. E 77 (1), 011907.

Doyeux, V., Guyot, Y., Chabannes, V., Prud-Homme, C., Ismail, M., 2013. Simulation of two-fluid flows using a finite element/level set method. Application to bubbles and vesicle dynamics. J. Comput. Appl. Math. 246, 251–259.

Du, Q., 2011. Phase field calculus, curvature-dependent energies, and vesicle membranes. Philos. Mag. 91 (1), 165–181.

Du, Q., Zhu, L., 2006. Analysis of a mixed finite element method for a phase field bending elasticity model of vesicle membrane deformation. J. Comput. Math. 24 (3), 265–280.

Du, Q., Liu, C., Wang, X., 2004. A phase field approach in the numerical study of the elastic bending energy for vesicle membranes. J. Comput. Phys. 198 (2), 450–468.

Du, Q., Liu, C., Ryham, R., Wang, X., 2005a. A phase field formulation of the Willmore problem. Nonlinearity 18, 1249–1267.

Du, Q., Liu, C., Ryham, R., Wang, X., 2005b. Modeling the spontaneous curvature effects in static cell membrane deformations by a phase field formulation. Commun. Pure Appl. Anal. 4 (3), 537–548.

Du, Q., Liu, C., Wang, X., 2006. Simulating the deformation of vesicle membranes under elastic bending energy in three dimensions. J. Comput. Phys. 212 (2), 757–777.

Du, Q., Li, M., Liu, C., 2007. Analysis of a phase field Navier–Stokes vesicle–fluid interaction model. Discrete Contin. Dyn. Syst. 8 (3), 539–556.

Elson, E.L., Fried, E., Dolbow, J.E., Genin, G.M., 2010. Phase separation in biological membranes: integration of theory and experiment. Annu. Rev. Biophys. 39, 207–226.

Emmerich, H., 2008. Advances of and by phase-field modelling in condensed-matter physics. Adv. Phys. 57 (1), 1–87.

Esedoglu, S., Ratz, A., Roger, M., 2013. Colliding interfaces in old and new diffuse-interface approximations of Willmore-flow. Commun. Math. Sci. 12 (1), 125–147.

Farshbaf-Shaker, M.H., Garcke, H., 2011. Thermodynamically consistent higher order phase field Navier–Stokes models with applications to biological membranes. Discrete Contin. Dyn. Syst., Ser. S 4 (2), 371–389.

Gao, L., Lipowsky, R., Shillcock, J., 2008. Tension-induced vesicle fusion: pathways and pore dynamics. Soft Matter 4 (6), 1208–1214.

Garcia-Saez, A.J., Chiantia, S., Schwille, P., 2007. Effect of line tension on the lateral organization of lipid membranes. J. Biol. Chem. 282 (46), 33537–33544.

Grafmüller, A., Shillcock, J., Lipowsky, R., 2007. Pathway of membrane fusion with two tension-dependent energy barriers. Phys. Rev. Lett. 98 (21), 218101.

Hallett, F.R., Marsh, J., Nickel, B.G., Wood, J.M., 1993. Mechanical properties of vesicles. Biophys. J. 64, 426–434.

Haußer, F., Li, S., Lowengrub, J., Marth, W., Ratz, A., Voigt, A., 2013. Thermodynamically consistent models for two-component vesicles. Int. J. Biomath. Biostat. 2 (1), 19–48.

Helfrich, W., 1973. Elastic properties of lipid bilayers: theory and possible experiments. Z. Naturforsch., C 28 (11), 693–703.

Hess, S.T., Kumar, M., Verma, A., Farrington, J., Kenworthy, A., Zimmerberg, J., 2005. Quantitative electron microscopy and fluorescence spectroscopy of the membrane distribution of influenza hemagglutinin. J. Cell Biol. 169 (6), 965–976.

Huang, K.C., Mukhopadhyay, R., Wingreen, N.S., 2006. A curvature-mediated mechanism for localization of lipids to bacterial poles. PLoS Comput. Biol. 2 (11), e151.

Kaoui, B., Krüger, T., Harting, J., 2012. How does confinement affect the dynamics of viscous vesicles and red blood cells? Soft Matter 8 (35), 9246–9252.

Kim, Y., Lai, M.-C., 2010. Simulating the dynamics of inextensible vesicles by the penalty immersed boundary method. J. Comput. Phys. 229 (12), 4840–4853.

Kraus, M., Wintz, W., Seifert, U., Lipowsky, R., 1996. Fluid vesicles in shear flow. Phys. Rev. Lett. 77 (17), 3685.

Krüger, T., 2016. Effect of tube diameter and capillary number on platelet margination and near-wall dynamics. Rheol. Acta 55, 511–526.

Kumar, P.S., Gompper, G., Lipowsky, R., 2001. Budding dynamics of multicomponent membranes. Phys. Rev. Lett. 86 (17), 3911.

Laadhari, A., Saramito, P., Misbah, C., 2014. Computing the dynamics of biomembranes by combining conservative level set and adaptive finite element methods. J. Comput. Phys. 263, 328–352.

Lalonde, M.S., Sundquist, W.I., 2012. How HIV finds the door. Proc. Natl. Acad. Sci. 109 (46), 18631–18632.

Laradji, M., Kumar, P.S., 2004. Dynamics of domain growth in self-assembled fluid vesicles. Phys. Rev. Lett. 93 (19), 198105.

Laradji, M., Kumar, P.S., 2005. Domain growth, budding, and fission in phase-separating self-assembled fluid bilayers. J. Chem. Phys. 123 (22), 224902.

Lázaro, G.R., Hernández-Machado, A., Pagonabarraga, I., 2014. Rheology of red blood cells under flow in highly confined microchannels: I. Effect of elasticity. Soft Matter 10 (37), 7195–7206.

Lázaro, G.R., Pagonabarraga, I., Hernández-Machado, A., 2015. Phase-field theories for mathematical modeling of biological membranes. Chem. Phys. Lipids 185, 46–60.

Liu, J., Sun, Y., Drubin, D.G., Oster, G.F., 2009. The mechanochemistry of endocytosis. PLoS Biol. 7 (9), e1000204.

Liu, Y., Takahashi, T., Tucsnak, M., 2012. Strong solutions for a phase field Navier–Stokes vesicle-fluid interaction model. J. Math. Fluid Mech. 14 (1), 177–195.

Löber, J., Ziebert, F., Aranson, I.S., 2015. Collisions of deformable cells lead to collective migration. Sci. Rep. 5.
Lowengrub, J., Allard, J., Aland, S., 2016. Numerical simulation of endocytosis: viscous flow driven by membranes with non-uniformly distributed curvature-inducing molecules. J. Comput. Phys. 309, 112–128.
Lowengrub, J.S., Ratz, A., Voigt, A., 2009. Phase-field modeling of the dynamics of multicomponent vesicles: spinodal decomposition, coarsening, budding and fission. Phys. Rev. E 79 (3), 31926.
Luo, Z.Y., Wang, S.Q., He, L., Xu, F., Bai, B.F., 2013. Inertia-dependent dynamics of three-dimensional vesicles and red blood cells in shear flow. Soft Matter 9 (40), 9651–9660.
Maitre, E., Misbah, C., Peyla, P., Raoult, A., 2012. Comparison between advected-field and level-set methods in the study of vesicle dynamics. Physica D 241 (13), 1146–1157.
Markvoort, A., Smeijers, A., Pieterse, K., van Santen, R., Hilbers, P., 2007. Lipid-based mechanisms for vesicle fission. J. Phys. Chem. B 111 (20), 5719–5725.
Marth, W., Voigt, A., 2013. Signaling networks and cell motility: a computational approach using a phase field description. J. Math. Biol., 91–112.
Marth, W., Aland, S., Voigt, A., 2016. Margination of white blood cells: a computational approach by a hydrodynamic phase field model. J. Fluid Mech. 790, 389–406.
van Meer, G., Sprong, H., 2004. Membrane lipids and vesicular traffic. Curr. Opin. Cell Biol. 16 (4), 373–378.
Noguchi, H., Gompper, G., 2005. Dynamics of fluid vesicles in shear flow: effect of membrane viscosity and thermal fluctuations. Phys. Rev. E 72 (1), 011901.
Nowak, S.A., Chou, T., 2008. Membrane lipid segregation in endocytosis. Phys. Rev. E 78 (2), 21908.
Peskin, C.S., 2002. The immersed boundary method. Acta Numer. 11, 479–517.
Pozrikidis, C., 1992. Boundary Integral and Singularity Methods for Linearized Viscous Flow. Cambridge University Press.
Ratz, A., Voigt, A., 2006. PDE's on surfaces—a diffuse interface approach. Commun. Math. Sci. 4 (3), 575–590.
Rauch, C., Farge, E., 2000. Endocytosis switch controlled by transmembrane osmotic pressure and phospholipid number asymmetry. Biophys. J. 78 (6), 3036–3047.
Salac, D., Miksis, M.J., 2012. Reynolds number effects on lipid vesicles. J. Fluid Mech. 711, 122–146.
Seifert, U., Berndl, K., Lipowsky, R., 1991. Shape transformations of vesicles: phase diagram for spontaneous- curvature and bilayer-coupling models. Phys. Rev. A 44 (2), 1182–1202.
Semrau, S., Idema, T., Holtzer, L., Schmidt, T., Storm, C., 2008. Accurate determination of elastic parameters for multicomponent membranes. Phys. Rev. Lett. 100 (8), 088101.
Shao, D., Rappel, W.-J., Levine, H., 2010. Computational model for cell morphodynamics. Phys. Rev. Lett. 105 (10), 108104.
Shao, D., Levine, H., Rappel, W.-J., 2012. Coupling actin flow, adhesion, and morphology in a computational cell motility model. Proc. Natl. Acad. Sci. 109 (18), 6851–6856.
Shillcock, J.C., Lipowsky, R., 2005. Tension-induced fusion of bilayer membranes and vesicles. Nat. Mater. 4 (3), 225–228.
Simons, K., Ikonen, E., et al., 1997. Functional rafts in cell membranes. Nature 387 (6633), 569–572.
Smith, K.A., Uspal, W.E., 2007. Shear-driven release of a bud from a multicomponent vesicle. J. Chem. Phys. 126 (7), 075102.
Sohn, J., Li, S., Li, X., Lowengrub, J., 2012. Axisymmetric multicomponent vesicles: a comparison of hydrodynamic and geometric models. Int. J. Numer. Methods Biomed. Eng., 346–368.
Sohn, J.S., Tseng, Y.-H.H., Li, S., Voigt, A., Lowengrub, J.S., 2010. Dynamics of multicomponent vesicles in a viscous fluid. J. Comput. Phys. 229 (1), 119–144.
Takeda, M., Leser, G.P., Russell, C.J., Lamb, R.A., 2003. Influenza virus hemagglutinin concentrates in lipid raft microdomains for efficient viral fusion. Proc. Natl. Acad. Sci. 100 (25), 14610–14617.
Tian, A., Johnson, C., Wang, W., Baumgart, T., 2007. Line tension at fluid membrane domain boundaries measured by micropipette aspiration. Phys. Rev. Lett. 98 (20), 208102.
Veatch, S.L., Keller, S.L., 2003. Separation of liquid phases in giant vesicles of ternary mixtures of phospholipids and cholesterol. Biophys. J. 85 (5), 3074–3083.
Wang, X., 2008. Asymptotic analysis of phase field formulations of bending elasticity models. SIAM J. Math. Anal. 39 (5), 1367–1401.
Wang, X., Du, Q., 2007. Modelling and simulations of multi-component lipid membranes and open membranes via diffuse interface approaches. J. Math. Biol. 56 (3), 347–371.

Watanabe, S., Rost, B.R., Camacho-Perez, M., Davis, M.W., Sohl-Kielczynski, B., Rosenmund, C., Jorgensen, E.M., 2013. Ultrafast endocytosis at mouse hippocampal synapses. Nature 504 (7479), 242–247.
Watanabe, S., Trimbuch, T., Camacho-Perez, M., Rost, B.R., Brokowski, B., Söhl-Kielczynski, B., Felies, A., Davis, M.W., Rosenmund, C., Jorgensen, E.M., 2014. Clathrin regenerates synaptic vesicles from endosomes. Nature 515.
Yazdani, A., Bagchi, P., 2013. Influence of membrane viscosity on capsule dynamics in shear flow. J. Fluid Mech. 718, 569–595.
Zhao, H., Shaqfeh, E.S., 2011. The dynamics of a vesicle in simple shear flow. J. Fluid Mech. 674, 578–604.

CHAPTER

10

Modeling and Experimental Analysis of Thermal Therapy during Short Pulse Laser Irradiation

S. Miller*, C. Gross Jones*, K. Mitra[†,#]

*Department of Mechanical and Aerospace Engineering, Florida Institute of Technology, Melbourne, FL, United States [†]Department of Biomedical Engineering, Department of Mechanical and Aerospace Engineering, Florida Institute of Technology, Melbourne, FL, United States

10.1 INTRODUCTION

Pulsed lasers are widely used for therapeutic medical applications. In order to help predict the best pulsing protocols to use in therapeutic laser treatments, it can be very beneficial to perform numerical modeling and theoretical analysis. Different models of varying complexity have been developed for such analyzes.

The peak temperatures achieved during laser thermal therapy may be affected by blood flow in living tissues, which convectively cools regions heated above the body's core temperature. The two dominant approaches to account for the effects of blood flow in computational models of heat transfer in tissue have been termed the continuum model and the vascular model (Stańczyk and Telega, 2002; Stańczyk et al., 2007; Ganguly et al., 2015b). While the continuum model accounts for blood flow through a bulk perfusion term that is applied to the entirety of a single-phase domain in the heat equation, the vascular model consists of a multi-phase domain of discrete vessels containing blood with an assigned velocity field within a solid tissue volume.

The Pennes bio-heat equation has been widely used in the development of continuum models, which is given by:

$$\rho_t c_t \frac{\partial T_t}{\partial t} = \nabla \cdot (k_t \nabla T_t) + \omega_b c_b (T_a - T_t) + q_{met}, \tag{10.1}$$

[#] Chapter coordinator.

where the subscripts t and b denote tissue and blood, respectively, ρ is density, c is constant pressure specific heat, k is thermal conductivity, ω is the volumetric flow rate of blood per volume of tissue, T is temperature, T_a is arterial blood temperature, and q_{met} is a metabolic heat source term. In Eq. (10.1), $\omega_b c_b (T_a - T_t)$ is a linear heat sink term that is proportional to both ω and the temperature difference between the blood and tissue. Adding a volumetric heat source term representing the addition of thermal energy by the laser, Q_L, to the right hand side of Eq. (10.1) provides a continuum model of laser heating of living tissues. The continuum model of laser heating of excised tissues or non-perfused tissue phantoms, Eq. (10.2), is obtained by removing the linear heat sink term and metabolic heat generation term from Eq. (10.1) and adding a volumetric heat source term:

$$\rho_t c_t \frac{\partial T_t}{\partial t} = \nabla \cdot (k_t \nabla T_t) + Q_L, \tag{10.2}$$

where the subscript i denotes the tissue layer (e.g., epidermis, dermis, hypodermis).

In the vascular model, heat transfer within the discrete vessels is dependent on vessel diameter and branching and blood flow velocity, which determine whether the flow is laminar or turbulent. These parameters vary depending on the particular physiological regions and vessels included in the model. In the case of laminar flow in discrete vessels with uniform average velocity, u, the governing equation for the heat transfer due to blood flow is given by (Ganguly et al., 2015a, 2015b):

$$\rho_b c_{p_b} \frac{\partial \mathbf{T}_b}{\partial t} + \rho_b c_{p_b} u \nabla \mathbf{T}_b = \nabla \cdot (k_b \nabla \mathbf{T}_b) + Q_L, \tag{10.3}$$

where the subscript b denotes blood. It is also possible to create hybrid models in which larger vessels are modeled discretely using Eq. (10.3) and small capillaries are considered in a bulk perfusion term using Eq. (10.1).

Numerical models of laser treatment of tissues yield different results for temperature rises achieved within the target tissues depending on whether a continuum or vascular model is used. In numerical models of temperature increases in a tissue volume due to a subsurface-focused laser heat source, continuum models predicted larger volumes of temperature rise than vascular models did (Ganguly et al., 2015b). Comparison of both models of subsurface-focused laser tissue irradiation to experimental results using live anesthetized mice demonstrated that vascular models showed better agreement with experimental data than did continuum models (Ganguly et al., 2015a). Since continuum models consider bulk properties, they cannot account for light reflection and scattering at vessel walls. Accurate modeling of light scattering by blood vessels requires consideration of the curvature and diffuse reflectance of the vessels (Milanič and Majaron, 2011; Majaron et al., 2015), strengthening the case for the inclusion of discrete vessels in models of laser thermal therapies.

The laser operating mode must also be taken into consideration in the models considered above. Continuous wave (CW) lasers deliver nearly uniform energy to the target tissue, resulting in a constant treatment of the tissue and steady heat diffusion to tissues surrounding the laser focal point. Short pulse lasers, however, provide short pulses of significantly higher power than CW lasers, and the spacing of these pulses in time is shorter than the thermal relaxation time of tissues. Consequently, each pulse of a short pulse laser results in an incremental temperature increase before the energy from the previous pulse is able to diffuse away from the laser focal point, resulting in a highly localized step-wise temperature rise at the site of irradiation (Ganguly et al., 2015a; Jaunich et al., 2008). For a given average power, shorter laser pulses therefore result in greater energy delivered per pulse and more localized rises in temperature.

A representation of the pulsed laser heat source term, Q_L [W m^{-3}], given by Ganguly et al. (2015a, 2015b), Kim and Guo (2007) that can be used in Eqs. (10.2) and (10.3) to model heat transfer in tissues due to short pulse laser irradiation is:

$$Q_L(x, y, z) = \frac{\mu_{a_i}}{4\pi} L_o \exp\left\{-4\ln 2 \times \left[\frac{(t_{rel} - \frac{z}{c_i})}{t_p} - 1.5\right]^2\right\} \times \exp\left(\frac{-2(x^2+y^2)}{\sigma^2}\right)\exp\left(-z\mu'_{e_i}\right), \tag{10.4}$$

where the absorption and reduced extinction coefficients of each medium are μ_{a_i} [m^{-1}] and μ'_{e_i} [m^{-1}], respectively, and c_i is the speed of light in each medium [m s^{-1}]. The laser pulse full width at half maximum (FWHM) is given by t_p, and t_{rel} is the relative time of the current time step of the model within the laser pulse width [s], which is considered for a duration of $3 \cdot t_p$ for each pulse. The $1/e$ beam radius is σ, and x, y, and z are the coordinates at each point within the model volume [m]. In Eq. (10.4), the tissue medium is oriented with the xy-plane normal to the incident laser beam, and the beam is focused at the tissue surface ($z = 0$ m). The reduced extinction coefficient of each medium is given by the sum of the absorption and reduced scattering coefficients, $\mu'_{e_i} = \mu_{a_i} + \mu'_{s_i}$ [1/m], where $\mu'_{s_i} = (1 - g_i)\mu_{s_i}$ [1/m]. The degree of anisotropy of the scattering phase function in each medium is given by g_i [dimensionless] and the scattering coefficient in each medium is given by μ_{s_i} [1/m]. The average intensity of the laser beam over the duration of each pulse at $z = 0$ m, L_o [W m^{-2}], is given by:

$$L_o = \frac{P}{\pi r_0^2}, \tag{10.5}$$

where the average laser power during each pulse, P, is given by the average laser power measured continually, P_{avg}, divided by the time that the laser is on, $P = \frac{P_{avg}}{t_p \cdot f}$. The radius of the laser spot size at the $z = 0$ m surface is given by r_0, and f is the laser repetition rate [1/s]. During the intervals between laser pulses, $Q_L = 0$. Removing the temporal term in Eq. (10.4) gives a CW laser heat source term:

$$Q_L(x, y, z) = \frac{\mu_{a_i}}{4\pi} L_o \exp\left(\frac{-2(x^2+y^2)}{\sigma^2}\right)\exp\left(-z\mu'_{e_i}\right). \tag{10.6}$$

In the CW case, $L_o = \frac{P_{avg}}{\pi r_0^2}$.

The ability of short pulse laser based thermal therapies to achieve treatment goals with minimal damage to surrounding healthy tissues is dependent upon appropriately selecting the laser parameters with consideration to the target tissue's optical and thermophysical properties. Tissue properties and vascularization affect tissue energy absorption, temperature rise, and volume of tissue heated.

Laser thermal therapies must raise the temperature of the target tissue above a threshold temperature in order to achieve the desired temperature rise and treatment goals within the target tissue. For human tissues, 43°C is commonly accepted to be the threshold temperature for calculating the rate of cell death due to thermal increases. Thermal therapies are designed to deliver a specific thermal dose, termed CEM_{43} or TD_{43}, which is the cumulative equivalent minutes at which the target volume is maintained at or above 43°C (Sapareto and Dewey, 1984; Dewhirst et al., 2003; Yarmolenko et al., 2011). Cell damage and death are not determined by the static maximum temperature achieved during a laser treatment, but instead are dependent on the time history of the temperature rise within the target tissue. The thermal dose for a system whose temperature changes in time, such as a tissue region during laser thermal therapy, is defined as:

$$CEM_{43} = \int_0^t R^{(43-T)} dt, \tag{10.7}$$

where t is the duration of the treatment in minutes, $T(x, y, z, t)$ is the temperature of the tissue at each location in the target volume for each instant in time, and $R = 0.25$ for $T < 43$°C or $R = 0.50$ for $T \geq 43$°C (Sapareto and Dewey, 1984; Dewhirst et al., 2003; Yarmolenko et al., 2011). Using numerical integration, CEM_{43} is evaluated as a summation over all time steps, Δt_i,

$$CEM_{43} = \sum_{i=0}^{n} \Delta t_i R^{(43-T)}, \tag{10.8}$$

where n is the total number of time steps. In the case in which the temperature is constant for all time steps,

$$CEM_{43} = \Delta t R^{(43-T)}. \tag{10.9}$$

Determining the CEM_{43} needed to achieve treatment goals requires analysis of tissue responses to treatments over time (Yarmolenko et al., 2011). Different treatment regimes, such as one longer duration treatment at lower sustained temperatures or many shorter treatments at higher sustained temperatures, may deliver the same total CEM_{43}. The specific range of CEM_{43} required for cell damage and death varies between different types of tissues. CEM_{43} ranges in all types of human tissues have not been fully characterized, and insufficient data have been published to account for the long term effects of heat on the nervous system (Yarmolenko et al., 2011). Numerical models of laser irradiation that include both the period of laser irradiation and subsequent period of thermal equilibration provide temperature time histories that can be used to predict CEM_{43}. Such predictions may aid in assessing whether a given laser treatment can be expected to achieve the desired CEM_{43}.

Some tissues may appear sufficiently affected to achieve treatment goals immediately following irradiation but may regain function over time (Yarmolenko et al., 2011). Other factors, including thermosensitivity, or the change in tissue thermal response with repeated exposure to hyperthermia, also play a role in the CEM_{43} required to achieve therapeutic goals (Dewhirst et al., 2003). While human studies are lacking, the effects of thermal dose fractionation was investigated in a study of canine sarcomas, which found that a single larger thermal dose per week reduced tumor volume more than smaller thermal doses delivered 3 to 4 times per week (Thrall et al., 2012). Accurate in vivo experimental characterization of the long term effects of laser irradiation and associated CEM_{43} on the full range of human tissues is needed in order to optimize treatment regimes.

Ex vivo experiments to predict the thermal dose resulting from laser irradiation may not be accurate in predicting thermal doses achieved by irradiation of living tissues. Thin excised tissues maintained at an ambient temperature of approximately 24°C cannot accurately account for the fact that increasing the temperature of in vivo human tissue with a core temperature of 37°C by 6°C can raise the tissue temperature to the 43°C threshold. Instead, temperature rises of nearly 20°C and unrealistically high average laser powers and/or lasing times are required to heat excised tissue to the 43°C threshold, providing an insufficient model for laser treatment optimization.

There is extensive literature on the development and use of optical tissue phantoms, which are objects that simulate the optical properties of biological tissues, and many results are reported in a comprehensive review paper (Pogue and Patterson, 2006). These phantoms mimic tissue optical properties at the wavelength of the light source, including absorption and scattering coefficients, scattering phase function, tissue anisotropy, and index of refraction. Experiments on optical tissue phantoms may aid in predicting laser radiant fluence (a measure of total incident radiant energy per surface area) within the target tissue. In addition to mimicking tissue optical properties, tissue phantoms may be designed to also match tissue thermophysical properties, including specific heat, thermal conductivity and density, mechanical properties, such as viscosity, modulus of elasticity and toughness, acoustic properties, and chemical composition.

The three primary functions of materials used to construct optical tissue phantoms are the matrix materials (which may be liquids, gels or solids), the scatterers, and the absorbers (Pogue and Patterson, 2006). Gel matrix materials, such as agar (Milanič et al., 2007), gelatin (Cook et al., 2011), polyvinyl alcohol gels, and polyvinyl chloride plastisol (PVCP) gels (Bohndiek et al., 2013; Vogt et al., 2016) best simulate the mechanical properties of soft tissue and can be biocompatible (Pogue and Patterson, 2006). The PVCP gels provide acoustic, mechanical and optical property tunability (Bohndiek et al., 2013; Vogt et al., 2016). Epoxy matrix phantoms (Jaunich et al., 2008) provide the advantage of long term stability and durability and can be used in instrument calibration (Pogue and Patterson, 2006). While the choice between gel and solid matrix materials often involves a trade-off between biological relevance and long-term stability and durability, PVCP has been characterized as both stable and biologically relevant (Bohndiek et al., 2013; Vogt et al., 2016). Common scattering materials include titanium dioxide (Bohndiek et al., 2013; Jaunich et al., 2008), aluminum oxide, Intralipid (Cook et al., 2011; Whelan et al., 2001; Di Ninni et al., 2011), and polymer or silica microspheres (Pogue and Patterson, 2006; Cook et al., 2011). Frequently used absorbers include dyes and dye blends using components such as India Ink (Cook et al., 2011; Jaunich et al., 2008), Evans Blue (Cook et al., 2011), Direct Red 81 (Cook et al., 2011), and various ProJet or AlexaFluor dyes (Pogue and Patterson, 2006). Absorbers are selected to have absorption peaks that match the wavelength(s) of the light sources. Biological materials have been used in cases in which phantom stability was not a primary concern. For example, collagen has been used as a scatterer and blood has been used as an absorber (Pogue and Patterson, 2006).

Phantoms have been used to study the optical and thermal effects of blood flow. One example of vascularized tissue phantoms was created by filling lucite boxes with a ground lean veal–gelatin blend and placing one 1.4 mm or 4 mm diameter polytetrafluoroethylene (PTFE) vessel inside the phantom (Whelan et al., 1995). Water was pumped through the vessel by a

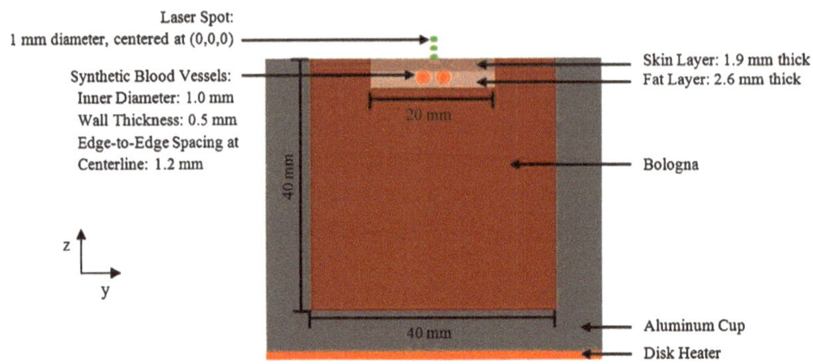

FIGURE 10.1 Cross-section (yz-plane) of phantom at $y = 0$ mm. Blood vessels are centered at $x = 1.6$ and $x = -1.6$ mm.

peristaltic pump and the lucite boxes were immersed in a water bath (Whelan et al., 1995). The phantoms were then irradiated with a fiber optic coupled CW Nd:YAG laser (Whelan et al., 1995). Additionally, 3D printed epoxy resin bone phantoms with hollow channels containing milk have been created to consider the optical effects of blood flow as measured by optical flowmeters (Binzoni et al., 2014).

In response to the many challenges inherent in laboratory experiments that characterize laser–tissue interactions, a novel temperature controlled synthetic tissue phantom with countercurrent synthetic blood flow was designed and developed. The phantoms were irradiated with a short pulse Nd:YAG laser and the measured surface temperature profiles were compared with numerical modeling results obtained using a COMSOL Multiphysics vascular model. The effects of blood flow on temperature rise during laser irradiation and subsequent heat diffusion following irradiation were analyzed. In order for the convective cooling effect of the subsurface countercurrent blood flow to be evident in surface temperature profiles, surface temperatures were required to rise above the 37°C temperature of the blood. The phantom volume was of sufficient size to fully attenuate all laser light as well as to prevent the boundary conditions for all boundaries other than the convective boundary at the surface incident to the laser from affecting the laser heat affected zone. This phantom design could be adapted for the experimental optimization of various thermal treatments.

10.2 METHODS

10.2.1 Vascularized Tissue Phantom Preparation

The vascularized tissue phantoms were constructed of materials mimicking soft tissues and contained discrete vessels with constant countercurrent blood flow. The PID temperature controllers were used to maintain the phantom interior at 37°C, representing core human body temperature. The $z = 0$ mm surface was exposed to ambient air at 24°C. A cross section of the phantom design is shown in Fig. 10.1.

A volume of a two-layer, 4.5 mm thick SynDaver synthetic tissue composed of a 1.9 mm thick synthetic skin layer and a 2.6 mm thick synthetic fat layer (SynDaver Labs, Tampa, FL) was inlayed in a volume of beef bologna. The bologna served to extend the length of the phantom volume in the z-direction beyond the 4.5 mm thickness of the synthetic skin and fat tissue in order to model a volume of tissue in which the laser light would be fully attenuated. Beef bologna has been shown to have thermophysical and optical properties similar to tissue and to demonstrate tissue-like responses to laser irradiation (Banerjee et al., 2005). Two 1 mm inner diameter (ID) synthetic blood vessels (SynDaver Labs, Tampa, FL) with a wall thickness of 0.5 mm and countercurrent flow were threaded through the top of the fat layer and bologna such that the top edge of the vessels were immediately below the synthetic skin layer, at $z = -1.9$ mm. The vessels ran parallel to each other and were spaced 1.2 mm apart, with outer edges at $x = 0.6$ mm and $x = -0.6$ mm, such that the outer diameters of the vessels were spaced 0.1 mm on either side of the laser irradiation spot (Fig. 10.1). The synthetic vessels were affixed with cyanoacrylate glue to 18 gauge Luer-Lok blunt tip needles trimmed to a length of 18 mm. The volume of synthetic skin, fat and bologna with synthetic vessels was then placed in a 40 mm diameter custom-machined aluminum cup affixed to a flexible silicone rubber disk heater (SRFR-3, Omega Engineering, Inc., Stamford, CT). The cup was mounted on an optical bench with the Nd:YAG laser beam incident at the center of the synthetic skin surface. A needle T-type thermocouple was inserted into the interior of the bologna and connected to a Watlow 96 PID temperature controller (Watlow, Winona, MN) to maintain the phantom interior at 37°C.

A length of amber rubber 6.35 mm ID tubing (Carolina Scientific, Burlington, NC) was affixed with cyanoacrylate glue onto male Luer adapters and screwed into the Luer-Lok needles to connect the rubber tubing to the synthetic vessels. Excess rubber tubing between connections to the synthetic vessels was placed in a water bath maintained at 37°C by a second Watlow 96 PID temperature controller and T-type thermocouple. The inlet end of the rubber tubing was inserted into a cup of synthetic blood (Carolina Scientific) anchored in the water bath and then threaded through a 3D printed peristaltic pump. The pump was driven by a stepper motor (RM57-102, Parker Compumotor, Petaluma, CA) controlled with custom software and printed circuit boards with a custom display. Fig. 10.2 shows a schematic of the blood flow circuit. The volumetric flow rate through the circuit was measured to be $0.077\ \mathrm{cm^3\,s^{-1}}$, which is on the same order of magnitude as the $8\ \mathrm{cm\,s^{-1}}$ velocity in a 1 mm inner diameter vessel, or $0.0628\ \mathrm{cm^3\,s^{-1}}$ volumetric flow rate, used in Sassaroli et al. (2012).

The optical and thermophysical properties of the materials used in the tissue phantom are given in Table 10.1.

10.2.2 Experimental Methods

An Nd:YAG (neodymium-doped yttrium aluminum garnet) laser rod (9500, Lasermetrics, Inc., Saddlebrook, NJ) was continually excited with a parallel-mounted krypton flash bulb (rated at 148.1 V @ 20 A) to produce 1064 nm radiation within the cavity that entered an intracavity acousto-optic Q-switch connected to a Q-switch driver (MVM Electronics, Melbourne, FL) and exited as a pulsed collimated laser beam of 500 ns FWHM duration pulses. A laser pulse repetition rate of 1 kHz was used. The collimated short pulse laser beam was then focused to a 1 mm diameter spot size by a UV fused silica plano-convex lens with a focal

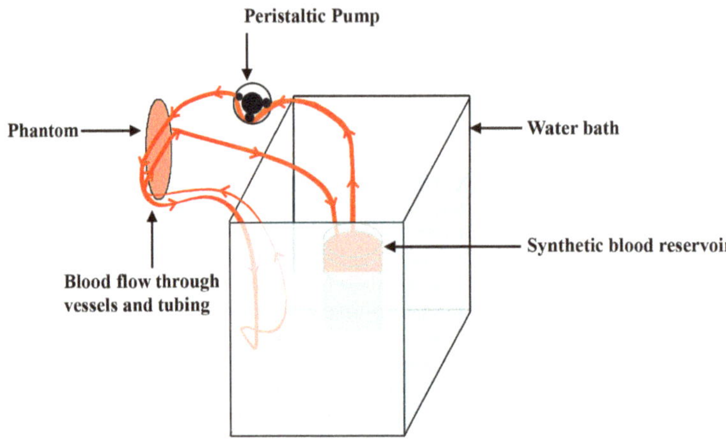

FIGURE 10.2 Blood flow circuit showing direction of blood flow through vessels and tubing. The blood flow was driven by a peristaltic pump connected to a stepper motor.

TABLE 10.1 Optical and thermophysical properties of materials used to create the vascularized synthetic tissue phantoms

Materials	Thermal conductivity k $[\mathrm{W\,m^{-1}\,K^{-1}}]$	Specific heat C $[\mathrm{J\,kg^{-1}\,K^{-1}}]$	Density ρ $[\mathrm{kg\,m^{-3}}]$	Reduced extinction coefficient μ_e' [1/m]	Absorption coefficient μ_a [1/m]	Index of refraction η	Speed of light in medium c [m/s]
Synthetic skin layer	0.2*	4000‖	1400*,‖	360‖	300‖	1.55§	1.93×10^{8}•
Synthetic fat layer	0.2*	4000‖	1400*,‖	110‖	50‖	1.37ǂ	2.19×10^{8}•
Bologna	0.82*	4600‡	950*	1000‡	400‡	N/A	N/A
Synthetic blood	1.3†	3600†	1000†	200†	100†	1.33ǂ	2.25×10^{8}•

* *Experimentally determined,* • *Calculated,* † *Ganguly et al. (2015a),* ‖ *Modified from Ganguly et al. (2015a) based on observations,* § *Modified from Milanič and Majaron (2011) based on observations,* ǂ *Majaron et al. (2015),* ‡ *Banerjee et al. (2005).*

length of 20 mm (LA4647, Thorlabs, Inc., Newton, NJ). The laser cavity and Q-switch were outfitted with a custom continuous flow water cooling system.

Pulses were triggered by the internal rate generator of a digital delay/pulse generator (DG535, Stanford Research Systems, Inc., Sunnyvale, CA) with a 25 ppm crystal oscillator timebase and a rate range of 0 Hz to 1 MHz connected to the Q-switch driver. Pulse shape was measured and monitored using a PIN diode (C1083, Hamamatsu, Hamamatsu City, Japan) with a 0.8 ns risetime and 400 to 1100 nm spectral response range at the back end of the laser set-up connected to an oscilloscope (7854, vertical module 7A26, horizontal module 7B53A, Tektronix, Beaverton, OR). The laser beam power output was measured from an approximately 4% split beamsampler (BSF10-C, Thorlabs Inc.) at the front end of the resonator using an optical power meter (1830-C, Newport, Irvine, CA) connected to a Newport 818-SL sili-

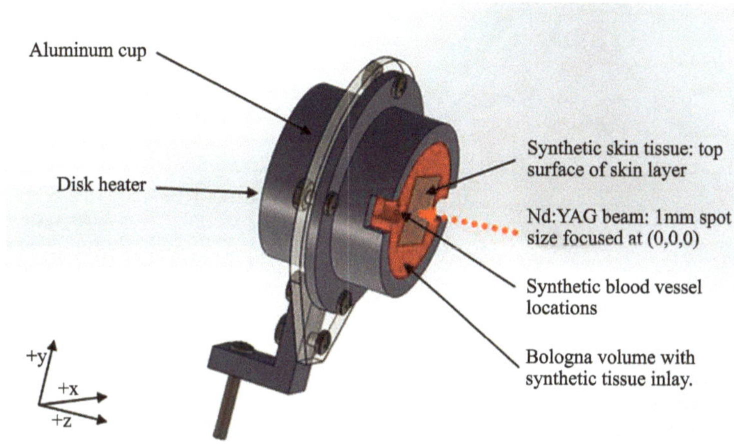

FIGURE 10.3 Experimental setup of vascularized tissue phantom in path of Nd:YAG beam.

con photodiode sensor with an 883-SL calibrated attenuator in order to avoid exceeding the maximum measurable power of 2 W. Power meter readings were continually recorded using a LabVIEW VI. The average laser power was 2.02 ± 0.03 W.

The Nd:YAG laser spot was focused at the center of the exposed $z = 0$ mm surface of the synthetic skin tissue as shown in Fig. 10.3. Different laser irradiation times of 1, 5, 10, and 20 s were tested.

Thermal images of the surface temperature of the irradiated volume during the laser irradiation and subsequent cooling periods were recorded using a FLIR A6703 SC thermal camera (FLIR, Wilsonville, OR) at a rate of 60 frames per second (fps). The spatial resolution of the FLIR images was 18 pixels (px) mm^{-1} in a 640 px by 512 px field of view.

10.2.3 Model Formulation

Models with and without blood flow were created using COMSOL Multiphysics 5.1. A physics-controlled mesh with refinement in the region immediately surrounding the laser focal point was used for both models. The model that included blood flow had 194,379 elements with maximum and minimum face areas of 7.56×10^{-6} and 4.00×10^{-8} m^2, respectively. This model consisted of a bologna cylinder with a diameter and height of 40 mm containing a 20 mm square inlay of a two-layer, 4.5 mm thick synthetic tissue as specified in Section 10.2.1. Two synthetic blood vessels (1 mm inner diameter, 0.5 mm wall thickness) with countercurrent flow ran through the bologna and fat layer of the synthetic tissue, with the top of the vessels immediately below the skin layer of the synthetic tissue surface as shown in the y-plane cut view in Fig. 10.1. The meshed geometry for the model that included blood flow is shown in Fig. 10.4.

The model that did not include blood flow did not contain any blood vessels but was otherwise identical to the model that included blood flow. The model without blood flow

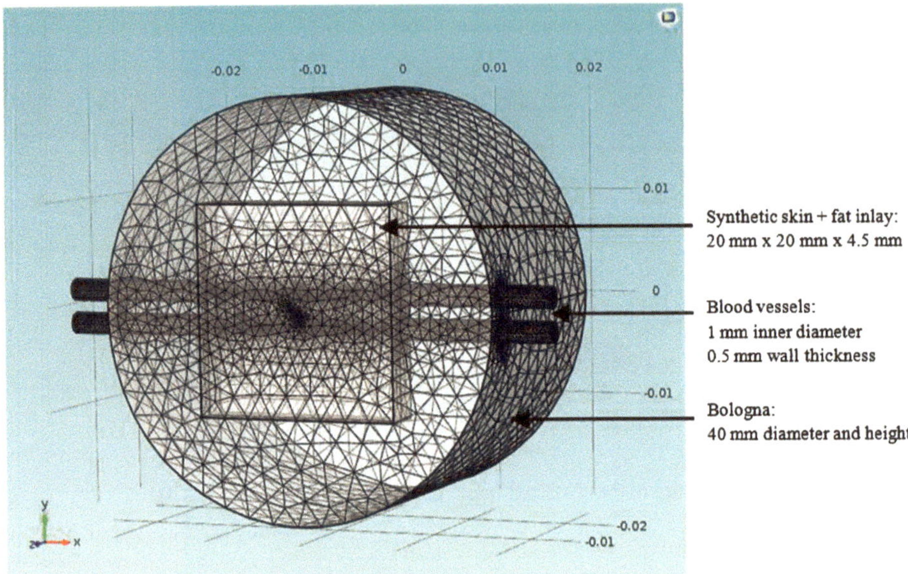

FIGURE 10.4 Mesh for model including blood flow.

had 53,570 elements with the same minimum and maximum face areas as the model that included blood flow.

In both the models with and without blood flow, the $z = -40$ mm and circumferential surfaces of the bologna were assigned constant temperature boundary conditions at 37°C. A convective boundary condition with a free stream temperature of 24°C, which was the measured room temperature, and $h = 10\,\mathrm{W\,m^{-2}\,K^{-2}}$ were assigned to the $z = 0$ mm surface of the bologna and synthetic skin tissue. The initial temperature of the interior of the model volume was set at the room temperature of 24°C. Before the laser heat source was applied to each model, the boundary conditions were run for 45 min of simulation time to yield a steady state temperature distribution within each model. In this way, the initial temperature distribution within the models before the laser heat source was applied closely approximated that in the experimental tissue phantoms.

The governing equation in the solid domains (synthetic skin and fat tissue, bologna) is given by Eq. (10.2), and the thermophysical properties used are listed in Table 10.1. Blood flow in the countercurrent blood vessel pair was assumed to be fully developed laminar flow with a uniform average velocity, u, of magnitude 98 $\mathrm{mm\,s^{-1}}$. This velocity was directly calculated from the measured experimental volumetric flow rate of 0.077 $\mathrm{cm^3\,s^{-1}}$. The governing equation for the heat transfer due to blood flow is given by Eq. (10.3).

The pulsed laser heat source term, Q_L, given by Eq. (10.4) was used in Eqs. (10.2) and (10.3) to consider $t_p = 500$ ns laser pulses beginning at time $t = 0$ s and repeating every 0.001 s, for a repetition rate of $f = 1$ kHz. At the laser focal point of $(x, y, z) = (0, 0, 0)$, $\sigma = 0.428$ mm, and $r_0 = 0.5$ mm. During the time that the laser was on, t_{on}, Q_L was assumed to be Gaussian in space and time and was included in the governing equation for each domain in the model

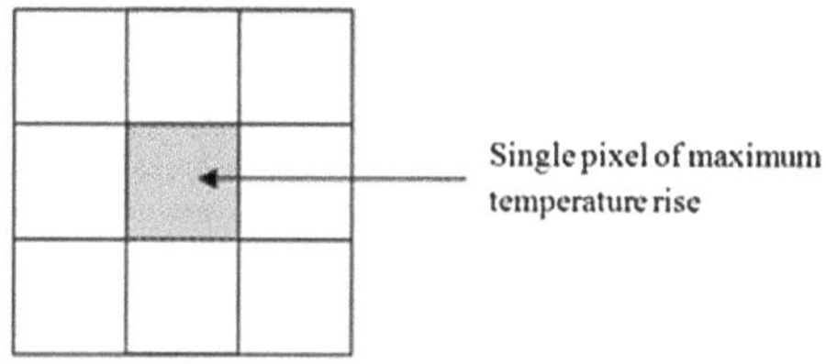

FIGURE 10.5 The pixels immediately surrounding the single pixel of maximum temperature rise were used to calculate 9 pixel averages. Each pixel corresponded to an area of 0.1667 mm × 0.1667 mm.

(skin, fat, blood, bologna). During the times that the laser was off between laser pulses, t_{off}, $Q_L = 0\ \mathrm{W\,m^{-3}}$. A CW laser heat source was also considered by incorporating Eq. (10.6) into Eqs. (10.2) and (10.3).

10.3 RESULTS AND DISCUSSION

In order to determine the effects of subsurface countercurrent vessels on the surface temperature rise in the phantoms due to short pulse laser irradiation, the maximum surface temperature rise was analyzed in phantoms with and without blood flow. Surface temperatures ($z = 0$ mm) were recorded with the thermal camera for cases with and without blood flow in the synthetic tissue phantoms for Nd:YAG irradiation of 1, 5, 10, and 20 s duration. All experiments were performed for a laser repetition rate of 1 kHz and average power of 2.02 ± 0.03 W. The thermal camera spatial calibration was 18 $\mathrm{px\,mm^{-1}}$ on a pixel side, with 1 px = 0.11 $\mathrm{mm^2}$. Temperature rise was analyzed using 9 px averages (3 px × 3 px, corresponding to an area of 0.17 mm × 0.17 mm) around the single pixel of maximum temperature at the laser focal point as well as using the single pixel maximum temperature rise (Fig. 10.5). The average noise floor of the images was 0.025 ± 0.027°C for 9 px averages and 0.069 ± 0.038°C for 1 px values.

The surface temperature of the synthetic vascularized tissue phantoms prior to irradiation averaged 29.2°C. At a room temperature of 24°C, the average surface temperature of human skin is around 31.5°C and that of human hands is around 30.5°C (Olesen, 1982). The surface temperatures of the synthetic tissue phantoms were maintained close to human skin temperature by using PID temperature controllers and a phantom volume large enough to prevent boundary conditions from affecting the surface temperature at the point of laser irradiation. This setup provided a more realistic surface temperature prior to laser irradiation than thin excised tissue samples maintained entirely at room temperature.

The diffusive effect of the subsurface countercurrent blood flow on the peak surface temperature was most pronounced after 20 s of irradiation, at which point the case without blood flow resulted in a 6.7% higher peak surface temperature than the case with blood flow (Fig. 10.9). A Mann–Whitney rank sum test identified that the experimentally measured temperature rises in phantoms with blood flow and without blood flow were significantly different throughout the duration of the 20 s irradiation ($p < 0.001$).

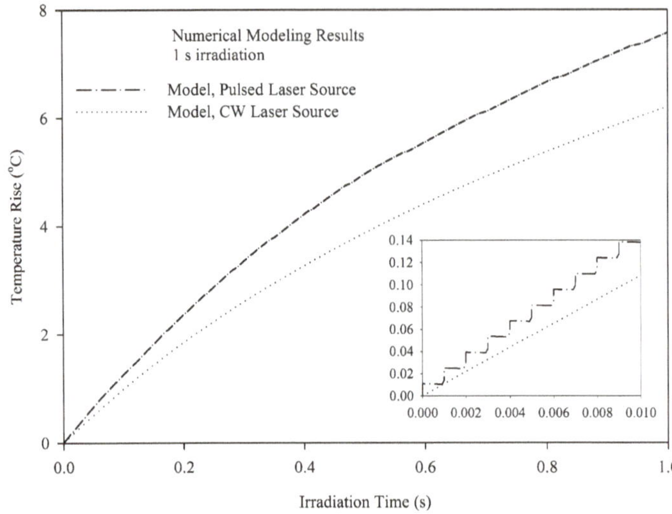

FIGURE 10.6 Comparison of the transient surface temperature rise at the laser focal point obtained numerically in tissue phantoms with blood perfusion between a pulsed laser and a CW laser source for 1 s irradiation time.

Surface temperature rises in the vascularized synthetic tissue phantoms due to 20 s short pulse laser irradiation with an irradiance of $257.19 \pm 3.82\ \mathrm{W\,cm^{-2}}$ were comparable to temperature rises in excised tissue samples due to 20 s of short pulse laser irradiation with an irradiance of $439.26 \pm 21.64\ \mathrm{W\,cm^{-2}}$ (Ganguly et al., 2015a). These results predict that a 71% greater irradiance would be required in order for excised tissue samples to exhibit equivalent surface temperature rises to those observed after 20 s of short pulse laser irradiation of the vascularized phantoms. The phantom design provides a viable alternative to live mouse experiments for analyzing surface temperature rises that may be achieved by laser irradiation. Additionally, the results from experiments with these phantoms are in closer agreement with results from experiments on live anesthetized mice than excised skin tissue samples maintained at room temperature (Ganguly et al., 2015a).

The COMSOL numerical models described in Section 10.2.3 were used to analyze heat transfer using both pulsed (Eq. (10.4)) and CW (Eq. (10.6)) laser heat source terms. A comparison of the transient surface temperature rise predicted by numerical models using a pulsed laser heat source term with that predicted by models using a CW laser heat source term for cases considering blood flow is shown in Fig. 10.6. The ratio of the temperature rise predicted by models using the pulsed source term to that predicted by models using the CW source term approached 1.2 at 1 s irradiation (Fig. 10.7). For longer irradiation durations, temperature rises modeled using the pulsed laser heat source term are well approximated using the CW laser heat source term (Ganguly et al., 2015a).

The surface temperature rises predicted by numerical models using the CW laser heat source term showed agreement with experimental temperature rises measured in the vascularized synthetic tissue phantoms for cases with blood flow (Fig. 10.8).

A Mann–Whitney rank sum test identified significant differences between the surface temperature rises predicted by numerical models for cases with and without blood flow at all

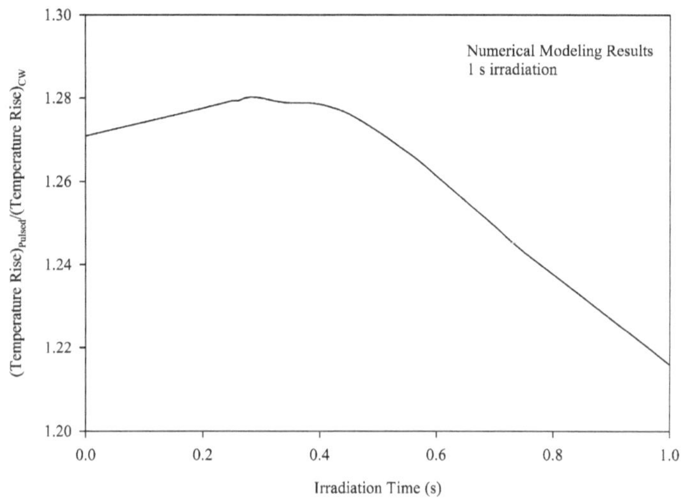

FIGURE 10.7 Transient ratio of the surface temperature rise for a pulsed laser source to that for a CW laser source obtained numerically in tissue phantoms with blood perfusion for 1 s irradiation time.

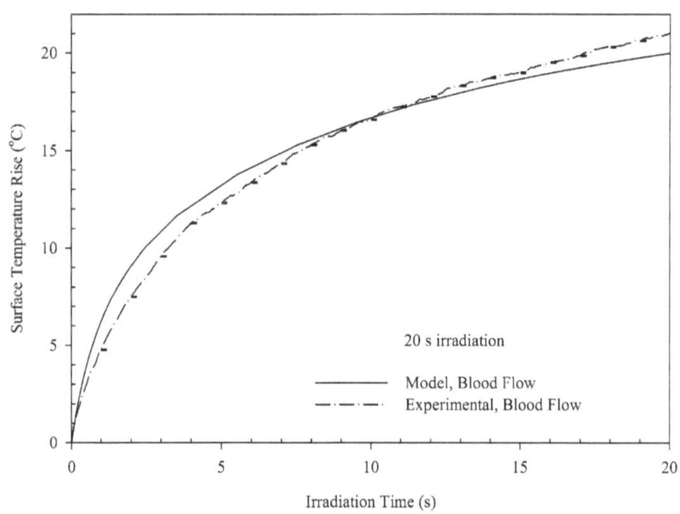

FIGURE 10.8 Comparison of transient surface temperature rises at the laser focal point in tissue phantoms with blood perfusion between numerical modeling and experimental measurements during 20 s laser irradiation.

times during a 20 s irradiation ($p = 0.044$). Fig. 10.9 shows a comparison of the transient temperature rises predicted by numerical models with experimental data for cases both with and without blood flow for irradiation durations of 20 s.

Since subsurface temperature rises due to laser irradiation were not measured experimentally, the numerical models were used to analyze the effect of blood flow on subsurface temperature rises due to short pulse laser irradiation. A Mann–Whitney rank-sum test in-

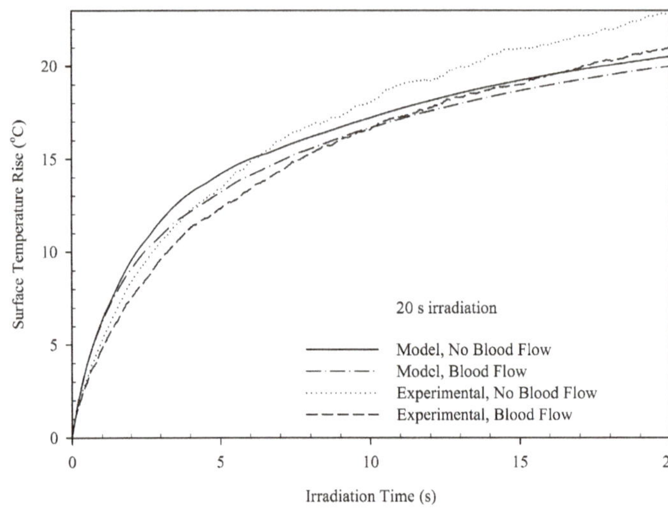

FIGURE 10.9 Comparison of transient surface temperature rise at the laser focal point during 20 s laser irradiation between numerical models and experimental measurements for cases with and without blood flow.

dicated that subsurface temperature rises at the synthetic skin-fat boundary ($z = -1.9$ mm) predicted by models with blood flow showed significant differences from those predicted by models without blood flow at all times during a 20 s laser irradiation duration ($p < 0.001$) (Fig. 10.10).

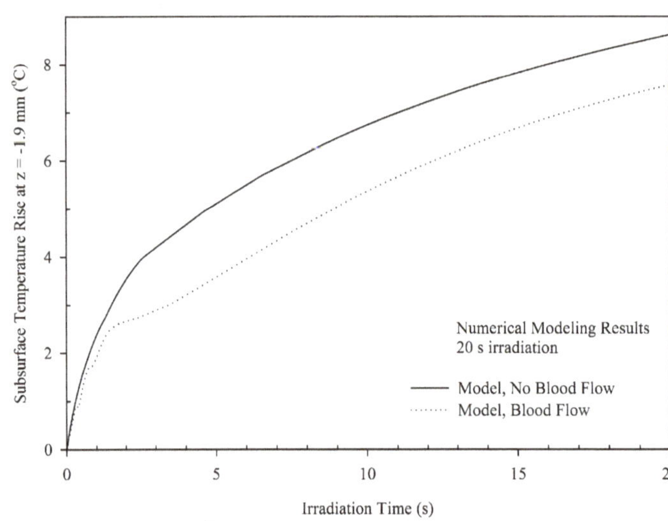

FIGURE 10.10 Comparison of the transient subsurface temperature rise at the synthetic skin-fat boundary ($z = -1.9$ mm) beneath the laser focal point between numerical modeling for cases with and without blood flow during 20 s laser irradiation.

Numerical models considering blood flow were used to determine the depth at which temperature rise due to laser irradiation was fully attenuated. The subsurface temperature rise profile beneath the laser focal point is shown in Fig. 10.11. Temperature rises were fully attenuated at a depth of 5 mm, demonstrating that the use of bologna to extend the phantom depth beyond the 4.5 mm thickness of the synthetic tissue was necessary in order to prevent boundary conditions from affecting subsurface temperature rises. Fig. 10.11 also demonstrates that the phantom volume was sufficiently large that subsurface temperature rises were not influenced by boundary conditions.

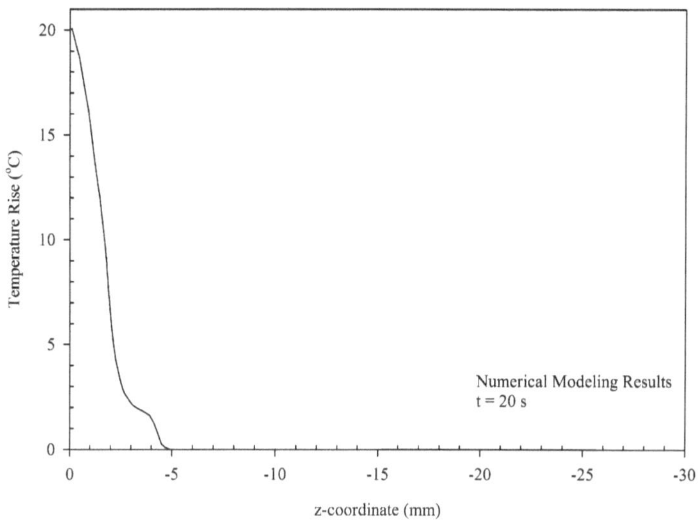

FIGURE 10.11 Subsurface temperature rise profile beneath the laser focal point obtained numerically in tissue phantom with blood flow at $t = 20$ s laser irradiation.

10.4 CONCLUSIONS

The numerical models presented are well suited for parametric analysis of the effects of thermophysical and optical properties of tissues on temperature rises due to short pulse laser irradiation for any given set of laser parameters (repetition rate, pulse width, average power, and irradiation duration). Numerical modeling results confirmed that the vascularized synthetic tissue phantoms were large enough to attenuate the temperature rises due to short pulse laser irradiation for irradiation durations of 1 to 20 s and an average laser power of 2.02 ± 0.03 W without introducing effects from the boundary conditions at the outer boundaries of the phantoms. Numerical models also demonstrated that blood flow had a significant effect on both the surface and subsurface transient temperature rise at all times during 20 s of laser irradiation. The effect of blood flow on the peak surface temperature rise during laser irradiation identified by the numerical models was validated using experimental measurements. The novel temperature controlled vascularized synthetic tissue phantom design could be adapted for use in the experimental optimization of various thermal treatments. Blood flow rates could easily be varied by varying the rate of the stepper motor, countercurrent blood vessel size and spacing could be modified, and additional pairs of countercurrent vessels could be added to the base phantom design. These phantoms present a viable alternative to live animals for use in analyzing the thermal effects of a given laser therapy.

References

Banerjee, A., Ogale, A.A., Das, C., Mitra, K., Subramanian, C., 2005. Temperature distribution in different materials due to short pulse laser irradiation. Heat Transf. Eng. 26 (8), 41–49.

Binzoni, T., Torricelli, A., Giust, R., Sanguinetti, B., Bernhard, P., Spinelli, L., 2014. Bone tissue phantoms for optical flowmeters at large interoptode spacing generated by 3D-stereolithography. Biomed. Opt. Express 5 (8), 2715–2725.

Bohndiek, S.E., Bodapati, S., Van de Sompel, D., Kothapalli, S.-R., Gambhir, S.S., 2013. Development and application of stable phantoms for the evaluation of photoacoustic imaging instruments. PLoS ONE 8 (9), e75533.

Cook, J.R., Bouchard, R.R., Emelianov, S.Y., 2011. Tissue-mimicking phantoms for photoacoustic and ultrasonic imaging. Biomed. Opt. Express 2 (11), 3193–3206.

Dewhirst, M.W., Viglianti, B.L., Lora-Michiels, M., Hanson, M., Hoopes, P.J., 2003. Basic principles of thermal dosimetry and thermal thresholds for tissue damage from hyperthermia. Int. J. Hyperth. 19, 267–294.

Di Ninni, P., Martelli, F., Zaccanti, G., 2011. Intralipid: towards a diffusive reference standard for optical tissue phantoms. Phys. Med. Biol. 56, N21–N28.

Ganguly, M., Miller, S., Mitra, K., 2015a. Model development and experimental validation for analyzing initial transients of irradiation of tissues during thermal therapy using short pulse lasers. Lasers Surg. Med. 47 (9), 711–722.

Ganguly, M., O'Flaherty, R., Mitra, K., Sajjadi, A., 2015b. Tissue response to short pulse laser irradiation. In: Becker, S.M., Kuznetsov, A.V. (Eds.), Heat Transfer and Fluid Flow in Biological Processes, vol. 1. Elsevier, Amsterdam, pp. 43–58.

Jaunich, M., Raje, S., Kim, K., Mitra, K., Guo, Z., 2008. Bio-heat transfer analysis during short pulse laser irradiation of tissues. Int. J. Heat Mass Transf. 51 (23–24), 5511–5521.

Kim, K., Guo, X., 2007. Multi-time-scale heat transfer modeling of turbid tissues exposed to short-pulsed irradiations. Comput. Methods Programs Biomed. 86, 112–123.

Majaron, B., Milanič, M., Premru, J., 2015. Monte Carlo simulation of radiation transport in human skin with rigorous treatment of curved tissue boundaries. J. Biomed. Opt. 20 (1), 015002.

Milanič, M., Majaron, B., 2011. Three-dimensional Monte Carlo model of pulsed laser treatment of cutaneous vascular lesions. J. Biomed. Opt. 16 (12), 128002.

Milanič, M., Majaron, B., Nelson, J.S., 2007. Pulsed photothermal temperature profiling of agar tissue phantoms. Lasers Med. Sci. 22, 279–284.

Olesen, B.W., 1982. Thermal Comfort Technical Review, vol. 2. Brüél and Kjær, Copenhagen, pp. 3–41.

Pogue, B.W., Patterson, M.S., 2006. Review of tissue simulating phantoms for optical spectroscopy, imaging and dosimetry. J. Biomed. Opt. 11 (4), 041102.

Sapareto, S.A., Dewey, W.C., 1984. Thermal dose determination in cancer therapy. Int. J. Radiat. Oncol. Biol. Phys. 10, 787–800.

Sassaroli, E., Li, K.C.P., O'Neill, B.E., 2012. Modeling focused ultrasound exposure for the optimal control of thermal dose distribution. Sci. World J., 252741.

Stańczyk, M., Telega, J.J., 2002. Modelling of heat transfer in biomechanics – a review: Part I. Soft tissues. Acta Bioeng. Biomech. 4, 31–61.

Stańczyk, M., Van Leeuwen, G.M.J., Van Steenhoven, A.A., 2007. Discrete vessel heat transfer in perfused tissue – model comparison. Phys. Med. Biol. 52, 2379–2391.

Thrall, D.E., Maccarini, P., Stauffer, P., Macfall, J., Hauck, M., Snyder, S., Case, B., Linder, K., Lan, L., Mccall, L., Dewhirst, M.W., 2012. Thermal dose fractionation affects tumor physiological response. Int. J. Hyperth. 28 (5), 431–440.

Vogt, W.C., Jia, C., Wear, K.A., Garra, B.S., Pfefer, T.J., 2016. Biologically relevant photoacoustic imaging phantoms with tunable optical and acoustic properties. J. Biomed. Opt. 21 (10), 101405.

Whelan, W.M., Chun, P., Chin, C.L., Sherar, M.D., Vitkin, I.A., 2001. Laser thermal therapy: utility of interstitial fluence monitoring for locating optical sensors. Phys. Med. Biol. 46, N91–N96.

Whelan, W.M., Wyman, D.R., Wilson, B.C., 1995. Investigations of large vessel cooling during interstitial laser heating. Med. Phys. 22 (1), 105–115.

Yarmolenko, P.S., Moon, E.J., Landon, C., Manzoor, A., Hochman, D.W., Viglianti, B.L., Dewhirst, M.W., 2011. Thresholds for thermal damage to normal tissues: an update. Int. J. Hyperth. 27 (4), 320–343.

CHAPTER

11

Micro-Scale Bio-Heat Diffusion Using Green's Functions

F. de Monte[*,#], A. Haji-Sheikh[†]*

[*]Department of Industrial and Information Engineering and Economics, University of L'Aquila, L'Aquila, Italy [†]Department of Mechanical and Aerospace Engineering, University of Texas at Arlington, Arlington, TX, United States

11.1 INTRODUCTION

When a living biological tissue is subject to a heating therapy, for example, hyperthermia therapy by laser irradiation, it is important to understand the evolution of its temperature as well as that for the blood within the vessels. For that reason, many mathematical models have been presented in the specialized literature, such as the bio-heat parabolic diffusive Pennes equation (Pennes, 1948), the hyperbolic (or single-phase lag, SPL) bio-heat equation (Zhang, 2009), the dual-phase lag (DPL) bio-heat equation (Zhang, 2009; Zhou et al., 2009; Liu, 2015; Liu and Chen, 2016) and, more recently, the two-step model that considers the biological tissue as a porous medium under Local Thermal Non-Equilibrium (LTNE) conditions (Nakayama and Kuwahara, 2008; Zhang, 2009). Among these models, the most reliable seems to be the two-step model based on two coupled partial differential equations, one for the tissue (solid phase) and the other for the blood (fluid phase). Furthermore, there is a lack of oscillations in Dual-Phase-Lagging heat conduction for a porous slab, as proven by Vadàsz (2005).

However, when dealing with the two-step model, by means of an adequate uncoupling procedure (Minkowycz et al., 1999; Vadàsz, 2005; Haji-Sheikh and Minkowycz, 2008), it is possible to derive a diffusive partial differential equation for the perfused biological tissue that is similar to the dual-phase lag bio-heat equation, as shown by Zhang (2009). However, there are substantial differences since, in these equations, the two phase-lag times are obtained as functions of the various thermal properties of blood and biological tissues. The solution to this equation is however not discussed by Zhang (2009) as well as is not approached earlier by Nakayama and Kuwahara (2008). While numerical solutions are given

[#] Chapter coordinator.

by Zhou et al. (2009), Liu (2015), and Liu and Chen (2016), where a hybrid discretization scheme based on the Laplace transform is proposed, no analytical solutions are available.

In this chapter, an exact analytical solution to the DPL bio-heat equation coming from local thermal non-equilibrium conditions is developed. Additionally, this equation is derived in a more complete way by removing some simplifying assumptions done by Zhang (2009), for example, keeping the time partial derivative of the blood velocity. Also, following the approach in Vadàsz (2005), the phase-lag time for the temperature gradient is derived accounting for the thermal properties of both blood and tissue.

To obtain the analytical solution, the bio-heat term appearing in the DPL diffusive equation is canceled out by means of an appropriate dependent-variable transformation. Then, following the approach by Hays-Stang and Haji-Sheikh (1999), a modified separation of variables (SOV) method is applied for solving the three-dimensional (3D) transient DPL equation for finite regular biological tissues. Once this is solved, DPL Greens functions are derived as well as the DPL Green's function solution equation. A relationship with the well-established parabolic Green's functions available in the book by Cole et al. (2011) is also provided.

An applicative example of the proposed procedure is given for laser-irradiated 1D rectangular biological tissues when the laser light is highly absorbed (Zhou et al., 2009; Liu, 2015). The results indicate that the temperature rise concerns only the first layers (2–3 mm) of living biological tissue for a laser power density of $q_{IR}^{(0)} = 2\ \mathrm{W/cm^2}$. Also, the maximum temperature rise is obtained with a lower perfusion rate depending on the type of biological tissue and related blood vessels.

11.2 BALANCE EQUATIONS

A biological tissue consists of three parts, that is, blood vessels (small capillaries), cells and interstitium, as shown in Fig. 11.1. Also, the interstitium can be further divided into the extracellular matrix (ECM), where the metabolic heat generation (S_m) takes place (Wright et al., 2012), and the extravascular (or interstitial) fluid (Nakayama and Kuwahara, 2008).

However, for sake of simplicity and following Khaled and Vafai's approach (2003), the whole anatomic structure of the biological tissue can be treated as a fluid-saturated porous medium through which the blood infiltrates and flows through small capillaries. In particular, the solid matrix phase is represented by the extravascular region (including cells and interstitium), in which metabolic reactions may take place; and the fluid phase is represented by the vascular region, that is, blood vessels. Additionally, according to earlier studies (Baish et al., 1986; Yuan, 2009; Zhang, 2009), the physical porous model assumes that all blood vessels are straight, merge into the tissue and are uniformly distributed. Therefore, the whole biological domain can be considered as an arrangement of repeated and staggered regular hexagonal units with the blood vessels at the center. A small portion of its cross-sectional area is shown in Fig. 11.2.

In particular, d_s is the diameter of an equivalent circular tissue, d_b is the inner diameter of vessels and L is the vessel distance or pitch between two adjacent vessels that may be taken as $L = \sqrt{3}d_s/2 \approx 0.866 \cdot d_s$. As the hexagonal units are regular, the longitudinal, transverse and diagonal pitches are equal. Under this ideal representation, the tissue porosity may

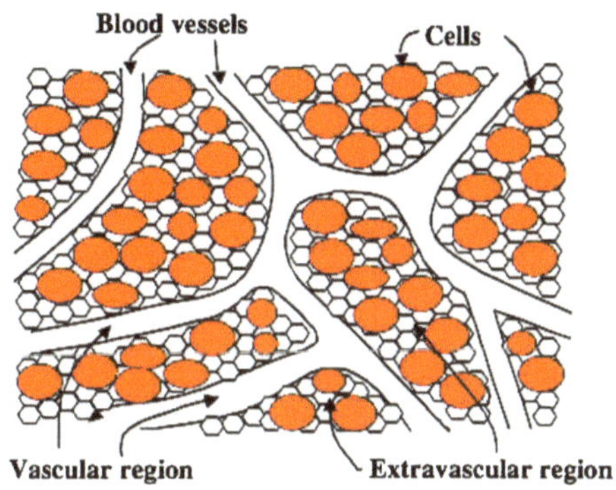

FIGURE 11.1 Schematic view of a biological tissue with vascular and extra-vascular regions (Nakayama and Kuwahara, 2008).

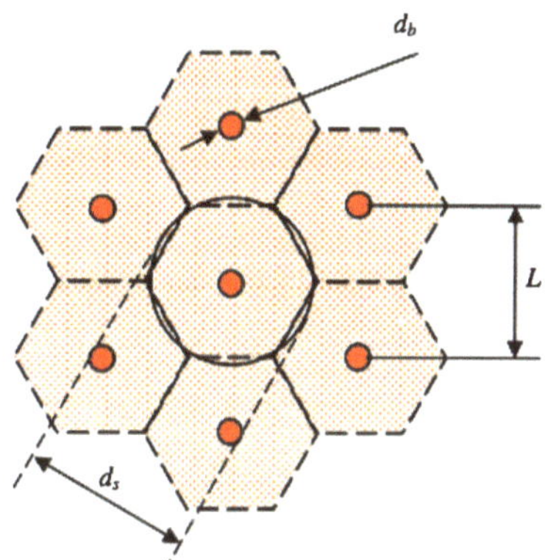

FIGURE 11.2 Biological domain as an assembly of repeated hexagonal units with blood vessels at the center (Zhang, 2009).

be approximated by $\varepsilon = (2\pi\sqrt{3}/9) \cdot (d_b/d_s)^2 \approx 1.2 \cdot (d_b/d_s)^2$. Some numerical values of these quantities are listed in Table 11.1.

The blood phase (liquid) and tissue phase (solid matrix) are here assumed to be isotropic and homogeneous. During hyperthermia therapy an external heat source (S_h) is applied (e.g., by a laser irradiation). The three-dimensional transient equations representing the physics (Nakayama and Kuwahara, 2008; Zhang, 2009) are presented here.

In the blood phase, the conservation of mass, the conservation of momentum, and the conservation of energy are represented as:

$$\nabla \cdot \langle \mathbf{v} \rangle^b = 0 \tag{11.1a}$$

$$\langle \mathbf{v} \rangle = -\frac{K}{\mu} \nabla \langle p \rangle^b, \tag{11.1b}$$

TABLE 11.1 Geometric parameters of the biological domain. The quantities d_s and d_b are given by Baish et al. (1986), Yuan (2009), and Zhang (2009)

Case	d_s [mm]	d_b [mm]	L [mm]	ε
1	17.83	1.14	15.44	0.0049
2	12.85	1.14	11.13	0.0095
3	10.75	1.14	9.31	0.0136
4	9.7	1.14	8.40	0.0167
5	8.65	1.14	7.49	0.0210
6	19.82	2.28	17.16	0.0160
7	14.42	2.28	12.49	0.0302
8	12.06	2.28	10.44	0.0432
9	10.48	2.28	9.08	0.0572
10	9.92	2.28	8.59	0.0639
11	20.98	4.56	18.17	0.0571
12	15.73	4.56	13.62	0.1016
13	13.58	4.56	11.76	0.1363
14	12.06	4.56	10.44	0.1729
15	11.27	4.56	9.76	0.1980

$$\underbrace{\varepsilon\rho_b c_b \frac{\partial \langle T_b\rangle^b}{\partial t}}_{\text{energy storage}} + \underbrace{\varepsilon\rho_b c_b \langle \mathbf{v}\rangle^b \cdot \nabla\langle T_b\rangle^b}_{\text{convective term}} = \underbrace{\varepsilon\nabla\left(k_b \nabla\langle T_b\rangle^b\right)}_{\text{diffusive term}} + \underbrace{\frac{4\varepsilon h_b}{d_b}\left(\langle T_s\rangle^s - \langle T_b\rangle^b\right)}_{\text{local heat transfer}}$$
$$+ \underbrace{\rho_b c_b \omega\left(\langle T_s\rangle^s - \langle T_b\rangle^b\right)}_{\text{perfusion term}} + \underbrace{\varepsilon S_h}_{\text{hyperthermia therapy}} . \tag{11.1c}$$

The conservation of energy in the solid tissue phase is represented as

$$\underbrace{(1-\varepsilon)\rho_s c_s \frac{\partial \langle T_s\rangle^s}{\partial t}}_{\text{energy storage}} = \underbrace{(1-\varepsilon)\nabla\left(k_s \nabla\langle T_s\rangle^s\right)}_{\text{diffusive term}} + \underbrace{\frac{4\varepsilon h_b}{d_b}\left(\langle T_b\rangle^b - \langle T_s\rangle^s\right)}_{\text{local heat transfer}}$$
$$+ \underbrace{\rho_b c_b \omega\left(\langle T_b\rangle^b - \langle T_s\rangle^s\right)}_{\text{perfusion term}} + \underbrace{(1-\varepsilon)S_m}_{\text{metabolic heating}} + \underbrace{(1-\varepsilon)S_h}_{\text{hyperthermia therapy}} \tag{11.1d}$$

where $\langle p\rangle^p$ and $\langle \mathbf{v}\rangle^b$ are the 'intrinsic' volume-averaged pressure and velocity of the blood within the small capillaries, respectively; and $\langle T_b\rangle^b$ and $\langle T_s\rangle^s$ are the 'intrinsic' volume-averaged temperatures of blood and biological tissue, respectively.

They are defined as

$$\langle p\rangle^b = \frac{1}{V_{rev}^b}\int_{V_{rev}^b} p\, dV_{rev}^b, \qquad \langle \mathbf{v}\rangle^b = \frac{1}{V_{rev}^b}\int_{V_{rev}^b} \mathbf{v}\, dV_{rev}^b, \tag{11.2a}$$

$$\langle T_b \rangle^b = \frac{1}{V^b_{rev}} \int_{V^b_{rev}} T_b \, dV^b_{rev}, \qquad \langle T_s \rangle^s = \frac{1}{V^s_{rev}} \int_{V^s_{rev}} T_s \, dV^s_{rev}. \tag{11.2b}$$

The above way of averaging over a volume is based on the volume of each phase contained in r.e.v. (representative elementary volume), that is, V^b_{rev} for the blood phase (which is a portion of the r.e.v., i.e., ε) and V^s_{rev} for the solid matrix phase (which is the complementary portion of the r.e.v., i.e., $1-\varepsilon$).

Another way of averaging over a volume is based on the total volume of the r.e.v. (incorporating both fluid and solid phases) given by $V_{rev} = V^b_{rev} + V^s_{rev}$. This leads to

$$\langle p \rangle = \frac{1}{V_{rev}} \int_{V_{rev}} p \, dV_{rev} = \frac{1}{V_{rev}} \underbrace{\int_{V^b_{rev}} p \, dV^b_{rev}}_{V^b_{rev} \langle p \rangle^b} = \varepsilon \langle p \rangle^b, \tag{11.3a}$$

$$\langle \mathbf{v} \rangle = \frac{1}{V_{rev}} \int_{V_{rev}} \mathbf{v} \, dV_{rev} = \frac{1}{V_{rev}} \underbrace{\int_{V^b_{rev}} \mathbf{v} \, dV^b_{rev}}_{V^b_{rev} \langle \mathbf{v} \rangle^b} = \varepsilon \langle \mathbf{v} \rangle^b, \tag{11.3b}$$

$$\langle T_b \rangle = \frac{1}{V_{rev}} \int_{V_{rev}} T_b \, dV_{rev} = \frac{1}{V_{rev}} \underbrace{\int_{V^b_{rev}} T_b \, dV^b_{rev}}_{V^b_{rev} \langle T_b \rangle^b} = \varepsilon \langle T_b \rangle^b, \tag{11.3c}$$

$$\langle T_s \rangle = \frac{1}{V_{rev}} \int_{V_{rev}} T_s \, dV_{rev} = \frac{1}{V_{rev}} \underbrace{\int_{V^s_{rev}} T_s \, dV^s_{rev}}_{V^s_{rev} \langle T_s \rangle^s} = (1-\varepsilon) \langle T_s \rangle^s. \tag{11.3d}$$

Both volume-averaged quantities, i.e., $\langle p \rangle$, $\langle \mathbf{v} \rangle$, $\langle T_b \rangle$, and $\langle T_s \rangle$, and 'intrinsic' volume-averaged quantities, $\langle p \rangle^b$, $\langle \mathbf{v} \rangle^b$, $\langle T_b \rangle^b$, and $\langle T_s \rangle^s$, depend on x, y, and z (as well as time for unsteady heat transfer), which are the rectangular coordinates of the centroid P of the r.e.v.. The volume-averaged velocity $\langle \mathbf{v} \rangle$ is the so-called Darcy velocity, also known as filtration or seepage velocity, and is the one appearing in the momentum equation (11.1b) or Darcy's law.

11.2.1 Heat Transfer and Blood Perfusion

In the conservation of energy of the blood and the tissue, i.e., Eqs. (11.1c) and (11.1d), the second and third terms appearing on the RHS of these equations are important and need further discussion.

The second term represents the contribution of the interfacial convective heat transfer between blood and its vessels (that is part of the solid matrix phase), which is due to the temperature difference between them. Specifically, $h_b = \mathrm{Nu}(k_b/d_b)$ is the heat transfer coefficient, k_b is the blood thermal conductivity and d_b is the inner diameter of the vascular tube. Sometimes (Nakayama and Kuwahara, 2008; Zhang, 2009), the concept of specific surface area $a_b = 4\varepsilon/d_b$ is employed in the convective heat transfer term in place of d_b. Some numerical values of h_b are given in Table 11.2 for $\mathrm{Nu} = 4.93$ (Roetzel and Xuan, 1998) and

TABLE 11.2 Parameters characterizing the energy exchanges between solid and fluid phases

Case	h_b [W/(m^2 K)]	ω_b [kg/(m^3 s)]	$4\varepsilon h_b/d_b$ [W/(m^3 K)]	$c_b\omega_b$ [W/(m^3 K)]	U [W/(m^3 K)]
1	2160	1	37176	3770	40946
2	2160	2	72076	7540	79616
3	2160	3	103183	11310	114493
4	2160	4	126702	15080	141782
5	2160	5	159326	18850	178176
6	1080	1	30348	3770	34118
7	1080	2	57281	7540	64821
8	1080	3	81939	11310	93249
9	1080	4	108493	15080	123573
10	1080	5	121202	18850	140052
11	540	1	27076	3770	30846
12	540	2	48177	7540	55717
13	540	3	64631	11310	75941
14	540	4	81986	15080	97066
15	540	5	93889	18850	112739

$k_b = 0.5$ W/(m K) (Yuan, 2009). Further values of Nusselt number are given by Nakayama et al. (2004) for more complex porous structures.

The third term gives the contribution of blood perfusion, that is a mass transfer process for nutritive purposes between the blood and the surrounding tissue to which is associated an energy exchange. In detail, ω is the perfusion rate whose order of magnitude is $\omega = 2 \cdot 10^{-4} \rightarrow 5 \cdot 10^{-4}$ m^3/(s m^3 tissue) (Nakayama and Kuwahara, 2008). Other researchers set a higher numerical value, such as $\omega = 1.87 \cdot 10^{-3}$ m^3/(s m^3 tissue) (Yamada et al., 1995). Also, Baish et al. (1986) and, then, Yuan (2009) and Zhang (2009) give the perfusion rate as $\omega_b = \rho_b \omega$, so having units of kg/(s m^3 tissue). Some numerical values of ω_b were obtained experimentally by Baish et al. (1986) and are listed in Table 11.2.

Now, the two-step model defined through Eqs. (11.1c) and (11.1d) can be rewritten in a more compact form as follows:

$$\varepsilon \rho_b c_b \left(\frac{\partial T_b}{\partial t} + \mathbf{v} \cdot \nabla T_b \right) = \varepsilon k_b \nabla^2 T_b + U(T_s - T_b) + \varepsilon S_h, \tag{11.4a}$$

$$(1-\varepsilon)\rho_s c_s \frac{\partial T_s}{\partial t} = (1-\varepsilon) k_s \nabla^2 T_s + U(T_b - T_s) + (1-\varepsilon) S_m + (1-\varepsilon) S_h \tag{11.4b}$$

where, for the sake of brevity, the following short notation is used: $\langle T_b \rangle^b = T_b$, $\langle T_s \rangle^s = T_s$ and $\langle \mathbf{v} \rangle^b = \mathbf{v}$. Also, the thermal conductivities of blood and biological tissue have been assumed temperature-independent and U is the coupling factor between solid (tissue) and fluid (blood) phases accounting for both convection and perfusion phenomena. It may be taken as

$$U = \frac{4\varepsilon h_b}{d_b} + c_b \rho_b \omega = \mathrm{Nu} \frac{4\varepsilon k_b}{d_b^2} + c_b \underbrace{\rho_b \omega}_{\omega_b} = 4.93 \frac{4\varepsilon k_b}{d_b^2} + c_b \omega_b \tag{11.5}$$

whose numerical values are listed in Table 11.2 for the same cases of Table 11.1. The contribution of convective heat transfer ($4\varepsilon h_b/d_b$) and blood perfusion ($c_b\omega_b$) to the coupling coefficient is also given in the same table. For that purpose, the blood specific heat is taken as $c_b = 3770$ J/(kg K) (Yuan, 2009).

Eqs. (11.4a) and (11.4b) are two coupled linear, non-homogeneous, parabolic partial differential equations of the second-order in two unknowns, i.e., T_s and T_b. To solve them, it is necessary to apply an uncoupling procedure whose starting point is to add together Eqs. (11.4a) and (11.4b) yielding

$$(1-\varepsilon)\rho_s c_s \frac{\partial T_s}{\partial t} + \varepsilon\rho_b c_b \frac{\partial T_b}{\partial t} + \varepsilon\rho_b c_b \mathbf{v}\cdot\nabla T_b = (1-\varepsilon)k_s\nabla^2 T_s + \varepsilon k_b \nabla^2 T_b + (1-\varepsilon)S_m + S_h \tag{11.6}$$

where the velocity distribution **v** may separately be derived by solving the equations of blood motion (11.1a) and (11.1b) where the above temperatures do not appear. However, the motion equations will not be solved here and a representative value for the velocity will be used as discussed in Section 11.3.

11.2.2 Exact Uncoupling Procedure

The tissue temperature T_s may be derived from Eq. (11.4a) as

$$T_s = T_b + \frac{\varepsilon\rho_b c_b}{U}\frac{\partial T_b}{\partial t} + \frac{\varepsilon\rho_b c_b}{U}\mathbf{v}\cdot\nabla T_b - \frac{\varepsilon k_b}{U}\nabla^2 T_b - \frac{\varepsilon}{U}S_h. \tag{11.7}$$

Substituting Eq. (11.7) in Eq. (11.6) yields

$$\frac{\varepsilon\rho_b c_b}{\overline{\rho c}}\left[\mathbf{v}\cdot\nabla T_b + \frac{(1-\varepsilon)\rho_s c_s}{U}\frac{\partial}{\partial t}(\mathbf{v}\cdot\nabla T_b) - \frac{(1-\varepsilon)k_s}{U}\nabla^2(\mathbf{v}\cdot\nabla T_b)\right] + \frac{\partial T_b}{\partial t} + \underbrace{\frac{\varepsilon(1-\varepsilon)\rho_s c_s \rho_b c_b}{U\overline{\rho c}}}_{\tau_q}\frac{\partial^2 T_b}{\partial t^2} = \overline{\alpha}\left[\nabla^2 T_b + \underbrace{\frac{\varepsilon(1-\varepsilon)(k_s\rho_b c_b + k_b\rho_s c_s)}{U\overline{k}}}_{\tau_T}\frac{\partial}{\partial t}(\nabla^2 T_b) - \frac{\varepsilon(1-\varepsilon)k_s k_b}{U\overline{k}}\nabla^4 T_b\right] + S_b \tag{11.8}$$

where τ_q and τ_T may be seen as phase-lag times for heat flux and temperature gradient, respectively. Also, other selected parameters:

$$\overline{k} = (1-\varepsilon)k_s + \varepsilon k_b = k_{s,eff} + k_{b,eff}, \tag{11.9a}$$

$$\overline{\rho c} = (1-\varepsilon)\rho_s c_s + \varepsilon\rho_b c_b = C_{s,eff} + C_{b,eff}, \tag{11.9b}$$

$$\overline{\alpha} = \frac{\overline{k}}{\overline{\rho c}} \tag{11.9c}$$

are the overall thermal conductivity, the overall volumetric heat capacity, and the overall thermal diffusivity of the porous medium (biological tissue filled by blood). Note that $k_{s,eff}$ and $k_{b,eff}$ are the effective thermal conductivities of the solid and blood phases, respectively, as well as $C_{s,eff}$ and $C_{b,eff}$ are the effective volumetric heat capacities of the corresponding phases. In addition,

$$S_b = \frac{(1-\varepsilon)S_m + S_h}{\overline{\rho c}} - \frac{\varepsilon(1-\varepsilon)}{U\overline{\rho c}}\left(k_s \nabla^2 S_h - \rho_s c_s \frac{\partial S_h}{\partial t}\right) \tag{11.9d}$$

that simplifies notably when S_h is uniform and time-independent.

Eq. (11.8) is a fourth-order partial differential equation where the blood temperature T_b is the sole unknown. Once this equation is solved and the blood temperature T_b is known, one can obtain the tissue temperature T_s by means of Eq. (11.7). Alternatively, by using a similar procedure, it is possible to get a partial differential equation where the tissue temperature T_s is the sole unknown.

11.2.3 Approximate Uncoupling Procedure

The exact uncoupling procedure described in the previous subsection leads to a partial differential equation, i.e., Eq. (11.8), which cannot be solved analytically. However, if the blood phase is approximated to behave using a lumped capacitance approach, as done by Minkowycz et al. (1999) and, then, by Zhang (2009), Eq. (11.7) may be represented as

$$T_s \approx T_b + \underbrace{\frac{\varepsilon \rho_b c_b}{U}}_{\tau_S} \frac{\partial T_b}{\partial t} \tag{11.10}$$

Thus, one obtains, in place of Eq. (11.8),

$$\underbrace{\frac{\varepsilon(1-\varepsilon)\rho_s c_s \rho_b c_b}{U\overline{\rho c}}}_{\tau_q} \frac{\partial^2 T_b}{\partial t^2} + \frac{\partial T_b}{\partial t} + \frac{\varepsilon \rho_b c_b}{\overline{\rho c}} \mathbf{v} \cdot \nabla T_b$$
$$= \overline{\alpha}\left[\nabla^2 T_b + \underbrace{\frac{\varepsilon(1-\varepsilon)k_s \rho_b c_b}{U\overline{k}}}_{\tau_T} \frac{\partial}{\partial t}\left(\nabla^2 T_b\right)\right] + \frac{(1-\varepsilon)S_m + S_h}{\overline{\rho c}} \tag{11.11}$$

where $\overline{k}$, $\overline{\rho c}$, and $\overline{\alpha}$ are as defined by Eqs. (11.9a), (11.9b), and (11.9c), respectively, and the blood temperature T_b is the sole unknown.

It is noteworthy to mention that Eq. (11.10) exhibits a time lag between tissue and blood temperatures and this time lag (or relaxation time) is given by $\tau_S = (\varepsilon \rho_b c_b)/U$. In fact, its RHS may be seen as a first-order approximation of $T_b(\mathbf{r}, t + \tau_S)$.

Once the blood temperature T_b has been established by solving Eq. (11.11), the tissue temperature T_s can also be derived by means of Eq. (11.10). Alternatively, one can obtain a partial differential equation where the tissue temperature T_s is the sole unknown. To get this, one can rewrite Eq. (11.10) as

$$T_b \approx T_s - \frac{\varepsilon \rho_b c_b}{U}\frac{\partial T_b}{\partial t}. \tag{11.12}$$

Substituting Eq. (11.12) into Eq. (11.11) yields

$$\begin{aligned}
&\frac{\varepsilon(1-\varepsilon)\rho_s c_s \rho_b c_b}{U\overline{\rho c}}\frac{\partial^2 T_s}{\partial t^2} + \frac{\partial T_s}{\partial t} + \frac{\varepsilon \rho_b c_b}{\overline{\rho c}}\mathbf{v}\cdot\nabla T_s = \overline{\alpha}\left[\nabla^2 T_s + \frac{\varepsilon(1-\varepsilon)k_s \rho_b c_b}{U\overline{k}}\frac{\partial}{\partial t}(\nabla^2 T_s)\right] \\
&+ \frac{(1-\varepsilon)S_m + S_h}{\overline{\rho c}} + \frac{\varepsilon \rho_b c_b}{U}\frac{\partial}{\partial t}\left\{\frac{\varepsilon(1-\varepsilon)\rho_s c_s \rho_b c_b}{U\overline{\rho c}}\frac{\partial^2 T_b}{\partial t^2} + \frac{\partial T_b}{\partial t} + \frac{\varepsilon \rho_b c_b}{\overline{\rho c}}\mathbf{v}\cdot\nabla T_b\right. \\
&\left. - \overline{\alpha}\left[\nabla^2 T_b + \frac{\varepsilon(1-\varepsilon)k_s \rho_b c_b}{U\overline{k}}\frac{\partial}{\partial t}(\nabla^2 T_b)\right]\right\} - \frac{(\varepsilon \rho_b c_b)^2}{U\overline{\rho c}}\left(\frac{\partial \mathbf{v}}{\partial t}\right)\cdot\nabla T_s
\end{aligned} \tag{11.13}$$

where the quantity between curly brackets on the RHS may be calculated by using Eq. (11.11). Also, because of Eq. (11.1a), the blood velocity may be considered uniform. Therefore, Eq. (11.13) reduces to

$$\begin{aligned}
&\underbrace{\frac{\varepsilon(1-\varepsilon)\rho_s c_s \rho_b c_b}{U\overline{\rho c}}}_{\tau_q}\frac{\partial^2 T_s}{\partial t^2} + \frac{\partial T_s}{\partial t} = \overline{\alpha}\left[\nabla^2 T_s + \underbrace{\frac{\varepsilon(1-\varepsilon)k_s \rho_b c_b}{U\overline{k}}}_{\tau_T}\frac{\partial}{\partial t}(\nabla^2 T_s)\right] \\
&+ \frac{(1-\varepsilon)S_m + S_h}{\overline{\rho c}} + \underbrace{\frac{\varepsilon \rho_b c_b}{U}}_{\tau_S}\frac{\partial}{\partial t}\left[\frac{(1-\varepsilon)S_m + S_h}{\overline{\rho c}}\right] - \frac{\varepsilon \rho_b c_b}{\overline{\rho c}}\left[\nabla T_s\cdot\left(\mathbf{v} + \underbrace{\frac{\varepsilon \rho_b c_b}{U}}_{\tau_S}\frac{d\mathbf{v}}{dt}\right)\right]
\end{aligned} \tag{11.14}$$

where the temperature T_s is the sole unknown.

Eq. (11.14) reduces exactly to the one derived by Zhang (2009, Eq. (26)) if the blood velocity is assumed time-independent, i.e., $d\mathbf{v}/dt \approx 0$. Additionally, the phase-lag time for heat flux τ_q appearing in Eqs. (11.11) and (11.14) is exactly the same as the one appearing in Eq. (11.8). But the time phase lag for temperature gradient τ_T is a bit different. In fact, it does not contain the term $k_b \rho_s c_s$. For this reason, a better approximate uncoupling procedure would be based on the following relation

$$T_s \approx T_b + \frac{\varepsilon \rho_b c_b}{U}\frac{\partial T_b}{\partial t} - \frac{\varepsilon k_b}{U}\nabla^2 T_b. \tag{11.15}$$

Using this in Eq. (11.6) provides the following two uncoupled partial differential equations for blood and tissue phases, respectively,

$$\begin{aligned}
&\underbrace{\frac{\varepsilon(1-\varepsilon)\rho_s c_s \rho_b c_b}{U\overline{\rho c}}}_{\tau_q}\frac{\partial^2 T_b}{\partial t^2} + \frac{\partial T_b}{\partial t} + \frac{\varepsilon \rho_b c_b}{\overline{\rho c}}\mathbf{v}\cdot\nabla T_b - \frac{(1-\varepsilon)S_m + S_h}{\overline{\rho c}} \\
&= \overline{\alpha}\left[\nabla^2 T_b + \underbrace{\frac{\varepsilon(1-\varepsilon)(k_s \rho_b c_b + k_b \rho_s c_s)}{U\overline{k}}}_{\tau_T}\frac{\partial}{\partial t}(\nabla^2 T_b) - \underbrace{\frac{\varepsilon(1-\varepsilon)k_s k_b}{U\overline{k}}}_{\approx 0}\nabla^4 T_b\right],
\end{aligned} \tag{11.16a}$$

$$\underbrace{\frac{\varepsilon(1-\varepsilon)\rho_s c_s \rho_b c_b}{U\overline{\rho c}}}_{\tau_q}\frac{\partial^2 T_s}{\partial t^2}+\frac{\partial T_s}{\partial t}+\frac{\varepsilon\rho_b c_b}{\overline{\rho c}}\left[\nabla T_s\cdot\left(\mathbf{v}+\underbrace{\frac{\varepsilon\rho_b c_b}{U}}_{\tau_S}\frac{d\mathbf{v}}{dt}\right)\right]$$
$$=\overline{\alpha}\left[\nabla^2 T_s+\underbrace{\frac{\varepsilon(1-\varepsilon)(k_s\rho_b c_b+k_b\rho_s c_s)}{U\overline{k}}}_{\tau_T}\frac{\partial}{\partial t}(\nabla^2 T_s)-\underbrace{\frac{\varepsilon(1-\varepsilon)k_s k_b}{U\overline{k}}\nabla^4 T_s}_{\approx 0}\right]+\frac{(1-\varepsilon)S_m+S_h}{\overline{\rho c}}$$
$$+\underbrace{\frac{\varepsilon\rho_b c_b}{U}}_{\tau_S}\frac{\partial}{\partial t}\left[\frac{(1-\varepsilon)S_m+S_h}{\overline{\rho c}}\right]$$
(11.16b)

where the bi-harmonic term can be neglected according to Tzou (1997) and Vadàsz (2005).

In this manner, the time phase lag for temperature gradient τ_T is exactly the same as the one appearing in Eq. (11.8) and in accordance with Vadàsz's work (2005). It is also interesting to observe that, for each term appearing in Eq. (11.16b), the corresponding first time derivative exists.

11.3 DUAL-PHASE LAG BIO-HEAT DIFFUSION EQUATION

The contribution of the blood flow to the tissue temperature distribution is represented by the third term on the LHS of Eq. (11.16b). If the blood velocity is assumed to be time-independent, i.e., $d\mathbf{v}/dt \approx 0$, it is reasonable to assume that

$$\varepsilon\rho_b c_b\mathbf{v}\cdot\nabla T_s \approx U(T_s - T_b) \qquad (11.17a)$$

as suggested first by Khaled and Vafai (2003) and, then, by Zhang (2009).

The above equation can also be obtained from the two original energy balances defined by Eqs. (11.4a) and (11.4b) under steady-state conditions, absence of heat sources and assuming that all heat conducted in the interior of the porous medium takes place within the solid portion (i.e., tissue) only. In fact, it follows that

$$\varepsilon\rho_b c_b\mathbf{v}\cdot\nabla T_s = U(T_s - T_b) = (1-\varepsilon)k_s\nabla^2 T_s \qquad (11.17b)$$

where equating the two end quantities gives the porous diffusion equation as derived by Schneider (though in a one-dimensional case) (1955, p. 219) under the same simplifying assumptions and with $T_b = T_s = T$ (Local Thermal Equilibrium, LTE for short), that is,

$$(1-\varepsilon)k_s\nabla^2 T = \varepsilon\rho_b c_b\mathbf{v}\cdot\nabla T. \qquad (11.17c)$$

Note that in the approximate equation (11.17a) the blood temperature may be considered uniform and constant according to Pennes's equation, for example, $T_b = T_{b,in} = 37°\text{C}$. This assumption is reasonable because T_b on the RHS of Eq. (11.17a) should represent only the interstitial blood (or extravascular fluid) that is different from the blood in the vessels (see Fig. 11.1). Concerning this, for the sake of simplicity, the biological tissue was divided into two

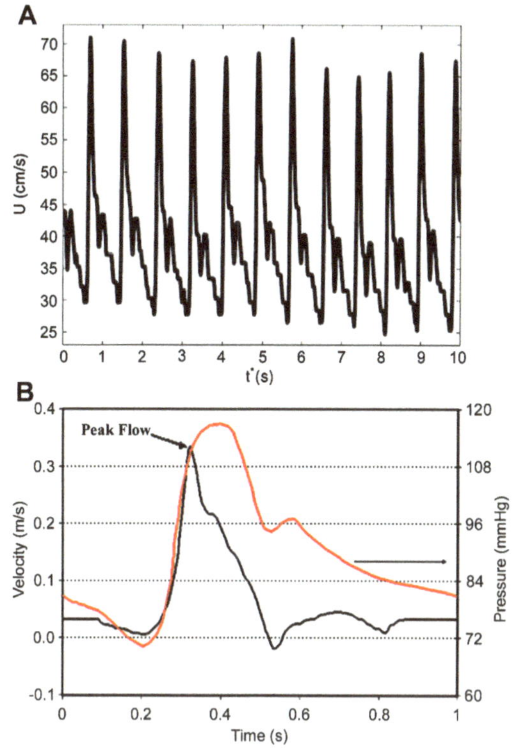

FIGURE 11.3 Pulsatile blood flow: (A) Mikhal et al. (2012), and (B) AlAmiri (2013).

distinctive regions (vascular and extravascular regions), as discussed in Section 11.2, though there are actually three compartments. Additionally, the LHS of Eq. (11.17a) depends only on $T_s(\mathbf{r}, t)$ and, hence, its RHS is also expected to depend only on $T_s(\mathbf{r}, t)$.

However, as the time variations of the blood velocity within blood vessels (arteries, arterioles, capillaries, venules) are not negligible (Lee et al., 1994; Mikhal et al., 2012; AlAmiri, 2013), as shown in Fig. 11.3, an extension of Eq. (11.17a) might be as follows:

$$\begin{aligned}
\varepsilon\rho_b c_b\left[\nabla T_s\cdot\left(\mathbf{v}+\underbrace{\frac{\varepsilon\rho_b c_b}{U}}_{\tau_S}\frac{d\mathbf{v}}{dt}\right)\right] &= \varepsilon\rho_b c_b\left[\nabla T_s\cdot\left(\mathbf{v}_{min}+g(t)\Delta\mathbf{v}+\tau_S\Delta\mathbf{v}\frac{dg}{dt}\right)\right]\\
&= \varepsilon\rho_b c_b\nabla T_s\cdot\mathbf{v}_{min}\left(1+\frac{|\Delta\mathbf{v}|}{|\mathbf{v}_{min}|}g(t)+\tau_S\frac{|\Delta\mathbf{v}|}{|\mathbf{v}_{min}|}\frac{dg}{dt}\right)\\
&\approx U(T_s-T_{b,in})\left[1+\underbrace{\frac{|\Delta\mathbf{v}|}{|\mathbf{v}_{min}|}\left(g(t)+\tau_S\frac{dg}{dt}\right)}_{\phi(t)}\right]=[1+\phi(t)]U(T_s-T_{b,in})
\end{aligned}\tag{11.17d}$$

where $g(t) = g(t + \Delta t_c)$ is a periodic function oscillating between 0 and 1 and having the cardiac cycle period Δt_c as a period. Also, $\mathbf{v}_{min}$ is the minimum velocity of the blood within its vessels and $\mathbf{v}_{min} + \Delta\mathbf{v} = \mathbf{v}_{max}$ is the peak velocity during the cardiac cycle. Hence, ϕ may be seen as a pulsatile parameter.

It is commonly believed that pulsatile flow generated by the pumping action of the heart is dampened out by the time it reaches the microcirculation. However, Lee et al. (1994) demonstrated, for a number of 24 perfused capillaries/mm, that pulsatility exists in the capillary bed and that it directly correlates with the pumping action of the heart. Specifically, the blood velocity has an average value of 0.8–1.2 mm/s with regular fluctuations that coincide with the period of the cardiac cycle (range of periods, 0.5–0.8 s). Also, the peak velocity amplitudes (range: 0.25–0.55 mm/s) are correlated directly with the amplitude of the pulse pressure and are about 40% of the average velocity.

Substituting Eq. (11.17d) into Eq. (11.16b) yields

$$\underbrace{\frac{\varepsilon(1-\varepsilon)\rho_s c_s \rho_b c_b}{U\overline{\rho c}}}_{\tau_q}\frac{\partial^2 T_s}{\partial t^2} + \frac{\partial T_s}{\partial t} = \overline{\alpha}\left[\nabla^2 T_s + \underbrace{\frac{\varepsilon(1-\varepsilon)(k_s\rho_b c_b + k_b\rho_s c_s)}{U\overline{k}}}_{\tau_T}\frac{\partial}{\partial t}(\nabla^2 T_s)\right]$$
$$-\frac{1}{\overline{\rho c}}[1+\phi(t)]U(T_s - T_{b,in}) + \frac{(1-\varepsilon)S_m + S_h}{\overline{\rho c}} + \underbrace{\frac{\varepsilon\rho_b c_b}{U}}_{\tau_S}\frac{\partial}{\partial t}\left[\frac{(1-\varepsilon)S_m + S_h}{\overline{\rho c}}\right] \tag{11.18}$$

which is similar to the well-known dual-phase lag (DPL) bio-heat equation; where now the phase lag times for the heat flux τ_q and for the temperature gradient τ_T may be taken as

$$\tau_q = \frac{\varepsilon(1-\varepsilon)\rho_s c_s \rho_b c_b}{U\overline{\rho c}}, \tag{11.19a}$$

$$\tau_T = \frac{\varepsilon(1-\varepsilon)(k_s\rho_b c_b + k_b\rho_s c_s)}{U\overline{k}}. \tag{11.19b}$$

According to the dual-phase lag model proposed by Tzou (1997), the heat flux and the temperature gradient are related each other through

$$\mathbf{q}_s(\mathbf{r}, t+\tau_q) = -\overline{k}\nabla T_s(\mathbf{r}, t+\tau_T) \tag{11.20a}$$

whose first-order approximation gives

$$\mathbf{q}_s + \tau_q\frac{\partial \mathbf{q}_s}{\partial t} = -\overline{k}\left[\nabla T_s + \tau_T\frac{\partial}{\partial t}(\nabla T_s)\right]. \tag{11.20b}$$

Finally, Eq. (11.18) may be rewritten as

$$\tau_q\frac{\partial^2 T_s}{\partial t^2} + \frac{\partial T_s}{\partial t} = \overline{\alpha}\left[\nabla^2 T_s + \tau_T\frac{\partial}{\partial t}(\nabla^2 T_s)\right] - \underbrace{\frac{1}{\overline{\rho c}}U[1+\phi(t)](T_s - T_{b,in})}_{\textit{bio-heat term}} + \frac{1}{\overline{\rho c}}\left(S + \tau_S\frac{\partial S}{\partial t}\right) \tag{11.21a}$$

where, for the sake of compactness, S and τ_S indicate

$$S = (1-\varepsilon)S_m + S_h, \qquad \tau_S = \frac{\varepsilon \rho_b c_b}{U} = \frac{C_{b,eff}}{U}. \tag{11.21b}$$

Eq. (11.21a) is the same as the one given by Zhang (2009, Eq. (28)) with the exception of 1) the formula for calculating the time phase lag for temperature gradient τ_T, i.e., Eq. (11.19b); and 2) the second term on the RHS where $\phi(t)$ defined in Eq. (11.17d) oscillates around zero. In the current case, the above phase lag is consistent with the one defined by Vadàsz (2005). Also, Eq. (11.21a) is similar to the classical DPL equation as given by Tzou (1997, Eq. (2.10)). The main difference (apart from the presence of the so-called "bio-heat" term) is that τ_q and τ_S are different. However, they are very close, i.e., $\tau_q \approx \tau_S$, as shown in the next subsection.

11.3.1 Phase Lag Times

The time phase lags listed before may be rewritten as

$$\tau_q = \frac{[(1-\varepsilon)\rho_s c_s]\cdot(\varepsilon\rho_b c_b)}{U[(1-\varepsilon)\rho_s c_s + \varepsilon\rho_b c_b]} = \frac{C_{s,eff}C_{b,eff}}{U(C_{s,eff}+C_{b,eff})} = \frac{1}{U}\left(\frac{1}{C_{s,eff}} + \frac{1}{C_{b,eff}}\right)^{-1}, \tag{11.22a}$$

$$\tau_T = \frac{[(1-\varepsilon)k_s]\cdot(\varepsilon\rho_b c_b) + (\varepsilon k_b)\cdot[(1-\varepsilon)\rho_s c_s]}{U[(1-\varepsilon)k_s + \varepsilon k_b]} = \frac{1}{U}\frac{k_{s,eff}C_{b,eff} + k_{b,eff}C_{s,eff}}{k_{s,eff}+k_{b,eff}}. \tag{11.22b}$$

Eq. (11.22a) states that the phase lag time for heat flux τ_q is affected by the effective volumetric heat capacities of the solid and blood phases (i.e., $C_{s,eff}$ and $C_{b,eff}$) which can be considered to behave as thermal capacitors connected in series, as already observed by Vadàsz (2005). Eq. (11.22b) says that the phase lag time for temperature gradient τ_T is influenced by the weighted average of the effective volumetric heat capacities of the solid and fluid phases, assuming as weighting factors the effective thermal conductivities of the complementary phases (i.e., $k_{s,eff}$ and $k_{b,eff}$).

Taking the ratio τ_T/τ_q gives

$$\frac{\tau_T}{\tau_q} = 1 + \frac{k_{s,eff}C_{b,eff}^2 + k_{b,eff}C_{s,eff}^2}{(k_{s,eff}+k_{b,eff})C_{s,eff}C_{b,eff}} > 1 \tag{11.23}$$

that is greater than 1 and independent of the coupling factor U defined by Eq. (11.5) between the two phases, as observed by Vadàsz (2005) too. Also, the case $\tau_T > \tau_q$ implies that, in Eq. (11.20), the heat flux vector $\mathbf{q}_s$ is the cause and the temperature gradient ∇T_s is the effect.

Now, by introducing $N_{C,eff} = C_{s,eff}/C_{b,eff}$ and $N_{k,eff} = k_{s,eff}/k_{b,eff}$ as the effective volumetric heat capacity ratio and the effective thermal conductivity ratio, respectively, Eq. (11.23) may be rewritten as

$$\frac{\tau_T}{\tau_q} = 1 + \underbrace{\frac{N_{k,eff} + N_{C,eff}^2}{(N_{k,eff}+1)N_{C,eff}}}_{\approx 1} \approx 2. \tag{11.24}$$

TABLE 11.3 Phase-lag times

Case	τ_q [s]	τ_T [s] (*)	τ_T [s]	τ_S [s]
1	0.471	0.464	0.943	0.474
2	0.468	0.460	0.936	0.472
3	0.464	0.457	0.928	0.470
4	0.458	0.452	0.917	0.466
5	0.457	0.451	0.914	0.467
6	1.827	1.756	3.653	1.856
7	1.789	1.752	3.577	1.844
8	1.755	1.733	3.509	1.834
9	1.728	1.723	3.455	1.832
10	1.691	1.663	3.381	1.806
11	6.909	6.825	13.819	7.328
12	6.485	6.449	12.970	7.218
13	6.136	6.127	12.273	7.105
14	5.832	5.866	11.664	7.051
15	5.576	5.630	11.151	6.952

(*) Zhang (2009).

Note that the result of Eq. (11.24) being equal to 2 can be explained by considering that

$$N_{C,eff} = \frac{C_{s,eff}}{C_{b,eff}} = \frac{(1-\varepsilon)\rho_s c_s}{\varepsilon \rho_b c_b} = \left(\frac{1}{\varepsilon} - 1\right) N_C \approx \frac{1}{\varepsilon} - 1, \tag{11.25a}$$

$$N_{k,eff} = \frac{k_{s,eff}}{k_{b,eff}} = \frac{(1-\varepsilon)k_s}{\varepsilon k_b} = \left(\frac{1}{\varepsilon} - 1\right) N_k \approx \frac{1}{\varepsilon} - 1 \tag{11.25b}$$

where the volumetric heat capacity ratio $N_C = (\rho_s c_s)/(\rho_b c_b)$ is nearly 1 as well as the thermal conductivity ratio $N_k = k_s/k_b$. In detail, Yuan (2009) considers N_C and N_k exactly equal to 1 while Zhang (2009) set them equal to 1.023 and 1.256, respectively. Therefore, $\tau_T \approx 2\tau_q$.

Now, bearing in mind Eq. (11.19a) and the second of the two Eqs. (11.21b), the ratio τ_S/τ_q gives

$$\frac{\tau_S}{\tau_q} = \frac{(1-\varepsilon)\rho_s c_s + \varepsilon \rho_b c_b}{(1-\varepsilon)\rho_s c_s} = 1 + \frac{C_{b,eff}}{C_{s,eff}} = 1 + \frac{1}{N_{C,eff}} \approx 1 + \frac{\varepsilon}{1-\varepsilon} \tag{11.26}$$

where Eq. (11.9b) and, then, Eq. (11.25a) have been used. Therefore, the ratio τ_S/τ_q is greater than 1 and independent of the coupling factor U. However, as $\varepsilon \in [0.0049, 0.1980]$ (see Table 11.1), $N_{C,eff} > 1$ and $\tau_S/\tau_q \in [1.005, 1.25]$, that is, $\tau_S \approx \tau_q$.

By using the numerical values given in Tables 11.1 and 11.2, the above phase-lag times are computed and listed in Table 11.3 for $k_b = k_s = 0.5$ W/(m K), $c_b = c_s = 3770$ J/(kg K) and $\rho_b = \rho_s = 1050$ kg/m^3 (Yuan, 2009). A comparison with Zhang's τ_T-values (2009) indicates that the phase lag time for temperature gradient is here nearly double according to Vadàsz (2005).

11.3.2 Transformations of the Dependent Variable

By defining a new variable $\theta_s(\mathbf{r},t) = T_s(\mathbf{r},t) - T_{b,in}$, Eq. (11.21a) becomes

$$\overline{\alpha}\nabla^2\theta_s + \overline{\alpha}\tau_T\frac{\partial}{\partial t}(\nabla^2\theta_s) + \frac{1}{\overline{\rho c}}\left(S + \tau_S\frac{\partial S}{\partial t}\right) - \underbrace{\frac{[1+\phi(t)]U}{\overline{\rho c}}\theta_s}_{\text{bio-heat term}} = \frac{\partial\theta_s}{\partial t} + \tau_q\frac{\partial^2\theta_s}{\partial t^2} \tag{11.27}$$

where $\theta_s(\mathbf{r},t)$ is the sole unknown.

In order to get an exact analytical solution, the pulsatile parameter $\phi(t)$ defined in Eq. (11.17d) is set equal to 0 afterwards. Thus, its time oscillation remains a subject for future research, and the "bio-heat" term may be canceled out from Eq. (11.27) by suitably defining a new dependent variable $\Psi(\mathbf{r},t)$ related to $\theta_s(\mathbf{r},t)$ (see Appendix A) by

$$\theta_s(\mathbf{r},t) = \Psi(\mathbf{r},t)e^{-\sigma t} \tag{11.28}$$

where σ is a constant defined as

$$\sigma = \frac{1-\sqrt{1-4(\tau_q U/\overline{\rho c})}}{2\tau_q}. \tag{11.29}$$

Now, substituting Eq. (11.28) into Eq. (11.27), bearing in mind that $\overline{\alpha} = \overline{k}/\overline{\rho c}$ and, then, multiplying the resulting equation by $\exp(\sigma t)$ yields

$$\frac{\overline{k}(1-\tau_T\sigma)}{\overline{\rho c}(1-2\tau_q\sigma)}\nabla^2\Psi + \frac{\overline{k}(1-\tau_T\sigma)}{\overline{\rho c}(1-2\tau_q\sigma)}\frac{\tau_T}{(1-2\tau_T\sigma)}\frac{\partial}{\partial t}(\nabla^2\Psi)$$
$$+\frac{1}{\overline{\rho c}(1-2\tau_q\sigma)}\left(S + \tau_S\frac{\partial S}{\partial t}\right)e^{\sigma t} = \frac{\partial\Psi}{\partial t} + \frac{\tau_q}{(1-2\tau_q\sigma)}\frac{\partial^2\Psi}{\partial t^2} \tag{11.30}$$

where the third term on the LHS has to be manipulated in order to get the classical DPL diffusion equation. The result is

$$\left(S + \tau_S\frac{\partial S}{\partial t}\right)e^{\sigma t} = Se^{\sigma t} + \tau_S\frac{\partial(Se^{\sigma t})}{\partial t} - \tau_S S\frac{\partial e^{\sigma t}}{\partial t} = Se^{\sigma t}(1-\sigma\tau_S) + \tau_S\frac{\partial Se^{\sigma t}}{\partial t}$$
$$= (1-\tau_S\sigma)\left(S_\sigma + \frac{\tau_S}{1-\tau_S\sigma}\frac{\partial S_\sigma}{\partial t}\right) \tag{11.31a}$$

where in this case

$$S_\sigma = Se^{\sigma t} = \left[(1-\varepsilon)S_m + S_h\right]e^{\sigma t} \tag{11.31b}$$

with S defined by the former of the two Eqs. (11.21b).

TABLE 11.4 Transformed phase-lag times

Case	σ [s^{-1}]	τ_q^* [s]	τ_T^* [s]	τ_S^* [s]
1	0.011	0.476	0.953	0.476
2	0.021	0.477	0.955	0.477
3	0.031	0.477	0.955	0.477
4	0.038	0.475	0.950	0.475
5	0.048	0.478	0.956	0.477
6	0.009	1.890	3.780	1.889
7	0.018	1.910	3.820	1.907
8	0.026	1.930	3.860	1.925
9	0.035	1.964	3.928	1.957
10	0.040	1.954	3.908	1.946
11	0.009	7.854	15.708	7.827
12	0.017	8.258	16.517	8.198
13	0.024	8.630	17.261	8.532
14	0.032	9.228	18.457	9.069
15	0.038	9.661	19.323	9.441

Substituting Eq. (11.31a) into Eq. (11.30) gives

$$\tilde{\alpha}\nabla^2\Psi + \tilde{\alpha}\tau_T^*\frac{\partial}{\partial t}(\nabla^2\Psi) + \frac{1}{\widetilde{\rho c}}\left(S_\sigma + \tau_S^*\frac{\partial S_\sigma}{\partial t}\right) = \frac{\partial \Psi}{\partial t} + \tau_q^*\frac{\partial^2\Psi}{\partial t^2} \tag{11.32}$$

where $\widetilde{\rho c}$, $\tilde{\alpha} = \tilde{k}/\widetilde{\rho c}$, τ_T^*, τ_q^* and τ_S^* are defined in Appendix A through Eqs. (A.7b)–(A.7f), respectively. Also, Eq. (A.7a) gives $\tilde{k}$.

Eq. (11.32) is the dual-phase lag (DPL) diffusion equation without the bio-heat term and where the term $\tau_S^*(\partial S_\sigma/\partial t)$ is the so-called *'apparent'* heating term introduced by Tzou (1997). Numerical values for the constant σ defined by Eq. (11.29) and the transformed phase-lag times appearing in the above DPL equation are listed in Table 11.4 for the same cases of the previous Tables 11.1–11.3.

Once the $\Psi(\mathbf{r}, t)$ function is known, the tissue temperature may be taken as

$$T_s(\mathbf{r}, t) = T_{b,in} + \theta_s(\mathbf{r}, t) = T_{b,in} + \Psi(\mathbf{r}, t)e^{-\sigma t}. \tag{11.33}$$

Then, the blood temperature $T_b(\mathbf{r}, t)$ can be derived by integrating Eq. (11.15) where the harmonic term can be neglected and $\varepsilon\rho_b c_b/U = \tau_S$, as defined by the second of the two Eqs. (11.21b). Following a standard integration procedure (Gradshteyn and Ryzhik, 2007; see p. 1096, #16.316), the result is

$$T_b(\mathbf{r}, t) = T_{b,in}e^{-t/\tau_S} + \frac{e^{-t/\tau_S}}{\tau_S}\int_{t'=0}^{t} e^{t'/\tau_S}\,T_s(\mathbf{r}, t')\,dt'. \tag{11.34a}$$

where t' is a dummy time variable.

Substituting Eq. (11.33) into Eq. (11.34a) yields

$$T_b(\mathbf{r},t) = T_{b,in} + \frac{e^{-t/\tau_S}}{\tau_S}\int_{t'=0}^{t} \Psi(\mathbf{r},t')e^{-(\sigma-1/\tau_S)t'}\,dt'. \tag{11.34b}$$

11.4 BOUNDARY AND INITIAL CONDITIONS

Boundary conditions for biological tissue and blood can take on various forms as with other heat conduction problems in porous media. A distinction between non-permeable and permeable boundary surfaces is necessary, as shown in the next subsections.

11.4.1 Impermeable Boundary Surfaces

If the ith boundary surface of the biological tissue of interest may be considered as an impermeable surface (for example, when the tissue is in contact with the wall of an artery), generalized non-homogeneous boundary conditions for biological tissue and blood may be taken, respectively, as

$$k_s\left(\frac{\partial T_s}{\partial n_i}\right)_{\mathbf{r}_i} = h_{w,i}\left[T_{\infty,i} - T_s(\mathbf{r}_i,t)\right] + q_{i,s}, \tag{11.35a}$$

$$k_b\left(\frac{\partial T_b}{\partial n_i}\right)_{\mathbf{r}_i} = h_{w,i}\left[T_{\infty,i} - T_b(\mathbf{r}_i,t)\right] + q_{i,b} \tag{11.35b}$$

where $T_{\infty,i} = T_{b,in} = 37°\mathrm{C}$ and $\mathbf{r}_i$ is the location of the ith boundary in a specific coordinate system. While the terms $q_{i,s}$ and $q_{i,b}$ are unknown, their combination

$$q_{w,i} = \overline{q}_i = (1-\varepsilon)q_{i,s} + \varepsilon q_{i,b} \tag{11.36}$$

is known. Note that $h_{w,i}$ is the heat transfer coefficient of the porous medium with the surrounding fluid (blood within the artery) at the ith impermeable boundary surface; and $q_{w,i} = \overline{q}_i$ is the overall surface heat flux applied to the same boundary of the porous medium.

When the heat transfer coefficient $h_{w,i}$ with the surrounding ambient is very large (as it approaches infinity), the result is a non-homogeneous boundary condition of the first kind for both phases, that is,

$$T_s(\mathbf{r}_i,t) = T_b(\mathbf{r}_i,t) = T_{b,in}. \tag{11.37}$$

On the contrary, when the heat transfer coefficient $h_{w,i}$ is very small (as it approaches zero), the result is a non-homogeneous boundary condition of the second kind for the tissue phase and for the blood phase

$$k_s\left(\frac{\partial T_s}{\partial n_i}\right)_{\mathbf{r}_i} = q_{i,s}, \tag{11.38a}$$

$$k_b\left(\frac{\partial T_b}{\partial n_i}\right)_{\mathbf{r}_i} = q_{i,b}. \tag{11.38b}$$

In such a case, as $q_{i,s}$ and $q_{i,b}$ are unknown while $q_{w,i}$ defined by Eq. (11.36) is prescribed, the actual boundary condition may be derived by multiplying Eq. (11.38a) by $(1-\varepsilon)$ and Eq. (11.38b) by ε, and then adding these together. This results in

$$(1-\varepsilon)k_s\left(\frac{\partial T_s}{\partial n_i}\right)_{\mathbf{r}_i}+\varepsilon k_b\left(\frac{\partial T_b}{\partial n_i}\right)_{\mathbf{r}_i}=k_{s,eff}\left(\frac{\partial T_s}{\partial n_i}\right)_{\mathbf{r}_i}+k_{b,eff}\left(\frac{\partial T_b}{\partial n_i}\right)_{\mathbf{r}_i}=q_{w,i} \tag{11.39}$$

where $q_{w,i}$ can in general be a function of both space and time.

Eq. (11.39) is a boundary condition of the second kind involving both tissue and blood temperature gradients at the ith impermeable boundary surface. This condition is in accordance with the first approach presented by Amiri et al. (1995) and, later, discussed by Lee and Vafai (1999).

Now, substituting Eq. (11.15) into Eq. (11.39) and neglecting the harmonic term yields

$$\overline{k}\left[\left(\frac{\partial T_b}{\partial n_i}\right)_{\mathbf{r}_i}+\overline{\tau}_T\frac{\partial}{\partial t}\left(\frac{\partial T_b}{\partial n_i}\right)_{\mathbf{r}_i}\right]=q_{w,i} \tag{11.40a}$$

where the sole blood temperature gradient is involved. Also, $\overline{k}$ is defined by Eq. (11.9a) and

$$\overline{\tau}_T=\frac{(1-\varepsilon)\varepsilon k_s\rho_b c_b}{\overline{k}U}=\frac{k_{s,eff}C_{b,eff}}{U(k_{s,eff}+k_{b,eff})}\approx(1-\varepsilon)\underbrace{\frac{\varepsilon\rho_b c_b}{U}}_{\tau_S}=(1-\varepsilon)\tau_S=\overline{\tau}_S \tag{11.40b}$$

that is different from τ_T defined by Eq. (11.19b) but is nearly equal to $\overline{\tau}_S$ as $k_s\approx\overline{k}$ because of $k_s\approx k_b$ (see Section 11.3.1).

Alternatively, one can obtain a boundary condition involving only the tissue temperature at the ith boundary. To obtain it, note that T_b can be determined from Eq. (11.15) neglecting the harmonic term; and then when this is substituted into Eq. (11.40a) the result is

$$\overline{k}\left[\left(\frac{\partial T_s}{\partial n_i}\right)_{\mathbf{r}_i}+\overline{\tau}_T\frac{\partial}{\partial t}\left(\frac{\partial T_s}{\partial n_i}\right)_{\mathbf{r}_i}\right]=q_{w,i}+\tau_S\frac{\partial}{\partial t}\left\{\overline{k}\left[\left(\frac{\partial T_b}{\partial n_i}\right)_{\mathbf{r}_i}+\overline{\tau}_T\frac{\partial}{\partial t}\left(\frac{\partial T_b}{\partial n_i}\right)_{\mathbf{r}_i}\right]\right\}. \tag{11.41}$$

The quantity between curly brackets on the RHS may be calculated by using Eq. (11.40a). Then, bearing in mind that $\overline{\tau}_T\approx\overline{\tau}_S$, it follows that

$$\overline{k}\left[\left(\frac{\partial T_s}{\partial n_i}\right)_{\mathbf{r}_i}+\overline{\tau}_S\frac{\partial}{\partial t}\left(\frac{\partial T_s}{\partial n_i}\right)_{\mathbf{r}_i}\right]=q_{w,i}+\tau_S\frac{\partial q_{w,i}}{\partial t} \tag{11.42}$$

where the tissue temperature T_s is the sole unknown.

Taking the ratio $\overline{\tau}_S/\tau_S$ gives $(1-\varepsilon)$ that is less than 1 and independent of the coupling factor U defined by Eq. (11.5) between the two phases. Consequently, Eq. (11.42) states that the temperature gradient at the ith boundary is the cause and the applied surface heat flux at the same boundary is the effect.

If $q_{w,i}=0$, the ith boundary surface is not only impermeable but also thermally insulated, and Eq. (11.42) reduces to

$$\left(\frac{\partial T_s}{\partial n_i}\right)_{\mathbf{r}_i}=0. \tag{11.43}$$

Similarly, multiplying Eq. (11.35a) by $(1-\varepsilon)$ and Eq. (11.35b) by ε, and then adding these together, the result is

$$(1-\varepsilon)\left[k_s\left(\frac{\partial T_s}{\partial n_i}\right)_{\mathbf{r}_i}+h_{w,i}T_s(\mathbf{r}_i,t)\right]+\varepsilon\left[k_b\left(\frac{\partial T_b}{\partial n_i}\right)_{\mathbf{r}_i}+h_{w,i}T_b(\mathbf{r}_i,t)\right]=h_{w,i}T_{b,in}+q_{w,i} \tag{11.44}$$

where $T_{\infty,i}=T_{b,in}$ has been used. Then, substituting Eq. (11.15) into Eq. (11.44) and neglecting the harmonic term yields

$$\left[\overline{k}\left(\frac{\partial T_b}{\partial n_i}\right)_{\mathbf{r}_i}+h_{w,i}T_b(\mathbf{r}_i,t)\right]+\overline{\tau}_S\frac{\partial}{\partial t}\left[k_s\left(\frac{\partial T_b}{\partial n_i}\right)_{\mathbf{r}_i}+h_{w,i}T_b(\mathbf{r}_i,t)\right]=h_{w,i}T_{b,in}+q_{w,i} \tag{11.45}$$

where the blood temperature is the sole unknown. Also, $k_s\approx\overline{k}$ in Eq. (11.45) as $k_s\approx k_b$ (see Section 11.3.1).

Alternatively, one can obtain a generalized boundary condition involving only the tissue temperature at the ith boundary. To obtain it, note that T_b can be determined from Eq. (11.15) neglecting the harmonic term; and then when this is substituted into Eq. (11.45) the result is

$$\begin{aligned}&\left[\overline{k}\left(\frac{\partial T_s}{\partial n_i}\right)_{\mathbf{r}_i}+h_{w,i}T_s(\mathbf{r}_i,t)\right]+\overline{\tau}_S\frac{\partial}{\partial t}\left[\overline{k}\left(\frac{\partial T_s}{\partial n_i}\right)_{\mathbf{r}_i}+h_{w,i}T_s(\mathbf{r}_i,t)\right]\\&=(h_{w,i}T_{b,in}+q_{w,i})+\tau_S\frac{\partial}{\partial t}(h_{w,i}T_{b,in}+q_{w,i})\end{aligned} \tag{11.46}$$

where the tissue temperature T_s is the sole unknown.

Note that the generalized boundary condition (11.46) was obtained by making use of Eq. (11.45). For $q_{w,i}=0$, Eq. (11.46) reduces to a simpler non-homogeneous boundary condition of the 3rd kind.

11.4.2 Permeable Boundary Surfaces

If the ith boundary surface of the porous medium is a permeable surface, that is, a surface crossed by a fluid supplied from (or received by) a reservoir at temperature $T_{\infty,i}$ and having $v_\infty^{(n_i)}=\mathbf{v}_\infty\cdot\mathbf{n}_i$ as undisturbed velocity normal to that boundary, generalized non-homogeneous boundary conditions for biological tissue and blood may be taken, respectively, as

$$k_s\left(\frac{\partial T_s}{\partial n_i}\right)_{\mathbf{r}_i}=h_{w,i}\left[T_{\infty,i}-T_s(\mathbf{r}_i,t)\right]+q_{i,s}, \tag{11.47a}$$

$$k_b\left(\frac{\partial T_b}{\partial n_i}\right)_{\mathbf{r}_i}=v_\infty^{(n_i)}\rho_b c_b\left[T_{\infty,i}-T_b(\mathbf{r}_i,t)\right]+q_{i,b} \tag{11.47b}$$

where $T_{\infty,i} = T_{b,in} = 37°C$; n_i is the direction of the outward pointing unit vector $\mathbf{n}_i$ normal to the surface at the same boundary, while $q_{i,s}$ and $q_{i,b}$ are unknown. Note that the thermal energy exchange in Eq. (11.47b) is due to a mass transfer process.

Now, multiplying Eqs. (11.47a) and (11.47b) by $(1-\varepsilon)$ and ε, respectively, and then adding together the resulting equations gives

$$(1-\varepsilon)\left[k_s\left(\frac{\partial T_s}{\partial n_i}\right)_{\mathbf{r}_i} + h_{w,i}T_s(\mathbf{r}_i,t)\right] + \varepsilon\left[k_b\left(\frac{\partial T_b}{\partial n_i}\right)_{\mathbf{r}_i} + v_\infty^{(n_i)}\rho_b c_b T_b(\mathbf{r}_i,t)\right]$$
$$= \left[(1-\varepsilon)h_{w,i} + \varepsilon v_\infty^{(n_i)}\rho_b c_b\right]T_{b,in} + q_{w,i} \tag{11.48}$$

where $q_{w,i}$ defined by Eq. (11.36) is known and $h_{w,i} \approx v_\infty^{(n_i)}\rho_b c_b$ according to Jiji (2009; see Eq. (5.10), p. 167), under local thermal equilibrium conditions between the two phases. Then, substituting Eq. (11.15) into Eq. (11.89) and neglecting the harmonic term yields

$$\left[\bar{k}\left(\frac{\partial T_b}{\partial n_i}\right)_{\mathbf{r}_i} + \left(v_\infty^{(n_i)}\rho_b c_b\right)T_b(\mathbf{r}_i,t)\right] + \bar{\tau}_S\frac{\partial}{\partial t}\left[\bar{k}\left(\frac{\partial T_b}{\partial n_i}\right)_{\mathbf{r}_i} + \left(v_\infty^{(n_i)}\rho_b c_b\right)T_b(\mathbf{r}_i,t)\right]$$
$$= \left(v_\infty^{(n_i)}\rho_b c_b\right)T_{b,in} + q_{w,i} \tag{11.49}$$

where the approximate relation $k_s \approx \bar{k}$ was used in the previous equation as $k_s \approx k_b$ (see Section 11.3.1).

Alternatively, one can obtain a generalized boundary condition involving only the tissue temperature at the ith boundary. To obtain it, note that T_b can be determined from Eq. (11.15) neglecting the harmonic term; and then when this is substituted into Eq. (11.49) the result is

$$\left[\bar{k}\left(\frac{\partial T_s}{\partial n_i}\right)_{\mathbf{r}_i} + \left(v_\infty^{(n_i)}\rho_b c_b\right)T_s(\mathbf{r}_i,t)\right] + \bar{\tau}_S\frac{\partial}{\partial t}\left[\bar{k}\left(\frac{\partial T_s}{\partial n_i}\right)_{\mathbf{r}_i} + \left(v_\infty^{(n_i)}\rho_b c_b\right)T_s(\mathbf{r}_i,t)\right]$$
$$= \left[\left(v_\infty^{(n_i)}\rho_b c_b\right)T_{b,in} + q_{w,i}\right] + \tau_S\frac{\partial}{\partial t}\left[\left(v_\infty^{(n_i)}\rho_b c_b\right)T_{b,in} + q_{w,i}\right] \tag{11.50}$$

where the tissue temperature T_s is the sole unknown.

Note that Eq. (11.49) was used to obtain the generalized boundary condition (11.50). It is relevant to observe that Eqs. (11.49) and (11.50) for permeable boundaries are exactly the same as Eqs. (11.45) and (11.46) for non-permeable ones. The only difference is that the term $(v_\infty^{(n_i)}\rho_b c_b)$ occurs in place of the term $h_{w,i}$. For $q_{w,i} = 0$, Eq. (11.50) reduces to a simpler non-homogeneous boundary condition of the 3rd kind.

11.4.3 Initial Conditions

The initial conditions for tissue and blood phases can in general be arbitrary functions of space as

$$T_s(\mathbf{r},0) = T_{s,in}(\mathbf{r}), \qquad T_b(\mathbf{r},0) = T_{b,in}(\mathbf{r}). \tag{11.51}$$

However, according to the assumption of constant and uniform blood temperature done for Eq. (11.27) in Section 11.3.2, it results in $T_b(\mathbf{r},0) = T_{b,in} = 37°C$.

11.4.4 Transformed Boundary and Initial Conditions

By using the temperature differences $\theta_s(\mathbf{r},t) = T_s(\mathbf{r},t) - T_{b,in}$ and $\theta_b(\mathbf{r},t) = T_b(\mathbf{r},t) - T_{b,in}$, Eqs. (11.44) and (11.48) (where $h_{w,i} \approx v_\infty^{(n_i)}\rho_b c_b$) may concisely be rewritten as

$$(1-\varepsilon)\left[k_s\left(\frac{\partial\theta_s}{\partial n_i}\right)_{\mathbf{r}_i} + P_i\theta_s(\mathbf{r}_i,t)\right] + \varepsilon\left[k_b\left(\frac{\partial\theta_b}{\partial n_i}\right)_{\mathbf{r}_i} + P_i\theta_b(\mathbf{r}_i,t)\right] = q_{w,i} \tag{11.52}$$

where $P_i = h_{w,i}$ for non-permeable boundaries (see Section 11.4.1), while $P_i = v_\infty^{(n_i)}\rho_b c_b$ for permeable boundary surfaces (see Section 11.4.2).

Then, by means of Eq. (11.15) rewritten as

$$\theta_s \approx \theta_b + \frac{\varepsilon\rho_b c_b}{U}\frac{\partial\theta_b}{\partial t}, \tag{11.53}$$

where the harmonic term has been neglected, Eqs. (11.45) and (11.49) may concisely be taken as

$$\left[\overline{k}\left(\frac{\partial\theta_b}{\partial n_i}\right)_{\mathbf{r}_i} + P_i\theta_b(\mathbf{r}_i,t)\right] + \overline{\tau}_S\frac{\partial}{\partial t}\left[\overline{k}\left(\frac{\partial\theta_b}{\partial n_i}\right)_{\mathbf{r}_i} + P_i\theta_b(\mathbf{r}_i,t)\right] = q_{w,i}. \tag{11.54}$$

Similarly, one can get a boundary condition where θ_s is the sole unknown

$$\left[\overline{k}\left(\frac{\partial\theta_s}{\partial n_i}\right)_{\mathbf{r}_i} + P_i\theta_s(\mathbf{r}_i,t)\right] + \overline{\tau}_S\frac{\partial}{\partial t}\left[\overline{k}\left(\frac{\partial\theta_s}{\partial n_i}\right)_{\mathbf{r}_i} + P_i\theta_s(\mathbf{r}_i,t)\right] = q_{w,i} + \tau_S\frac{\partial q_{w,i}}{\partial t}. \tag{11.55}$$

Then, using Eq. (11.28) in the previous equation yields

$$\begin{aligned}&\left[\overline{k}\left(\frac{\partial\Psi}{\partial n_i}\right)_{\mathbf{r}_i} + P_i\Psi(\mathbf{r}_i,t)\right] + \frac{\overline{\tau}_S}{1-\overline{\tau}_S\sigma}\frac{\partial}{\partial t}\left[\overline{k}\left(\frac{\partial\Psi}{\partial n_i}\right)_{\mathbf{r}_i} + P_i\Psi(\mathbf{r}_i,t)\right]\\ &= \frac{1-\sigma\tau_S}{1-\sigma\overline{\tau}_S}\left[\left(q_{w,i}e^{\sigma t}\right) + \tau_S^*\frac{\partial(q_{w,i}e^{\sigma t})}{\partial t}\right]\end{aligned} \tag{11.56}$$

where τ_S^* is defined in Appendix A by Eq. (A.7f).

By substituting Eq. (A.7a) in Eq. (11.56), it results in

$$\left[\tilde{k}\left(\frac{\partial\Psi}{\partial n_i}\right)_{\mathbf{r}_i} + \tilde{P}_i\Psi(\mathbf{r}_i,t)\right] + \overline{\tau}_T^*\frac{\partial}{\partial t}\left[\tilde{k}\left(\frac{\partial\Psi}{\partial n_i}\right)_{\mathbf{r}_i} + \tilde{P}_i\Psi(\mathbf{r}_i,t)\right] = q_{w,i}^{(\sigma)} + \tau_S^*\frac{\partial q_{w,i}^{(\sigma)}}{\partial t} \tag{11.57a}$$

or, by considering both sides as first-order expansions of Ψ gradient and heat flux at the ith boundary,

$$\left[\tilde{k}\left(\frac{\partial\Psi}{\partial n_i}\right)_{\mathbf{r}_i} + \tilde{P}_i\Psi\right](\mathbf{r}_i,t+\overline{\tau}_T^*) = q_{w,i}^{(\sigma)}(\mathbf{r}_i,t+\tau_S^*) \tag{11.57b}$$

where $\tilde{k}$ is defined by Eq. (A.7a). Also, other selected parameters are:

$$\tilde{P}_i = \frac{1-\tau_T\sigma}{1-\tau_S\sigma}P_i, \tag{11.58a}$$

$$q_{w,i}^{(\sigma)} = q_{w,i} \frac{1-\tau_T\sigma}{1-\overline{\tau}_S\sigma} e^{\sigma t}, \tag{11.58b}$$

$$\overline{\tau}_T^* = \frac{\overline{\tau}_S}{1-\sigma\overline{\tau}_S} \frac{1-\tau_S\sigma}{1-\tau_T\sigma} \approx \left(\frac{1-\tau_S\sigma}{1-\overline{\tau}_S\sigma}\right) \frac{\overline{\tau}_T}{1-\tau_T\sigma}. \tag{11.58c}$$

For $\tilde{P}_i \to \infty$ and $\tilde{P}_i \to 0$, Eq. (11.57a) reduces, respectively, to

$$\Psi(\mathbf{r}_i, t) = 0, \qquad \underbrace{\tilde{k}\left(\frac{\partial \Psi}{\partial n_i}\right)_{\mathbf{r}_i} + \overline{\tau}_T^* \frac{\partial}{\partial t} \tilde{k}\left(\frac{\partial \Psi}{\partial n_i}\right)_{\mathbf{r}_i}}_{\approx \tilde{k}(\frac{\partial \Psi}{\partial n_i})_{\mathbf{r}_i}(\mathbf{r}_i, t+\overline{\tau}_T^*)} = \underbrace{q_{w,i}^{(\sigma)} + \tau_S^* \frac{\partial q_{w,i}^{(\sigma)}}{\partial t}}_{\approx q_{w,i}^{(\sigma)}(\mathbf{r}_i, t+\tau_S^*)} \tag{11.59}$$

where the former is a homogeneous boundary condition of the 1st kind while the latter is a non-homogeneous boundary condition of the 2nd kind.

For $q_{w,i} = 0$, Eq. (11.57a) reduces to a simpler homogeneous boundary condition of the 3rd kind, that is,

$$\tilde{k}\left(\frac{\partial \Psi}{\partial n_i}\right)_{\mathbf{r}_i} + \tilde{P}_i \Psi(\mathbf{r}_i, t) = 0. \tag{11.60a}$$

Then, for $\tilde{P}_i \to \infty$ and $\tilde{P}_i \to 0$, Eq. (11.60a) reduces, respectively, to

$$\Psi(\mathbf{r}_i, t) = 0, \qquad \left(\frac{\partial \Psi}{\partial n_i}\right)_{\mathbf{r}_i} = 0. \tag{11.60b}$$

It is interesting to observe that the homogeneous boundary conditions for the DPL diffusion equation are exactly the same as the ones for the parabolic diffusion.

As regards the initial condition for the tissue phase, it is given by the first of the two Eqs. (11.51). In the current case, its transformation gives

$$\Psi(\mathbf{r}, 0) = \Psi_{in}(\mathbf{r}) = T_{s,in}(\mathbf{r}) - T_{b,in}, \tag{11.61}$$

where $T_{b,in} = 37°\text{C}$.

However, regarding the dual-phase lag diffusion equation (11.32) another initial condition is needed, that is, $\dot{\Psi}_{in}(\mathbf{r}) = (\partial\Psi/\partial t)_{t=0}$. This can be derived from Eq. (11.33) by taking its time partial derivative as follows:

$$\frac{\partial T_s}{\partial t} = e^{-\sigma t} \frac{\partial \Psi}{\partial t} - \sigma e^{-\sigma t} \Psi(\mathbf{r}, t). \tag{11.62}$$

For $t = 0$, the previous equation becomes

$$\left[\frac{\partial \Psi(\mathbf{r}, t)}{\partial t}\right]_{t=0} = \dot{\Psi}_{in}(\mathbf{r}) = \sigma\Psi(\mathbf{r}, 0) + \underbrace{\left[\frac{\partial T_s(\mathbf{r}, t)}{\partial t}\right]_{t=0}}_{=0} = \sigma\Psi_{in}(\mathbf{r}), \tag{11.63}$$

that is, the second initial condition for Eq. (11.32).

11.5 TEMPERATURE SOLUTION IN FINITE REGULAR TISSUES WITH HOMOGENEOUS BOUNDARY CONDITIONS

Consideration is given first to the solution of Eq. (11.32) subject to the homogeneous boundary conditions of Eqs. (11.60a) and (11.60b). For homogeneous boundary conditions, an exact analytical solution was given by Hays-Stang and Haji-Sheikh (1999). They indicated the possibility to use a superposition technique to get two separate solutions: one for the known initial conditions and the other for the contribution of volumetric heat source.

Following here the same approach, one can define a temperature solution to the dual-phase lag diffusion equation (11.32) subject to homogeneous boundary conditions as

$$\Psi(\mathbf{r},t) = \Psi_I(\mathbf{r},t) + \Psi_S(\mathbf{r},t) \tag{11.64}$$

where the function $\Psi_I(\mathbf{r},t)$ represents a solution using the initial conditions, while $\Psi_S(\mathbf{r},t)$ is a solution that defines the contribution of volumetric heat source. The former of these two solutions is presented in Section 11.5.1 and the latter in Section 11.5.2. Both the solutions use a classical Fourier series technique.

Consider the classical Fourier-type heat conduction equation

$$\tilde{\alpha}\nabla^2\hat{\Psi} = \frac{\partial\hat{\Psi}}{\partial t} \tag{11.65}$$

in a finite regular body subject to homogeneous boundary conditions. The fundamental solution to Eq. (11.65) can be determined using the separation of variables (SOV) to be

$$\hat{\Psi}(\mathbf{r},t) = \sum_{n=1}^{\infty} c_n F_n(\mathbf{r}) e^{-\gamma_n t} \tag{11.66}$$

where the constant c_n accounts for the initial condition.

The space-variable function $F_n(\mathbf{r}) = F_n(x,y,z)$ is the nth eigenfunction (corresponding to the nth eigenvalue γ_n and accounting for the thermal diffusivity) that satisfies the relation

$$\tilde{\alpha}\nabla^2 F_n(\mathbf{r}) = -\gamma_n F_n(\mathbf{r}) \tag{11.67}$$

for regular geometries. In accordance with the Sturm–Liouville problem, the function $F_n(\mathbf{r})$ also satisfies the orthogonality property

$$\int_V F_n(\mathbf{r}') F_m(\mathbf{r}')\, dV' = \begin{cases} 0 & \text{when } n \neq m \\ N_n & \text{when } n = m \end{cases} \tag{11.68}$$

where N_n is the so-called norm.

11.5.1 Solution Due to Initial Conditions

The analytical solution of Eq. (11.32) having the initial condition as only driving term (hence, $S_\sigma = 0$) can be obtained by modifying Eq. (11.66) as

$$\Psi_I(\mathbf{r},t) = \sum_{n=1}^{\infty} \psi_{I,n}(t) F_n(\mathbf{r}) e^{-\gamma_n t} \tag{11.69}$$

where $\psi_{I,n}(t)$ is an unknown function of time that accounts for the phase lags appearing in Eq. (11.32) where now $S_\sigma = 0$.

The evaluation of the function $\psi_{I,n}(t)$ may be obtained by placing $\Psi_I(\mathbf{r},t)$ into Eq. (11.32) with $S_\sigma = 0$ and using first Eq. (11.67) and, then, Eq. (11.68). This results in a linear, homogeneous, second-order ordinary differential equation for the determination of the time function $\psi_{I,n}(t)$ as

$$\frac{d^2\psi_{I,n}(t)}{dt^2} - 2\beta_n \frac{d\psi_{I,n}(t)}{dt} + \lambda_n^2 \psi_{I,n}(t) = 0 \tag{11.70}$$

where

$$\beta_n = \gamma_n \left(1 - \frac{\tau_T^*}{2\tau_q^*} - \frac{1}{2\gamma_n \tau_q^*} \right) \tag{11.71a}$$

and

$$\lambda_n = \gamma_n \left(1 - \frac{\tau_T^*}{\tau_q^*} \right)^{1/2}. \tag{11.71b}$$

The solution of Eq. (11.70) is

$$\begin{aligned} \psi_{I,n}(t) &= B_{I,1n} \exp\left(\beta_n t + \sqrt{\beta_n^2 - \lambda_n^2 t}\right) + B_{I,2n} \exp\left(\beta_n t - \sqrt{\beta_n^2 - \lambda_n^2 t}\right) \\ &= e^{\beta_n t} \left[D_{I,1n} \sinh\left(\sqrt{\beta_n^2 - \lambda_n^2 t}\right) + D_{I,2n} \cosh\left(\sqrt{\beta_n^2 - \lambda_n^2 t}\right) \right] \end{aligned} \tag{11.72}$$

where $D_{I,1n} = B_{I,1n} - B_{I,2n}$ and $D_{2n} = B_{I,1n} + B_{I,2n}$.

The two initial conditions to be used for the determination of coefficients $D_{I,1n}$ and $D_{I,2n}$ are

$$\Psi_I(\mathbf{r},0) = \Psi_{in}(\mathbf{r}), \qquad \left[\frac{\partial \Psi_I(\mathbf{r},t)}{\partial t} \right]_{t=0} = \dot{\Psi}_{in}(\mathbf{r}) \tag{11.73}$$

where $\Psi_{in}(\mathbf{r})$ and $\dot{\Psi}_{in}(\mathbf{r})$ are given by Eqs. (11.61) and (11.63), respectively. Thus, by imposing that Eq. (11.69), with the $\psi_{I,n}(t)$ function defined in Eq. (11.72), satisfies the two initial conditions listed before and bearing in mind the orthogonality property (11.68), the result is

$$D_{I,2n} = \frac{1}{N_n} \int_V \Psi_{in}(\mathbf{r}') F_n(\mathbf{r}') \, dV', \tag{11.74a}$$

$$D_{I,1n} = \frac{1}{\sqrt{\beta_n^2 - \lambda_n^2}}\left[(\gamma_n - \beta_n)D_{I,2n} + \frac{1}{N_n}\int_V \dot{\Psi}_{in}(\mathbf{r}')F_n(\mathbf{r}')\,dV'\right]. \tag{11.74b}$$

After determination of these constants, the function $\psi_{I,n}(t)$ defined by Eq. (11.72) becomes

$$\psi_{I,n}(t) = e^{\beta_n t}\frac{\sinh(\sqrt{\beta_n^2 - \lambda_n^2}t)}{N_n\sqrt{\beta_n^2 - \lambda_n^2}}\left[(\gamma_n - \beta_n)\int_V \Psi_{in}(\mathbf{r}')F_n(\mathbf{r}')\,dV'\right.$$
$$\left. + \int_V \dot{\Psi}_{in}(\mathbf{r}')F_n(\mathbf{r}')\,dV'\right] + e^{\beta_n t}\frac{\cosh(\sqrt{\beta_n^2 - \lambda_n^2}t)}{N_n}\left[\int_V \Psi_{in}(\mathbf{r}')F_n(\mathbf{r}')\,dV'\right]. \tag{11.75}$$

Therefore, the solution $\Psi_I(\mathbf{r}, t)$ is

$$\Psi_I(\mathbf{r}, t) = \sum_{n=1}^{\infty}\frac{F_n(\mathbf{r})}{N_n}e^{-(\gamma_n-\beta_n)t}\left\{\frac{\sinh(\sqrt{\beta_n^2 - \lambda_n^2}t)}{\sqrt{\beta_n^2 - \lambda_n^2}}\right.$$
$$\cdot\left[(\gamma_n - \beta_n)\int_V F_n(\mathbf{r}')\Psi_{in}(\mathbf{r}')\,dV' + \int_V F_n(\mathbf{r}')\dot{\Psi}_{in}(\mathbf{r}')\,dV'\right]$$
$$\left. + \cosh\left(\sqrt{\beta_n^2 - \lambda_n^2}t\right)\int_V F_n(\mathbf{r}')\Psi_{in}(\mathbf{r}')\,dV'\right\} \tag{11.76}$$

where N_n is the norm as given by Eq. (11.68).

It is to be noted that the relative magnitude of β_n and λ_n defined by Eqs. (11.71a) and (11.71b), respectively, affects the quantity $\sqrt{\beta_n^2 - \lambda_n^2}$ within Eq. (11.75) and, hence, within Eq. (11.76). This quantity can be real or imaginary while the function $\psi_{I,n}(t)$ is always real. Whenever $\beta_n^2 - \lambda_n^2 < 0$, the $\Psi_I(\mathbf{r}, t)$ solution can exhibit a wave-like phenomenon and the convergence of Eq. (11.76) should be fast (exponential convergence).

11.5.2 Solution Due to a Heating Source

Next, a similar mathematical procedure can provide a solution that includes the contribution of the volumetric heat source; that is, using the relation

$$\Psi_S(\mathbf{r}, t) = \sum_{n=1}^{\infty}\psi_{S,n}(t)F_n(\mathbf{r})e^{-\gamma_n t}. \tag{11.77}$$

The substitution of $\Psi_S(\mathbf{r}, t)$ into the DPL diffusion equation (11.32) and, then, the application of Eqs. (11.67) and (11.68) lead to an ordinary differential for the determination of the time function $\psi_{S,n}(t)$, that is,

$$\frac{d^2\psi_{S,n}(t)}{dt^2} - 2\beta_n\frac{d\psi_{S,n}(t)}{dt} + \lambda_n^2\psi_{S,n}(t) = S_n^*(t) \tag{11.78}$$

where β_n and λ_n are as in the previous subsection and, hence, defined by Eqs. (11.71). Also, the non-homogeneous term $S_n^*(t)$ may be taken as

$$S_n^*(t) = \frac{e^{\gamma_n t}}{\tau_q^* N_n \widetilde{\rho c}} \int_V F_n(\mathbf{r}') \left[S_\sigma(\mathbf{r}', t) + \tau_S^* \frac{\partial S_\sigma(\mathbf{r}', t)}{\partial t} \right] dV' \tag{11.79}$$

where S_σ is defined by Eq. (11.31b) and $\widetilde{\rho c}$ by Eq. (A.7b).

The solution for Eq. (11.78) is

$$\psi_{S,n}(t) = \psi_{S,n,g}(t) + \psi_{S,n,p}(t) \tag{11.80}$$

where $\psi_{S,n,g}(t)$ is the general solution to the homogeneous differential equation associated with Eq. (11.78), which is exactly the same as Eq. (11.70). Therefore, $\psi_{S,n,g}(t) = \psi_{I,n}(t)$ as given by Eq. (11.72). As far as the particular integral $\psi_{S,n,p}(t)$ to Eq. (11.78) is concerned, it may be calculated as (see Gradshteyn and Ryzhik, 2007; Sect. 16.516, p. 1100)

$$\psi_{S,n,p}(t) = \int_{\tau=0}^{t} \frac{e^{\beta_n(t+\tau)}[e^{-\sqrt{\beta_n^2-\lambda_n^2}(t-\tau)} - e^{\sqrt{\beta_n^2-\lambda_n^2}(t-\tau)}]}{W(\tau)} S_n^*(\tau)\, d\tau \tag{11.81}$$

where the Wronskian $W(\tau) = -2\sqrt{\beta_n^2 - \lambda_n^2} e^{2\beta_n \tau}$. By some algebra, this may be expressed as

$$\psi_{S,n,p}(t) = \int_{\tau=0}^{t} \frac{e^{\beta_n(t-\tau)} \sinh[\sqrt{\beta_n^2 - \lambda_n^2}(t-\tau)]}{\sqrt{\beta_n^2 - \lambda_n^2}} S_n^*(\tau)\, d\tau. \tag{11.82}$$

An alternative way to derive $\psi_{S,n,p}(t)$ is based on the Laplace Transform Method and has been shown by Haji-Sheikh and Beck (1994). Therefore, Eq. (11.80) becomes

$$\begin{aligned} \psi_{S,n}(t) = {} & e^{\beta_n t}\left[D_{S,1n} \sinh\left(\sqrt{\beta_n^2 - \lambda_n^2}\, t\right) + D_{S,2n} \cosh\left(\sqrt{\beta_n^2 - \lambda_n^2}\, t\right) \right] \\ & + \int_{\tau=0}^{t} \frac{e^{\beta_n(t-\tau)} \sinh[\sqrt{\beta_n^2 - \lambda_n^2}(t-\tau)]}{\sqrt{\beta_n^2 - \lambda_n^2}} S_n^*(\tau)\, d\tau. \end{aligned} \tag{11.83}$$

Using the initial conditions $\Psi_{S,in}(\mathbf{r}) = 0$ and $[\partial \Psi_S(\mathbf{r}, t)/\partial t]_{t=0} = 0$ leads to the determination of coefficients $D_{S,1n}$ and $D_{S,2n}$. This is done by requiring that Eq. (11.77), with the $\psi_{S,n}(t)$ function defined in Eq. (11.83), satisfy these initial conditions and, then, bearing in mind the orthogonality property (11.68). The result is that $D_{S,1n} = 0$ and $D_{S,2n} = 0$. Thus, the $\Psi_S(\mathbf{r}, t)$ solution is

$$\begin{aligned} \Psi_S(\mathbf{r}, t) = {} & \frac{1}{\tau_q^* \widetilde{\rho c}} \sum_{n=1}^{\infty} \frac{F_n(\mathbf{r})}{N_n} \int_{\tau=0}^{t} \frac{e^{(\beta_n - \gamma_n)(t-\tau)} \sinh[\sqrt{\beta_n^2 - \lambda_n^2}(t-\tau)]}{\sqrt{\beta_n^2 - \lambda_n^2}} \\ & \times \left\{ \int_V F_n(\mathbf{r}') \left[S_\sigma(\mathbf{r}', \tau) + \tau_S^* \frac{\partial S_\sigma(\mathbf{r}', \tau)}{\partial \tau} \right] dV' \right\} d\tau. \end{aligned} \tag{11.84}$$

As previously noted, the quantity $\sqrt{\beta_n^2 - \lambda_n^2}$ can be real or imaginary while the function $\psi_{S,n}(t)$ is always real. Whenever $\beta_n^2 - \lambda_n^2 < 0$, the $\Psi_S(\mathbf{r}, t)$ solution can exhibit a wave-like phenomenon.

11.6 TEMPERATURE SOLUTION IN FINITE REGULAR TISSUES WITH NON-HOMOGENEOUS BOUNDARY CONDITIONS

For non-homogeneous boundary conditions associate with Eq. (11.32), such as Eq. (11.57a) or the second of the two Eqs. (11.59), the temperature solution may be derived by applying the principle of superposition for linear problems, as done in Section 11.5.

If the boundary conditions are time-independent, the function Ψ_B (accounting for only these conditions) has two parts, a steady-state solution and what is called a 'complementary' transient solution, that is,

$$\Psi_B(\mathbf{r},t) = \Psi_{B,ss}(\mathbf{r}) + \Psi_{B,ct}(\mathbf{r},t) \tag{11.85a}$$

where

$$\Psi_{B,ss}(\mathbf{r}) = \lim_{t\to\infty} \Psi_B(\mathbf{r},t). \tag{11.85b}$$

In the current case, because of the variable transformation due to Eq. (11.28), the modified applied heat flux $q_{w,i}^{(\sigma)}$ appearing in Eq. (11.57a) and defined through Eq. (11.58b) is always time-dependent, even if the original surface heat flux $q_{w,i}$ is constant. For this reason, the function Ψ_B will have a quasi-steady part and the usual 'complementary' transient part, that is,

$$\Psi_B(\mathbf{r},t) = \Psi_{B,qs}(\mathbf{r},t) + \Psi_{B,ct}(\mathbf{r},t) \tag{11.86a}$$

where

$$\Psi_{B,qs}(\mathbf{r},t) = \lim_{t\to\infty} \Psi_B(\mathbf{r},t). \tag{11.86b}$$

Note that this might also occur in the particular case of time-independent boundary conditions when they are all of the second kind. For example, a slab insulated on one side and heated by a constant heat flux on the other side (see Example 6.6 given by Cole et al., 2011).

Therefore, the complete solution may be taken as

$$\Psi(\mathbf{r},t) = \Psi_I(\mathbf{r},t) + \Psi_S(\mathbf{r},t) + \Psi_B(\mathbf{r},t) \tag{11.87}$$

where the function $\Psi_I(\mathbf{r},t)$ and $\Psi_S(\mathbf{r},t)$ are still given by Eqs. (11.76) and (11.84), respectively, while Ψ_B is the solution to the following governing equations

$$\tilde{\alpha}\nabla^2\Psi_B + \tilde{\alpha}\tau_T^*\frac{\partial}{\partial t}(\nabla^2\Psi_B) = \frac{\partial \Psi_B}{\partial t} + \tau_q^*\frac{\partial^2\Psi_B}{\partial t^2}, \tag{11.88a}$$

$$\left[\tilde{k}\left(\frac{\partial \Psi_B}{\partial n_i}\right)_{\mathbf{r}_i} + \tilde{P}_i\Psi_B(\mathbf{r}_i,t)\right] + \overline{\tau}_T^*\frac{\partial}{\partial t}\left[\tilde{k}\left(\frac{\partial \Psi_B}{\partial n_i}\right)_{\mathbf{r}_i} + \tilde{P}_i\Psi_B(\mathbf{r}_i,t)\right] = q_{w,i}^{(\sigma)} + \tau_S^*\frac{\partial q_{w,i}^{(\sigma)}}{\partial t}, \tag{11.88b}$$

$$\Psi_B(\mathbf{r},0) = 0, \tag{11.88c}$$

$$\left[\frac{\partial \Psi_B(\mathbf{r},t)}{\partial t}\right]_{t=0} = \dot{\Psi}_{B,in}(\mathbf{r}) = 0 \tag{11.88d}$$

where the initial conditions are zero and the heating source is absent.

11.6.1 Alternative Solution

The solution to the defining equations listed before may be obtained by using a technique similar to the alternative solution proposed by Cole et al. (2011) (see p. 164, Section 5.3.2) for improving convergence of series solution involving non-homogeneous boundary conditions. Following this technique, the function Ψ_B may be written as

$$\Psi_B(\mathbf{r},t)=\Psi_B^*(\mathbf{r},t)+\Psi_B'(\mathbf{r},t) \tag{11.89}$$

where $\Psi_B^*(\mathbf{r},t)$ satisfies the original non-homogeneous boundary conditions (but does not need to satisfy the initial conditions) and, hence, $\Psi_B'(\mathbf{r},t)$ will be freed from this constraint.

The function Ψ_B^* satisfies the following equations

$$\nabla^2\Psi_B^*=0, \tag{11.90a}$$

$$\left[\tilde{k}\left(\frac{\partial\Psi_B^*}{\partial n_i}\right)_{\mathbf{r}_i}+\tilde{P}_i\Psi_B^*(\mathbf{r}_i,t)\right]+\tilde{\tau}_T^*\frac{\partial}{\partial t}\left[\tilde{k}\left(\frac{\partial\Psi_B^*}{\partial n_i}\right)_{\mathbf{r}_i}+\tilde{P}_i\Psi_B^*(\mathbf{r}_i,t)\right]=q_{w,i}^{(\sigma)}+\tau_S^*\frac{\partial q_{w,i}^{(\sigma)}}{\partial t}. \tag{11.90b}$$

Once $\Psi_B^*(\mathbf{r},t)$ is determined (see Example 5.3 given by Cole et al., 2011), it is possible to find the equations that define $\Psi_B'(\mathbf{r},t)$ by replacing Eq. (11.89) into the original initial boundary value problem (IBVP) for Ψ_B, i.e., Eqs. (11.88a)–(11.88d). The result is

$$\tilde{\alpha}\nabla^2\Psi_B'+\tilde{\alpha}\tau_T^*\frac{\partial}{\partial t}(\nabla^2\Psi_B')+\frac{1}{\widetilde{\rho c}}S_B^*(\mathbf{r},t)=\frac{\partial\Psi_B'}{\partial t}+\tau_q^*\frac{\partial^2\Psi_B'}{\partial t^2}, \tag{11.91a}$$

$$\tilde{k}\left(\frac{\partial\Psi_B'}{\partial n_i}\right)_{\mathbf{r}_i}+\tilde{P}_i\Psi_B'(\mathbf{r}_i,t)=0, \tag{11.91b}$$

$$\Psi_B'(\mathbf{r},0)=-\Psi_B^*(\mathbf{r},0)=-\Psi_{B,in}^*(\mathbf{r}), \tag{11.91c}$$

$$\left[\frac{\partial\Psi_B'(\mathbf{r},t)}{\partial t}\right]_{t=0}=\dot{\Psi}_{B,in}'(\mathbf{r})=-\left[\frac{\partial\Psi_B^*(\mathbf{r},t)}{\partial t}\right]_{t=0}=-\dot{\Psi}_{B,in}^*(\mathbf{r}) \tag{11.91d}$$

where the auxiliary source term is

$$S_B^*(\mathbf{r},t)=-\widetilde{\rho c}\left(\frac{\partial\Psi_B^*}{\partial t}+\tau_q^*\frac{\partial^2\Psi_B^*}{\partial t^2}\right). \tag{11.92}$$

Because $\Psi_B^*(\mathbf{r},t)$ satisfies the original boundary conditions, note that the boundary conditions for $\Psi_B'(\mathbf{r},t)$ are completely homogeneous. Therefore, the solution technique proposed in Section 11.5 can be applied to find Ψ_B' in the form

$$\Psi_B'(\mathbf{r},t)=\Psi_{B,I}'(\mathbf{r},t)+\Psi_{B,S}'(\mathbf{r},t) \tag{11.93}$$

where $\Psi'_{B,I}(\mathbf{r},t)$ and $\Psi'_{B,S}(\mathbf{r},t)$ may be calculated by using Eqs. (11.76) and (11.84), respectively. The result is

$$\Psi'_{B,I}(\mathbf{r},t) = -\sum_{n=1}^{\infty} \frac{F_n(\mathbf{r})}{N_n} e^{-(\gamma_n-\beta_n)t} \left\{ \frac{\sinh(\sqrt{\beta_n^2-\lambda_n^2}\,t)}{\sqrt{\beta_n^2-\lambda_n^2}} \times \left[(\gamma_n-\beta_n)\int_V F_n(\mathbf{r}')\Psi^*_{B,in}(\mathbf{r}')\,dV' + \int_V F_n(\mathbf{r}')\dot{\Psi}^*_{B,in}(\mathbf{r}')\,dV' \right] + \cosh\left(\sqrt{\beta_n^2-\lambda_n^2}\,t\right)\int_V F_n(\mathbf{r}')\Psi^*_{B,in}(\mathbf{r}')\,dV' \right\}, \tag{11.94a}$$

$$\Psi'_{B,S}(\mathbf{r},t) = -\frac{1}{\tau_q^*}\sum_{n=1}^{\infty} \frac{F_n(\mathbf{r})}{N_n} \int_{\tau=0}^{t} \frac{e^{(\beta_n-\gamma_n)(t-\tau)}\sinh[\sqrt{\beta_n^2-\lambda_n^2}(t-\tau)]}{\sqrt{\beta_n^2-\lambda_n^2}} \times \left[\int_V F_n(\mathbf{r}')\left(\frac{\partial \Psi_B^*}{\partial t} + \tau_q^* \frac{\partial^2 \Psi_B^*}{\partial t^2} \right) dV' \right] d\tau. \tag{11.94b}$$

Substituting Eq. (11.93) into Eq. (11.89) yields

$$\Psi_B(\mathbf{r},t) = \Psi_B^*(\mathbf{r},t) + \Psi'_{B,I}(\mathbf{r},t) + \Psi'_{B,S}(\mathbf{r},t) \tag{11.95}$$

where Ψ_B^* is the solution to Eqs. (11.90a)–(11.90b).

11.7 GREEN'S FUNCTIONS FOR FINITE REGULAR TISSUES

The solution procedure employed in the two previous sections leads to the development of a methodology that is based on the Green's function (GF) solution method. The GF technique is a very powerful method to solve in an integral form a dual-phase lag diffusion problem, in particular when non-homogeneous terms are involved.

Two different approaches can be used to derive the DPL Green's functions (GFs) and, hence, the related DPL Green's function solution equation (GFSE).

- One way is to derive first the GFs by using the temperature solution obtained in Sections 11.5 and 11.6 and, then, the DPL GFSE. As a matter of fact, in the current case, the so-called DPL 'alternative' GFSE is derived as dealing with the alternative solution of Section 11.6.1.
- Another way is to determine first the DPL GFSE and then to determine the GFs by still employing the solution derived in Sections 11.5 and 11.6.

The former is discussed in the following subsections while the latter is subject for future research and should lead to the same result, so indicating symbolic intrinsic verification (Cole et al., 2014), apart from $\Psi_B(\mathbf{r},t)$ due to non-homogeneous boundary conditions as here dealing with the alternative solution of Section 11.6.1.

11.7.1 Dual-Phase Lag Green's Functions

By definition, the Green's function $G(\mathbf{r},t|\mathbf{r}',\tau)$, in its general form, describes the temperature response (at location $\mathbf{r}$ and at time t) to a local instantaneous heat source (at a location $\mathbf{r}'$ at a time τ) having the volumetric heat capacity as strength. This definition applies to the classical Fourier-type heat conduction (Cole et al., 2011), thermal wave (or single-phase lag) conduction (Haji-Sheikh and Beck, 1994; Haji-Sheikh et al., 2013) and dual-phase lag conduction (Hays-Stang and Haji-Sheikh, 1999).

Then, for the Green's function $G^{DPL}(\mathbf{r},t|\mathbf{r}^*,\tau^*)$ related to the dual-phase lag diffusion equation (such as the one defined by Eq. (11.32)), the local instantaneous heat source is

$$S_\sigma(\mathbf{r},t)=\widetilde{\rho c}\,\delta\left(\mathbf{r}-\mathbf{r}^*\right)\delta\left(t-\tau^*\right) \tag{11.96}$$

where $\delta(\cdot)$ is the Dirac delta function.

Thus, Eq. (11.84) provides the Green's function when $S_\sigma(\mathbf{r},t)$ is given by Eq. (11.96), that is,

$$\begin{aligned}G^{DPL}\left(\mathbf{r},t|\mathbf{r}^*,\tau^*\right)=&\frac{1}{\tau_q^*}\sum_{n=1}^{\infty}\frac{F_n(\mathbf{r})}{N_n}\left[\int_V F_n\left(\mathbf{r}'\right)\delta\left(\mathbf{r}'-\mathbf{r}^*\right)dV'\right]\\&\cdot\left\{\int_{\tau=0}^{t}\frac{e^{(\beta_n-\gamma_n)(t-\tau)}\sinh[\sqrt{\beta_n^2-\lambda_n^2}(t-\tau)]}{\sqrt{\beta_n^2-\lambda_n^2}}\delta\left(\tau-\tau^*\right)d\tau\right.\\&\left.-\tau_S^*\frac{\partial}{\partial\tau^*}\int_{\tau=0}^{t}\frac{e^{(\beta_n-\gamma_n)(t-\tau)}\sinh[\sqrt{\beta_n^2-\lambda_n^2}(t-\tau)]}{\sqrt{\beta_n^2-\lambda_n^2}}\delta\left(\tau-\tau^*\right)d\tau\right\}\end{aligned} \tag{11.97}$$

where the following relation has been used:

$$\frac{\partial\delta(\tau-\tau^*)}{\partial\tau}=-\frac{\partial\delta(\tau-\tau^*)}{\partial\tau^*}. \tag{11.98}$$

Now, by using the integration properties of the Dirac delta function (see Appendix D.1 of Cole et al., 2011), the result is

$$\begin{aligned}G^{DPL}\left(\mathbf{r},t|\mathbf{r}^*,\tau^*\right)=&\sum_{n=1}^{\infty}\frac{F_n(\mathbf{r})F_n(\mathbf{r}^*)}{N_n}\cdot\frac{e^{(\beta_n-\gamma_n)(t-\tau^*)}\sinh[\sqrt{\beta_n^2-\lambda_n^2}(t-\tau^*)]}{\tau_q^*\sqrt{\beta_n^2-\lambda_n^2}}\\&-\tau_S^*\frac{\partial}{\partial\tau^*}\left\{\sum_{n=1}^{\infty}\frac{F_n(\mathbf{r})F_n(\mathbf{r}^*)}{N_n}\cdot\frac{e^{(\beta_n-\gamma_n)(t-\tau^*)}\sinh[\sqrt{\beta_n^2-\lambda_n^2}(t-\tau^*)]}{\tau_q^*\sqrt{\beta_n^2-\lambda_n^2}}\right\}.\end{aligned} \tag{11.99}$$

Bearing in mind that $-\partial f(t-\tau^*)/\partial\tau^*=\partial f(t-\tau^*)/\partial t$ and after replacing τ^* by τ and $\mathbf{r}^*$ by $\mathbf{r}'$, the DPL Green's function is

$$\begin{aligned}G^{DPL}\left(\mathbf{r},t|\mathbf{r}',\tau\right)&=G_1^{DPL}\left(\mathbf{r},t|\mathbf{r}',\tau\right)+\tau_S^*\frac{\partial}{\partial t}G_1^{DPL}\left(\mathbf{r},t|\mathbf{r}',\tau\right)\\&=G_1^{DPL}\left(\mathbf{r},t|\mathbf{r}',\tau\right)+G_2^{DPL}\left(\mathbf{r},t|\mathbf{r}',\tau\right)\end{aligned} \tag{11.100}$$

where

$$G_1^{DPL}(\mathbf{r},t|\mathbf{r}',\tau)=\sum_{n=1}^{\infty}\frac{F_n(\mathbf{r})F_n(\mathbf{r}')e^{-\gamma_n(t-\tau)}}{N_n}\left\{\frac{e^{\beta_n(t-\tau)}\sinh[\sqrt{\beta_n^2-\lambda_n^2}(t-\tau)]}{\tau_q^*\sqrt{\beta_n^2-\lambda_n^2}}\right\}, \tag{11.101a}$$

$$G_2^{DPL}(\mathbf{r},t|\mathbf{r}',\tau)=\frac{\tau_S^*}{\tau_q^*}\left[G_{2a}^{DPL}(\mathbf{r},t|\mathbf{r}',\tau)-G_{2b}^{DPL}(\mathbf{r},t|\mathbf{r}',\tau)\right] \tag{11.101b}$$

with

$$G_{2a}^{DPL}(\mathbf{r},t|\mathbf{r}',\tau)=\sum_{n=1}^{\infty}\frac{F_n(\mathbf{r})F_n(\mathbf{r}')e^{-\gamma_n(t-\tau)}}{N_n}\left\{e^{\beta_n(t-\tau)}\cosh\left[\sqrt{\beta_n^2-\lambda_n^2}(t-\tau)\right]\right\}, \tag{11.102a}$$

$$G_{2b}^{DPL}(\mathbf{r},t|\mathbf{r}',\tau)=\sum_{n=1}^{\infty}\frac{F_n(\mathbf{r})F_n(\mathbf{r}')e^{-\gamma_n(t-\tau)}}{N_n}\cdot\left\{(\gamma_n-\beta_n)\frac{e^{\beta_n(t-\tau)}\sinh[\sqrt{\beta_n^2-\lambda_n^2}(t-\tau)]}{\sqrt{\beta_n^2-\lambda_n^2}}\right\}. \tag{11.102b}$$

These two Green's functions may be manipulated as shown in Appendix B to give

$$G_{2a}^{DPL}(\mathbf{r},t|\mathbf{r}',\tau)=\frac{\tau_q^*}{\tau_S^*}G_2^{DPL}(\mathbf{r},t|\mathbf{r}',\tau)-\tilde{\alpha}\frac{\tau_T^*}{2}\nabla_0^2G_1^{DPL}(\mathbf{r},t|\mathbf{r}',\tau)+\frac{1}{2}G_1^{DPL}(\mathbf{r},t|\mathbf{r}',\tau) \tag{11.103a}$$

$$G_{2b}^{DPL}(\mathbf{r},t|\mathbf{r}',\tau)=-\tilde{\alpha}\frac{\tau_T^*}{2}\nabla_0^2G_1^{DPL}(\mathbf{r},t|\mathbf{r}',\tau)+\frac{1}{2}G_1^{DPL}(\mathbf{r},t|\mathbf{r}',\tau) \tag{11.103b}$$

11.7.2 DPL and Fourier-Type Green's Functions

Bearing in mind Eqs. (11.101a)–(11.102b), the complete DPL Green's function defined through Eq. (11.100) may be rewritten as

$$G^{DPL}(\mathbf{r},t|\mathbf{r}',\tau)=\sum_{n=1}^{\infty}\frac{F_n(\mathbf{r})F_n(\mathbf{r}')e^{-\gamma_n(t-\tau)}}{N_n}\Phi_n(t-\tau) \tag{11.104a}$$

where the function

$$\Phi_n(t-\tau)=\frac{e^{\beta_n(t-\tau)}}{\tau_q^*}\left\{\frac{1+\tau_S^*(\beta_n-\gamma_n)}{\sqrt{\beta_n^2-\lambda_n^2}}\sinh\left[\sqrt{\beta_n^2-\lambda_n^2}(t-\tau)\right]+\tau_S^*\cosh\left[\sqrt{\beta_n^2-\lambda_n^2}(t-\tau)\right]\right\} \tag{11.104b}$$

accounts for the dual-phase lag coming from local thermal non-equilibrium (LTNE) conditions of the living biological tissue. In fact, when $\Phi_n = 1$, the classical transient Green's function for the Fourier-type diffusion equation is obtained, that is,

$$G(\mathbf{r}, t|\mathbf{r}', \tau) = \sum_{n=1}^{\infty} \frac{F_n(\mathbf{r})F_n(\mathbf{r}')e^{-\gamma_n(t-\tau)}}{N_n}. \tag{11.105}$$

This is available in the book by Cole et al. (2011) for any kind of boundary conditions and different systems of coordinates (rectangular, cylindrical and spherical).

11.7.3 DPL Alternative Green's Function Solution Equation

Eq. (11.84) derived by the SOV method may be rewritten by using the Green's function $G_1^{DPL}(\mathbf{r}, t|\mathbf{r}', \tau)$ defined by Eq. (11.101a). This results in

$$\Psi_S(\mathbf{r}, t) = \frac{1}{\widetilde{\rho c}} \int_{\tau=0}^{t} \int_V G_1^{DPL}(\mathbf{r}, t|\mathbf{r}', \tau)\left[S_\sigma(\mathbf{r}', \tau) + \tau_S^* \frac{\partial S_\sigma(\mathbf{r}', \tau)}{\partial \tau}\right] dV'\, d\tau. \tag{11.106}$$

Similarly, Eq. (11.76) derived by the SOV method may be rewritten by using the DPL Green's functions listed in Section 11.7.1 so that

$$\begin{aligned}\Psi_I(\mathbf{r}, t) = &\int_V \left[G_{2a}^{DPL}(\mathbf{r}, t|\mathbf{r}', 0) + G_{2b}^{DPL}(\mathbf{r}, t|\mathbf{r}', 0)\right]\Psi_{in}(\mathbf{r}')\, dV' \\ &+ \tau_q^* \int_V G_1^{DPL}(\mathbf{r}, t|\mathbf{r}', 0)\dot{\Psi}_{in}(\mathbf{r}')\, dV'\end{aligned} \tag{11.107}$$

where, bearing in mind Eqs. (11.103a) and (11.103b), the quantity between square brackets within the first integral may be rewritten as

$$\begin{aligned}&\left[G_{2a}^{DPL}(\mathbf{r}, t|\mathbf{r}', 0) + G_{2b}^{DPL}(\mathbf{r}, t|\mathbf{r}', 0)\right] \\ &= G_1^{DPL}(\mathbf{r}, t|\mathbf{r}', \tau) + \frac{\tau_q^*}{\tau_S^*} G_2^{DPL}(\mathbf{r}, t|\mathbf{r}', \tau) - \tilde{\alpha}\tau_T^* \nabla_0^2 G_1^{DPL}(\mathbf{r}, t|\mathbf{r}', \tau).\end{aligned} \tag{11.108}$$

Substituting it in Eq. (11.107) gives

$$\begin{aligned}\Psi_I(\mathbf{r}, t) = &\int_V \left(G_1^{DPL}\big|_{\tau=0} + \frac{\tau_q^*}{\tau_S^*} G_2^{DPL}\Big|_{\tau=0} - \tilde{\alpha}\tau_T^* \nabla_0^2 G_1^{DPL}\big|_{\tau=0}\right)\Psi_{in}(\mathbf{r}')\, dV' \\ &+ \tau_q^* \int_V G_1^{DPL}\big|_{\tau=0}\dot{\Psi}_{in}(\mathbf{r}')\, dV'.\end{aligned} \tag{11.109}$$

Finally, Eq. (11.95) along with Eqs. (11.94a) and (11.94b) may be rewritten as

$$\begin{aligned}\Psi_B(\mathbf{r},t) =& \Psi_B^*(\mathbf{r},t) - \int_{\tau=0}^{t}\int_V G_1^{DPL}(\mathbf{r},t|\mathbf{r}',\tau)\left(\frac{\partial \Psi_B^*}{\partial t} + \tau_q^*\frac{\partial^2 \Psi_B^*}{\partial t^2}\right)dV'd\tau \\ &- \int_V \left(G_1^{DPL}\big|_{\tau=0} + \frac{\tau_q^*}{\tau_S^*}G_2^{DPL}\Big|_{\tau=0} - \tilde{\alpha}\tau_T^*\nabla_0^2 G_1^{DPL}\big|_{\tau=0}\right)\Psi_{B,in}^*(\mathbf{r}')\,dV' \\ &- \tau_q^*\int_V G_1^{DPL}\big|_{\tau=0}\dot{\Psi}_{B,in}^*(\mathbf{r}')\,dV' \end{aligned} \tag{11.110}$$

where Ψ_B^* is the solution to Eqs. (11.90a)–(11.90b).

Summing up Eqs. (11.106), (11.109), and (11.110) yields the so-called DPL alternative Green's function solution equation (AGFSE). This equation leads to a much faster convergence of the series-solution due to non-homogeneous boundary conditions than the DPL GFSE (here not derived).

11.8 TEMPERATURE DISTRIBUTION IN A LASER-IRRADIATED BIOLOGICAL TISSUE

For effective laser treatment processes, such as laser hyperthermia, it is important to understand the evolution of temperature in the biological tissues as a result of material and heat transfer parameters that affect the laser treatment itself. As the laser light can be highly absorbed or strongly scattered (Zhou et al., 2009; Liu, 2015), two cases can be considered according to these situations. In both cases, a broad laser beam with a uniform incident irradiance (q_{IR}) is applied at time $t = 0$ up to t_{IR} normally to a finite slab of perfused biological tissue with a thickness of L_s and a uniform initial temperature of $T_{s,in} = T_{b,in} = 37°C$. As the spot size of the broad laser beam is much larger than the thickness of the thermally affected zone for the time interval of interest, a one-dimensional model with the $x = L_s$ boundary considered thermally insulated would be sufficient for analyzing the thermal response of the heated porous medium (Zhou et al., 2009; Liu, 2015), as shown in Fig. 11.4. The boundary condition at the front (irradiated) surface $x = 0$ will be discussed in the next subsection, where only the case of highly absorbed laser radiation will be analyzed.

11.8.1 Defining Equations for Highly Absorbed Laser Light

When laser irradiation is highly absorbed by a tissue (e.g., for some UV and IR wave lengths), the laser light is absorbed within a very small depth, a few μm. In this case, laser heating can reasonably be approximated as a surface heat flux at $x = 0$. Alternatively, the irradiated surface at $x = 0$ may be regarded as being thermally insulated and, hence, the laser heating can be considered as a local (plane) heat source located at $x = 0$, that is,

$$S_h(x,t) = (1 - R_d)q_{IR}(t)\delta(x) \tag{11.111a}$$

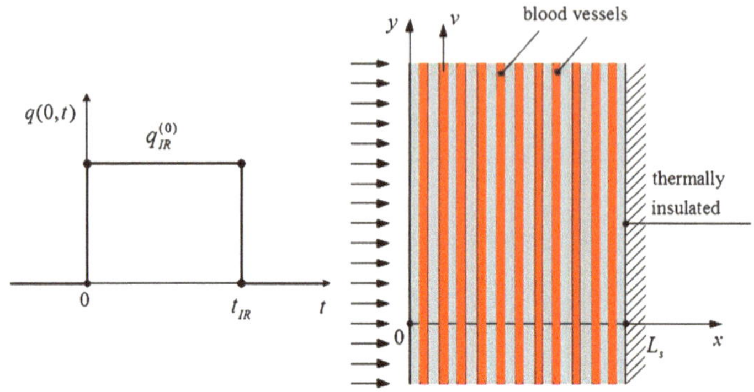

FIGURE 11.4 Schematic of a biological tissue subject to a laser irradiation at the boundary $x = 0$ for a finite time period of t_{IR}. The blood flows along y having v as a velocity.

where R_d is the diffuse reflectance of light at the irradiated surface. Also,

$$q_{IR}(t) = q(0,t) = q_{IR}^{(0)}\left[H(t) - H(t - t_{IR})\right] \tag{11.111b}$$

where $q_{IR}^{(0)}$ is the time-independent incident laser heat flux, t_{IR} is the laser exposure time and $H(.)$ is the so-called Heaviside or unit step function, as depicted in Fig. 11.4.

The one-dimensional transient governing equations for this heating problem can be derived from the more general ones defined in the previous sections, namely Eqs. (11.32), (11.57a), (11.61), and (11.63). Therefore,

$$\tilde{\alpha}\frac{\partial^2\Psi}{\partial x^2} + \tilde{\alpha}\tau_T^*\frac{\partial}{\partial t}\left(\frac{\partial^2\Psi}{\partial x^2}\right) + \frac{1}{\widetilde{\rho c}}\left(S_\sigma + \tau_S^*\frac{\partial S_\sigma}{\partial t}\right) = \frac{\partial\Psi}{\partial t} + \tau_q^*\frac{\partial^2\Psi}{\partial t^2}, \tag{11.112a}$$

$$-\tilde{k}\left(\frac{\partial\Psi}{\partial x}\right)_{x=0} + \tilde{P}_0\Psi(0,t) = 0, \tag{11.112b}$$

$$\tilde{k}\left(\frac{\partial\Psi}{\partial x}\right)_{x=L_s} + \tilde{P}_L\Psi(L_s,t) = 0, \tag{11.112c}$$

$$\Psi(x,0) = \Psi_{in}(x) = 0, \tag{11.112d}$$

$$\left[\frac{\partial\Psi(x,t)}{\partial t}\right]_{t=0} = \dot{\Psi}_{in}(x) = \sigma\Psi_{in}(x) = 0 \tag{11.112e}$$

where, for permeable boundaries of biological tissues (such as in this case), $P_0 = P_L = u_\infty\rho_b c_b$, as shown in Section 11.4.2.

If the blood vessels can be considered as a bundle of staggered tubes placed along the direction normal to the incident laser radiation, as shown in Fig. 11.4, the blood velocity component along x vanishes and, hence, $P_0 \approx P_L \approx 0$. In other words, two homogeneous boundary conditions of the second kind are obtained. Therefore, Eqs. (11.112b) and (11.112c) reduce, respectively, to

$$\left(\frac{\partial \Psi}{\partial x}\right)_{x=0} = 0, \qquad \left(\frac{\partial \Psi}{\partial x}\right)_{x=L_s} = 0. \tag{11.113}$$

11.8.2 Temperature Solutions

As the initial and boundary conditions are homogeneous, the only driving term is due to metabolic and laser heating appearing in Eq. (11.112a) through $S_\sigma = Se^{\sigma t}$ defined by Eq. (11.31b). Thus, the solution given by Eq. (11.106) is readapted for the current 1D problem so that

$$\Psi(x,t) = \frac{1}{\widetilde{\rho c}} \int_{\tau=0}^{t} \int_{x'=0}^{L_s} G_{1,X22}^{DPL}(x,t|x',\tau)\left[S_\sigma(x',\tau) + \tau_S^* \frac{\partial S_\sigma(x',\tau)}{\partial \tau}\right] dx'\, d\tau \tag{11.114}$$

where the one-dimensional Green's function $G_{1,X22}^{DPL}(x,t|x',\tau)$ comes from the more general GF defined by Eq. (11.101a). In this case

$$G_{1,X22}^{DPL}(x,t|x',\tau) = \sum_{n=1}^{\infty} \frac{F_n(x)F_n(x')e^{-\gamma_n(t-\tau)}}{N_n} \left\{\frac{e^{\beta_n(t-\tau)}\sinh[\sqrt{\beta_n^2-\lambda_n^2}(t-\tau)]}{\tau_q^*\sqrt{\beta_n^2-\lambda_n^2}}\right\} \tag{11.115}$$

where the subscript "X22" indicates that the diffusion of heat occurs along x and the boundary conditions are both of the 2nd kind (Cole et al., 2011; Cole et al., 2014). For the associate eigenvalue (or Sturm–Liouville) problem, the result is

$$\gamma_n = (n\pi)^2 \frac{\tilde{\alpha}}{L_s^2} \quad (n = 0, 1, 2, \ldots), \tag{11.116a}$$

$$F_n(x) = \cos(\sqrt{\gamma_n/\tilde{\alpha}}\,x) = \cos(n\pi x/L_s) \quad (n = 0, 1, 2, \ldots), \tag{11.116b}$$

$$N_n = \begin{cases} L_s & \text{for } n = 0 \\ L_s/2 & \text{for } n = 1, 2, \ldots. \end{cases} \tag{11.116c}$$

Note that, for $n = 0$, it results in $\gamma_0 = 0$ and $F_0(x) = 1$. Therefore, this case has to be included in the series solution expressed by Eq. (11.115). Also, bearing in mind Eqs. (11.71a) and (11.71b), the constants are $\beta_0 = -1/(2\tau_q^*)$ and $\lambda_0 = 0$. Thus, Eq. (11.115) may be rewritten as

$$G_{1,X22}^{DPL}(x,t|x',\tau) = \frac{1 - e^{-\frac{t-\tau}{\tau_q^*}}}{L_s} + \frac{2}{L_s}\sum_{n=1}^{\infty} \cos\left(\frac{n\pi x}{L_s}\right)\cos\left(\frac{n\pi x'}{L_s}\right)e^{-\gamma_n(t-\tau)} \cdot \left\{\frac{e^{\beta_n(t-\tau)}\sinh[\sqrt{\beta_n^2-\lambda_n^2}(t-\tau)]}{\tau_q^*\sqrt{\beta_n^2-\lambda_n^2}}\right\}. \tag{11.117}$$

Note that while there are two values for the constant $\sqrt{\beta_0^2} = \pm 1/(2\tau_q^*)$, for either case the result is always Eq. (11.117).

Then, bearing in mind Eqs. (11.31b), (11.111a), and (11.111b), the quantity between square brackets in Eq. (11.114) becomes

$$S_\sigma(x',\tau) + \tau_S^* \frac{\partial S_\sigma(x',\tau)}{\partial \tau} = \tau_S^*(1-R_d)q_{IR}^{(0)}\delta(x')\cdot\left[\delta(\tau) - \delta(\tau - t_{IR})e^{\sigma t_{IR}}\right] + \left(1-\sigma\tau_S^*\right)(1-R_d)q_{IR}^{(0)}\delta(x')\left[H(\tau) - H(\tau - t_{IR})\right]e^{\sigma\tau} \quad (11.118)$$

where appropriate Dirac delta function properties have been used (Cole et al., 2011, Appendix D), such as $\delta(\tau)e^{\sigma\tau} = \delta(\tau)$ and $\delta(\tau - t_{IR})e^{\sigma\tau} = \delta(\tau - t_{IR})e^{\sigma t_{IR}}$. Also, note that the metabolic heating term S_m does not appear in Eq. (11.118). The reason is that the initial tissue temperature of $T_{s,in} = 37°C$ is due to the metabolic heating S_m and this term should not be accounted for twice.

Substituting Eq. (11.118) in Eq. (11.114), and with some algebraic manipulation, the result is

$$\Psi(x,t) = \frac{\tau_S^*(1-R_d)q_{IR}^{(0)}}{\widetilde{\rho c}}\left[G_{1,X22}^{DPL}(x,t|0,0) - e^{\sigma t_{IR}}H(t-t_{IR})G_{1,X22}^{DPL}(x,t|0,t_{IR})\right] + \left(1-\tau_S^*\sigma\right)\frac{(1-R_d)q_{IR}^{(0)}}{\widetilde{\rho c}}\left[\int_{\tau=0}^{t} G_{1,X22}^{DPL}(x,t|0,\tau)e^{\sigma\tau}d\tau - H(t-t_{IR})\int_{\tau=t_{IR}}^{t} G_{1,X22}^{DPL}(x,t|0,\tau)e^{\sigma\tau}d\tau\right] \quad (11.119)$$

where $G_{1,X22}^{DPL}(x,t|0,t_{IR}) = 0$ for $t < t_{IR}$ by the Green's function causality relation (Cole et al., 2011; see p. 37); the effect can in fact not begin before the instantaneous source. Similarly, the second integral between square brackets in Eq. (11.119) vanishes for $t < t_{IR}$, as the effect at time t cannot begin before the instantaneous source at $\tau \in [t_{IR}, t]$. For this reason, the above GF and the second integral involving GF have been multiplied by the unit step function $H(t - t_{IR})$ in Eq. (11.119).

Then, by using Eq. (11.117), the result is

$$G_{1,X22}^{DPL}(x,t|0,0) = \frac{1 - e^{-\frac{t}{\tau_q^*}}}{L_s} + \frac{2}{\tau_q^* L_s}\sum_{n=1}^{\infty}\cos\left(\frac{n\pi x}{L_s}\right)\frac{e^{(\beta_n - \gamma_n)t}}{\sqrt{\beta_n^2 - \lambda_n^2}}\cdot\sinh\left(\sqrt{\beta_n^2 - \lambda_n^2}\,t\right), \quad (11.120a)$$

$$G_{1,X22}^{DPL}(x,t|0,t_{IR}) = \frac{1 - e^{-\frac{t-t_{IR}}{\tau_q^*}}}{L_s} + \frac{2}{\tau_q^* L_s}\sum_{n=1}^{\infty}\cos\left(\frac{n\pi x}{L_s}\right)\frac{e^{(\beta_n - \gamma_n)(t-t_{IR})}}{\sqrt{\beta_n^2 - \lambda_n^2}}\cdot\sinh\left[\sqrt{\beta_n^2 - \lambda_n^2}(t - t_{IR})\right], \quad (11.120b)$$

$$\int_{\tau=0}^{t} G_{1,X22}^{DPL}(x,t|0,\tau)e^{\sigma\tau}d\tau = \frac{e^{\sigma t}-1}{\sigma L_s} + e^{-\frac{t}{\tau_q^*}}\frac{\tau_q^*}{(\tau_q^*\sigma+1)L_s}\left[1-e^{(\sigma+1/\tau_q^*)t}\right]$$
$$+\frac{2e^{\sigma t}}{\tau_q^* L_s}\sum_{n=1}^{\infty}\frac{1}{\sqrt{\beta_n^2-\lambda_n^2}}\cos\left(\frac{n\pi x}{L_s}\right)$$
$$\cdot\int_{u=0}^{t} e^{(\beta_n-\sigma-\gamma_n)u}\sinh\left(\sqrt{\beta_n^2-\lambda_n^2}\,u\right)du, \tag{11.120c}$$

$$\int_{\tau=t_{IR}}^{t} G_{1,X22}^{DPL}(x,t|0,\tau)e^{\sigma\tau}d\tau = \frac{e^{\sigma t}-e^{\sigma t_{IR}}}{\sigma L_s} + e^{-\frac{t}{\tau_q^*}}\frac{\tau_q^*}{(\tau_q^*\sigma+1)L_s}\left[e^{(\sigma+1/\tau_q^*)t_{IR}}-e^{(\sigma+1/\tau_q^*)t}\right]$$
$$+\frac{2e^{\sigma t}}{\tau_q^* L_s}\sum_{n=1}^{\infty}\frac{1}{\sqrt{\beta_n^2-\lambda_n^2}}\cos\left(\frac{n\pi x}{L_s}\right)$$
$$\cdot\int_{u=0}^{t-t_{IR}} e^{(\beta_n-\sigma-\gamma_n)u}\sinh\left(\sqrt{\beta_n^2-\lambda_n^2}\,u\right)du. \tag{11.120d}$$

Note that the dummy variable τ of the integral on the RHS of the last two equations has been replaced by $u = t - \tau$. Once these two integrals are solved (see Section 11.8.3), one can substitute Eqs. (11.120) in Eq. (11.119) and get the transformed tissue temperature $\Psi(x,t)$ in an exact analytical closed-form.

Additionally, when $t \to \infty$, the steady-state transformed solution is obtained and Eq. (11.119) reduces to

$$\Psi(x, t\to\infty) = \frac{(1-R_d)q_{IR}^{(0)}}{\sigma L_s\widetilde{\rho c}}\left(e^{\sigma t_{IR}}-1\right) \approx \frac{(1-R_d)q_{IR}^{(0)}}{L_s\widetilde{\rho c}}t_{IR} \tag{11.121a}$$

where the latter expression is valid for $\sigma t_{IR} << 1$. The correctness of this equation may be verified by applying the first law of thermodynamics to the finite biological tissue of Fig. 11.4 during its heating treatment, that is,

$$\int_{t'=0}^{\infty}(1-R_d)q_{IR}^{(0)}e^{\sigma t'}dt' = (1-R_d)q_{IR}^{(0)}\int_{t'=0}^{t_{IR}}e^{\sigma t'}dt' = L_s\widetilde{\rho c}\big[\Psi(x,\infty)-\underbrace{\Psi_{in}(x)}_{=0}\big] \tag{11.121b}$$

which leads exactly to the same result as given by Eq. (11.121a). This is an indication of symbolic intrinsic verification (Cole et al., 2014).

Then, once $\Psi(x,t)$ is known, the tissue temperature $T_s(x,t)$ can be derived by using Eq. (11.33) readapted for the current 1D problem. As far as the blood temperature $T_b(x,t)$ is concerned, it can be derived by means of Eq. (11.34b) rewritten for a one-dimensional problem as follows:

$$T_b(x,t) = T_{b,in} + \frac{e^{-t/\tau_S}}{\tau_S}\int_{t'=0}^{t}\Psi(x,t')e^{-(\sigma-1/\tau_S)t'}dt' \tag{11.122}$$

where $\Psi(x,t)$ is given by Eq. (11.119).

11.8.3 Convergence of the Temperature Series-Solutions

Computational difficulties might arise during the evaluation of the two GFs defined by Eqs. (11.120a) and (11.120b), and appearing in Eq. (11.119), because of the positive argument of the hyperbolic sine. For the case of $k_s = k_b$, $\rho_s = \rho_b$ and $c_s = c_b$, bearing in mind Eqs. (11.71a) and (11.71b), the above argument simplifies to

$$\sqrt{\beta_n^2 - \lambda_n^2} = \sqrt{\left(\gamma_n + \frac{1}{2\tau_q^*}\right)^2 - \frac{\gamma_n}{\tau_q^*}} > 0 \quad (n = 0, 1, 2, \ldots). \tag{11.123a}$$

Therefore, when n increases, the above argument can be very large and the computation of the hyperbolic sine can lead to an overflow error.

In contrast, the argument of the exponential is always negative and, for the same special case, simplifies to

$$(\beta_n - \gamma_n) = -\left(\gamma_n + \frac{1}{2\tau_q^*}\right) < 0 \quad (n = 0, 1, 2, \ldots). \tag{11.123b}$$

However, summing up Eqs. (11.123a) and (11.123b) yields

$$(\beta_n - \gamma_n) + \sqrt{\beta_n^2 - \lambda_n^2} = -\left[\left(\gamma_n + \frac{1}{2\tau_q^*}\right) - \sqrt{\left(\gamma_n + \frac{1}{2\tau_q^*}\right)^2 - \frac{\gamma_n}{\tau_q^*}}\right] < 0 \quad (n = 0, 1, 2, \ldots) \tag{11.123c}$$

that is always negative. This result suggests writing out the hyperbolic sine in Eqs. (11.120a) and (11.120b) and combining its argument with the exponential argument. In detail,

$$\begin{aligned} G_{1,X22}^{DPL}(x, t|0, 0) = & \frac{1 - e^{-\frac{t}{\tau_q^*}}}{L_s} + \frac{1}{\tau_q^* L_s} \sum_{n=1}^{\infty} \cos\left(\frac{n\pi x}{L_s}\right) \frac{e^{[(\beta_n - \gamma_n) + \sqrt{\beta_n^2 - \lambda_n^2}]t}}{\sqrt{\beta_n^2 - \lambda_n^2}} \\ & - \frac{1}{\tau_q^* L_s} \sum_{n=1}^{\infty} \cos\left(\frac{n\pi x}{L_s}\right) \frac{e^{[(\beta_n - \gamma_n) - \sqrt{\beta_n^2 - \lambda_n^2}]t}}{\sqrt{\beta_n^2 - \lambda_n^2}}, \end{aligned} \tag{11.124a}$$

$$\begin{aligned} G_{1,X22}^{DPL}(x, t|0, t_{IR}) = & \frac{1 - e^{-\frac{t - t_{IR}}{\tau_q^*}}}{L_s} + \frac{1}{\tau_q^* L_s} \sum_{n=1}^{\infty} \cos\left(\frac{n\pi x}{L_s}\right) \frac{e^{[(\beta_n - \gamma_n) + \sqrt{\beta_n^2 - \lambda_n^2}](t - t_{IR})}}{\sqrt{\beta_n^2 - \lambda_n^2}} \\ & - \frac{1}{\tau_q^* L_s} \sum_{n=1}^{\infty} \cos\left(\frac{n\pi x}{L_s}\right) \frac{e^{[(\beta_n - \gamma_n) - \sqrt{\beta_n^2 - \lambda_n^2}](t - t_{IR})}}{\sqrt{\beta_n^2 - \lambda_n^2}} \end{aligned} \tag{11.124b}$$

where either summation exhibits an exponential convergence, so avoiding the computational difficulties mentioned above. However, those summations having the expression (11.123c) as an argument will have a slow convergence for short times.

Similarly, the integrals appearing on the RHS of Eqs. (11.120c) and (11.120d) may conveniently be rewritten as follows:

$$\begin{aligned}
&\int_{u=0}^{t} e^{(\beta_n-\sigma-\gamma_n)u}\sinh\left(\sqrt{\beta_n^2-\lambda_n^2}u\right)du \\
&= \frac{1}{2}\int_{u=0}^{t} e^{[(\beta_n-\sigma-\gamma_n)+\sqrt{\beta_n^2-\lambda_n^2}]u}du - \frac{1}{2}\int_{u=0}^{t} e^{[(\beta_n-\sigma-\gamma_n)-\sqrt{\beta_n^2-\lambda_n^2}]u}du \\
&= \frac{e^{[(\beta_n-\sigma-\gamma_n)+\sqrt{\beta_n^2-\lambda_n^2}]t}-1}{2[(\beta_n-\sigma-\gamma_n)+\sqrt{\beta_n^2-\lambda_n^2}]} - \frac{e^{[(\beta_n-\sigma-\gamma_n)-\sqrt{\beta_n^2-\lambda_n^2}]t}-1}{2[(\beta_n-\sigma-\gamma_n)-\sqrt{\beta_n^2-\lambda_n^2}]},
\end{aligned} \tag{11.125a}$$

$$\begin{aligned}
&\int_{u=0}^{t-t_{IR}} e^{(\beta_n-\sigma-\gamma_n)u}\sinh\left(\sqrt{\beta_n^2-\lambda_n^2}u\right)du \\
&= \frac{1}{2}\int_{u=0}^{t-t_{IR}} e^{[(\beta_n-\sigma-\gamma_n)+\sqrt{\beta_n^2-\lambda_n^2}]u}du - \frac{1}{2}\int_{u=0}^{t-t_{IR}} e^{[(\beta_n-\sigma-\gamma_n)-\sqrt{\beta_n^2-\lambda_n^2}]u}du \\
&= \frac{e^{[(\beta_n-\sigma-\gamma_n)+\sqrt{\beta_n^2-\lambda_n^2}](t-t_{IR})}-1}{2[(\beta_n-\sigma-\gamma_n)+\sqrt{\beta_n^2-\lambda_n^2}]} - \frac{e^{[(\beta_n-\sigma-\gamma_n)-\sqrt{\beta_n^2-\lambda_n^2}](t-t_{IR})}-1}{2[(\beta_n-\sigma-\gamma_n)-\sqrt{\beta_n^2-\lambda_n^2}]}.
\end{aligned} \tag{11.125b}$$

Therefore, Eqs. (11.120c) and (11.120d) become

$$\begin{aligned}
&\int_{\tau=0}^{t} G_{1,X22}^{DPL}(x,t|0,\tau)e^{\sigma\tau}d\tau \\
&= \frac{e^{\sigma t}-1}{\sigma L_s} + e^{-\frac{t}{\tau_q^*}}\frac{\tau_q^*}{(\tau_q^*\sigma+1)L_s}\left[1-e^{(\sigma+1/\tau_q^*)t}\right] \\
&\quad + \frac{1}{\tau_q^* L_s}\sum_{n=1}^{\infty}\cos\left(\frac{n\pi x}{L_s}\right)\frac{e^{[(\beta_n-\gamma_n)+\sqrt{\beta_n^2-\lambda_n^2}]t}}{(\beta_n-\sigma-\gamma_n)\sqrt{\beta_n^2-\lambda_n^2}+(\beta_n^2-\lambda_n^2)} \\
&\quad - \frac{e^{\sigma t}}{\tau_q^* L_s}\sum_{n=1}^{\infty}\cos\left(\frac{n\pi x}{L_s}\right)\frac{1}{(\beta_n-\sigma-\gamma_n)\sqrt{\beta_n^2-\lambda_n^2}+(\beta_n^2-\lambda_n^2)} \\
&\quad - \frac{1}{\tau_q^* L_s}\sum_{n=1}^{\infty}\cos\left(\frac{n\pi x}{L_s}\right)\frac{e^{[(\beta_n-\gamma_n)-\sqrt{\beta_n^2-\lambda_n^2}]t}}{(\beta_n-\sigma-\gamma_n)\sqrt{\beta_n^2-\lambda_n^2}-(\beta_n^2-\lambda_n^2)} \\
&\quad + \frac{e^{\sigma t}}{\tau_q^* L_s}\sum_{n=1}^{\infty}\cos\left(\frac{n\pi x}{L_s}\right)\frac{1}{(\beta_n-\sigma-\gamma_n)\sqrt{\beta_n^2-\lambda_n^2}-(\beta_n^2-\lambda_n^2)},
\end{aligned} \tag{11.126a}$$

$$\begin{aligned}
&\int_{\tau=t_{IR}}^{t} G_{1,X22}^{DPL}(x,t|0,\tau)e^{\sigma\tau}\,d\tau \\
&= \frac{e^{\sigma t}-e^{\sigma t_{IR}}}{\sigma L_s} + e^{-\frac{t}{\tau_q^*}} \frac{\tau_q^*}{(\tau_q^*\sigma+1)L_s}\left[e^{(\sigma+1/\tau_q^*)t_{IR}} - e^{(\sigma+1/\tau_q^*)t}\right] \\
&+ \frac{e^{\sigma t}}{\tau_q^* L_s}\sum_{n=1}^{\infty}\cos\left(\frac{n\pi x}{L_s}\right)\frac{e^{[(\beta_n-\sigma-\gamma_n)+\sqrt{\beta_n^2-\lambda_n^2}](t-t_{IR})}}{(\beta_n-\sigma-\gamma_n)\sqrt{\beta_n^2-\lambda_n^2}+(\beta_n^2-\lambda_n^2)} \\
&- \frac{e^{\sigma t}}{\tau_q^* L_s}\sum_{n=1}^{\infty}\cos\left(\frac{n\pi x}{L_s}\right)\frac{1}{(\beta_n-\sigma-\gamma_n)\sqrt{\beta_n^2-\lambda_n^2}+(\beta_n^2-\lambda_n^2)} \\
&- \frac{e^{\sigma t}}{\tau_q^* L_s}\sum_{n=1}^{\infty}\cos\left(\frac{n\pi x}{L_s}\right)\frac{e^{[(\beta_n-\sigma-\gamma_n)-\sqrt{\beta_n^2-\lambda_n^2}](t-t_{IR})}}{(\beta_n-\sigma-\gamma_n)\sqrt{\beta_n^2-\lambda_n^2}-(\beta_n^2-\lambda_n^2)} \\
&+ \frac{e^{\sigma t}}{\tau_q^* L_s}\sum_{n=1}^{\infty}\cos\left(\frac{n\pi x}{L_s}\right)\frac{1}{(\beta_n-\sigma-\gamma_n)\sqrt{\beta_n^2-\lambda_n^2}-(\beta_n^2-\lambda_n^2)}
\end{aligned} \tag{11.126b}$$

where some of the sums listed before exhibit an exponential convergence; while others exhibit an algebraic convergence requiring many terms.

11.8.4 Numerical Results

By implementing Eq. (11.119) and related expressions, i.e., Eqs. (11.124a), (11.124b), (11.126a), and (11.126b), in Matlab ambient, plots of tissue temperature as a function of both space and time could be derived. These plots are shown in Figs. 11.5–11.8 and refer to cases 1, 5, 6, and 10 of Tables 11.1–11.4. Additionally, they were drawn for a tissue of $L_s = 5$ cm as a thickness, a laser power density of $q_{IR}^{(0)} = 2\ \text{W/cm}^2$, a diffuse reflectance of light at the irradiated surface $R_d = 0.05$ and a laser exposure time $t_{IR} = 5$ s (Zhou et al., 2009; Liu, 2015).

Both Figs. 11.5 and 11.6 consider blood vessels having a diameter of 1.14 mm but with different porosity and perfusion rate (Cases 1 and 5). The higher perfusion rate assumed in Fig. 11.6 leads to a lower temperature with respect to the one plotted in Fig. 11.5. The same difference occurs in Figs. 11.7 and 11.8 where a diameter of 2.28 mm for the blood vessels was considered.

Then, either figure provides the tissue temperature as a function of time and space. In the latter case, the heating (laser "on") and cooling (laser "off") processes of the perfused tissue are put into evidence assuming the time as a parameter. Only a few mm of tissue are affected by the laser therapy. In fact, at a distance of 3 mm from the irradiated surface, the tissue heating is practically negligible.

The above value of 3 mm is in accordance with the so-called penetration distance $d_p = \sqrt{10A\bar{\alpha}t}$ (de Monte et al., 2008; see Eq. (9); de Monte et al., 2012) under local thermal equilibrium conditions. It is defined as the conservative distance from the heated boundary for a given time t at which the temperature just begins to change the fraction of 10^{-A} ($A = 1, 2, \ldots, 10$) of the maximum heated surface temperature rise up to and including that

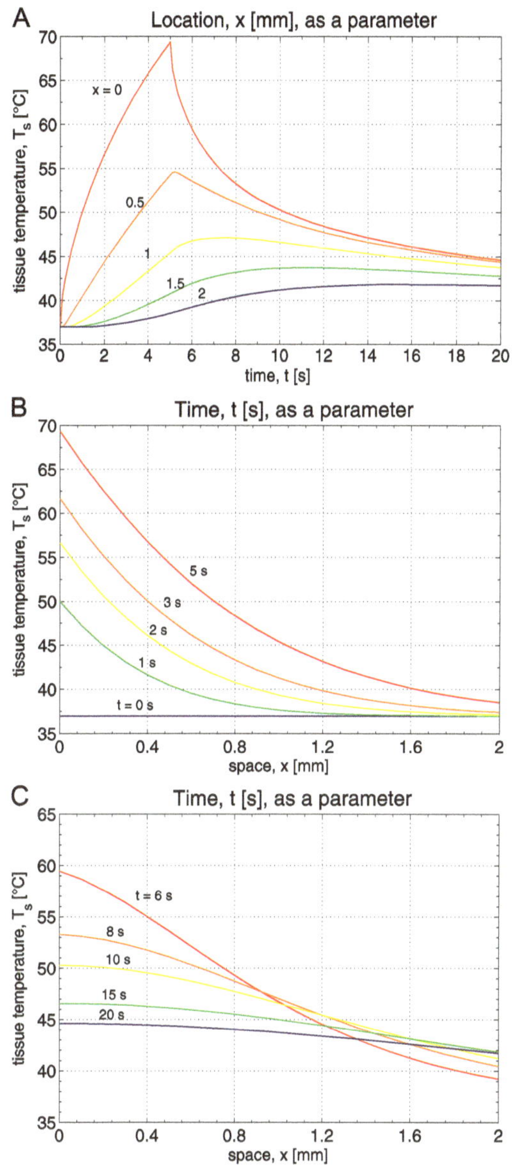

FIGURE 11.5 Plots of tissue temperature for a laser power density of $q_{IR}^{(0)} = 2\ \mathrm{W/cm^2}$ ("**Case 1**" of Tables 11.1–11.4): (A) as a function of time; (B) versus space with laser "on" (heating); and (C) versus space with laser "off" (cooling).

time t. Setting $t = t_{IR} = 5$ s, $A = 2$ (i.e., 1%) and applying Eq. (11.9c) for evaluating the overall thermal diffusivity $\bar{\alpha} = 0.125\ \mathrm{mm^2/s}$ of the porous medium (biological tissue filled by blood), it results in $d_p = 3.5$ mm, that is a conservative estimate of the previous distance of 3 mm.

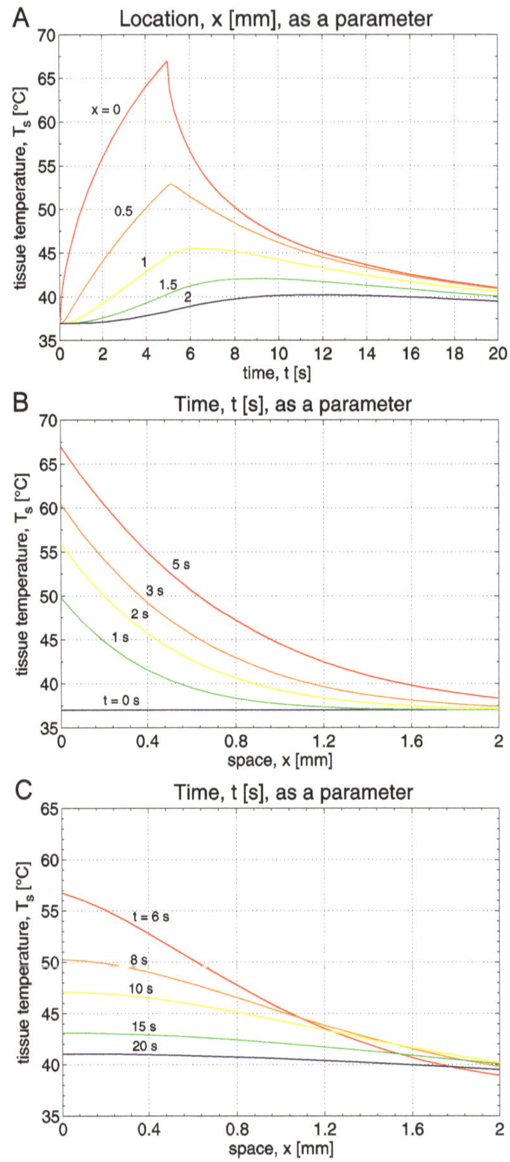

FIGURE 11.6 Plots of tissue temperature for a laser power density of $q_{IR}^{(0)} = 2\,\mathrm{W/cm^2}$ (**"Case 5"** of Tables 11.1–11.4): (A) as a function of time; (B) versus space with laser "on" (heating); and (C) versus space with laser "off" (cooling).

11.9 CONCLUSIONS

The dual-phase lag bio-heat diffusion equation under local thermal non-equilibrium conditions has been derived. By using an appropriate dependent-variable transformation followed

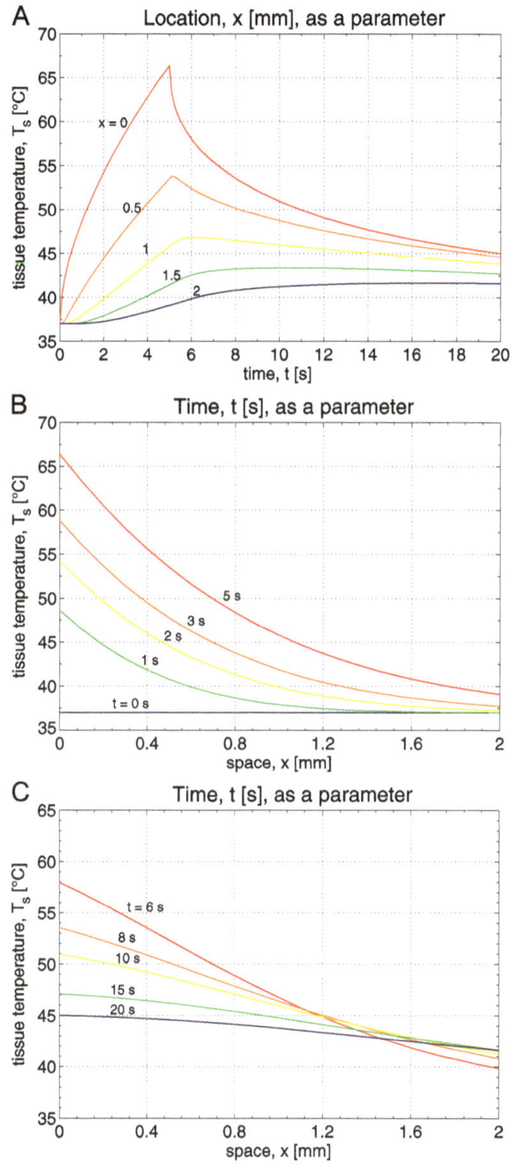

FIGURE 11.7 Plots of tissue temperature for a laser power density of $q_{IR}^{(0)} = 2\ \text{W/cm}^2$ ("**Case 6**" of Tables 11.1–11.4): (A) as a function of time; (B) versus space with laser "on" (heating); and (C) versus space with laser "off" (cooling).

by a modified separation of variables method, an exact analytical solution to this equation was obtained. Dual-phase lag Green's functions and related Green's function solution equation were also obtained for finite and regular biological tissues. An application to four types

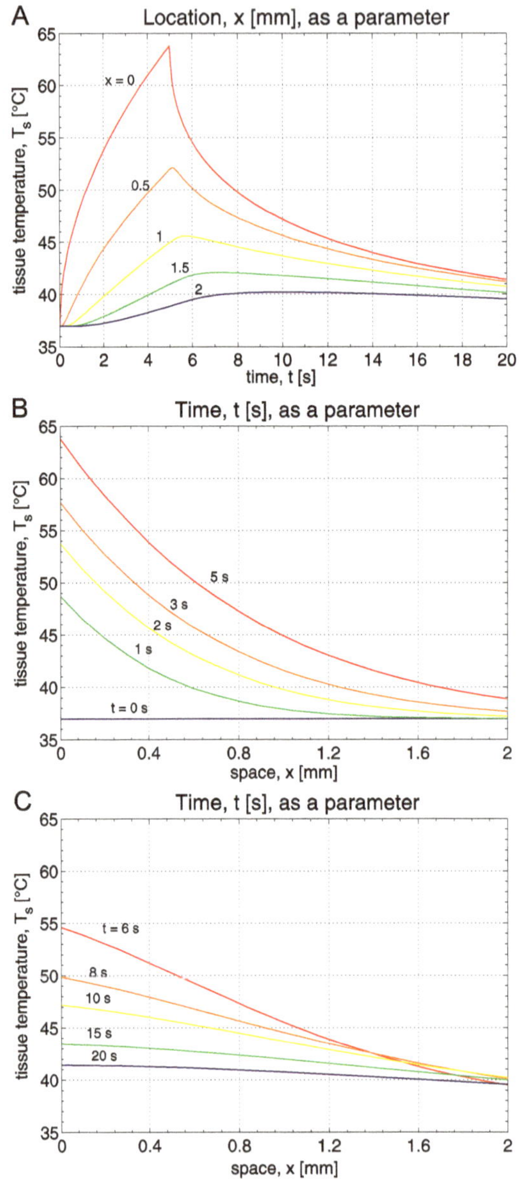

FIGURE 11.8 Plots of tissue temperature for a laser power density of $q_{IR}^{(0)} = 2\ \mathrm{W/cm^2}$ ("**Case 10**" of Tables 11.1–11.4): (A) as a function of time; (B) versus space with laser "on" (heating); and (C) versus space with laser "off" (cooling).

of biological tissues subject to thermal therapy when the laser light is highly absorbed from the same tissues was given. A computer code written in Matlab ambient allowed the temperature distribution within the tissue to be promptly obtained as a function of both space and time. Then, the concept of penetration distance could a priori give insight about the thickness of biological tissue (only a few mm) affected by the laser therapy.

APPENDIX A

The transformation of the dependent variable as defined by Eq. (11.28) may be obtained by placing $\theta_s(\mathbf{r},t) = \Psi(\mathbf{r},t) f(t)$ in Eq. (11.27) and, then, by searching for the unknown function $f(t)$ in such a way as to cancel out the "bio-heat" term from the same equation.

Thus, substituting $\theta_s(\mathbf{r},t) = \Psi(\mathbf{r},t) f(t)$ in Eq. (11.27) yields

$$\Psi\left[\tau_q \frac{d^2 f}{dt^2} + \frac{df}{dt} + \frac{U}{\overline{\rho c}} f(t)\right] + \left(2\tau_q \frac{\partial \Psi}{\partial t} - \overline{\alpha}\tau_T \nabla^2 \Psi\right)\frac{df}{dt}$$
$$+ f(t)\left[\tau_q \frac{\partial^2 \Psi}{\partial t^2} + \frac{\partial \Psi}{\partial t} - \overline{\alpha}\nabla^2\Psi - \overline{\alpha}\tau_T \frac{\partial}{\partial t}\left(\nabla^2\Psi\right)\right] - \frac{1}{\overline{\rho c}}\left(S + \tau_S \frac{\partial S}{\partial t}\right) = 0. \qquad \text{(A.1)}$$

So that the function $f(t)$ can cancel out the "bio-heat" term, it has to satisfy the following linear, homogeneous, second-order ordinary differential equation:

$$\tau_q \frac{d^2 f}{dt^2} + \frac{df}{dt} + \frac{U}{\overline{\rho c}} f(t) = 0, \qquad \text{(A.2)}$$

that is, the quantity between square brackets appearing on the LHS of Eq. (A.1) as first term.

Solving Eq. (A.2) provides

$$f(t) = c_1 f_1(t) + c_2 f_2(t), \qquad \text{(A.3)}$$

that is, a linear combination of two solutions, i.e., $f_1(t) = e^{-\sigma_1 t}$ and $f_2(t) = e^{-\sigma_2 t}$, where

$$\sigma_1 = \frac{1 - \sqrt{1 - 4\tau_q U/\overline{\rho c}}}{2\tau_q}, \qquad \text{(A.4a)}$$

$$\sigma_2 = \frac{1 + \sqrt{1 - 4\tau_q U/\overline{\rho c}}}{2\tau_q}. \qquad \text{(A.4b)}$$

It is relevant to observe that the argument of the square root appearing in these two equations is always positive or equal to zero as $\tau_q U/\overline{\rho c} \le 1/4$. In fact, bearing in mind Eqs. (11.9b) and (11.22a), the result is

$$\frac{\tau_q U}{\overline{\rho c}} = \frac{C_{s,eff} C_{b,eff}}{(C_{s,eff} + C_{b,eff})^2} = \frac{N_{C,eff}}{(N_{C,eff} + 1)^2}. \qquad \text{(A.5)}$$

For $N_{C,eff} = C_{s,eff}/C_{b,eff} = 1$, the quantity $\tau_q U/\overline{\rho c}$ has its maximum value, and this maximum value is equal to $1/4$. Consequently, σ_1 and σ_2 are positive real numbers. The former is always less than or equal to $1/(2\tau_q)$, while the latter is always greater than or equal to $1/(2\tau_q)$.

Also, as $N_{C,eff} > 1$ (see Section 11.3.1), we have $\tau_q U/\overline{\rho c} < 1/4$. In detail, by using Eq. (11.25a), we obtain $\tau_q U/\overline{\rho c} \approx \varepsilon(1-\varepsilon)$ where $\varepsilon \in [0.0049, 0.1980]$ (see Table 11.1).

Now, substituting $f(t) = e^{-\sigma t}$ (where σ can be equal to σ_1 or σ_2) into Eq. (A.1), by some algebra we obtain

$$\tilde{\alpha}\nabla^2\Phi + \tilde{\alpha}\tau_T^*\frac{\partial}{\partial t}(\nabla^2\Phi) + \frac{1}{\widetilde{\rho c}}\left(S_\sigma + \tau_S^*\frac{\partial S_\sigma}{\partial t}\right) = \frac{\partial \Phi}{\partial t} + \tau_q^*\frac{\partial^2\Phi}{\partial t^2} \tag{A.6}$$

where

$$\tilde{k} = \bar{k}\frac{1-\tau_T\sigma}{1-\tau_S\sigma}, \tag{A.7a}$$

$$\widetilde{\rho c} = \overline{\rho c}\frac{1-2\tau_q\sigma}{1-\tau_S\sigma}, \tag{A.7b}$$

$$\tilde{\alpha} = \frac{\tilde{k}}{\widetilde{\rho c}} = \bar{\alpha}\frac{1-\tau_T\sigma}{1-2\tau_q\sigma}, \tag{A.7c}$$

$$\tau_T^* = \frac{\tau_T}{1-\tau_T\sigma}, \tag{A.7d}$$

$$\tau_q^* = \frac{\tau_q}{1-2\tau_q\sigma} = \frac{\tau_q}{\sqrt{1-4\tau_q U/\overline{\rho c}}}, \tag{A.7e}$$

$$\tau_S^* = \frac{\tau_S}{1-\tau_S\sigma}. \tag{A.7f}$$

Bearing in mind that $\tau_T \approx 2\tau_q$ and $\tau_q \approx \tau_S$ (see Section 11.3.1), all the quantities listed before are positive when $\sigma = \sigma_1$ and are negative when $\sigma = \sigma_2$. But, from a physical point of view, all the constants appearing in Eq. (A.6) have to be positive in order to represent the dual-phase lag diffusion equation. Therefore, the case of σ_2 cannot be accepted.

APPENDIX B

The Green's function $G_{2a}^{DPL}(\mathbf{r}, t|\mathbf{r}', \tau)$ and $G_{2b}^{DPL}(\mathbf{r}, t|\mathbf{r}', \tau)$ defined by Eqs. (11.102a) and (11.102b), respectively, may be manipulated as follows.

By using Eq. (11.71a), the quantity $(\gamma_n - \beta_n)$ appearing in $G_{2b}^{DPL}(\mathbf{r}, t|\mathbf{r}', \tau)$ becomes

$$(\gamma_n - \beta_n) = \gamma_n\frac{\tau_T^*}{2\tau_q^*} + \frac{1}{2\tau_q^*}. \tag{B.1}$$

Consequently, Eq. (11.102b) may be split into two parts as

$$G_{2b}^{DPL}(\mathbf{r},t|\mathbf{r}',\tau)=\frac{\tau_T^*}{2}\underbrace{\sum_{n=1}^{\infty}\frac{F_n(\mathbf{r})F_n(\mathbf{r}')e^{-\gamma_n(t-\tau)}}{N_n}\left\{\gamma_n\frac{e^{\beta_n(t-\tau)}\sinh[\sqrt{\beta_n^2-\lambda_n^2}(t-\tau)]}{\tau_q^*\sqrt{\beta_n^2-\lambda_n^2}}\right\}}_{G_{2c}^{DPL}(\mathbf{r},t|\mathbf{r}',\tau)}$$
$$+\frac{1}{2}G_1^{DPL}(\mathbf{r},t|\mathbf{r}',\tau)=\frac{\tau_T^*}{2}G_{2c}^{DPL}(\mathbf{r},t|\mathbf{r}',\tau)+\frac{1}{2}G_1^{DPL}(\mathbf{r},t|\mathbf{r}',\tau). \tag{B.2}$$

In addition, using Eq. (11.67) in the $\mathbf{r}'$ space results in the relation $\gamma_n F_n(\mathbf{r}')=-\tilde{\alpha}\nabla_0^2 F_n(\mathbf{r}')$, where ∇_0 stands for the "grad" operator in the $\mathbf{r}'$ space. Thus,

$$G_{2c}^{DPL}(\mathbf{r},t|\mathbf{r}',\tau)=-\tilde{\alpha}\sum_{n=1}^{\infty}\frac{F_n(\mathbf{r})\nabla_0^2F_n(\mathbf{r}')e^{-\gamma_n(t-\tau)}}{N_n}\left\{\frac{e^{\beta_n(t-\tau)}\sinh[\sqrt{\beta_n^2-\lambda_n^2}(t-\tau)]}{\tau_q^*\sqrt{\beta_n^2-\lambda_n^2}}\right\}$$
$$=-\tilde{\alpha}\nabla_0^2\underbrace{\sum_{n=1}^{\infty}\frac{F_n(\mathbf{r})F_n(\mathbf{r}')e^{-\gamma_n(t-\tau)}}{N_n}\left\{\frac{e^{\beta_n(t-\tau)}\sinh[\sqrt{\beta_n^2-\lambda_n^2}(t-\tau)]}{\tau_q^*\sqrt{\beta_n^2-\lambda_n^2}}\right\}}_{G_1^{DPL}(\mathbf{r},t|\mathbf{r}',\tau)}$$
$$=-\tilde{\alpha}\nabla_0^2G_1^{DPL}(\mathbf{r},t|\mathbf{r}',\tau). \tag{B.3}$$

Substituting Eq. (B.3) in Eq. (B.2) gives

$$G_{2b}^{DPL}(\mathbf{r},t|\mathbf{r}',\tau)=-\tilde{\alpha}\frac{\tau_T^*}{2}\nabla_0^2G_1^{DPL}(\mathbf{r},t|\mathbf{r}',\tau)+\frac{1}{2}G_1^{DPL}(\mathbf{r},t|\mathbf{r}',\tau). \tag{B.4}$$

By replacing Eq. (B.4) in Eq. (11.101b), the result is

$$G_2^{DPL}(\mathbf{r},t|\mathbf{r}',\tau)=\frac{\tau_S^*}{\tau_q^*}\left[G_{2a}^{DPL}(\mathbf{r},t|\mathbf{r}',\tau)+\tilde{\alpha}\frac{\tau_T^*}{2}\nabla_0^2G_1^{DPL}(\mathbf{r},t|\mathbf{r}',\tau)-\frac{1}{2}G_1^{DPL}(\mathbf{r},t|\mathbf{r}',\tau)\right]. \tag{B.5}$$

Also from the previous equation, it can be shown that $G_{2a}^{DPL}(\mathbf{r},t|\mathbf{r}',\tau)$ may be represented as

$$G_{2a}^{DPL}(\mathbf{r},t|\mathbf{r}',\tau)=\frac{\tau_q^*}{\tau_S^*}G_2^{DPL}(\mathbf{r},t|\mathbf{r}',\tau)-\tilde{\alpha}\frac{\tau_T^*}{2}\nabla_0^2G_1^{DPL}(\mathbf{r},t|\mathbf{r}',\tau)$$
$$+\frac{1}{2}G_1^{DPL}(\mathbf{r},t|\mathbf{r}',\tau). \tag{B.6}$$

References

AlAmiri, A.M., 2013. Fluid–structure interaction analysis of pulsatile blood flow and heat transfer in living tissues during thermal therapy. J. Fluids Eng. 135 (4), 041103.

Amiri, A., Vafai, K., Kuzay, T.M., 1995. Effects of boundary conditions on non-Darcian heat transfer through porous media and experimental comparisons. Numer. Heat Transf., Part A 27, 651–664.

Baish, J.W., Foster, K.R., Ayyaswamy, P.S., 1986. Perfused phantom models of microwave irradiated tissues. J. Biomech. Eng. 108, 239–245.
Cole, K., Beck, J.V., Haji-Sheikh, A., Litkouhi, B., 2011. Heat Conduction Using Green's Functions, 2nd ed. Taylor and Francis, Boca Raton, FL.
Cole, K., Beck, J.V., Woodbury, K.A., de Monte, F., 2014. Intrinsic verification and a heat conduction database. Int. J. Therm. Sci. 78, 36–47.
de Monte, F., Beck, J.V., Amos, D.E., 2008. Diffusion of thermal disturbances in two-dimensional Cartesian transient heat conduction. Int. J. Heat Mass Transf. 51, 5931–5941.
de Monte, F., Beck, J.V., Amos, D.E., 2012. Solving two-dimensional Cartesian unsteady heat conduction problems for small values of the time. Int. J. Therm. Sci. 60, 106–113.
Gradshteyn, I.S., Ryzhik, I.M., 2007. Table of Integrals, Series, and Products, 7th edition. Academic Press.
Haji-Sheikh, A., Beck, J.V., 1994. Green's function solution for thermal wave equation in finite bodies. Int. J. Heat Mass Transf. 37, 2615–2626.
Haji-Sheikh, A., Minkowycz, W.J., 2008. Heat transfer analysis under local thermal non-equilibrium conditions. In: Vadàsz, P. (Ed.), Emerging Topics in Heat and Mass Transfer in Porous Media: From Bioengineering and Microelectronics to Nanotechnology. Springer Science+Business Media B.V.
Haji-Sheikh, A., De Monte, F., Beck, J.V., 2013. Temperature solutions in thin films using thermal wave Green's function solution equation. Int. J. Heat Mass Transf. 62, 78–86.
Hays-Stang, K.J., Haji-Sheikh, A., 1999. A unified solution for heat conduction in thin films. Int. J. Heat Mass Transf. 42, 455–465.
Jiji, L.M., 2009. Heat Conduction, 3rd edition. Springer.
Khaled, A.R.A., Vafai, K., 2003. The role of porous media in modeling flow and heat transfer in biological tissues. Int. J. Heat Mass Transf. 46, 4989–5003.
Lee, D.-Y., Vafai, K., 1999. Analytical characterization and conceptual assessment of solid and fluid temperature differentials in porous media. Int. J. Heat Mass Transf. 42, 423–435.
Lee, J.J., Tyml, K., Menkis, A.H., Novick, R.J., Mckenzie, F.N., 1994. Evaluation of pulsatile and nonpulsatile flow in capillaries of goat skeletal muscle using intravital microscopy. Microvasc. Res. 48, 316–327.
Liu, K.-C., 2015. Analysis for high-order effects in thermal lagging to thermal responses in biological tissue. Int. J. Heat Mass Transf. 81, 347–354.
Liu, K.-C., Chen, Y.-S., 2016. Analysis of heat transfer and burn damage in a laser irradiated living tissue with the generalized dual-phase-lag model. Int. J. Therm. Sci. 103, 1–9.
Mikhal, J., Slump, C.H., Geurts, B.J., 2012. Simulation of pulsatile flow in cerebral aneurysms: from medical images to flow and forces. In: Murai, Y. (Ed.), Aneurysm. InTech. ISBN 978-953-51-0730-9.
Minkowycz, W.J., Haji-Sheikh, A., Vafai, K., 1999. On departure from local thermal equilibrium in porous media due to a rapidly changing heat source: the Sparrow number. Int. J. Heat Mass Transf. 42, 3373–3385.
Nakayama, A., Kuwahara, F., 2008. A general bioheat transfer model based on the theory of porous media. Int. J. Heat Mass Transf. 51, 3190–3199.
Nakayama, A., Kuwahara, F., Hayashi, T., 2004. Numerical modelling for three-dimensional heat and fluid flow through a bank of cylinders in yaw. J. Fluid Mech. 498, 139–159.
Pennes, H.H., 1948. Analysis of tissue and arterial blood temperatures in the resting human forearm. J. Appl. Physiol. 1, 93–122.
Roetzel, W., Xuan, Y., 1998. Transient response of the human limb to an external stimulus. Int. J. Heat Mass Transf. 41, 229–239.
Schneider, P.J., 1955. Conduction Heat Transfer. Addison-Wesley Publishing Company.
Tzou, D.Y., 1997. Macro- to Microscale Heat Transfer: The Lagging Behaviour. Taylor & Francis, Washington.
Vadàsz, P., 2005. Lack of oscillations in dual-phase-lagging heat conduction for a porous slab subject to imposed heat flux and temperature. Int. J. Heat Mass Transf. 48, 2822–2828.
Wright, J., Han, B., Chuong, C.J., 2012. Biphasic investigation of tissue mechanical response during freezing front propagation. J. Biomech. Eng. 134, 061005.
Yamada, Y., Tien, T., Ohta, M., 1995. Theoretical analysis of temperature variation of biological tissues irradiated by light. In: ASME/JSME Joint Thermal Engineering Conference, vol. 4, pp. 575–581.
Yuan, P., 2009. Numerical analysis of an equivalent heat transfer coefficient in a porous model for simulating a biological tissue in a hyperthermia therapy. Int. J. Heat Mass Transf. 52, 1734–1740.
Zhang, Y., 2009. Generalized dual-phase lag bioheat equations based on nonequilibrium heat transfer in living biological tissues. Int. J. Heat Mass Transf. 52, 4829–4834.
Zhou, J., Chen, J.K., Zhang, Y., 2009. Dual-phase lag effects on thermal damage to biological tissues caused by laser irradiation. Comput. Biol. Med. 39, 286–293.

Nomenclature

c	Specific heat, J kg^{-1} K^{-1}
C	Volumetric heat capacity (ρc), J m^{-3} K^{-1}
d_b	Inner diameter of the blood vessels (see Fig. 11.2), m
d_s	Diameter of an equivalent circular biological tissue (see Fig. 11.2), m
F_n	nth eigenfunction
G	Green's function, m^{-3} (for 3D cases) and m^{-1} (for 1D cases)
h	Heat transfer coefficient, W m^{-2} K^{-1}
H	Heaviside (or unit step) function
k	Thermal conductivity, W m^{-1} K^{-1}
K	Permeability, m s^{-1}
L	Distance between two adjacent blood vessels (see Fig. 11.2), m
L_s	Tissue thickness (see Fig. 11.4), m
$\mathbf{n}_i$	Unit vector normal to the ith permeable boundary surface
N_C	Volumetric heat capacity ratio (C_s/C_b)
N_k	Thermal conductivity ratio (k_s/k_b)
N_n	nth norm, m^3 (for 3D cases) and m (for 1D cases)
Nu	Nusselt number
p	Pressure, Pa
P_i	Dimensional quantity at the ith boundary surface ($h_{w,i}$ or $v_\infty^{(n_i)}\rho_b c_b$), W m^{-2} K^{-1}
q	Heat flux per unit area, W m^{-2}
$\mathbf{q}$	Heat flux vector per unit area, W m^{-2}
$\mathbf{r}$	Position vector, m
$\mathbf{r}'$	Position vector appearing in Green's functions, m
S	Volumetric heating source, W m^{-3}
t	Time, s
T	Temperature, °C
U	Coupling factor between solid and fluid phases defined by Eq. (11.5), W m^{-3} K^{-1}
v	Velocity component along y (see Fig. 11.4), m s^{-1}
$v^{(n_i)}$	Velocity component normal to the ith permeable boundary surface ($\mathbf{v}\cdot\mathbf{n}_i$), m s^{-1}
$\mathbf{v}$	Velocity vector, m s^{-1}
x, y, z	Rectangular space coordinates, m

Greek symbols

α	Thermal diffusivity, m^2 s
β_n	Constant defined by Eq. (11.71a) and related to γ_n, s^{-1}
γ_n	nth eigenvalue, s^{-1}
δ	Dirac delta function, m^{-1} (for local source) and s^{-1} (for instantaneous source)
ε	Porosity
θ	Temperature difference ($T - T_{b,in}$), °C
λ_n	Constant defined by Eq. (11.71b) and related to γ_n, s^{-1}
μ	Dynamic viscosity, kg m^{-1} s^{-1}
ρ	Density, kg m^{-3}
σ	Constant defined by Eq. (11.29), s^{-1}
τ	Phase-lag time. Also, time appearing in Green's functions, s
ϕ	Pulsatile parameter (dimensionless) defined in Eq. (11.17d)
ψ	nth function of time (dimensionless) accounting for the phase lags in the DPL equation
Ψ	Transformed temperature difference for tissue phase ($\Psi = \theta_s e^{\sigma t}$), °C
ω	Perfusion rate, m^3 s^{-1} m^{-3} tissue
ω_b	Mass perfusion rate ($\rho_b\omega$), kg s^{-1} m^{-3} tissue

Subscripts

0	Boundary at $x=0$
∞	Undisturbed fluid
b	Blood (fluid phase)
B	Boundary (due to boundary conditions)
eff	Effective
h	Heating therapy
i	ith boundary surface
I	Initial (due to initial condition)
IR	Irradiance (due to laser beam)
in	Initial ($t=0$)
L	Boundary at $x=L_s$
m	Metabolism
n	Counting integer for eigenvalues
q	Heat flux
rev	Representative elementary volume
s	Tissue (solid phase)
S	Source (due to heat source)
T	Temperature gradient
w	Wall

Superscripts

DPL	Dual-phase lag
b	Blood (fluid phase)
n_i	Direction of the outward pointing unit vector $\mathbf{n}_i$
s	Tissue (solid phase)

Further symbols

$\langle\ \rangle$	Volume-averaged
$\langle\ \rangle^j$	Intrinsic volume-averaged, $j=b$ (blood or fluid) or s (tissue or solid)
∇	Vector differential operator in the $\mathbf{r}$ space
∇_0	Vector differential operator in the $\mathbf{r}'$ space
∇^2	Laplacian (differential operator given by the divergent of the gradient of a scalar function)

CHAPTER

12

Microstructural Influences on Growth and Transport in Biological Tissue—A Multiscale Description

L. Irons, J. Collis†, R.D. O'Dea*,#*

*Centre for Mathematical Medicine and Biology, School of Mathematical Sciences, University of Nottingham, Nottingham, United Kingdom †School of Mathematical Sciences, University of Nottingham, Nottingham, United Kingdom

12.1 INTRODUCTION

A fundamental barrier to advancing the understanding of biological tissue growth lies in its inherently multiscale nature: interactions between processes acting at disparate scales can profoundly influence the emergent dynamics. A unified description of such phenomena requires reconciling insight obtained by theoretical or experimental study at one scale with observations at another. For example, diseases that manifest at the organ scale often arise through the interaction of microscopic events at the cellular scale; moreover, the resulting macroscale changes influence the microscopic dynamics.

This complexity makes multiscale experimental observation difficult so that additional support by theoretical means is necessary. However, the theoretical challenges are substantial: to comprehensively model the diverse contributory processes (from intra- and inter-cell signaling dynamics and cell biomechanics, to tissue-level transport patterning and mechanics) leads inevitably to formulations that are analytically and computationally intractable. As a result, a significant area of research concentrates on developing various techniques for coupling different dynamics occurring on multiple scales (see, e.g. Alarcón et al., 2006; Frieboes et al., 2007; Macklin et al., 2009; Van Leeuwen et al., 2009; Osborne et al., 2010; Powathil et al., 2015, and citations therein).

An approach that addresses this is to seek 'effective' descriptions (obtained via multiple-scale asymptotic homogenization) that provide a coarse-scale (macroscale) description of tissue dynamics whilst still incorporating aspects of the microscale physics. Such meth-

Chapter coordinator.

ods have been widely used in industrial/applied settings in order to provide descriptions of porous and poroelastic materials (Burridge and Keller, 1981; Mei and Auriault, 1991; Lee and Mei, 1997). More recently, these methods have been used to analyze problems in biological settings to consider motion and pattern formation in cell monolayers (Fozard et al., 2010; O'Dea and King, 2011, 2012), transport in biological tissues (Matzavinos and Ptashnyk, 2015) and the regulated transport of auxin in plants (Band and King, 2012; Chavarría-Krauser and Ptashnyk, 2010). Of specific relevance to the current work are recent studies that consider tumor growth and tissue engineering applications, a key feature of both being the interplay between growth, transport, microstructure and mechanics. For example, the efficacy of chemotherapeutic treatments is crucially dependent on drug transport through the microvasculature. This inspired Shipley and Chapman (2010) to consider the homogenization of flow and transport within a solid tumor, and to form a 'double porous medium' description at the macroscale. Their description allows for the microscopic structure and the permeability of the leaky vasculature to be explicitly considered. The dependence of macroscale transport on the structure of the microvasculature within this model was subsequently explored in Penta and Ambrosi (2015), highlighting the crucial influence of vascular tortuosity on the anti-tumor drug delivery. A simple description of tissue growth in a multiscale context was analyzed in O'Dea et al. (2015), in which a macroscale description of growth and transport in a porous material was obtained. Growth was represented by nutrient-limited microscale accretion of a rigid solid structure. Similar methods were used in Penta et al. (2014) to describe a growing poroelastic material in the absence of nutrient. A more sophisticated underlying growth model employing a multiphase formulation has been considered in Collis et al. (2016a, 2016b).

In this chapter we extend the analyses of Shipley and Chapman (2010), Penta and Ambrosi (2015), O'Dea et al. (2015) by considering the nutrient-limited appositional growth of a (microscale) porous medium as a model system that couples flow, nutrient transport and growth processes. The macroscale model that we derive comprises a double porous medium description for the flow (with additional contributions pertaining to growth), and an advection–reaction equation governing nutrient transport and consumption. This macroscale description is coupled to the underlying microscale dynamics via Laplace and Stokes problems defined in a prototypical 'unit cell'. The cell problems characterize the material's microstructure and flow physics—the solution to the Laplace problem influences flow in the porous tissue region, and the Stokes problem governs the flow of the surrounding fluid (such as the flow within the microvasculature). Suitable slip and leakage conditions are enforced between the porous tissue and fluid domains at the microscale. Slip in the microscale flow problem was neglected in the closely-related studies of Shipley and Chapman (2010) and Penta and Ambrosi (2015) on the basis of parameter estimates from the literature. Here, however, we choose to retain this feature in view of the strong variation of relevant parameters with tissue type (Netti et al., 2000), the inherent difficulty of parameterizing accurately models of this type (Zhang et al., 2000), and because of our aim to provide an exposition of multiscale analyses of rather broad relevance to tissue growth processes. Our numerical experiments identify how macroscale dynamics are influenced by the slip within the microstructure, as well as the growth and tortuosity of the microstructure. In particular, we highlight that for a large range of slip parameters (ranging over several orders of magnitude), the macroscale permeability depends strongly on both microscale tortuosity and slip.

The remainder of this chapter is organized as follows. In Section 12.2, we introduce the governing equations describing microscale flow, transport, and growth. In Section 12.3 we derive a corresponding macroscale description via multiscale homogenization. In Section 12.4, numerical results are presented, indicating in detail the dependence of the microscale permeability on vascular tortuosity and slip and indicating their influence (together with growth) on the macroscale model behavior. Finally, we summarize our results and indicate future directions in Section 12.5.

12.2 FORMULATION: NUTRIENT-LIMITED MICROSCALE GROWTH OF A POROUS MEDIUM

In this section we present the microscale model describing flow, transport, and growth. The underlying formulation is due to Shipley and Chapman (2010) and so we forgo a detailed discussion and motivation; in what follows it suffices to state the governing equations, together with a brief description of the underlying model assumptions. We highlight that the formulation that we study here is to be viewed as a model system within which we may highlight the importance of flow and transport phenomena in growing structures, and, as such, the presentation is intentionally generic. However, this model is broadly applicable to the study of solid tumors, in which the transport between the leaky neovasculature and interstitium (comprising cells and extracellular space) is of particular interest in the context of the efficacy of chemotherapy. The reader is directed to Shipley and Chapman (2010), Penta and Ambrosi (2015) and references therein for comprehensive motivation.

We consider a representation of biological tissue as an idealized porous structure, represented as a highly connected material with spatially periodic microstructure that comprises a growing 'tissue' domain, saturated with a viscous Newtonian fluid. As in Shipley and Chapman (2010), the tissue domain is itself assumed to be an isotropic porous medium and fluid flow within it is described by Darcy's law. We note in passing that such a choice is supported by related multiscale analyses of the microscale tissue domain itself (see, e.g. Mei and Auriault, 1989 and O'Dea et al., 2015), but such concerns are beyond the scope of this chapter. A passive diffusible nutrient is transported by the fluid and regulates microscale appositional tissue growth, which is modeled by a phase transition on the singular surface separating the two domains. This idealized description of growth is appropriate in the 'thin-rim' (or 'fast-consumption') limit (King and Franks, 2007) whereby nutrient consumption in the tissue dominates over diffusive transport. In particular, denoting the characteristic microscale lengthscale, tissue diffusivity, and nutrient consumption by l^*, D^* and λ^*, respectively, our description applies when $\beta = D^*/l^{*2}\lambda^* \ll 1$. In this case, nutrient consumption and associated growth occurs in a boundary layer of width $\mathcal{O}(\sqrt{\beta})$ with the remainder of the tissue remaining nutrient starved.

The microscale domain is denoted Ω, with boundary $\partial\Omega$, and has characteristic length l^*. Throughout, asterisks distinguish dimensional quantities from their dimensionless equivalents. We further partition Ω into a fluid domain Ω_f, and a tissue domain Ω_t, which is itself modeled as a porous material. Boundaries of the two domains are denoted by $\partial\Omega_f$ and $\partial\Omega_t$, which may include parts of the exterior boundary $\partial\Omega$ and the (rigid) interface between Ω_f

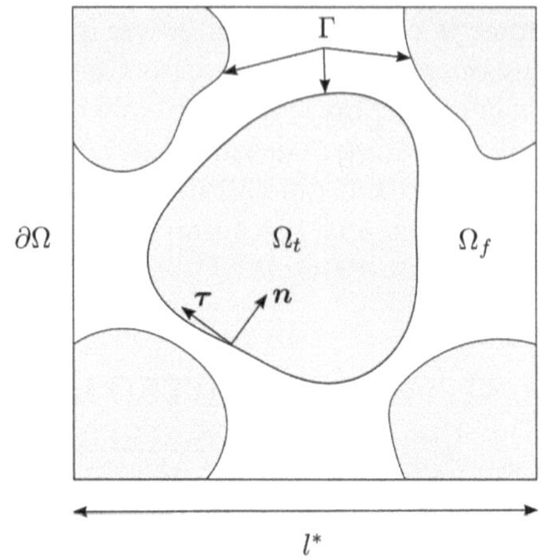

FIGURE 12.1 Schematic diagram of a single unit of the periodic microscale problem. Diagram adapted from Collis et al. (2016b, Fig. 2.1). The fluid and porous tissue domains are denoted by Ω_f and Ω_t, respectively, with interface Γ. The tangent(s) and inward normal to Ω_t on Γ are denoted $\boldsymbol{\tau}$ and $\boldsymbol{n}$.

and Ω_t, denoted Γ. The tangent(s) and inward normal to Ω_t on Γ are denoted $\boldsymbol{\tau}$ and $\boldsymbol{n}$. Fig. 12.1 illustrates the representation of the combined material Ω (that forms our representation of a tissue) via such a periodic cell geometry. The total porosity of the combined material is then given by $\Phi = |\Omega_f|/|\Omega|$. The macroscale characteristic length of the tissue is denoted L^*, and we assume that the lengthscales are well-separated. Correspondingly, we introduce a dimensionless parameter ε such that

$$0 < \varepsilon = \frac{l^*}{L^*} \ll 1. \tag{12.1}$$

12.2.1 Microscale Governing Equations and Boundary Conditions

We denote the fluid velocity and pressure by $\boldsymbol{v}^*$ and p^* (with restriction to Ω_f or Ω_t by the subscript f or t, respectively), and the nutrient concentration by c^*. The flow and nutrient transport in Ω_f, in dimensional form, are then governed by

$$\rho_f^* \left(\frac{\partial \boldsymbol{v}_f^*}{\partial t^*} + (\boldsymbol{v}_f^* \cdot \nabla^*) \boldsymbol{v}_f^* \right) = -\nabla^* p_f^* + \mu^* \nabla^{*2} \boldsymbol{v}_f^*, \tag{12.2}$$

$$\nabla^* \cdot \boldsymbol{v}_f^* = 0, \tag{12.3}$$

$$\frac{\partial c^*}{\partial t^*} + \boldsymbol{v}_f^* \cdot \nabla^* c^* = D^* \nabla^{*2} c^*, \tag{12.4}$$

in which ρ_f^* and μ^* are the density and dynamic viscosity of the fluid, respectively, D^* denotes the nutrient diffusivity, and t^* denotes time.

In the porous tissue domain, Ω_t, the flow is governed by Darcy's law so that

$$\boldsymbol{v}_t^* = -\frac{k^*}{\mu^*}\nabla^* p_t^*, \tag{12.5}$$

$$\nabla^* \cdot \boldsymbol{v}_t^* = 0, \tag{12.6}$$

for isotropic tissue permeability k^*.

The absence of nutrient in Ω_t is consistent with the thin-rim/fast-consumption limit that we consider; however, the inclusion of other diffusible substances that are present in both domains (and for which our boundary-consumption assumptions need not apply) is straightforward (see, e.g. Shipley and Chapman, 2010; Penta and Ambrosi, 2015; Collis et al., 2016b). To represent appositional growth in this limit, we consider mass conservation on the interface $\boldsymbol{x}^* \in \Gamma$, thereby obtaining the following conditions describing fluid exchange between Ω_f and Ω_t, nutrient consumption, and associated interfacial growth:

$$c^*\left(\boldsymbol{v}_f^* - \boldsymbol{v}_\Gamma^*\right)\cdot \boldsymbol{n} - D^*\nabla^* c^* \cdot \boldsymbol{n} = Q^*, \tag{12.7}$$

$$\left(\boldsymbol{v}_f^* - \boldsymbol{v}_\Gamma^*\right)\cdot \boldsymbol{n} - \frac{S^*(Q^*)}{\rho_f^*} = \left(\boldsymbol{v}_t^* - \boldsymbol{v}_\Gamma^*\right)\cdot \boldsymbol{n} - \frac{S^*(Q^*)}{\rho_s^*} = T^*\left(p_f^* - p_t^*\right), \tag{12.8}$$

$$\boldsymbol{v}_\Gamma^* \cdot \boldsymbol{n} = -\frac{S^*(Q^*)}{\phi_s \rho_s^*}, \tag{12.9}$$

in which $\boldsymbol{v}_\Gamma^* \cdot \boldsymbol{n}$ is the normal velocity of the interface (whose motion arises due to growth), T^* is the hydraulic permeability of the interface, and ρ_s^* and ϕ_s are the density and volume fraction of the solid constituents within the porous tissue domain, Ω_t. Note that ρ_s^* and ϕ_s are characteristic of Ω_t only—the porous domain Ω_t has an intrinsic porosity denoted $\phi = 1 - \phi_s$; cf. the combined porosity, Φ, which arises from considering $\Omega = \Omega_f \cup \Omega_t$.

Eq. (12.7) describes the nutrient flux, modified to account for the interfacial movement and nutrient consumption Q^*, and in (12.8) the function $S^*(Q^*)$ denotes mass conversion from fluid to solid at a rate dependent upon nutrient consumption. This process leads to growth of Ω_t, embodied by the normal interfacial motion given by (12.9). Eqs. (12.8) and (12.9) arise from considering the porous material in Ω_t to be a two phase mixture of a fluid and an immobile solid (with volume fraction ϕ_s), while Ω_f contains only fluid, and with phase change occurring on the interface between the two domains (see, e.g. Ateshian, 2011). Following Shipley and Chapman (2010), the pressure-driven fluid exchange between the two domains is assumed to obey a membrane law.

The functional forms for S^* and Q^* representing growth and consumption are to be constitutively specified. For this example, we adopt a simple growth and consumption model (following O'Dea et al., 2015) and assume that nutrient uptake occurs at a rate proportional to the nutrient concentration at the interface, coupled to interfacial growth:

$$Q^* = R^* c^* \quad \text{and} \quad S^* = \alpha^* Q^*, \tag{12.10}$$

where the constant R^* regulates the rate of nutrient consumption at the interface Γ, and α^* dictates the influence of nutrient uptake upon subsequent tissue growth.

With (12.2) being of higher order than (12.5), a further condition is required to determine the fluid velocity in Ω_f. As such, we constrain the flow along the interface Γ with the Beavers–Joseph–Saffman condition

$$\left[(\boldsymbol{n}\cdot\nabla^*)\boldsymbol{v}_f^*\right]\cdot\boldsymbol{\tau} = -\frac{\gamma}{\sqrt{k^*}}\boldsymbol{v}_f^*\cdot\boldsymbol{\tau}, \tag{12.11}$$

where γ is a dimensionless constant dependent on the interfacial properties. In the context of tumor growth, this will reflect microvascular rheology including, say, the influence of the endothelial glycocalyx (Shipley and Chapman, 2010). Lastly, to complete the system (and in keeping with our assumption of strong scale separation between micro- and macro-scale descriptions), we require that all solutions are periodic on the microscale domain Ω. Macroscale variation in Ω is of course permitted, thereby enabling spatial variation of material structure within this formulation.

We highlight that despite the simplified growth and consumption model that we have adopted, the dynamics in each domain remain fully-coupled: the flow coupling between Ω_t and Ω_f embodied by (12.8) and (12.11) influences the transport of nutrient to Γ and thereby its growth.

12.2.2 Non-dimensionalization

We non-dimensionalize as follows:

$$\left.\begin{array}{llll} \boldsymbol{x}^* = l^*\boldsymbol{x}, & \boldsymbol{v}^* = V^*\boldsymbol{v}, & \boldsymbol{v}_\Gamma^* = V^*\boldsymbol{v}_\Gamma, & c^* = C^*c, \\ p^* = \dfrac{\mu^* L^* V^*}{l^{*2}}p, & t^* = \dfrac{L^*}{V^*}t, & Q^* = \dfrac{l^* C^* V^*}{L^*}Q, & S^* = \dfrac{l^*\rho_s^* V^*}{L^*}S \end{array}\right\} \tag{12.12}$$

in which V^* and C^* are characteristic macroscale flow velocity and nutrient concentration scales.

Of particular significance is the timescale of macroscale advection, L^*/V^*, and in keeping with this we scale the growth and consumption terms with the dimensionless parameter, ε, of Eq. (12.1). A key impact of these choices is to ensure that the leading order problems that we develop are steady and that the free interface is quasi-stationary, thereby simplifying the analysis. A treatment of $\mathcal{O}(1)$ growth dynamics, through a timescale based on microscale growth, is possible (see, e.g. Penta et al., 2014) but leads to a computationally-intensive description in which a separate microscale problem must be solved at each macroscale point due to microscale growth. In the problem we develop below, we seek instead to obtain a more tractable description which permits coupling between microscale and macroscale dynamics.

In their dimensionless form, the equations governing flow and transport in Ω_f are given by

$$\varepsilon^2 \mathrm{Re}\left(\varepsilon\frac{\partial \boldsymbol{v}_f}{\partial t} + (\boldsymbol{v}_f \cdot \nabla)\boldsymbol{v}_f\right) = -\nabla p_f + \varepsilon \nabla^2 \boldsymbol{v}_f, \tag{12.13}$$

$$\nabla \cdot \boldsymbol{v}_f = 0, \tag{12.14}$$

$$\varepsilon\frac{\partial c}{\partial t} + \boldsymbol{v}_f \cdot \nabla c = \frac{1}{\mathrm{Pe}}\nabla^2 c. \tag{12.15}$$

In Ω_t the flow is described by

$$\varepsilon \boldsymbol{v}_t = -k\nabla p_t, \tag{12.16}$$

$$\nabla \cdot \boldsymbol{v}_t = 0. \tag{12.17}$$

Finally, the boundary conditions on the interface Γ are

$$c\left(\boldsymbol{v}_f - \boldsymbol{v}_\Gamma\right)\cdot \boldsymbol{n} - \frac{1}{\mathrm{Pe}}\nabla c \cdot \boldsymbol{n} = \varepsilon Q, \tag{12.18}$$

$$(\boldsymbol{v}_f - \boldsymbol{v}_\Gamma)\cdot \boldsymbol{n} - \varepsilon\rho S(Q) = (\boldsymbol{v}_t - \boldsymbol{v}_\Gamma)\cdot \boldsymbol{n} - \varepsilon S(Q) = \varepsilon T(p_f - p_t), \tag{12.19}$$

$$\left[(\boldsymbol{n}\cdot\nabla)\boldsymbol{v}_f\right]\cdot \boldsymbol{\tau} = -\theta \boldsymbol{v}_f \cdot \boldsymbol{\tau}, \tag{12.20}$$

where the dimensionless boundary velocity is

$$\boldsymbol{v}_\Gamma \cdot \boldsymbol{n} = -\varepsilon\frac{S(Q)}{\phi_s}, \tag{12.21}$$

and the Reynolds number and Péclet number, appropriate for macroscale advective flow, are defined by

$$\mathrm{Re} = \frac{\rho_f^* V^* L^*}{\mu^*} \quad \text{and} \quad \mathrm{Pe} = \frac{V^* l^*}{D^*}. \tag{12.22}$$

The remaining dimensionless groupings appearing in (12.13)–(12.20) are then given by

$$k = \frac{k^*}{l^{*2}}, \quad \rho = \frac{\rho_s^*}{\rho_f^*}, \quad T = \frac{T^*\mu^* L^{*2}}{l^{*3}}, \quad \text{and} \quad \theta = \frac{\gamma l^*}{\sqrt{k^*}}, \tag{12.23}$$

in which k describes the isotropic permeability of the porous tissue, ρ the relative density of the solid and fluid constituents, and T and θ govern the permeability (in terms of leakage) and slip properties of the interface, respectively. Motivated by biologically relevant parameters (investigated in detail in Shipley and Chapman, 2010), a suitable permeability scaling results in the interfacial pressure-driven flux being relegated to $\mathcal{O}(\varepsilon)$ in (12.19). Such a choice additionally ensures that the problem we consider remains quasi-steady, consistent with our aim to obtain a simplified macroscale description.

12.3 MULTIPLE SCALES ANALYSIS

With the aim of including microscopic structure and growth into a macroscale description, we employ a multiple-scales method to (12.13)–(12.21). Consistent with our assumption of well-separated spatial scales, we define dimensionless micro- and macroscale coordinates $\boldsymbol{x}$ and $\boldsymbol{X}$, respectively, related by $\boldsymbol{x} = \varepsilon^{-1}\boldsymbol{X}$. Under such a choice, the first and second order spatial derivatives are transformed as

$$\nabla = \nabla_{\boldsymbol{x}} + \varepsilon \nabla_{\boldsymbol{X}} \quad \text{and} \quad \nabla^2 = \nabla_{\boldsymbol{x}}^2 + 2\varepsilon \nabla_{\boldsymbol{X}} \cdot \nabla_{\boldsymbol{x}} + \varepsilon^2 \nabla_{\boldsymbol{X}}^2. \tag{12.24}$$

We now consider two-scale expansions of the form

$$\psi(\boldsymbol{x}, \boldsymbol{X}, t; \varepsilon) = \psi^{(0)}(\boldsymbol{x}, \boldsymbol{X}, t) + \varepsilon \psi^{(1)}(\boldsymbol{x}, \boldsymbol{X}, t) + \cdots \tag{12.25}$$

to obtain a hierarchy of problems in increasing orders of ε, for which the limiting behavior as $\varepsilon \to 0$ is of interest. In what follows it suffices to consider the problems at $\mathcal{O}(1)$ and $\mathcal{O}(\varepsilon)$, from which a unit cell problem in Ω may be derived that dictates the influence of the microstructure on the flow dynamics, and which parameterizes a macroscale description of the coupled flow and transport dynamics. Central to the construction of such a macroscale description from an underlying description are the following extrinsic volume averages over the relevant microscale domain:

$$\langle \psi \rangle_f = \frac{1}{|\Omega|} \int_{\Omega_f} \psi \, dV, \qquad \langle \psi \rangle_t = \frac{1}{|\Omega|} \int_{\Omega_t} \psi \, dV, \quad \text{and} \quad \langle \psi \rangle_\Gamma = \frac{1}{|\Omega|} \int_{\Gamma} \psi \, dS.$$

12.3.1 Microscale Flow and Transport

The microscale $\mathcal{O}(1)$ flow in each domain is governed by

$$\boldsymbol{0} = -\nabla_{\boldsymbol{x}} p_f^{(0)} \qquad \forall \boldsymbol{x} \in \Omega_f, \tag{12.26}$$

$$0 = \nabla_{\boldsymbol{x}} \cdot \boldsymbol{v}_f^{(0)} \qquad \forall \boldsymbol{x} \in \Omega_f, \tag{12.27}$$

$$\boldsymbol{0} = -k \nabla_{\boldsymbol{x}} p_t^{(0)} \qquad \forall \boldsymbol{x} \in \Omega_t, \tag{12.28}$$

$$0 = \nabla_{\boldsymbol{x}} \cdot \boldsymbol{v}_t^{(0)} \qquad \forall \boldsymbol{x} \in \Omega_t, \tag{12.29}$$

and the $\mathcal{O}(\varepsilon)$ flow by

$$\boldsymbol{0} = -\nabla_{\boldsymbol{X}} p_f^{(0)} - \nabla_{\boldsymbol{x}} p_f^{(1)} + \nabla_{\boldsymbol{x}}^2 \boldsymbol{v}_f^{(0)} \qquad \forall \boldsymbol{x} \in \Omega_f, \tag{12.30}$$

$$0 = \nabla_{\boldsymbol{X}} \cdot \boldsymbol{v}_f^{(0)} + \nabla_{\boldsymbol{x}} \cdot \boldsymbol{v}_f^{(1)} \qquad \forall \boldsymbol{x} \in \Omega_f, \tag{12.31}$$

$$\boldsymbol{v}_t^{(0)} = -k \left(\nabla_{\boldsymbol{x}} p_t^{(1)} + \nabla_{\boldsymbol{X}} p_t^{(0)} \right) \qquad \forall \boldsymbol{x} \in \Omega_t, \tag{12.32}$$

$$0 = \nabla_{\boldsymbol{x}} \cdot \boldsymbol{v}_t^{(1)} + \nabla_{\boldsymbol{X}} \cdot \boldsymbol{v}_t^{(0)} \qquad \forall \boldsymbol{x} \in \Omega_t. \tag{12.33}$$

From (12.26) and (12.28) we observe that the leading order pressures, $p_f^{(0)}$ and $p_t^{(0)}$, are constant over the periodic cell. We also highlight that from (12.21) the rigid interface is stationary at leading order $\left(\boldsymbol{v}_\Gamma^{(0)}=\mathbf{0}\right)$, with $\mathcal{O}(\varepsilon)$ motion given by $\boldsymbol{v}_\Gamma^{(1)}\cdot\boldsymbol{n}=-S^{(0)}/\phi_s$, where $S^{(0)}$ denotes the leading order contribution from S. The leading order boundary conditions are therefore given by

$$\left(\boldsymbol{n}\cdot\nabla_x\boldsymbol{v}_f^{(0)}\right)\cdot\boldsymbol{\tau}=-\theta\boldsymbol{v}_f^{(0)}\cdot\boldsymbol{\tau}\qquad\forall\boldsymbol{x}\in\Gamma,\tag{12.34}$$

$$\boldsymbol{v}_f^{(0)}\cdot\boldsymbol{n}=\boldsymbol{v}_t^{(0)}\cdot\boldsymbol{n}=0\qquad\forall\boldsymbol{x}\in\Gamma,\tag{12.35}$$

which are applied on the quasi-stationary interface Γ. At $\mathcal{O}(\varepsilon)$, we obtain:

$$\left[\boldsymbol{n}\cdot\left(\nabla_X\boldsymbol{v}_f^{(0)}+\nabla_x\boldsymbol{v}_f^{(1)}\right)\right]\cdot\boldsymbol{\tau}=-\theta\boldsymbol{v}_f^{(1)}\cdot\boldsymbol{\tau}\qquad\forall\boldsymbol{x}\in\Gamma,\tag{12.36}$$

$$\boldsymbol{v}_f^{(1)}\cdot\boldsymbol{n}=\left(\rho-\frac{1}{\phi_s}\right)S^{(0)}+T\left(p_f^{(0)}-p_t^{(0)}\right)\qquad\forall\boldsymbol{x}\in\Gamma,\tag{12.37}$$

$$\boldsymbol{v}_t^{(1)}\cdot\boldsymbol{n}=\left(1-\frac{1}{\phi_s}\right)S^{(0)}+T\left(p_f^{(0)}-p_t^{(0)}\right)\qquad\forall\boldsymbol{x}\in\Gamma.\tag{12.38}$$

The flows in each microscale domain, governed by (12.26)–(12.38), are now considered in turn. In Ω_f, the linearity of (12.27) and (12.30) together with the microscale invariance of $p_f^{(0)}$ suggest the use of the assumed representation:

$$\boldsymbol{v}_f^{(0)}=-\boldsymbol{K}_f\nabla_X p_f^{(0)},\tag{12.39}$$

$$p_f^{(1)}=-\boldsymbol{a}_f\cdot\nabla_X p_f^{(0)}+\overline{p_f}_1,\tag{12.40}$$

for an unknown permeability tensor $\boldsymbol{K}_f(\boldsymbol{x},\boldsymbol{X})$, vector $\boldsymbol{a}_f(\boldsymbol{x},\boldsymbol{X})$ and function $\overline{p_f}_1=\overline{p_f}_1(\boldsymbol{X},t)$. The function $\overline{p_f}_1(\boldsymbol{X},t)$ can be interpreted as a pressure average, varying only on the macroscale, whilst the vector $\boldsymbol{a}_f$ accounts for microscale $\mathcal{O}(\varepsilon)$ pressure variations within Ω_f. Substitution into (12.27), (12.30), (12.34) and (12.35) yields a cell problem, by which the microscale dependence of $\boldsymbol{v}_f^{(0)}$ and $p_f^{(1)}$ may be quantified. That is, $\boldsymbol{K}_f$ and $\boldsymbol{a}_f$ satisfy

$$\nabla_x\cdot\boldsymbol{K}_f=\mathbf{0}\qquad\forall\boldsymbol{x}\in\Omega_f,\tag{12.41}$$

$$-\nabla_x\boldsymbol{a}_f^T+\nabla_x^2\boldsymbol{K}_f=-\boldsymbol{I}\qquad\forall\boldsymbol{x}\in\Omega_f,\tag{12.42}$$

$$\boldsymbol{K}_f\cdot\boldsymbol{n}=\mathbf{0}\qquad\forall\boldsymbol{x}\in\Gamma,\tag{12.43}$$

$$(\boldsymbol{n}\cdot\nabla_x\boldsymbol{K}_f)\cdot\boldsymbol{\tau}=-\theta\boldsymbol{K}_f\cdot\boldsymbol{\tau}\qquad\forall\boldsymbol{x}\in\Gamma,\tag{12.44}$$

where $\boldsymbol{K}_f$ and $\boldsymbol{a}_f$ are periodic on Ω. To ensure uniqueness, we additionally specify $\langle\boldsymbol{a}_f\rangle_f=\mathbf{0}$. The above equations form a Stokes flow boundary value problem to be solved for a specific periodic cell geometry, through which structural features of the microscale cell, as well as the slip properties of the interface, are imparted to the flow.

A corresponding cell problem in Ω_t is obtained by substituting (12.32) into (12.29) and (12.35), and appealing to the microscale invariance of $p_t^{(0)}$, yielding

$$\nabla_x^2 p_t^{(1)} = 0 \qquad \forall \boldsymbol{x} \in \Omega_t, \tag{12.45}$$

$$\nabla_x p_t^{(1)} \cdot \boldsymbol{n} = -\nabla_X p_t^{(0)} \cdot \boldsymbol{n} \qquad \forall \boldsymbol{x} \in \Gamma, \tag{12.46}$$

for which the associated solvability condition is trivially satisfied. Eqs. (12.45) and (12.46) admit an ansatz of identical form to (12.40), providing the following microscale cell problem:

$$\nabla_x^2 \boldsymbol{a}_t = \boldsymbol{0} \qquad \forall \boldsymbol{x} \in \Omega_t, \tag{12.47}$$

$$(\nabla_x \boldsymbol{a}_t) \cdot \boldsymbol{n} = \boldsymbol{I} \cdot \boldsymbol{n} \qquad \forall \boldsymbol{x} \in \Gamma, \tag{12.48}$$

with the periodic function $\boldsymbol{a}_t$ having corresponding meaning in Ω_t to that of $\boldsymbol{a}_f$, and adhering to $\langle \boldsymbol{a}_t \rangle_t = \boldsymbol{0}$.

The leading order problem for the nutrient transport is given by

$$\boldsymbol{v}_f^{(0)} \cdot \nabla_x c^{(0)} - \frac{1}{\mathrm{Pe}} \nabla_x^2 c^{(0)} = 0 \qquad \forall \boldsymbol{x} \in \Omega_f, \tag{12.49}$$

together with the condition

$$c^{(0)} \boldsymbol{v}_f^{(0)} \cdot \boldsymbol{n} - \frac{1}{\mathrm{Pe}} \nabla_x c^{(0)} \cdot \boldsymbol{n} = 0 \qquad \forall \boldsymbol{x} \in \Gamma, \tag{12.50}$$

from which, together with (12.35), a straightforward argument shows that $c^{(0)} = c^{(0)}(\boldsymbol{X}, t)$. In view of this, the $\mathcal{O}(\varepsilon)$ equations comprise

$$\frac{\partial c^{(0)}}{\partial t} + \boldsymbol{v}_f^{(0)} \cdot \left(\nabla_X c^{(0)} + \nabla_x c^{(1)} \right) = \frac{1}{\mathrm{Pe}} \nabla_x^2 c^{(1)} \qquad \forall \boldsymbol{x} \in \Omega_f, \tag{12.51}$$

and

$$c^{(0)} \boldsymbol{v}_f^{(1)} \cdot \boldsymbol{n} - \frac{1}{\mathrm{Pe}} \left(\nabla_x c^{(1)} + \nabla_X c^{(0)} \right) \cdot \boldsymbol{n} = Q^{(0)} - \frac{S^{(0)} c^{(0)}}{\phi_s} \qquad \forall \boldsymbol{x} \in \Gamma, \tag{12.52}$$

together with the requirement that $c^{(1)}$ is periodic on Ω. The expansion of Eq. (12.21) has been exploited to eliminate the interfacial velocity in (12.52).

As is typical in these kinds of analyses (Shipley and Chapman, 2010; O'Dea et al., 2015; Collis et al., 2016b), the microscale description of nutrient transport is underdetermined; however, as we show below, the flow and transport problems developed in this section may be exploited to obtain a description of coupled flow and transport dynamics at the macroscale.

12.3.2 Macroscale Flow and Transport

The macroscale flow contributions from Ω_f and Ω_t are obtained by averaging (12.39) and (12.32) over the appropriate domains, yielding

$$\left\langle \boldsymbol{v}_f^{(0)} \right\rangle_f = -\langle \boldsymbol{K}_f \rangle_f \nabla_X p_f^{(0)}, \tag{12.53}$$

$$\left\langle \boldsymbol{v}_t^{(0)} \right\rangle_t = -\boldsymbol{K}_t \nabla_X p_t^{(0)}, \tag{12.54}$$

with $\boldsymbol{K}_t = k\left((1-\Phi)\boldsymbol{I} - \langle (\nabla_x \boldsymbol{a}_t)^T \rangle_t\right)$. We observe in each case a macroscale Darcy flow with permeability determined from the underlying microstructural and slip properties of the tissue, and obtained from the cell problems (12.41)–(12.44) and (12.47)–(12.48).

Equations governing the leading order macroscale pressures are obtained by averaging (12.31) and (12.33) over the relevant domains; in view of (12.37) and (12.38) we obtain

$$\nabla_X \cdot \left(\langle \boldsymbol{K}_f \rangle_f \nabla_X p_f^{(0)} \right) = \left(\rho - \frac{1}{\phi_s} \right) \left\langle S^{(0)} \right\rangle_\Gamma + \frac{\bar{A}T}{|\Omega|} \left(p_f^{(0)} - p_t^{(0)} \right), \tag{12.55}$$

$$\nabla_X \cdot \left(\boldsymbol{K}_t \nabla_X p_t^{(0)} \right) = \left(\frac{1}{\phi_s} - 1 \right) \left\langle S^{(0)} \right\rangle_\Gamma - \frac{\bar{A}T}{|\Omega|} \left(p_f^{(0)} - p_t^{(0)} \right), \tag{12.56}$$

wherein the constant $\bar{A}$ represents the surface area of the boundaries within the cell, defined by $\bar{A} = \int_\Gamma dS$.

The macroscale nutrient transport is obtained by averaging (12.51) which, via (12.52), (12.37) and the microscale independence of $c^{(0)}$, gives

$$\Phi_f \frac{\partial c^{(0)}}{\partial t} + \left\langle \boldsymbol{v}_f^{(0)} \right\rangle_f \cdot \nabla_X c^{(0)} = \rho c^{(0)} \left\langle S^{(0)} \right\rangle_\Gamma - \left\langle Q^{(0)} \right\rangle_\Gamma + \frac{\bar{A}T}{|\Omega|} \left(c^{(0)} \left(p_f^{(0)} - p_t^{(0)} \right) \right). \tag{12.57}$$

In summary, we obtain a flow description corresponding to a 'double porous medium' with both the porous tissue and surrounding fluid domains behaving as a porous medium, coupled by fluid transport between domains and with additional contributions due to growth. Moreover, Eqs. (12.53) and (12.54) highlight that the macroscale flow is influenced directly by the microscale geometry via the permeability tensors $\langle \boldsymbol{K}_f \rangle_f$ and $\boldsymbol{K}_t$. Although on the timescale of interest considered here, microscale growth enters the problem at $\mathcal{O}(\varepsilon)$, it induces an $\mathcal{O}(1)$ macroscale flow via contributions to the pressure. We remark that, consistent with our representation of growth as a phase transition, the magnitude of the pressure contributions are controlled by both the relative density of fluid and solid components, and the intrinsic porosity (through the solid volume fraction) of the growing porous material. In particular, pressure perturbations arise even in the absence of density differences (cf. O'Dea et al., 2015) since for a given rate of mass exchange, the interfacial motion is increased proportionally according to the solid content of the porous material being created (see (12.21)) thereby requiring additional flow to conserve mass. The final term in (12.55) and (12.56) acts to equalize the pressure between domains, with a rate influenced by the microscale boundary geometry and resulting surface area.

The nutrient transport at the macroscale, which determines the growth, is given by an advection–reaction equation, with reaction terms arising due to growth, consumption and pressure-driven fluid exchange between the domains. As in O'Dea et al. (2015), depending on the model parameter values, interfacial growth may provide an apparent source of nutrient in the fluid domain; in particular, the density and volume fraction of the solid component of Ω_t play a crucial role.

We reiterate the similarity of the underlying model to that of Shipley and Chapman (2010) and Penta and Ambrosi (2015), whose macroscale formulation for flow and pressure is therefore recovered in the absence of growth and associated consumption. However, the inclusion of nutrient-limited growth and the resulting boundary movement are important features to consider when modeling processes in biological tissue, and their presence is seen to induce important additional coupling between nutrient transport, fluid flow and pressure fields. Lastly, we note that the simpler growth model of O'Dea et al. (2015) is recovered by treating Ω_t as a solid.

12.4 RESULTS

We now present illustrative numerical simulations of the microscale cell problems (12.41)–(12.44) and (12.47)–(12.48), and the related macroscale description (12.55)–(12.57). Our aim is to indicate how the microscale flow depends on the microstructure (including its slip properties), and its inclusion within macroscale simulations of relevance to biological applications of broad interest.

12.4.1 Microscale Numerical Experiments

In this subsection, we present a series of numerical experiments demonstrating the numerical approximation to the solution of the microscale system derived in Section 12.3.1. As noted earlier, our focus is to consider a model highlighting how growth and transport mechanisms at the microscale influence macroscale dynamics, so we do not consider appropriate techniques associated with the reliable incorporation of relevant physiological microscale data. Instead, we choose to consider a representative, academic geometry chosen in the main as a test case with which to demonstrate suitable means of computing the microscale information that is required to parameterize the macroscale system. However, we additionally provide numerical experiments highlighting in detail the joint role of the microstructure (concentrating on a simple model of tortuosity, of relevance to tumor microvasculature, say) and slip on the microscale dynamics. In this aspect we extend the closely-related study of Penta and Ambrosi (2015), in which no-slip was imposed in place of (12.44).

12.4.1.1 *Parameterization of the Macroscale System*

For computational simplicity, we would, ideally, consider only two-dimensional calculations. However, in order to obtain meaningful average permeabilities to parameterize a macroscale problem, we are required to consider a three-dimensional problem. We there-

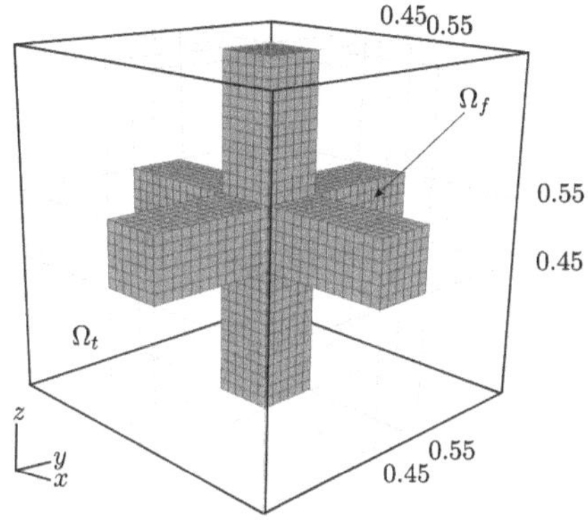

FIGURE 12.2 The periodic microscale geometry, Ω.

fore consider the microscale geometry $\Omega = (0,1)^3$, and further define three subsets of Ω, $\mathcal{A}_1 = (0.45, 0.55) \times (0.45, 0.55) \times (0,1)$, $\mathcal{A}_2 = (0.45, 0.55) \times (0,1) \times (0.45, 0.55)$, and $\mathcal{A}_3 = (0,1) \times (0.45, 0.55) \times (0.45, 0.55)$; and subsequently define $\Omega_f = \bigcup_{i=1}^{3} \mathcal{A}_i$ and $\Omega_t = \Omega \setminus \Omega_f$. From this definition of the geometry we may compute the total porosity $\Phi = 0.028$. A diagram of the microscale geometry is shown in Fig. 12.2.

It is beyond the scope of this chapter to present full details of the discretization of the microscale PDEs. It suffices to note that the choice of discretization follows the motivation provided in Collis et al. (2016a) which considered the numerical solution of a cell problem, arising from a similar multiscale analysis (Collis et al., 2016b), that comprises a pair of Stokes problems coupled across a singular interface by appropriate normal stress and tangential velocity continuity conditions. In Collis et al. (2016a), the authors derive a discontinuous Galerkin (dG) finite element (FE) method, closely following that presented in Toselli (2002); we follow the same process here and derive a suitable dG scheme for the solution of the cell problem for $\boldsymbol{K}_f$. For the cell problem governing $\boldsymbol{a}_t$, we employ a symmetric interior penalty dG scheme as described in Arnold et al. (2001) and the references cited therein. The key features distinguishing the current work from Collis et al. (2016a) are that there is no explicit coupling between the cell problems obtained in Ω_t and Ω_f across the interface Γ, and the slip conditions obtained on Γ take a different form to those presented in Collis et al. (2016b). Moreover, one of the cell problems corresponds to Laplace's equation, rather than a pair of coupled Stokes flows. All numerical experiments were carried out with the AptoFEM finite element package (Antonietti et al., 2015).

Fig. 12.3 shows representative numerical solutions of the microscale system for the fluid permeability $\boldsymbol{K}_f$ obtained with $\theta = 1$. We note that due to the symmetry of Ω and Ω_f, the computation of $K_{f,13}$, $K_{f,23}$ and $K_{f,33}$ is sufficient to specify $\boldsymbol{K}_f$ in Ω_f. Fig. 12.4 indicates the solution of the $\mathcal{O}(\varepsilon)$ pressure correction $\boldsymbol{a}_t$. Once more, the symmetry of Ω and Ω_t implies that computation of $a_{t,3}$ is sufficient to specify $\boldsymbol{a}_t$ in Ω_t.

12.4.1.2 The Influence of Microscale Flow Physics on Macroscale Transport

We now investigate the dependence of the fluid flow on the tissue microstructure and its slip properties. As noted above, in order to employ such microscale calculations to parameterize

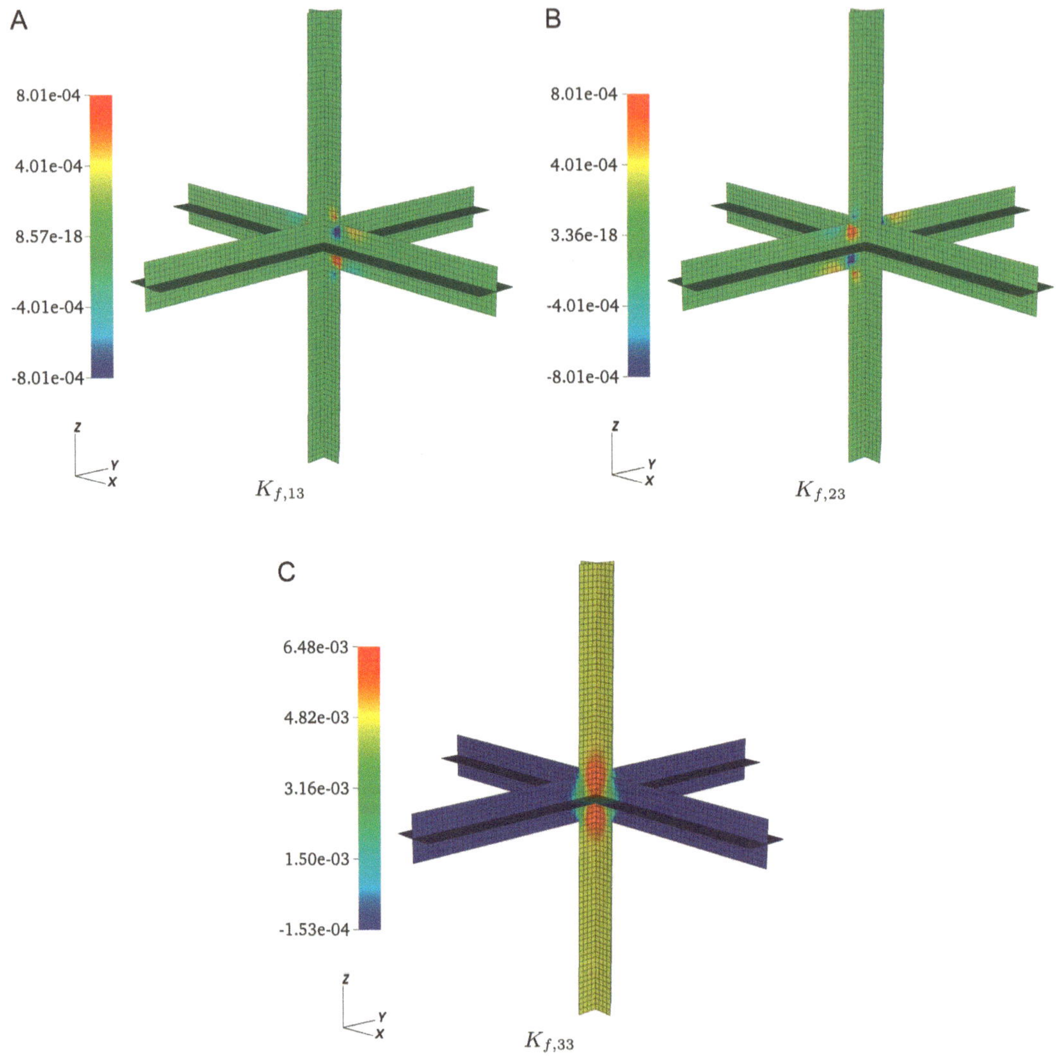

FIGURE 12.3 Components of the isotropic permeability tensor $\boldsymbol{K}_f$.

a macroscale problem, 3D calculations are necessary since Ω_t would otherwise be unconnected. In this subsection, however, our focus is to assess the dependence of the permeability of the material on its microscale properties, concentrating on the flow in Ω_f (which, moreover, is decoupled from Ω_t in the relevant cell problem) and so we consider 2D numerical calculations for computational ease. We anticipate that the results will be qualitatively similar in 3D.

In order to effect this study, we perturb an underlying 2D geometry that is analogous to that shown in Fig. 12.2 (for 3D) so that the centerline of each branch of the fluid domain Ω_f

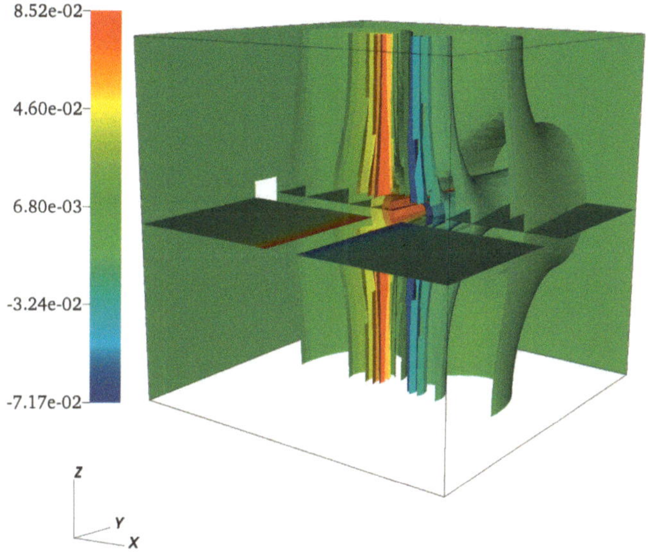

FIGURE 12.4 10 isosurfaces of $a_{t,3}$, indicating the $\mathcal{O}(\varepsilon)$ pressure variation in Ω_t.

is parameterized as

$$\mathscr{C}(z) = A \sin\left(\frac{2\pi \omega z}{\mathscr{L}}\right), \tag{12.58}$$

where A denotes an amplitude, ω denotes a spatial frequency, $\mathscr{L}$ denotes the length of the branch, and z is a parameterization of the centerline of the branch from $\partial\Omega$ to a central square given by $[0.45, 0.55]^2$. The values of the parameters in (12.58) that we employ in numerical simulations are given in Table 12.1.

Fig. 12.5 shows representative simulations highlighting the variation of $\boldsymbol{K}_f$ with the slip parameter θ, for an example 2D tortuous geometry. These results indicate clearly the strong influence of slip and tortuosity on the fluid flow. In particular, for a fixed microstructure, note the transition from a plug-type flow that arises in the case of small θ (Fig. 12.5A) to a Poiseuille-type flow (Fig. 12.5C) in the case of large θ. Quantitative comparison of the results in Fig. 12.5 allows us to observe an order of magnitude reduction in permeability over this range of θ. Fig. 12.6 quantifies this in more detail, highlighting that for small θ, large averaged permeability is obtained, in comparison to that at large θ. In each case, strong dependence on the microscale tortuosity is observed: at $\theta = 10^{-4}$ permeability is increased by two orders of magnitude between the least and most tortuous geometries; at $\theta = 10^5$, the variation is reduced but remains significant (we observe an 8-fold increase). We note, therefore, that the small absolute value changes in permeability observed at large θ are nevertheless expected to have a significant influence on macroscale transport. In each of these regimes, we additionally note that the permeability is insensitive to changes in θ—for $\theta \leqslant 10^{-2}$ and for $\theta \geqslant 10^3$ we obtain constant permeability (to within the discretization error or our computations). Importantly, however, for a large intermediate range of θ extending from $\theta = \mathcal{O}(1)$ over several

TABLE 12.1 Parameter values employed in the generation of Fig. 12.6. Here, $\mathscr{S}$ denotes the length of the parameterized branch centerline, given by $\mathscr{S} = \int_0^{\mathscr{L}} \sqrt{1+\left(\frac{d\mathscr{C}}{dz}\right)^2}\, dz$, an indicator of the tortuosity of a given geometry

ω	A	$\mathscr{S}$
0	–	0.45
1	0.025	0.46
1	0.050	0.50
2	0.050	0.62
3	0.050	0.78

orders of magnitude, the macroscale permeability exhibits very strong dependence on both the microstructure and the slip.

In Penta and Ambrosi (2015), slip was neglected on the basis of parameter estimates from the literature relevant to the tumor application under consideration (colon adenocarcinoma; Boucher et al., 1998); similar arguments were exploited in Shipley and Chapman (2010). These estimates provide $\theta = \mathcal{O}(10^3)$ in the current context, corresponding to the approximate Poiseuille-flow regime identified above. Conversely, estimates relating to murine mammary carcinoma given in Netti et al. (2000) yield $\theta \sim 400$ for which we observe significant θ-dependence. More generally, accurate parametrization of such models is highly problematic. In particular, the value of hydraulic conductivity κ^* (from which we compute $k^* = \kappa^* \mu^*$ in (12.23)) reported in the literature varies over several orders of magnitude depending on, for example, tissue type and deformation (Netti et al., 2000; Zhang et al., 2000), with the latter being of particular relevance in the context of tissue growth modeling.

Lastly, we note that θ (and its equivalent in Shipley and Chapman, 2010; Penta and Ambrosi, 2015) includes a dimensionless parameter γ that reflects interfacial properties, such as microvascular rheology in the context of tumor growth. In the absence of suitable data, this is set to unity in the cited studies. Our results indicate that this rather arbitrary choice could be significant, in that any reduction may place θ in the range for which slip dependence has a non-trivial effect on the macroscale dynamics.

12.4.2 Macroscale Dynamics

In this section, we return to the isotropic three dimensional microstructure depicted in Fig. 12.2 and present numerical experiments to demonstrate the solution of the macroscale system defined in (12.53)–(12.57). Representative parameter values, boundary and initial data, and geometry are employed, as well as microscale data obtained from the numerical experiments detailed in Section 12.4.1.1.

We emphasize that our focus here is on highlighting the macroscale model dynamics that arise from suitable consideration of coupled multiscale growth and transport processes, and suitable means of their computation. To this end, we consider the following macroscale model set-up (with broad relevance to transport dynamics and growth in vascularized tumors,

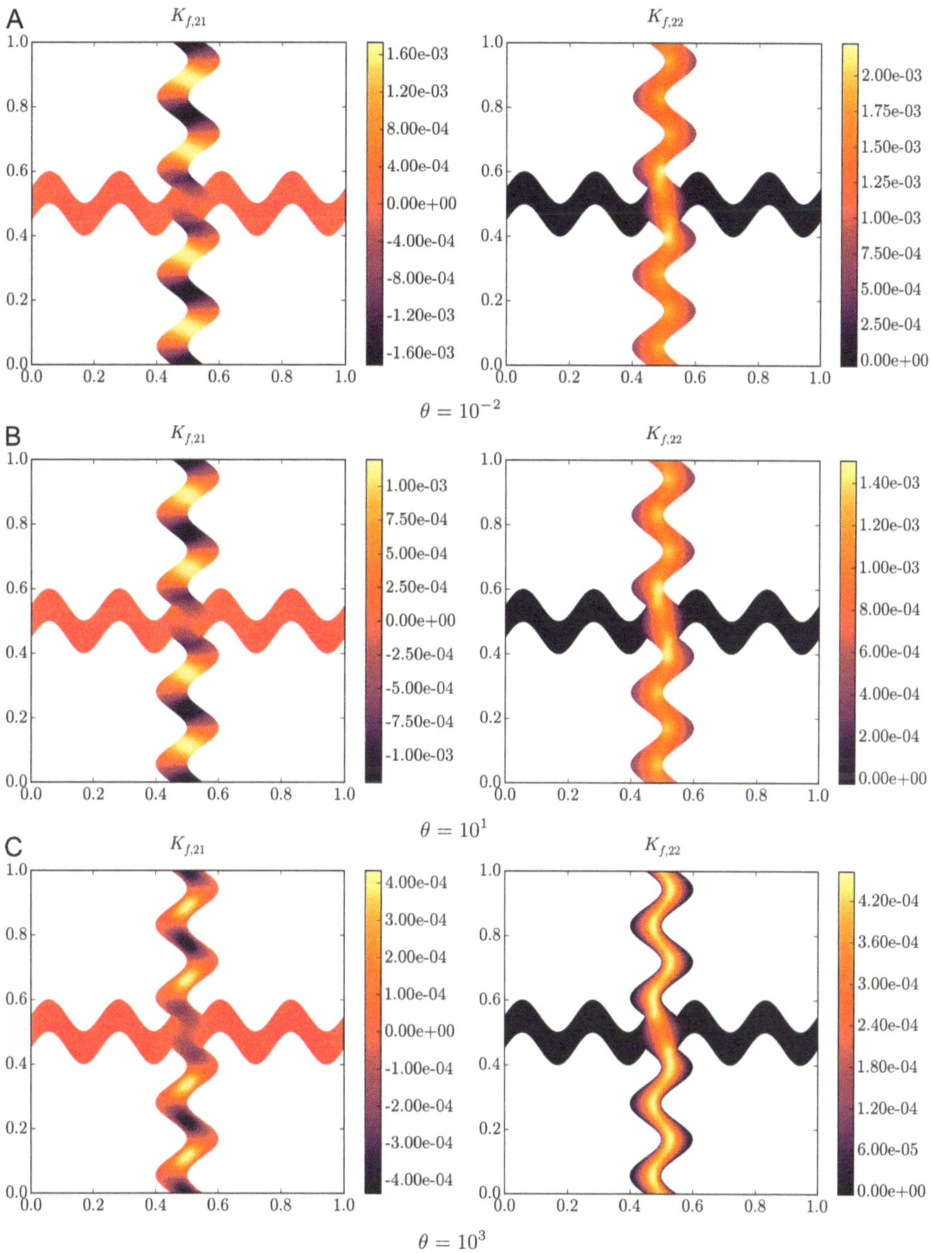

FIGURE 12.5 Illustrative plots of $\mathbf{K}_f$ for different values of the slip coefficient θ within a tortuous microscale geometry characterized by $A = 0.05$ and $\omega = 2$ in (12.58).

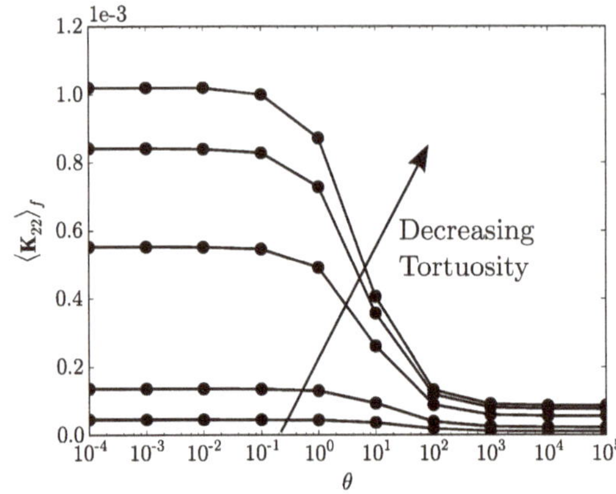

FIGURE 12.6 The variation of $\langle \boldsymbol{K}_{f,22} \rangle_f$ in response to changes in both slip and tortuosity. The microscale geometries employed here are characterized by the parameter values given in Table 12.1.

or tissue engineering applications) and investigate how the transport of a passive nutrient through the system is influenced by the nutrient-dependent microscale growth of the solid component. In addition to the macroscale system of equations given by (12.55)–(12.57) we are required to specify appropriate initial and boundary conditions on the macroscale domain and its boundary, denoted Ω_L and $\partial\Omega_L$, respectively. To this end, we specify boundary conditions

$$p_t^{(0)} = 0 \qquad \forall \boldsymbol{X} \in \partial\Omega_L, \tag{12.59}$$

$$p_f^{(0)} = 1 \qquad \forall \boldsymbol{X} \in \partial\Omega_L, \tag{12.60}$$

$$c^{(0)} = c_\partial(t) \qquad \forall \boldsymbol{X} \in \Lambda^-\left(\left\langle \boldsymbol{v}_f^{(0)} \right\rangle_f\right), \tag{12.61}$$

where $\Lambda^-\left(\left\langle \boldsymbol{v}_f^{(0)} \right\rangle_f\right) \subset \partial\Omega_L$ denotes the inflow portion of the boundary, and $c_\partial(t)$ denotes the boundary data for nutrient concentration. Furthermore, we specify the initial condition for $c^{(0)}$ as

$$c^{(0)}(0) = 0 \qquad \forall \boldsymbol{X} \in \Omega_L.$$

In this work, we choose not to consider specific uptake/consumption parameters or microscale/macroscale geometries appropriate for individual cancer growth or tissue engineering applications. However, the formulation that we have derived, and the computational method we employ, admits such considerations without difficulty.

In the numerical simulations that follow, we employ non-dimensional parameter values as described in Table 12.2, where given the symmetries of the microscale geometry we are able to write

$$\langle \boldsymbol{K}_f \rangle_f = K_f \boldsymbol{I} \qquad \text{and} \qquad \boldsymbol{K}_t = K_t \boldsymbol{I}.$$

TABLE 12.2 Non-dimensional parameters employed in the numerical experiments

Parameter	Value	Description
Φ	2.80×10^{-2}	Total porosity
ϕ_s	9.72×10^{-1}	Volume fraction of solid in (microscale) porous medium
θ	1.00×10^{1}	Slip coefficient
K_f	4.62×10^{-5}	Fluid permeability
K_t	9.70×10^{-1}	Tumor permeability
ρ	1.20×10^{1}	Relative density of fluid and solid constituents
$\bar{A}$	1.08×10^{1}	Surface area of boundary between Ω_f and Ω_t
$\widehat{T}$	1.00×10^{-3}	Permeability of interface
α	1.00×10^{-1}	Rate of surface accretion
R	1.00×10^{-3}	Rate of nutrient consumption

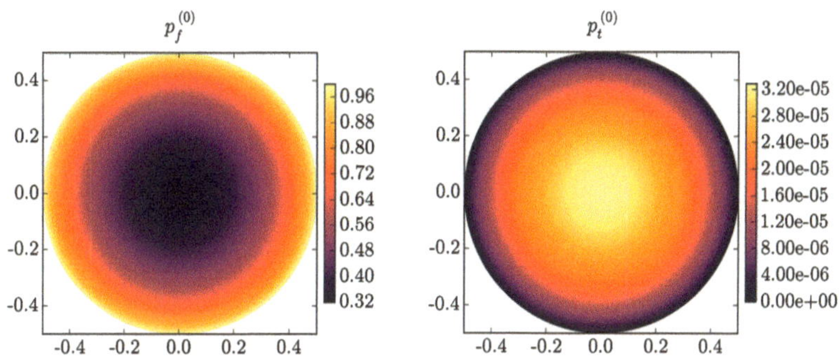

FIGURE 12.7 Numerical approximations of the pressure in the fluid and tissue domains, $p_f^{(0)}$ and $p_t^{(0)}$, at a representative time point $t = 1000$. These correspond to approximations to the solutions of (12.55) and (12.56), subject to (12.60) and (12.59), respectively.

In order to compute numerical approximations of the macroscale flow system we employ a symmetric interior penalty dG FE method (see, e.g. Arnold et al., 2001 and the references cited therein). For the transport equation, we follow the discretization described in Collis et al. (2016a), which employs a dG FE method based on that presented in Houston and Süli (2001), with an explicit Euler time stepper.

Fig. 12.7 shows the macroscale pressure in the fluid and tissue domains at a representative time $t = 1000$, indicating (in general) inward fluid flow from the boundary in the vasculature, and outward flow in the tissue domain. Figs. 12.8 and 12.9 show the time-evolution of the macroscale nutrient in the 2D domain for $c_\partial(t) = 1$, and along $y = 0$, $x \geqslant 0$ (note that under the choice of domain and boundary condition, these profiles are radially symmetric, so that the behavior indicated in Fig. 12.9 is reproduced throughout the domain). These highlight that nutrient is transported into the interior of the tissue under the influence of the pressure-induced flow described above, but that the extent of nutrient advection is limited. Eventually transport from the exterior achieves a quasi-steady configuration (as also observed in O'Dea et al., 2015) in which regions distant from the source exhibit nutrient depletion so that, due to (12.10), uptake and associated growth is restricted to tissue in the vicinity of the boundary.

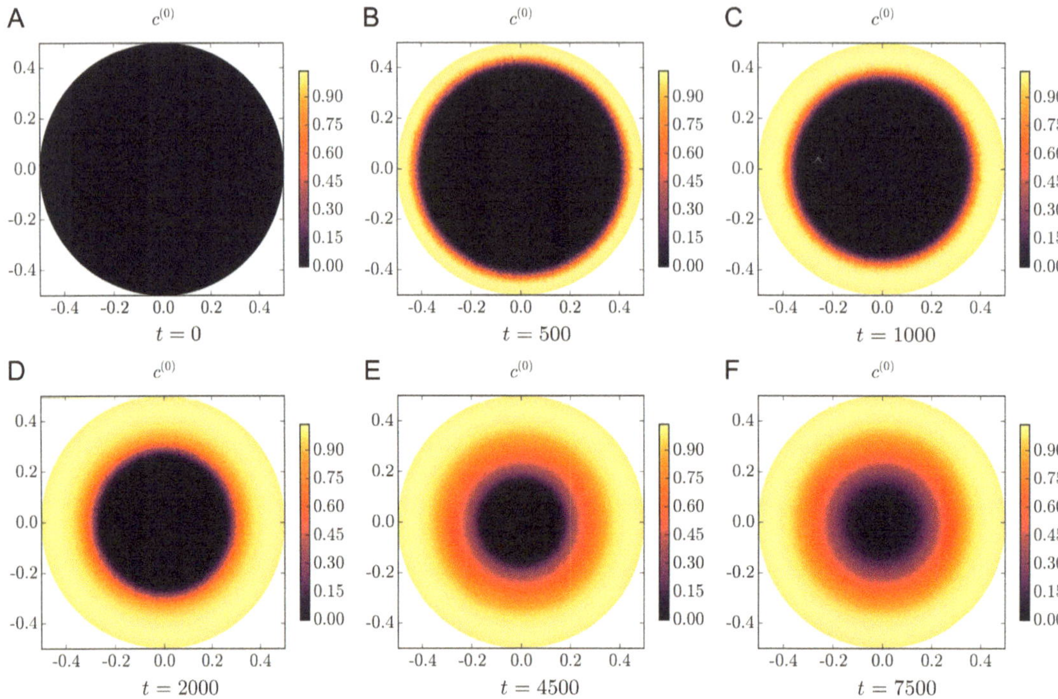

FIGURE 12.8 Numerical approximations of the nutrient concentration at $t = 0$, 500, 1000, 2000, 4500, 7500. These correspond to approximations to the solution of (12.57), subject to the boundary condition (12.61) for $c_\partial = 1$.

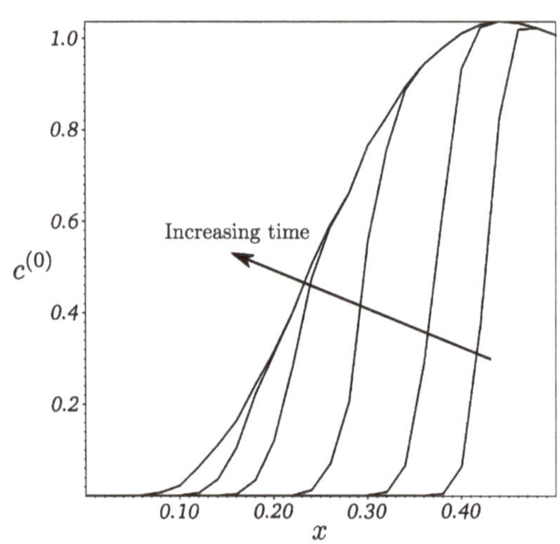

FIGURE 12.9 Evolution of the nutrient concentration at $y = 0$, $x \geqslant 0$. This corresponds to the approximation to the solutions of (12.57) subject to the boundary condition (12.61) for $c_\partial(t) = 1$.

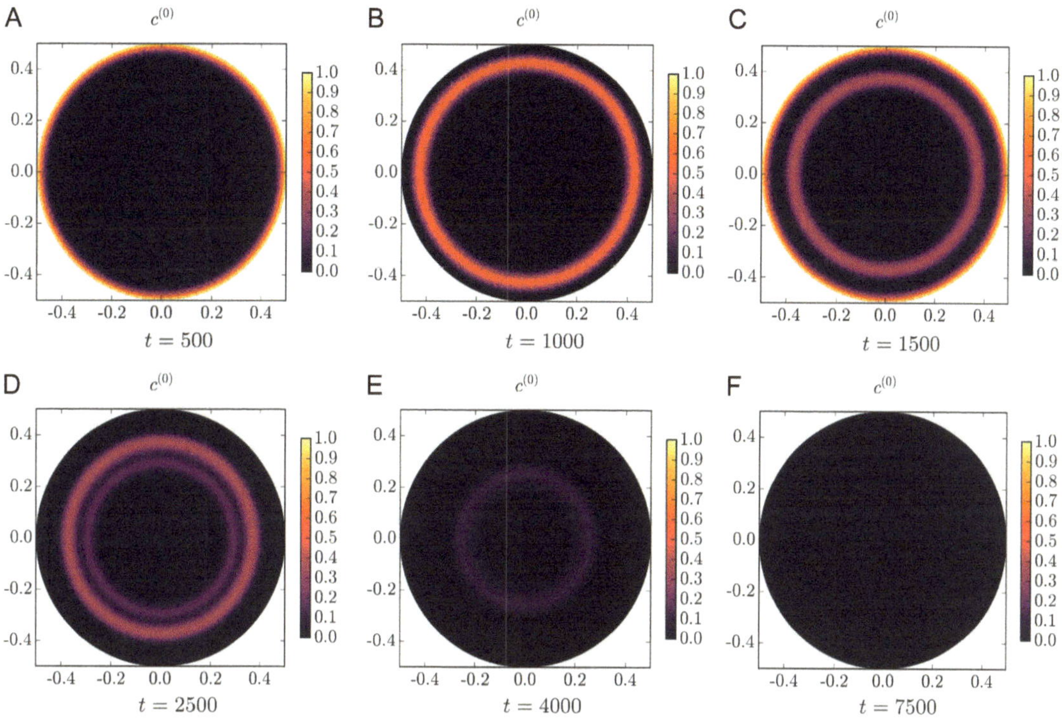

FIGURE 12.10 Numerical approximations of the nutrient concentration at $t = 500, 1000, 1500, 2500, 4000, 7500$. These correspond to approximations to the solution of (12.57), subject to the boundary condition (12.61) for c_∂ given by (12.62).

Fig. 12.10 indicates how a time-varying dose of nutrient is transported from the boundary for the boundary data

$$c_\partial(t) = \exp\left(-5 \times 10^{-5}(t - 500)^2\right) + \exp\left(-5 \times 10^{-5}(t - 1500)^2\right), \tag{12.62}$$

which produces two Gaussian pulses in the nutrient concentration, peaking at $t = 500$ and $t = 1500$. We include such a set up to indicate how considerations of relevance to, for example, the study of glycolytic pathways in the treatment of cancer may be easily investigated within this formulation. In this context, c is to be interpreted as glucose, whose concentration in plasma is naturally time-varying. While the overall dynamic is similar to that presented in Fig. 12.8 (c is transported inwards from the exterior of the tissue at a rate that decreases as it approaches the center), the flow within the tissue is, in fact, influenced by this new regime. The pulse of c traveling across the domain will induce an additional corresponding structure in p due to the dependence on c in the growth terms that appear in Eqs. (12.55) and (12.56).

12.5 DISCUSSION

In this chapter, we have considered the problem of deriving effective descriptions of relevance to tissue growth phenomena. From an underlying description of the microscale flow physics we derive, via a multiscale analysis, the governing equations suitable to describe fluid flow and nutrient transport within a growing tissue at the macroscale, which nevertheless accommodate explicitly the influence of dynamics occurring at the microscale. The multiscale homogenization technique that we employ has been widely applied in studies of porous and poroelastic materials (see, e.g. Burridge and Keller, 1981 and many others), and more recently in models of growing structures in a solid (Penta et al., 2014; O'Dea et al., 2015) and multiphase context (Collis et al., 2016b). Here we consider applying these ideas to a model of vascularized tumor tissue due to Shipley and Chapman (2010), thereby proposing a macroscale model suitable for the study of transport and growth of tumors (and other related applications).

Our microscale system comprises a growing porous medium (whose growth occurs at a singular interface separating the porous tissue domain from a surrounding fluid domain) and whose evolution is nutrient-limited. This description of growth is appropriate in the case for which nutrient consumption in the tissue dominates over diffusive transport. At the macroscale we obtain a flow description corresponding to a 'double porous medium' with both the porous tissue and the surrounding fluid domains behaving as a porous medium, and with fluid transport occurring between them. Additional contributions arise due to growth, and permeability in each domain is parameterized by microscale 'cell problems'. Nutrient transport is governed by an advection–reaction equation with dependencies on growth and consumption. As a consequence of the timescale on which we perform our analysis, microscale growth and transport processes are relegated to $\mathcal{O}(\varepsilon)$ so that the domain is quasi-stationary. Nevertheless, the effective description that we derive highlights that such growth and transport provide an $\mathcal{O}(1)$ contribution on the macroscale. The formulation that we obtain is a combination of those derived in Shipley and Chapman (2010) and O'Dea et al. (2015), and these formulations are recovered in the appropriate limits.

We demonstrate the behavior of the derived macroscale system by indicative numerical examples, employing academic examples of micro- and macro-scale geometry (full details of the numerical approach for a closely-related problem is given in Collis et al., 2016a). These results indicate that nutrient transport through the porous structure is a result of a balance between growth, uptake and advection. In the case of a constant external nutrient source, this eventually achieves a quasi-steady configuration (as observed in O'Dea et al., 2015) in which regions distant from the source exhibit nutrient depletion. These results are therefore of relevance to, say optimizing transport in tissue engineering applications, or drug delivery to tumors.

The microscale cell problems (and hence the macroscale behavior) are dependent on both the microscopic structure and physics of the material under consideration, through the geometry of the periodic cell, and the slip properties (via the parameter θ) of the interface between microscale fluid and tissue domains. However, in both Shipley and Chapman (2010) and Penta and Ambrosi (2015) the latter aspect is neglected on the basis of parameter estimates and scaling arguments. Here, we provide an indication of its importance, highlighting a wide range of slip parameter values for which the permeability is strongly dependent on both the microstructure and the slip. In view of the difficulty of accurate model parameterization, our results indicate that this feature is likely to be of importance in some applications of biological relevance.

In this work we have analyzed a model system with the aim of highlighting how microscale growth and transport processes of rather broad relevance to biological applications may be analyzed in a multiscale framework. A key assumption of the methodology that we employ is that the material may be described by a locally-periodic structure; while such an approximation may be suitable in certain tissue engineering applications (where scaffold production techniques may admit a highly regular microstructure), it is somewhat more problematic in more general settings. Obtaining suitable effective representations in non-periodic media remains a key area of research in this field. Furthermore, we have considered a simplified growth model, concentrating on appositional growth at a non-material interface—application of multiscale analyses to more comprehensive growth models (such as those accommodating interstitial growth of soft tissue) is of particular interest; the recent study of Collis et al. (2016b) seeks to make some progress in this regard; however, there are many open questions to be resolved.

Acknowledgements

J. Collis acknowledges support from the Engineering and Physical Sciences Research Council (under grant number EP/K039342/1).

References

Alarcón, T., Owen, M.R., Byrne, H.M., Maini, P.K., 2006. Multiscale modelling of tumour growth and therapy: the influence of vessel normalisation on chemotherapy. Comput. Math. Methods Med. 7, 85–119.

Antonietti, P., Giani, S., Hall, E., Houston, P., Krahl, R., 2015. AptoFEM documentation. URL http://www.aptofem.com.

Arnold, D.N., Brezzi, F., Cockburn, B., Marini, D.L., 2001. Unified analysis of discontinuous Galerkin methods for elliptic problems. SIAM J. Numer. Anal. (ISSN 0036-1429) 39, 1749–1779.

Ateshian, G.A., 2011. The role of mass balance equations in growth mechanics illustrated in surface and volume dissolution. J. Biomech. Eng. 133, 011010.

Band, L.R., King, J.R., 2012. Multiscale modelling of auxin transport in the plant-root elongation zone. J. Math. Biol. 65, 743–785.

Boucher, Y., Brekken, C., Netti, P.A., Baxter, L.T., Jain, R.K., 1998. Intratumoral infusion of fluid: estimation of hydraulic conductivity and implications for the delivery of therapeutic agents. Br. J. Cancer 78, 1442.

Burridge, R., Keller, J.B., 1981. Poroelasticity equations derived from microstructure. J. Acoust. Soc. Am. 70, 1140.

Chavarría-Krauser, A., Ptashnyk, M., 2010. Homogenization of long-range auxin transport in plant tissues. Nonlinear Anal., Real World Appl. 11, 4524–4532.

Collis, J., Hubbard, M.E., O'Dea, R.D., 2016a. Computational modelling of multiscale, multiphase fluid mixtures with application to tumour growth. Comput. Methods Appl. Mech. Eng. 309, 554–578.

Collis, J., Hubbard, M.E., O'Dea, R.D., 2016b. A multiscale analysis of drug transport and response for a multiphase tumor model. Submitted for publication. http://dx.doi.org/10.1017/S0956792516000413.

Fozard, J.A., Byrne, H.M., Jensen, O.E., King, J.R., 2010. Continuum approximations of individual-based models for epithelial monolayers. Math. Med. Biol. 27, 39–74.

Frieboes, H.B., Lowengrub, J.S., Wise, S., Zheng, X., Macklin, P., Bearer, E.L., Cristini, V., 2007. Computer simulation of glioma growth and morphology. NeuroImage 37, S59–S70.

Houston, P., Süli, E., 2001. hp-adaptive discontinuous Galerkin finite element methods for first-order hyperbolic problems. SIAM J. Sci. Comput. 23, 1226–1252.
King, J.R., Franks, S.J., 2007. Mathematical modelling of nutrient-limited tissue growth. In: Free Boundary Problems. Springer, pp. 273–282.
Lee, Cheo K., Mei, Chiang C., 1997. Re-examination of the equations of poroelasticity. Int. J. Eng. Sci. 35, 329–352.
Macklin, P., McDougall, S., Anderson, A.R.A., Chaplain, M.A.J., Cristini, V., Lowengrub, J., 2009. Multiscale modelling and nonlinear simulation of vascular tumour growth. J. Math. Biol. 58, 765–798.
Matzavinos, A., Ptashnyk, M., 2015. Homogenization of oxygen transport in biological tissues. Appl. Anal., 1–37.
Mei, C.C., Auriault, J-L., 1989. Mechanics of heterogeneous porous media with several spatial scales. Proc. R. Soc. Lond. A 426, 391–423.
Mei, C.C., Auriault, J.L., 1991. The effect of weak inertia on flow through a porous medium. J. Fluid Mech. 222, 647–663.
Netti, P.A., Berk, D.A., Swartz, M.A., Grodzinsky, A.J., Jain, R.K., 2000. Role of extracellular matrix assembly in interstitial transport in solid tumors. Cancer Res. 60, 2497–2503.
O'Dea, R.D., King, J.R., 2011. Multiscale analysis of pattern formation via intercellular signalling. Math. Biosci. 231, 172–185.
O'Dea, R.D., King, J.R., 2012. Continuum limits of pattern formation in hexagonal-cell monolayers. J. Math. Biol. 64, 579–610.
O'Dea, R.D., Nelson, M.R., El Haj, A.J., Waters, S.L., Byrne, H.M., 2015. A multiscale analysis of nutrient transport and biological tissue growth in vitro. Math. Med. Biol. 32, 345–366.
Osborne, J.M., Walter, A., Kershaw, S.K., Mirams, G.R., Fletcher, A.G., Pathmanathan, Pras, Gavaghan, D., Jensen, O.E., Maini, P.K., Byrne, H.M., 2010. A hybrid approach to multi-scale modelling of cancer. Philos. Trans. R. Soc. Lond. A 368, 5013–5028.
Penta, R., Ambrosi, D., 2015. The role of the microvascular tortuosity in tumor transport phenomena. J. Theor. Biol. 364, 80–97.
Penta, R., Ambrosi, D., Shipley, R.J., 2014. Effective governing equations for poroelastic growing media. Q. J. Mech. Appl. Math. 67, 69–91.
Powathil, G.G., Swat, M., Chaplain, M.A.J., 2015. Systems Oncology: Towards Patient-Specific Treatment Regimes Informed by Multiscale Mathematical Modelling. Semin. Cancer Biol., vol. 30. Elsevier, pp. 13–20.
Shipley, R.J., Chapman, S.J., 2010. Multiscale modelling of fluid and drug transport in vascular tumours. Bull. Math. Biol. 72, 1464–1491.
Toselli, A., 2002. hp-finite element discontinuous Galerkin approximations for the Stokes problem. Math. Models Methods Appl. Sci. 12, 1565–1616.
Van Leeuwen, I.M.M., Mirams, G.R., Walter, A., Fletcher, A., Murray, P., Osborne, J., Varma, S., Young, S.J., Cooper, J., Doyle, B., et al., 2009. An integrative computational model for intestinal tissue renewal. Cell Prolif. 42, 617–636.
Zhang, X-Y., Luck, J., Dewhirst, M.W., Yuan, F., 2000. Interstitial hydraulic conductivity in a fibrosarcoma. Am. J. Physiol., Heart Circ. Physiol. 279, H2726–H2734.

CHAPTER

13

How Dense Core Vesicles Are Delivered to Axon Terminals – A Review of Modeling Approaches

I.A. Kuznetsov[*,†], *A.V. Kuznetsov*[‡,#]

[*]Perelman School of Medicine, University of Pennsylvania, Philadelphia, PA, United States

[†]Department of Bioengineering, University of Pennsylvania, Philadelphia, PA, United States

[‡]Department of Mechanical and Aerospace Engineering, North Carolina State University, Raleigh, NC, United States

13.1 INTRODUCTION

Dense core vesicles (DCVs) are neuronal organelles that transport diverse cargos important for neuron growth and function, such as neuropeptides, monoamines, and neurotrophic factors. Dysregulation of DCV transport is linked to numerous disorders. Neuropeptides, one of the key components transported by DCVs, are synthesized in the neuron soma. They are then packaged in DCVs and transported toward nerve terminals. Published experimental results suggest that DCVs circulate in nerve terminals, undergoing repeated anterograde and retrograde transport cycles. Some DCVs are captured as they travel through en passant boutons, which are sequential varicose sites in axon terminals. Other DCVs that were not captured while passing the boutons switch direction when they leave the terminal and re-enter circulation. The chapter reviews recent modeling approaches to simulating DCV transport in axon terminals. In particular, differences in modeling DCV transport in terminals with type Ib and type III boutons are discussed. The effects of varying the rate of DCV synthesis in the soma are investigated. The ability of the model to predict the trends observed in experiments is also discussed. Testable predictions made by the model are examined.

[#] Chapter coordinator.

13.2 REVIEW OF RELEVANT LITERATURE

13.2.1 Dense Core Vesicles and Their Cargos

DCVs are membrane-bound organelles in which neural cells package diverse cargos, such as neuropeptides, hormones, monoamines, neurotrophic factors, and preassembled components that need to be delivered to synaptic active zones. DCVs deliver these cargos to axon terminals (Huang and Reichardt, 2001; Poo, 2001; Samson and Medcalf, 2006; van de Bospoort et al., 2012; Berg et al., 2000; Li and Cline, 2010; Sorra et al., 2006). The abovementioned cargos are typically packaged into DCVs in the soma and transported together (Bury and Sabo, 2011).

Neuropeptides, one of the major cargos transported in DCVs, belong to a large class of signaling molecules that are found in the brain and spinal cord (Merighi et al., 2011; Zhang et al., 2011). Neuropeptides are extracellular messengers that play a role in regulating immune system response, cell proliferation and tissue repair, food intake and obesity, mood, sexual behavior, and development (Jonsdottir, 2000; Schaffer et al., 1998; Arora and Anubhuti, 2006; Bos et al., 2012; Werner and Covenas, 2010; Argiolas and Melis, 2013). More than a hundred different neuropeptides have been identified to date (Gondre-Lewis et al., 2012). Neuropeptide-containing DCVs range in diameter between 100–300 nm (Sorra et al., 2006; Gondre-Lewis et al., 2012). Because neuropeptides must be constantly replenished, a continuous supply of DCVs to the axon terminal is required (Maday et al., 2014).

DCVs are synthesized in the soma. The synthesis occurs in stages. First immature secretory granules are formed in the trans-Golgi network; this is followed by coalescence of the granules and their maturation (Morvan and Tooze, 2008; May et al., 2014). High resolution images of DCVs, obtained by means of photoactivated localization microscopy, are reported in Scalettar et al. (2014).

Transport of DCVs is a microtubule-dependent process (Bittins et al., 2010). Anterograde transport of DCVs is driven by kinesin-3 (KIF1A) molecular motors (Barkus et al., 2008; Maday et al., 2014) and retrograde transport of DCVs is driven by dynein motors (Moughamian and Holzbaur, 2012; Wong, 2012; Goldstein and Yang, 2000; Kwinter et al., 2009). Recent data suggest that anterograde and retrograde transport of DCVs may be less interdependent than was previously thought (Cavolo et al., 2015).

13.2.2 Accumulation and Release of DCVs

In axon terminals, DCVs accumulate and release their content from sequential varicose sites called en passant boutons (hereafter boutons) (Shakiryanova et al., 2006; Holtmaat and Svoboda, 2009). In experiments conducted in mice, it was found that boutons exhibit structural plasticity; they can appear and disappear over a period of days (De Paola et al., 2006). Two possible modes of DCV fusion in boutons are full fusion and kiss-and-run (Zhang et al., 2011; Chiang et al., 2014). Full fusion results in the full collapse of the vesicle and the expulsion of all of its content. Kiss-and-run results in a partial release of vesicle content, which occurs through a fusion pore. Data obtained in Wong et al. (2015) suggest that a single DCV is not emptied at once. Rather, the release from a DCV may occur repeatedly through the kiss-and-run mechanism.

There is still no complete understanding of how DCV docking and release in the axons are regulated (Hoover et al., 2014; Nurrish, 2014; Walter et al., 2014). DCV exocytic discharge involves such events as tethering, direct docking, and priming for fusion (D'Alessandro and Meldolesi, 2013). Exocytic fusion machinery involves many proteins and the release of DCVs in neurons is controlled in a complicated manner. For example, van de Bospoort et al. (2012) investigated the effects of lack/overexpression of Munc13 proteins on DCV release. They found that the lack of Munc13 reduces and delays the release of DCVs from synapses while overexpression of Munc13 boosts the efficiency of DCV release.

13.2.3 DCV Transport in Axon Terminals

The conventional understanding of neuropeptide transport in axons was that boutons await passively for the arrival of the DCVs from the soma. Utilizing a novel method, called simultaneous photo bleaching and imaging (SPAIM), Wong et al. (2012) investigated DCV transport and accumulation in axons with type Ib terminals and challenged the conventional understanding of DCV delivery. Their research eliminated potential transport models such as delivery in order, stochastic delivery, and parallel address based delivery. Instead, they established that DCVs circulate in type Ib terminals. Initially DCVs accumulate at the fastest rate in the most distal bouton, but later they are distributed uniformly among other boutons due to sporadic capture, which occurs as DCVs pass boutons moving anterogradely or retrogradely.

In addition, experimental results obtained in Wong et al. (2012) indicate that in order to switch directions, DCVs may switch from anterograde to retrograde motors in the most distal bouton, and that it takes ~300 s to switch motors (Levitan, 2015). However, it should be pointed out that the change of motors is only a hypothesis and how exactly DCVs switch directions in the most distal bouton is not completely understood.

13.2.4 Link to Neurodegenerative Disorders

Understanding neuropeptide transport is important for understanding the underlying reasons for a number of neurological disorders, which are related to defects in DCV budding or transport. Dysregulations of neuropeptides and neurotransmitters are linked to pathogenesis of schizophrenia. Defects in DCV transport are related to many diseases, such as Fragile X syndrome (an inherited intellectual disorder), cognitive and mood disorders, and diabetes and obesity. There is a connection between deficiencies in cargo packaging into DCVs and Smith–Lemli–Opitz syndrome (a developmental disorder). Finally, vesicular deficiencies (of not only DCVs) are linked to Alzheimer's disease (Gondre-Lewis et al., 2012; Meyer-Lindenberg et al., 2011).

13.2.5 Importance of Mathematical Modeling for Better Understanding of Biological Issues Related to DCV Transport and Release

Ecdysis, the molding of the cuticle in insects (Loveall and Deitcher, 2010), is associated with massive neuropeptide release. After ecdysis in *Drosophila*, type III boutons become almost completely depleted of neuropeptides (Bulgari et al., 2014). Unlike for many neurotransmitters, there are no reuptake mechanisms for neuropeptides. Each neuropeptide molecule that had been released has to be replaced with a new molecule synthesized in the soma (Hokfelt et al., 2012). After neuropeptides are released, it may take hours or even longer to replenish them by transport of newly synthesized peptides from the soma (van den Pol, 2012).

The typical duration of a fluorescence recovery after photobleaching experiment reported in Wong et al. (2012) was ~5 min. This duration is much shorter than the time it would take to replenish neuropeptide stores after release. One benefit of modeling is thus the ability to simulate a much longer timeframe than that which is currently accessible to experiments.

Mathematical modeling can also help investigate physical phenomena that may lead to faster replenishment of boutons with neuropeptides. For example, it is conceivable that DCV production in the soma increases in the period preceding ecdysis, which may lead to the temporary appearance of DCV circulation in type III terminals (Kuznetsov and Kuznetsov, 2015b). This circulation may help replenishing neuropeptide stores in boutons because DCVs can be captured from the circulating pool.

13.2.6 Modeling of DCV Transport

Through innovative experimental techniques, significant progress has been made in recent years in understanding dynamics of DCV transport and capture in axon terminals (Wong et al., 2012; Bulgari et al., 2014; Wong et al., 2015; Cavolo et al., 2015). However, mathematical modeling of DCV transport is still in its infancy.

A large amount of information about a system can be obtained by considering conservations of mass, energy, and momentum. The models of DCV transport reviewed in this chapter extend this approach and treat the number of DCVs as a conserved quantity. The first model of DCV transport was developed in Kuznetsov and Kuznetsov (2013). It was applicable to axons with type Ib terminals only. This model was able to predict the development of circulation of DCVs in type Ib boutons, which supported the hypothesis of neuropeptide delivery to synapses by long range vesicle circulation and sporadic capture (Wong et al., 2012). The second model (Kuznetsov and Kuznetsov, 2015a) was able to simulate both terminals with type Ib and type III boutons and also investigated the effect of the transcription factor Dimmed (DIMM). The second model successfully simulated differences in DCV transport between terminals with type Ib and type III boutons found in Bulgari et al. (2014). The results showed that the increase of DCV capture efficiency in boutons coupled with varying bouton capacity can lead to a qualitative change in DCV accumulation in boutons. The third model from the same group included the capability to simulate changes in the DCV production rate in the soma (Kuznetsov and Kuznetsov, 2015b). The normal rate of DCV production was defined as the soma production rate that is just enough to compensate for the DCV destruction in the axon and boutons at steady state. Two rates of DCV production were investigated: the normal rate and a 10% increase. Effects of DCV production rate on DCV circulation were studied.

Predictions of the three models were testable and falsifiable. The models were obtained using certain assumptions about DCV transport in axon terminals. The implications for the

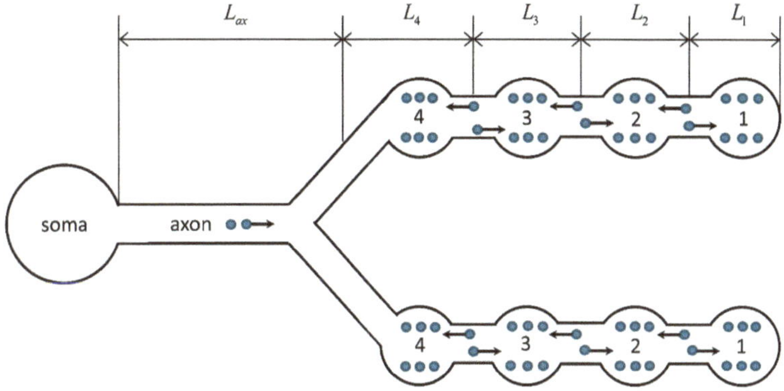

FIGURE 13.1 A schematic diagram of a nerve terminal with two branches. Each branch contains four en passant boutons, which are numbered as in Wong et al. (2012). The figure also displays transiting and resident DCVs.

understanding of transport in the terminals if some of the models' predictions would be found to be false in future experiments were discussed.

13.3 MATHEMATICAL MODELS OF DCV TRANSPORT AND ACCUMULATION IN AXON TERMINALS

13.3.1 Morphology of Axon Terminals

Although synapses at dendrites can occur in the absence of spines, many synapses are formed between boutons and dendritic spines (Anderson and Martin, 2001). Understanding transport and vesicle capture in boutons is thus important for understanding how the signaling machinery in neurons works. Boutons are divided into different types according to their morphology (Atwood et al., 1993; Jia et al., 1993). Here we review modeling of DCV transport and accumulation in axon terminals with type Ib and type III boutons in *Drosophila*. DCV transport in these two types of terminals was studied experimentally in Wong et al. (2012) and Bulgari et al. (2014).

At the terminal the axon may split into several branches (Fig. 13.1). Axon terminals with type Ib boutons in *Drosophila* motoneurons are characterized by continuous DCV circulation, when DCVs move through the boutons in the anterograde direction, turn around at the most distal bouton, move retrogradely through the boutons, and then turn around again once they reached the axon to re-enter the terminal. In type Ib boutons only ~10% of DCVs are captured as they pass a bouton either in anterograde or retrograde direction (Wong et al., 2012). The situation in type III boutons is quite different due to a much larger capture efficiency of DCVs in boutons. The largest bouton (which in the experiments of Bulgari et al. (2014) was the most proximal bouton, see Fig. 1A in their paper) now captures ~65% of DCVs. By the time DCVs passed the terminal, first in anterograde and then in retrograde motion, almost all DCVs are captured. Hence in axons with type III terminals and a normal rate of DCV production in the soma there is no DCV circulation.

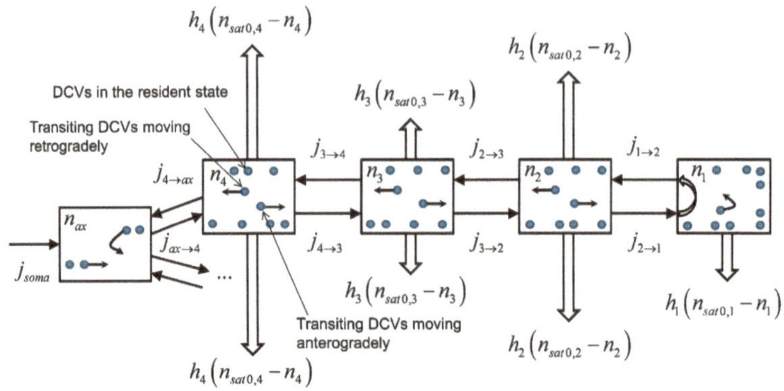

FIGURE 13.2 A kinetic diagram showing DCV fluxes between the compartments, $j_{ax\to4}$, $j_{4\to3}$, etc. (regular arrows) and fluxes resulting from capture of DCVs in boutons, $h(n_{sat0,4}-n_4)$, $h(n_{sat0,3}-n_3)$, etc. (block arrows). The diagram also shows anterogradely and retrogradely moving DCVs and DCVs in the resident state in boutons.

13.3.2 Simulated Geometry and Major Assumptions of the Model

A sketch of the problem is displayed in Fig. 13.1, which shows a neuron terminal with two branches, each of which contains four boutons. Since the two branches are identical, only one branch is simulated. The boutons are numbered the same way as in Wong et al. (2012). Fig. 13.2 displays DCV fluxes between the axon and four boutons; these fluxes are shown by regular arrows. Fluxes resulting from DCV capture in boutons are displayed by block arrows.

The following assumptions concerning the DCV fluxes were made in the model:

1. The DCV flux from the axon into the terminal is proportional to the DCV concentration in the axon.
2. Loss of DCVs in the resident state in boutons occurs due to a finite half-life of DCVs.
3. The rate of DCV capture into the resident state in a bouton is proportional to the difference between the saturated DCV concentration at infinite DCV half-life and the current DCV concentration in the resident state in the bouton. The utilization of the saturated DCV concentration at infinite DCV half-life allows simulating DCV capture even at steady-state, which is necessary to compensate for the DCV loss due to a finite half-life.
4. The DCV flux leaving a bouton is equal to the difference between the DCV flux coming to the bouton and the rate of DCV capture in the bouton.
5. Uncaptured DCVs that move anterogradely reverse their direction in the most distal bouton and start moving retrogradely, toward the axon. This requires switching molecular motors from kinesin to dynein. There is a time delay associated with this motor switching process.
6. Uncaptured retrogradely moving DCVs switch back from dynein to kinesin motors in the axon, which allows them to re-enter circulation.
7. The presence of transiting vesicles is neglected in the model as the number of transiting DCVs in a bouton is much less compared to the number of resident DCVs. The effect of transiting vesicles is accounted for only in terms of the DCV fluxes between the boutons.

13.3.3 Governing Equations

Here we follow the third, latest version of the model developed in Kuznetsov and Kuznetsov (2015b). Concentrations of DCVs in the axon and boutons are expressed by their linear number densities, vesicles/μm. The governing equations, expressing conservation of DCVs in the four boutons and in the axon, are

$$L_1\frac{dn_1}{dt}=\min\left[h_1(n_{sat0,1}-n_1),\, j_{2\to 1}\right]-L_1\frac{n_1\ln(2)}{T_{1/2}}, \tag{13.1}$$

$$L_2\frac{dn_2}{dt}=\min\left[h_2(n_{sat0,2}-n_2),\, j_{3\to 2}\right]+H[t-t_s]\min\left[h_2(n_{sat0,2}-n_2),\, j_{1\to 2}\right]-L_2\frac{n_2\ln(2)}{T_{1/2}}, \tag{13.2}$$

$$L_3\frac{dn_3}{dt}=\min\left[h_3(n_{sat0,3}-n_3),\, j_{4\to 3}\right]+H[t-t_s]\min\left[h_3(n_{sat0,3}-n_3),\, j_{2\to 3}\right]-L_3\frac{n_3\ln(2)}{T_{1/2}}, \tag{13.3}$$

$$L_4\frac{dn_4}{dt}=\min\left[h_4(n_{sat0,4}-n_4),\, j_{ax\to 4}\right]+H[t-t_s]\min\left[h_4(n_{sat0,4}-n_4),\, j_{3\to 4}\right]-L_4\frac{n_4\ln(2)}{T_{1/2}}, \tag{13.4}$$

$$L_{ax}\frac{dn_{ax}}{dt}=j_{soma}+2j_{4\to ax}-2j_{ax\to 4}-L_{ax}\frac{n_{ax}\ln(2)}{T_{1/2,ax}}, \tag{13.5}$$

where h_i is the mass transfer coefficient characterizing the rate of DCV capture into the resident state in bouton i ($i=1,\ldots,4$), H is the Heaviside step function, $j_{a\to b}$ is the flux of DCVs from compartment "a" to compartment "b" (see Fig. 13.2), j_{soma} is the rate of DCV production in the soma, L_{ax} is the length of the axon (see Fig. 13.1), L_i is the length of the compartment occupied by bouton i ($i=1,\ldots,4$) (see Fig. 13.1), n_{ax} is the concentration of DCVs in the axon, n_i is the concentration of resident DCVs in bouton i ($i=1,\ldots,4$), $n_{sat0,i}$ is the saturated concentration of DCVs in bouton i ($i=1,\ldots,4$) at infinite DCV half-life, t is the time, $T_{1/2}$ is the half-life of resident DCVs in boutons, and $T_{1/2,ax}$ is the half-life of DCVs in the axon.

Eqs. (13.1)–(13.4) involve functions that take the minimum of two quantities. The function $\min[h_1(n_{sat0,1}-n_1), j_{2\to 1}]$ on the right-hand side of Eq. (13.1) simulates the fact that DCV capture in a bouton cannot exceed the flux of DCVs that enter the bouton. If the flux of DCVs entering the bouton is large, than the capture rate is $h_1(n_{sat0,1}-n_1)$. Otherwise, all DCVs entering the bouton are captured, but not more, and the capture rate is $j_{2\to 1}$. Eqs. (13.2), (13.3), and (13.4) contain two of such functions because in boutons 2, 3, and 4 DCVs can be captured when they pass the bouton either in anterograde or in retrograde motion. The multiplier $H[t-t_s]$ in the term modeling retrograde capture simulates the assumption that it takes a time t_s (300 s) for DCVs to switch from anterograde to retrograde motors in the most distal bouton (bouton 1). For this reason, no retrograde capture is possible before $t=300$ s.

The flux from the axon to the most proximal bouton is assumed to be proportional to the concentration of DCVs in the axon:

$$j_{ax\to 4}=h_{in}n_{ax}, \tag{13.6}$$

where h_{in} is the mass transfer coefficient characterizing the rate at which DCVs enter bouton 4 from the axon.

A flux of DCVs leaving bouton 4 and continuing to move anterogradely is assumed to be equal to the flux of DCVs entering bouton 4 less the rate of DCV capture into the resident state in bouton 4:

$$j_{4\to3} = H_1\left[j_{ax\to4} - \min\left[h_4(n_{sat0,4} - n_4), j_{ax\to4}\right]\right]. \tag{13.7}$$

In Eq. (13.7) the function H_1 simulates the fact that the number of DCVs entering a bouton cannot be negative

$$H_1[\xi] = \xi H[\xi]. \tag{13.8}$$

The other fluxes are

$$j_{3\to2} = H_1\left[j_{4\to3} - \min\left[h_3(n_{sat0,3} - n_3), j_{4\to3}\right]\right], \tag{13.9}$$
$$j_{2\to1} = H_1\left[j_{3\to2} - \left[h_2(n_{sat0,2} - n_2), j_{3\to2}\right]\right], \tag{13.10}$$
$$j_{1\to2} = H[t - t_s]H_1\left[j_{2\to1} - \min\left[h_1(n_{sat0,1} - n_1), j_{2\to1}\right]\right], \tag{13.11}$$
$$j_{2\to3} = H[t - t_s]H_1\left[j_{1\to2} - \min\left[h_2(n_{sat0,2} - n_2), j_{1\to2}\right]\right], \tag{13.12}$$
$$j_{3\to4} = H[t - t_s]H_1\left[j_{2\to3} - \min\left[h_3(n_{sat0,3} - n_3), j_{2\to3}\right]\right], \tag{13.13}$$
$$j_{4\to ax} = H[t - t_s]H_1\left[j_{3\to4} - \min\left[h_4(n_{sat0,4} - n_4), j_{3\to4}\right]\right]. \tag{13.14}$$

Since the model is one-dimensional, the fluxes in Eqs. (13.6), (13.7), (13.9)–(13.14) have dimensions of vesicles/s.

Eqs. (13.1)–(13.5) are first-order ordinary differential equations, thus only initial conditions must be specified. The assumption that initially there were no resident DCVs in the boutons, but that the axon was initially filled with DCVs, results in the following initial conditions:

$$n_1(0) = 0, \quad n_2(0) = 0, \quad n_3(0) = 0, \quad n_4(0) = 0, \quad n_{ax}(0) = n_{sat,ax}, \tag{13.15}$$

where $n_{sat,ax}$ is the saturated concentration of DCVs in the axon.

13.3.4 Estimation of Parameter Values

Some model parameters can be estimated using published experimental data. Based on Wong et al. (2012), lengths of compartments occupied by the boutons, L_1, L_2, L_3, and L_4, are estimated to be 5 μm. The length of the axon used in experiments by Wong et al. (2012) was ~500 μm (Levitan, 2015). The flux from the axon to the most proximal bouton, $j_{ax\to4}$, is estimated to be 4 vesicles/min or 0.0667 vesicles/s (Bulgari et al., 2014). The half-life of resident DCVs, $T_{1/2}$, is estimated to be 6 hours or 2.16×10^4 s (Shakiryanova et al., 2006). Based on the fact that DCVs have to survive lengthy transport to the axon terminal (Levitan, 2015), the half-life of DCVs in the axon is assumed to be longer and is estimated to be 30 hours or 1.08×10^5 s. The capture efficiencies in type Ib boutons, ϖ_1, ϖ_2, ϖ_3, and ϖ_4, are estimated to be 10, 10, 10, and 40%, respectively, while in type III boutons they are estimated to be 40, 40, 65, and 20%, respectively (see the estimates in Kuznetsov and Kuznetsov, 2015b, which are based on the data reported in Wong et al., 2012 and Bulgari et al., 2014).

In type III terminals boutons have different sizes. The following equations have been used to quantify the relative bouton sizes:

$$n_{sat,2} = a_2 n_{sat,1}, \qquad n_{sat,3} = a_3 n_{sat,1}, \qquad n_{sat,4} = a_4 n_{sat,1}. \tag{13.16}$$

A representative type III terminal is described by the following values of morphological parameters: $a_2 = 3$, $a_3 = 9$, and $a_4 = 3$. This corresponds to the situation when the most distal bouton (bouton 1) is the smallest and the bouton located next to the proximal bouton (bouton 3) is the largest, see Fig. 13.1 for how boutons are numbered.

The rest of model parameters are estimated by using simple physical arguments. Saturated DCV concentrations in boutons, $n_{sat,i}$ $(i = 1, \ldots, 4)$, are estimated as follows. Since at steady-state all type Ib boutons have approximately the same number of DCVs (~200 vesicles, see Shakiryanova et al., 2006), the saturated concentration in any of Ib boutons is $n_{sat,i} = 200/L_i = 40$ vesicles/μm $(i = 1, \ldots, 4)$.

According to Bulgari et al. (2014), in axons with type III boutons there is almost no DCV circulation. This means that in type III terminals all DCVs that enter the terminal are captured before they leave the terminal. Captured DCVs compensate for DCV degradation in boutons. The conservation of DCVs in a type III terminal at steady state at the normal rate of DCV production in the soma gives the following:

$$j_{ax\to 4} = (L_1 n_{sat,1} + L_2 n_{sat,2} + L_3 n_{sat,3} + L_4 n_{sat,4}) \frac{\ln(2)}{T_{1/2}}. \tag{13.17}$$

Eqs. (13.16) and (13.17) result in $n_{sat,1} = 26.0$, $n_{sat,2} = 77.9$, $n_{sat,3} = 233.8$, and $n_{sat,4} = 77.9$ vesicles/μm.

Kuznetsov and Kuznetsov (2015b) estimated the saturated DCV concentration in the axon, $n_{sat,ax}$, to be 10% of the saturated DCV concentrations in type Ib boutons. This estimate was used for axons with type Ib and type III terminals. This means that $n_{sat,ax} = 0.1 \times 40$ vesicles/μm $= 4$ vesicles/μm.

A mass transfer coefficient characterizing DCV transport from the axon into the terminal was estimated as follows:

$$h_{in} = j_{ax\to 4} / n_{sat,ax}, \tag{13.18}$$

which gave $h_{in} = 0.0167$ μm/s.

Initially (at $t = 0$) there are no resident DCVs in the boutons. After passing each bouton the DCV flux drops by a certain percentage, characterized by the capture efficiency in that bouton. This fact is expressed as:

$$\varpi_4 j_{ax\to 4} = h_4 (n_{sat0,4} - 0), \tag{13.19}$$

$$\varpi_3 j_{ax\to 4} (1 - \varpi_4) = h_3 (n_{sat0,3} - 0), \tag{13.20}$$

$$\varpi_2 j_{ax\to 4} (1 - \varpi_4)(1 - \varpi_3) = h_2 (n_{sat0,2} - 0), \tag{13.21}$$

$$\varpi_1 j_{ax\to 4} (1 - \varpi_4)(1 - \varpi_3)(1 - \varpi_2) = h_1 (n_{sat0,1} - 0). \tag{13.22}$$

On the other hand, at steady-state the rate of DCV capture in a bouton equals to the rate of DCV loss in the bouton due to a finite DCV half-life:

$$2h_i(n_{sat0,i} - n_{sat,i}) = L_i \frac{n_{sat,i} \ln(2)}{T_{1/2}} \quad (i = 2, \ldots, 4), \tag{13.23}$$

$$h_1(n_{sat0,1} - n_{sat,1}) = L_1 \frac{n_{sat,1} \ln(2)}{T_{1/2}}. \tag{13.24}$$

There is no coefficient 2 on the left-hand side of Eq. (13.24) because DCVs pass the most distal bouton only once (they pass all other boutons twice, in anterograde and in retrograde motion).

The solution of Eqs. (13.19)–(13.24) for type Ib boutons results in the following:

$$\begin{aligned} h_4 &= 8.65 \times 10^{-5}, & h_3 &= 6.99 \times 10^{-5}, \\ h_2 &= 5.48 \times 10^{-5}, & h_1 &= 3.26 \times 10^{-4}\ \mu\text{m/s}, \end{aligned} \tag{13.25}$$

The units at the end of Eq. (13.25) apply to each parameter in Eq. (13.25). The same applies to Eqs. (13.26)–(13.28).

$$n_{sat0,4} = 77.1, \quad n_{sat0,3} = 85.9, \quad n_{sat0,2} = 98.5, \quad n_{sat0,1} = 59.7 \text{ vesicles/}\mu\text{m}. \tag{13.26}$$

The solution of Eqs. (13.19)–(13.24) for type III boutons results in the following:

$$\begin{aligned} h_4 &= 9.09 \times 10^{-5}, & h_3 &= 6.81 \times 10^{-5}, \\ h_2 &= 1.56 \times 10^{-5}, & h_1 &= 1.21 \times 10^{-5}\ \mu\text{m/s}, \end{aligned} \tag{13.27}$$

$$n_{sat0,4} = 147, \quad n_{sat0,3} = 509, \quad n_{sat0,2} = 478, \quad n_{sat0,1} = 371 \text{ vesicles/}\mu\text{m}. \tag{13.28}$$

The rate of DCV production in the soma (a parameter involved in Eq. (13.5)) is calculated as follows:

$$j_{soma} = \kappa \left[(2L_1 n_{sat,1} + 2L_2 n_{sat,2} + 2L_3 n_{sat,3} + 2L_4 n_{sat,4}) \frac{\ln(2)}{T_{1/2}} + L_{ax} n_{sat,ax} \frac{\ln(2)}{T_{1/2,ax}} \right], \tag{13.29}$$

where $\kappa = 1$ for the normal rate of DCV production in the soma, $\kappa = 1.25$ for the rate of DCV production increased by 25%, and $\kappa = 1.5$ for the rate of DCV production increased by 50%. If $\kappa = 1$, the rate of DCV production in the soma exactly equals the rate at which DCVs are destroyed in the axon and boutons. The value of j_{soma} should be understood as the net rate of DCV production in the soma. In fact, some DCVs may re-enter the soma for degradation (Wong et al., 2012). In this case, j_{soma} in Eq. (13.5) should be understood as the rate of DCV synthesis in the soma minus the rate at which DCVs re-enter the soma from the axon.

13.4 RESULTS AND DISCUSSION

Eqs. (13.1)–(13.5) were solved numerically by using Matlab's solver ODE45 (Matlab R2014b, MathWorks, Natick, MA, USA).

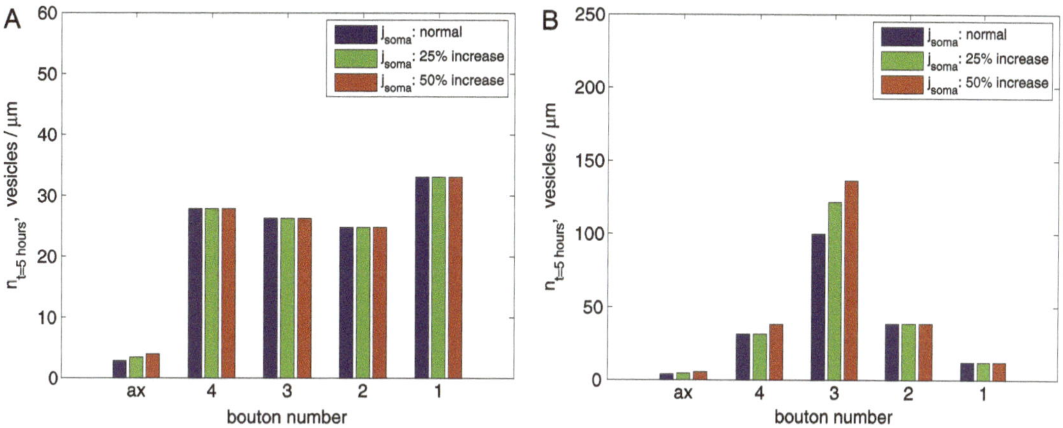

FIGURE 13.3 Concentrations of DCVs in the axon and in the resident state in various boutons at $t = 5$ hours after complete release of DCVs, at different rates of DCV production in the soma. (A) Terminal with type Ib boutons. (B) Terminal with type III boutons.

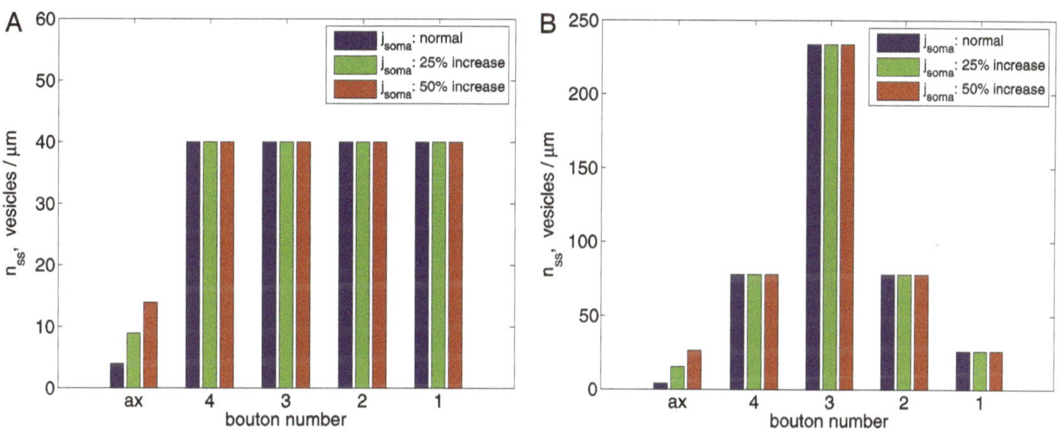

FIGURE 13.4 Concentrations of DCVs in the axon and in the resident state in various boutons at steady-state, at different rates of DCV production in the soma. (A) Terminal with type Ib boutons. (B) Terminal with type III boutons.

Figs. 13.3–13.7 show results for three different DCV production rates in the soma (j_{soma}) while Figs. 13.8–13.10 show results for three different half-lives of DCVs in the resident state ($T_{1/2}$). All other parameters in the model (including h_1, h_2, h_3, h_4, and $n_{sat0,1}$, $n_{sat0,2}$, $n_{sat0,3}$, $n_{sat0,4}$) are assumed to be unaffected by changing the values of j_{soma} and $T_{1/2}$; the values of these and other parameters of the model are given in Section 13.3.4.

Figs. 13.3–13.7, which show results for different DCV production rates in the soma, were computed assuming that $T_{1/2} = 6$ hours. The model shows a significant difference in the initial dynamics of DCV accumulation in type Ib and type III terminals. Indeed, in type Ib terminals the most distal bouton (bouton 1) fills with DCVs faster than other boutons (Fig. 13.3A), which is in agreement with experimental observations of Wong et al. (2012). On the other

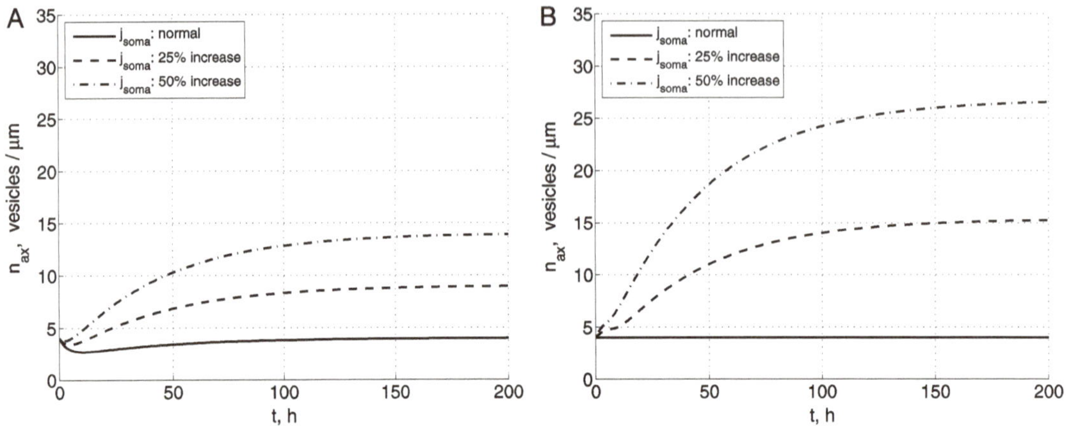

FIGURE 13.5 The buildup toward steady-state: concentrations of DCVs in the axon, at different rates of DCV production in the soma. (A) Terminal with type Ib boutons. (B) Terminal with type III boutons.

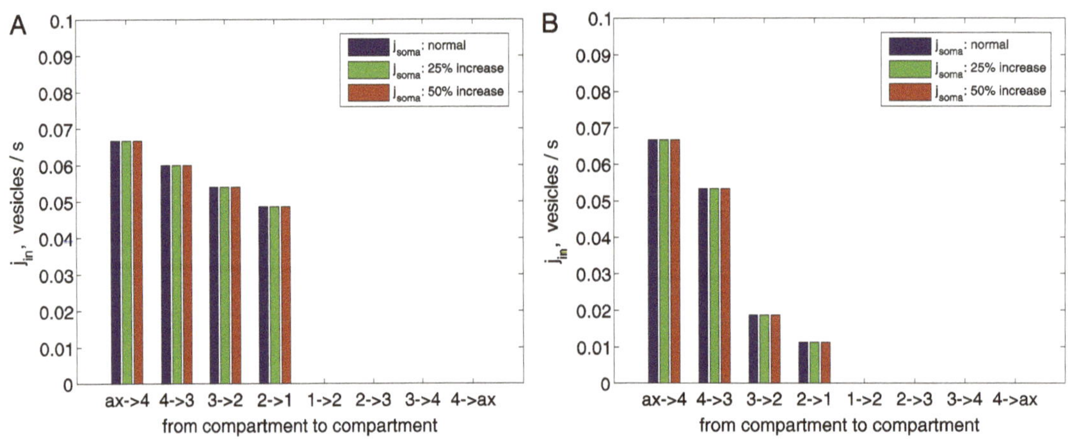

FIGURE 13.6 Fluxes between the axon and the most proximal bouton and between various boutons at the initial state ($t = 0$), at different rates of DCV production in the soma. (A) Terminal with type Ib boutons. (B) Terminal with type III boutons.

hand, in type III boutons the largest bouton (bouton 3) fills up first (Fig. 13.3B). This is consistent with Bulgari et al. (2014) who found that in type III terminals neuropeptide accumulation was initially most marked in proximal boutons.

Steady-state DCV concentrations in the resident state in boutons, $n_{ss,i}$, do not depend on the rate of DCV production in the soma (Figs. 13.4A and 13.4B); they are determined by bouton capacities, which are expressed by saturated DCV concentrations in boutons, $n_{sat,i}$. Despite the fact that in type Ib terminals in the beginning of the process DCVs preferentially accumulate in the most distal bouton, at steady-state DCVs are uniformly distributed between the boutons. In type III terminals at steady-state the DCV distribution is non-uniform,

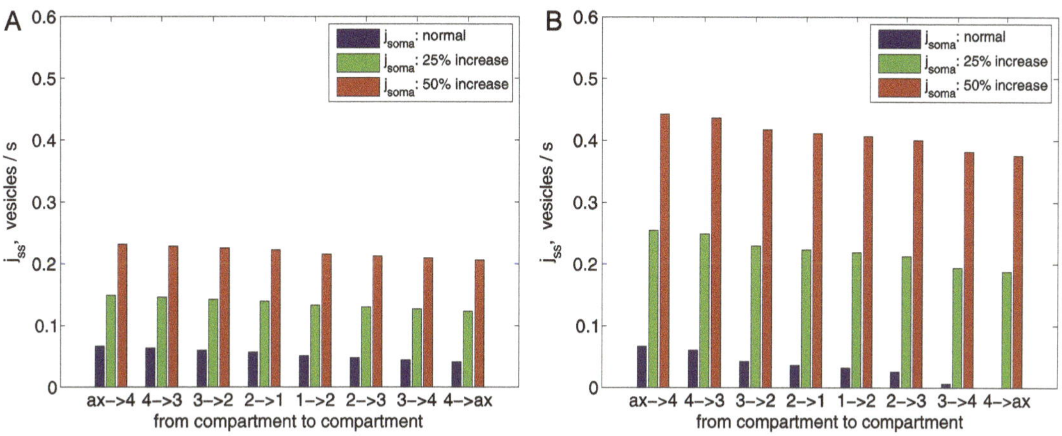

FIGURE 13.7 Fluxes between the axon and the most proximal bouton and between various boutons at steady-state, at different rates of DCV production in the soma. (A) Terminal with type Ib boutons. (B) Terminal with type III boutons.

with the greatest number of DCVs accumulating in the largest bouton (in our example it is bouton 3), and the smallest number of DCVs accumulating in the most distal bouton (bouton 1).

The DCV concentration in the axon builds up to larger steady-state values for larger values of j_{soma} (Figs. 13.5A and 13.5B). It is interesting that in axons with type III terminals, for a normal value of DCV production rate, $n_{ax}(t)$ stays constant (Fig. 13.5B, solid line). This is because in this case there is no DCV circulation, and DCVs do not return to the axon. On the other hand, the normal value of j_{soma} is calculated such that the production of DCVs would exactly compensate for DCV destruction in the axon and boutons (see Eq. (13.29) with $\kappa = 1$).

The initial fluxes from the axon to bouton 4 and between the boutons (Figs. 13.6A and 13.6B) do not depend on the production rate of DCVs, they are controlled by the initial DCV concentration in the axon (which determines the DCV flux from the axon into the terminal) and by the initial DCV capture rate in boutons (which determines what portion of DCVs passes the bouton). Initially there are only anterograde fluxes because it is assumed that it takes 300 s to change from anterograde to retrograde motors in the most distal bouton. The steady-state fluxes from the axon to bouton 4 and between the boutons (Figs. 13.7A and 13.7B) determine whether there is DCV circulation in the terminal. Indeed, for circulation to exist, $j_{4 \to ax}$ must be larger than zero. If $j_{4 \to ax}$ is not zero, the returning DCVs can switch their direction in the axon and move back into the terminal. In type Ib terminals there is circulation even for the normal rate of DCV production (Fig. 13.7A). In type III terminals there is no DCV circulation at the normal DCV production rate; all DCVs are captured before they reach the axon. However, if the rate of DCV production in the soma is increased, either by 25 or by 50%, the circulation develops, since there are uncaptured DCVs that leave the terminal and that can turn around in the axon.

Figs. 13.8–13.10 display results for various durations of the DCV half-life in boutons; these figures are computed assuming that the rate of DCV production in the soma in increased by 25% compared to the normal rate. This allows us to investigate whether the reduced DCV

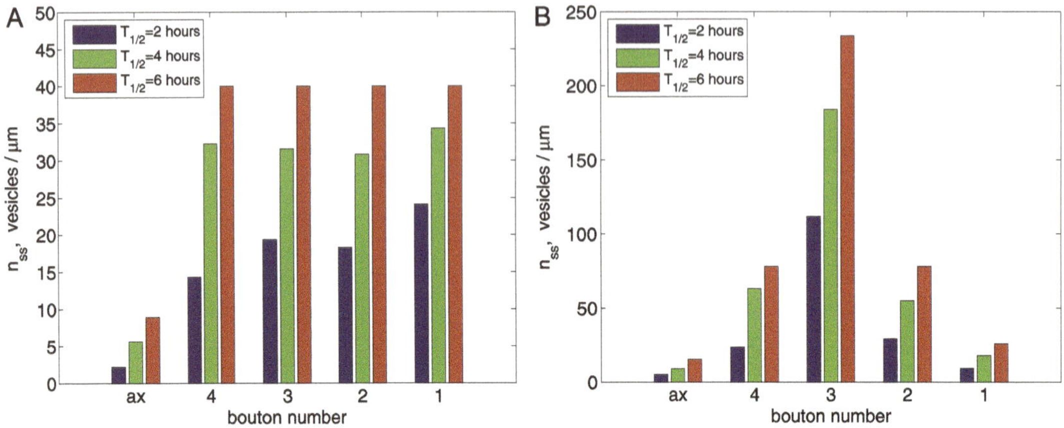

FIGURE 13.8 Concentrations of DCVs in the axon and in the resident state in various boutons at steady-state, at different DCV half-lives. (A) Terminal with type Ib boutons. (B) Terminal with type III boutons.

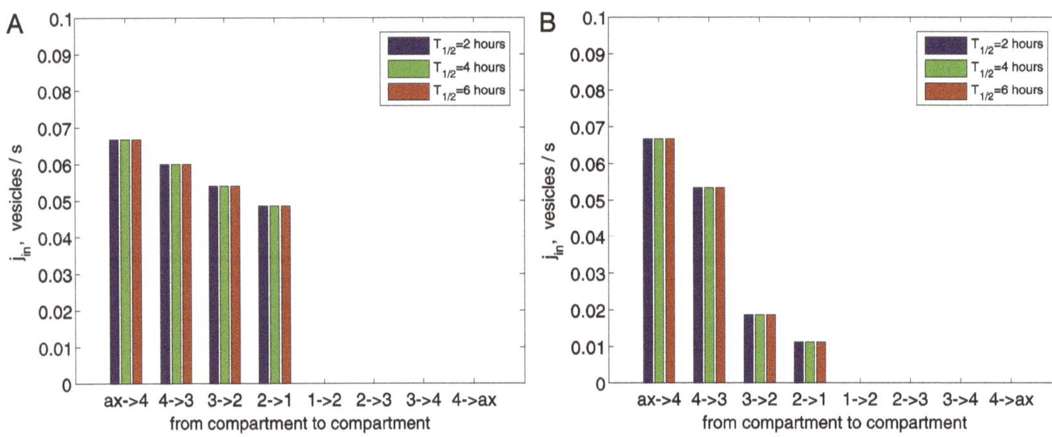

FIGURE 13.9 Fluxes between the axon and the most proximal bouton and between various boutons at the initial state ($t = 0$), at different DCV half-lives. (A) Terminal with type Ib boutons. (B) Terminal with type III boutons.

half-life can compensate for the increased DCV production rate. The steady-state DCV concentrations in boutons depend on the DCV half-life: for a smaller half-life there are less DCVs at steady-state in boutons (Figs. 13.8A and 13.8B).

The initial DCV fluxes do not depend on the DCV half-life, as expected (Figs. 13.9A and 13.9B). The steady-state fluxes strongly depend on the DCV half-life. For $T_{1/2} = 6$ hours there is DCV circulation in type III boutons (Fig. 13.10B); this is related to the fact that Fig. 13.10 displays the results for the increased DCV production rate. This effect was originally noted in Kuznetsov and Kuznetsov (2015b). Now we can see that this circulation becomes weaker when $T_{1/2} = 4$ hours, and completely disappears when $T_{1/2} = 2$ hours (Fig. 13.10B). Even more interesting is that DCV circulation can disappear in type Ib boutons if DCV half-life is

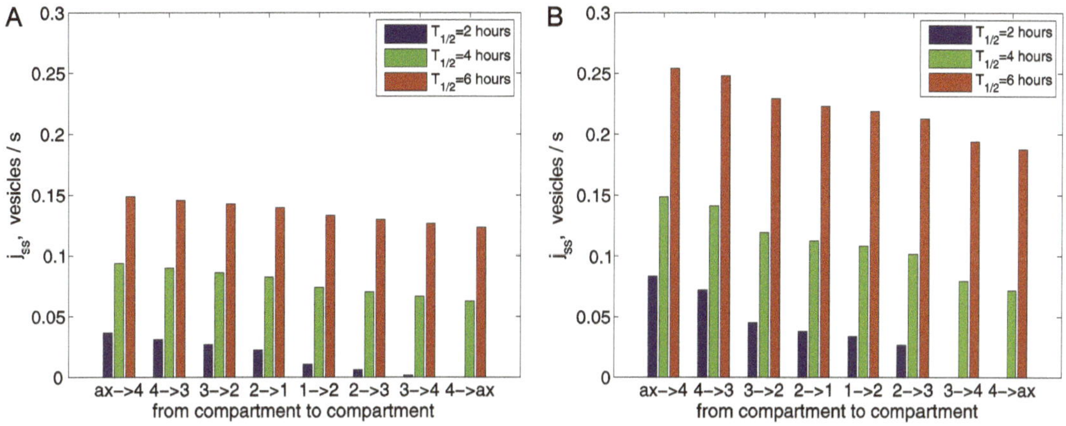

FIGURE 13.10 Fluxes between the axon and the most proximal bouton and between various boutons at steady-state, at different DCV half-lives. (A) Terminal with type Ib boutons. (B) Terminal with type III boutons.

decreased (Fig. 13.10A, results for $T_{1/2} = 2$ hours). These computational predictions provide potentially significant insight into DCV circulation in axon terminals and should be checked in future experiments.

13.5 FUTURE WORK

The longest axon in the human body is ~1 m in length. It takes DCVs days to travel this distance. Therefore, if neurons would rely only on DCVs synthesized in the soma to replenish the pool of resident DCVs in boutons following release, it would take many days to fully replenish boutons. Bulgari et al. (2014) suggested that a pool of circulating DCVs can be tapped into to replenish DCV stores. Simulating this effect would require including a pool of transiting DCVs into the model. Therefore, a two-concentration model that would account for separate pools of transiting and resident DCVs needs to be developed in future research. The two-concentration model must collapse to the one-concentration model under certain conditions (which must be established) for which the transiting vesicle concentration is negligible.

The importance of validating mathematical models by comparisons with experimental data cannot be overstated.

13.6 CONCLUSIONS

The developed model has the following capabilities. It simulates DCV transport in axon terminals and capture into the resident state in boutons. The model accounts for a finite half-life of resident DCVs. The model simulates the differences between DCV transport in type Ib and type III terminals. It exhibits initial distal DCV accumulation followed by an equal distribution of DCVs between the boutons in type Ib terminals which was observed during the experiments of Wong et al. (2012). It also simulates the initial proximal DCV accumulation followed by an unequal distribution of DCVs between the boutons in type III terminals and is hence in agreement with Bulgari et al. (2014). It also reproduces the fact that at a normal rate of DCV production in the soma there is DCV circulation in type Ib terminals but no DCV circulation in type III terminals (Wong et al., 2012; Bulgari et al., 2014).

The model makes some testable predictions. The most important prediction concerns the possibility of DCV circulation in type III boutons if the rate of DCV production in the soma is increased. It would be interesting to investigate whether this strategy is indeed used by neurons (whether the DCV production rate increases, for example, before a massive DCV release to create a circulation that can be used to replenish DCV stores). The model also predicts that the effect of increasing DCV production can be counter-acted by decreasing the DCV half-life in boutons. It also predicts that the reduction of DCV half-life in type Ib boutons can lead to the disappearance of DCV circulation in type Ib terminals.

Acknowledgement

AVK gratefully acknowledges the support of the Alexander von Humboldt Foundation through the Humboldt Research Award.

References

Anderson, J.C., Martin, K.A.C., 2001. Does bouton morphology optimize axon length? Nat. Neurosci. 4, 1166–1167.

Argiolas, A., Melis, M.R., 2013. Neuropeptides and central control of sexual behaviour from the past to the present: a review. Prog. Neurobiol. 108, 80–107.

Arora, S., Anubhuti, 2006. Role of neuropeptides in appetite regulation and obesity – a review. Neuropeptides 40, 375–401.

Atwood, H., Govind, C., Wu, C., 1993. Differential ultrastructure of synaptic terminals on ventral longitudinal abdominal muscles in *Drosophila* larvae. J. Neurobiol. 24, 1008–1024.

Barkus, R.V., Klyachko, O., Horiuchi, D., Dickson, B.J., Saxton, W.M., 2008. Identification of an axonal kinesin-3 motor for fast anterograde vesicle transport that facilitates retrograde transport of neuropeptides. Mol. Biol. Cell 19, 274–283.

Berg, E.A., Johnson, R.J., Leeman, S.E., Boyd, N., Kimerer, L., Fine, R.E., 2000. Isolation and characterization of substance P-containing dense core vesicles from rabbit optic nerve and termini. J. Neurosci. Res. 62, 830–839.

Bittins, C.M., Eichler, T.W., Hammer III, J.A., Gerdes, H., 2010. Dominant-negative myosin Va impairs retrograde but not anterograde axonal transport of large dense core vesicles. Cell. Mol. Neurobiol. 30, 369–379.

Bos, P.A., Panksepp, J., Bluthe, R., Van Honk, J., 2012. Acute effects of steroid hormones and neuropeptides on human social–emotional behavior: a review of single administration studies. Front. Neuroendocrinol. 33, 17–35.

Bulgari, D., Zhou, C., Hewes, R.S., Deitcher, D.L., Levitan, E.S., 2014. Vesicle capture, not delivery, scales up neuropeptide storage in neuroendocrine terminals. Proc. Natl. Acad. Sci. USA 111, 3597–3601.

Bury, L.A.D., Sabo, S.L., 2011. Coordinated trafficking of synaptic vesicle and active zone proteins prior to synapse formation. Neural Dev. 6, 24.

Cavolo, S.L., Zhou, C., Ketcham, S.A., Suzuki, M.M., Ukalovic, K., Silverman, M.A., Schroer, T.A., Levitan, E.S., 2015. Mycalolide B dissociates dynactin and abolishes retrograde axonal transport of dense-core vesicles. Mol. Biol. Cell 26, 2664–2672.

Chiang, H., Shin, W., Zhao, W., Hamid, E., Sheng, J., Baydyuk, M., Wen, P.J., Jin, A., Momboisse, F., Wu, L., 2014. Post-fusion structural changes and their roles in exocytosis and endocytosis of dense-core vesicles. Nat. Commun. 5, 3356.

D'Alessandro, R., Meldolesi, J., 2013. Expression and function of the dense-core vesicle membranes are governed by the transcription repressor REST. FEBS Lett. 587, 1915–1922.

De Paola, V., Holtmaat, A., Knott, G., Song, S., Wilbrecht, L., Caroni, P., Svoboda, K., 2006. Cell type-specific structural plasticity of axonal branches and boutons in the adult neocortex. Neuron 49, 861–875.
Goldstein, L.S.B., Yang, Z.H., 2000. Microtubule-based transport systems in neurons: the roles of kinesins and dyneins. Annu. Rev. Neurosci. 23, 39–71.
Gondre-Lewis, M.C., Park, J.J., Loh, Y.P., 2012. Cellular mechanisms for the biogenesis and transport of synaptic and dense-core vesicles. Int. Rev. Cell Mol. Biol. 299, 27–115.
Hokfelt, T., Xu, Zhi-Qingdavid, Shi, T., Adori, C., Zheng, K., Barde, S., Zhang, M., 2012. Neuropeptide systems – some basic concepts. In: Nyberg, F.G. (Ed.), Neuropeptides in Neuroprotection and Neuroregeneration. CRC Press, Boca Raton, FL.
Holtmaat, A., Svoboda, K., 2009. Experience-dependent structural synaptic plasticity in the mammalian brain. Nat. Rev. Neurosci. 10, 647–658.
Hoover, C.M., Edwards, S.L., Yu, S., Kittelmann, M., Richmond, J.E., Eimer, S., Yorks, R.M., Miller, K.G., 2014. A novel CaM kinase II pathway controls the location of neuropeptide release from *Caenorhabditis elegans* motor neurons. Genetics 196, 745–765.
Huang, E., Reichardt, L., 2001. Neurotrophins: roles in neuronal development and function. Annu. Rev. Neurosci. 24, 677–736.
Jia, X., Gorczyca, M., Budnik, V., 1993. Ultrastructure of neuromuscular-junctions in *Drosophila* – comparison of wild-type and mutants with increased excitability. J. Neurobiol. 24, 1025–1044.
Jonsdottir, I.H., 2000. Neuropeptides and their interaction with exercise and immune function. Immunol. Cell Biol. 78, 562–570.
Kuznetsov, I.A., Kuznetsov, A.V., 2013. A compartmental model of neuropeptide circulation and capture between the axon soma and nerve terminals. Int. J. Numer. Methods Biomed. Eng. 29, 574–585.
Kuznetsov, I.A., Kuznetsov, A.V., 2015a. Modeling neuropeptide transport in various types of nerve terminals containing en passant boutons. Math. Biosci. 261, 27–36.
Kuznetsov, I.A., Kuznetsov, A.V., 2015b. Can an increase in neuropeptide production in the soma lead to DCV circulation in axon terminals with type III en passant boutons? Math. Biosci. 267, 61–78.
Kwinter, D.M., Lo, K., Mafi, P., Silverman, M.A., 2009. Dynactin regulates bidirectional transport of dense-core vesicles in the axon and dendrites of cultured hippocampal neurons. Neuroscience 162, 1001–1010.
Levitan, E.S., 2015. Personal communication.
Li, J., Cline, H.T., 2010. Visual deprivation increases accumulation of dense core vesicles in developing optic tectal synapses in *Xenopus laevis*. J. Comp. Neurol. 518, 2365–2381.
Loveall, B.J., Deitcher, D.L., 2010. The essential role of bursicon during *Drosophila* development. BMC Dev. Biol. 10, 92.
Maday, S., Twelvetrees, A.E., Moughamian, A.J., Holzbaur, E.L.F., 2014. Axonal transport: cargo-specific mechanisms of motility and regulation. Neuron 84, 292–309.
May, R.M., Tabatadze, N., Czech, M.M., Woolley, C.S., 2014. Estradiol regulates large dense core vesicles in the hippocampus of adult female rats. Brain Struct. Funct. 219, 1947–1954.
Merighi, A., Salio, C., Ferrini, F., Lossi, L., 2011. Neuromodulatory function of neuropeptides in the normal CNS. J. Chem. Neuroanat. 42, 276–287.
Meyer-Lindenberg, A., Domes, G., Kirsch, P., Heinrichs, M., 2011. Oxytocin and vasopressin in the human brain: social neuropeptides for translational medicine. Nat. Rev. Neurosci. 12, 524–538.
Morvan, J., Tooze, S.A., 2008. Discovery and progress in our understanding of the regulated secretory pathway in neuroendocrine cells. Histochem. Cell Biol. 129, 243–252.
Moughamian, A.J., Holzbaur, E.L.F., 2012. Synaptic vesicle distribution by conveyor belt. Cell 148, 849–851.
Nurrish, S., 2014. Dense core vesicle release: controlling the where as well as the when. Genetics 196, 601–604.
Poo, M., 2001. Neurotrophins as synaptic modulators. Nat. Rev. Neurosci. 2, 24–32.
Samson, A.L., Medcalf, R.L., 2006. Tissue-type plasminogen activator: a multifaceted modulator of neurotransmission and synaptic plasticity. Neuron 50, 673–678.
Scalettar, B.A., Shaver, D., Kaech, S., Lochner, J.E., 2014. Super-resolution imaging of neuronal dense-core vesicles. J. Vis. Exp. 89, e51394.
Schaffer, M., Beiter, T., Becker, H.D., Hunt, T.K., 1998. Neuropeptides – mediators of inflammation and tissue repair? Arch. Surg. 133, 1107–1116.
Shakiryanova, D., Tully, A., Levitan, E.S., 2006. Activity-dependent synaptic capture of transiting peptidergic vesicles. Nat. Neurosci. 9, 896–900.

Sorra, K.E., Mishra, A., Kirov, S.A., Harris, K.M., 2006. Dense core vesicles resemble active-zone transport vesicles and are diminished following synaptogenesis in mature hippocampal slices. Neuroscience 141, 2097–2106.
van de Bospoort, R., Farina, M., Schmitz, S.K., De Jong, A., De Wit, H., Verhage, M., Toonen, R.F., 2012. Munc13 controls the location and efficiency of dense-core vesicle release in neurons. J. Cell Biol. 199, 883–891.
van den Pol, A.N., 2012. Neuropeptide transmission in brain circuits. Neuron 76, 98–115.
Walter, A.M., Kurps, J., De Wit, H., Schoening, S., Toft-Bertelsen, T.L., Lauks, J., Ziomkiewicz, I., Weiss, A.N., Schulz, A., Von Mollard, G.F., Verhage, M., Sorensen, J.B., 2014. The SNARE protein vti1a functions in dense-core vesicle biogenesis. EMBO J. 33, 1681–1697.
Werner, F., Covenas, R., 2010. Classical neurotransmitters and neuropeptides involved in major depression: a review. Int. J. Neurosci. 120, 455–470.
Wong, M.Y., 2012. Delivery and exocytosis of neuropeptide vesicles at the nerve terminal. PhD thesis. University of Pittsburgh.
Wong, M.Y., Cavolo, S.L., Levitan, E.S., 2015. Synaptic neuropeptide release by dynamin-dependent partial release from circulating vesicles. Mol. Biol. Cell 26, 2466–2474.
Wong, M.Y., Zhou, C., Shakiryanova, D., Lloyd, T.E., Deitcher, D.L., Levitan, E.S., 2012. Neuropeptide delivery to synapses by long-range vesicle circulation and sporadic capture. Cell 148, 1029–1038.
Zhang, Z., Wu, Y., Wang, Z., Dunning, F.M., Rehfuss, J., Ramanan, D., Chapman, E.R., Jackson, M.B., 2011. Release mode of large and small dense-core vesicles specified by different synaptotagmin isoforms in PC12 cells. Mol. Biol. Cell 22, 2324–2336.

Nomenclature

a_i Morphological parameters characterizing the ratio of the number of DCVs in bouton i ($i = 2, \ldots, 4$) to that in bouton 1 for type III boutons
h_{in} Mass transfer coefficient characterizing the rate at which DCVs enter bouton 4 from the axon (μm/s)
h_i Mass transfer coefficient characterizing the rate of DCV capture into the resident state in bouton i ($i = 1, \ldots, 4$) (μm/s)
H Heaviside step function
H_1 Function defined in Eq. (13.8)
$j_{a \to b}$ Flux of DCVs from compartment "a" to compartment "b" (vesicles/s)
j_{soma} Rate of DCV production in the soma (vesicles/s)
L_{ax} Length of the axon (μm)
L_i Length of the compartment occupied by bouton i ($i = 1, \ldots, 4$) (μm)
n_{ax} Concentration of DCVs in the axon (vesicles/μm)
n_i Concentration of resident DCVs in bouton i ($i = 1, \ldots, 4$) (vesicles/μm)
$n_{sat,i}$ Saturated concentration of DCVs in the resident state in bouton i ($i = 1, \ldots, 4$) (vesicles/μm)
$n_{sat0,i}$ Saturated concentration of DCVs in the resident state in bouton i ($i = 1, \ldots, 4$) at infinite DCV half-life (vesicles/μm)
$n_{sat,ax}$ Saturated concentration of DCVs in the axon (vesicles/μm)
t Time (s)
t_s Time it takes for DCVs to switch from anterograde to retrograde motors in the most distal bouton (s)
$T_{1/2}$ Half-life of resident DCVs in boutons (s)
$T_{1/2,ax}$ Half-life of DCVs in the axon (s)

Greek Symbols

κ Magnification coefficient defined in Eq. (13.29)
ϖ_i Efficiency of DCV capture in bouton i ($i = 1, \ldots, 4$)

Abbreviations

DCV Dense core vesicle

CHAPTER

14

Modeling of Food Digestion

S. Marze[#]

Biopolymères Interactions Assemblages, INRA, Nantes, France

14.1 INTRODUCTION

As digestion represents an essential process for life, it has been studied by distinguished philosophers and scientists. The historical research advances were reported in a fascinating article (Bloch, 1987). It highlighted the diversity of concepts about digestion, which was formerly seen as a succession of either chemical or physical processes, or studied through its biological effects. The understanding that these processes act together, and with a dependence on food structure and composition, made it necessary to develop new investigation methods. Early on in the field, modeling approaches were introduced to account for the main nutrients and processes. On the other hand, the food features have only recently been taken into account.

In this chapter, the various processes occurring within the digestive tract will be described to set the frame of the basic knowledge. Then, the development of the first models will be reported in their contexts, which were originally the understanding of drug absorption and of animal feed digestion. Finally, the recent microscale modeling approaches focusing on the fate of foods in some parts of the human digestive tract will be detailed.

14.2 THE COMPLEXITY OF FOOD DIGESTION AND ABSORPTION

Although digestion may be described as a biological process yielding nutrient and waste from food, it is actually a complex combination of various processes. In this section, the main processes are reviewed within the digestive tract and up to absorption, leaving the biology of hormonal, neural, and electrophysiological regulations out of the focus. The distinct roles of the colon and its microbiota are also not described here. This summary constitutes the basic knowledge, which can be found in any recent volume about gastrointestinal physiology (see, for example, Johnson, 2014).

[#] Chapter coordinator.

TABLE 14.1 The main products of hydrolysis at each step of the digestive tract

	Starches	Triglycerides	Proteins
Mouth	Dextrins	Diglycerides Fatty acids	
Stomach		Diglycerides Fatty acids	Polypeptides
Small intestine	Maltose	Monoglycerides Fatty acids	Tripeptides Dipeptides Amino acids
Intestinal brush border	Glucose		Amino acids

14.2.1 Chemical Processes

The major chemical reaction that converts food into nutrients is hydrolysis. This consists in the rupture of one or several bonds of a molecule by water. The digestive juices comprise various enzymes catalyzing this reaction, with a high selectivity in substrate and bond. The main digestive enzymes are amylases, lipases and proteases, responsible for the hydrolysis of starch (amylolysis), lipid (lipolysis) and protein (proteolysis), respectively.

In saliva produced in the mouth, ptyalin (an α-amylase) starts the hydrolysis of starch, and lingual lipase, to a minor extent, starts the hydrolysis of triglyceride. Then, in the stomach, the gastric juice contains pepsin, a protease that starts the hydrolysis of protein, and also contains gastric lipase that continues the hydrolysis of triglyceride. Lastly, in the small intestine, the digestive juice from pancreas is secreted into the duodenum to finalize food hydrolysis with pancreatic α-amylase, pancreatic lipase, and trypsin/chymotrypsin (two proteases). Ultimately, more enzymes from the intestinal brush border are required to complete the hydrolysis of starch and protein before absorption. Note that some vitamins also need specific enzymes to produce their absorbable form. The main end products of hydrolysis are glucose for starch, monoglyceride and fatty acid for triglyceride, and amino acid for protein. The intermediate products of hydrolysis are reported in Table 14.1.

It is important to note that the activity of all enzymes depends on pH and temperature. The passage of food through the digestive tract modifies pH, which is regulated by the secretion of electrolytes. While pH in the mouth after food ingestion is relatively stable around 7.0 (because the transit time is short), pH in the stomach may vary within the range 6.5 (fed state) to 1.5 (fasted state), and in the small intestine pH can range from 5.5 (duodenum) to 7.5 (ileum). Temperature in the stomach can also be modified by food, and usually returns to 37 °C within 20 min.

These changes affect the physicochemical conditions in the digestive tract. Temperature, ionic strength and pH modulate the electrostatic charge, conformation and solubility of molecules. Moreover, the various smaller molecules produced by hydrolysis also present specific properties.

TABLE 14.2 Ranges for some mechanical and temporal quantities in the digestive tube. The stomach transit times were evaluated from the literature, being too abundant to be cited systematically. The references are: (a) Anderson (1956), (b) Dejak et al. (2003), (c) Zhang and Hui (2015), (d) Harrison et al. (2014a), (e) Harrison and Cleary (2014), (f) Ono et al. (2004), (g) Haruma et al. (1994), (h) Indireshkumar et al. (2000), (i) Hveem et al. (2001), (j) Hausken et al. (2002), (k) Desipio et al. (2007), (l) Kwiatek et al. (2009), (m) Shama and Sherman (1973), (n) Ahmed et al. (2009), (o) Ferrua et al. (2014), (p) Ehrlein and Stockmann (1998), (q) de Loubens et al. (2013), (r) Lentle et al. (2013), (s) Lim et al. (2015), (t) Palmer et al. (1992), (u) Hiiemae (2004), (v) Read et al. (1986), (w) Camilleri et al. (1991), (x) Degen and Phillips (1996), (y) Bennink et al. (1999), (z) Graff et al. (2001)

	Pressure or stress (kPa unless otherwise specified)	Shear rate (s^{-1})	Transit time (min unless otherwise specified)
Mouth	Teeth 4–17 MPa (a), 8–86 MPa (b), 4–16 MPa (c), 40–440 (d), 23–55 (e) Tongue-palate 20–25 (f)	Tongue-palate 10–1000 (m)	Liquid 2–3.2 s (t) Solid 6.6–12 s (t) Soft solid 3.8–11.6 s (u) Hard solid 15.2–23 s (u)
Stomach	Antrum 5.9 (g), 0.8–1.6 (h), 0.9–2.1 (i), 3.5 (j), 1.3–1.7 (k), 0.7–1.1 (l) Pylorus 0.8–4.3 (h), 7.1–8.1 (k)	Antrum 0.02–0.8 (n), 0.01–0.04 (o) Pylorus 0.27–0.45 (o)	Solid 150–270 (average 210) Liquid 60–180 (average 110)
Small intestine	Duodenum 3.9 (g), 4.8 (j)	Lumen 0.7–5.8 (p), 0.02–0.9 (q), 0.02–1.2 (r) Villi 1–3 (s)	Solid 160–320 (v), 150–290 (w), 180–210 (x), 260–300 (y), 180–230 (z) Liquid 270–320 (y), 120–250 (z)

14.2.2 Physical Processes

As a consequence of the physicochemical conditions, initial food structures tend to alter under the action of physical phenomena such as crystallization, aggregation, flocculation, coalescence, and mass transfer. At the macroscale, this results in phase separation, particle size change, and evolution of rheological properties (elasticity, viscosity). Furthermore, assemblies are created from the association of nutrients and physiological molecules. This is the case for the intestinal mixed micelle composed of bile salts and lipids, or for various micronutrient/protein binding.

At the body scale, the muscles of the digestive tract contract and relax in order to separate, mix and propel food. This generates stress, strain, and determines transit times. These fundamental quantities are reported in Table 14.2, where the strain rate is characterized in the shear mode. The greatest mechanical breakdown is achieved in the mouth, with high stress and shear rate levels. Shear rate remains high in the pharynx and esophagus regions, whereas pressure decreases markedly (not reported in Table 14.2, as not directly related to digestion processes). Then, both pressure and shear rate decrease to lower levels as food progresses through the stomach and the small intestine, where transit times are greatly increased to compensate. Nevertheless, there is still some mechanical breakdown of food in the antrum region (referred to as grinding), linked to a selective gastric emptying based on particle size. Then, mixing is produced by a segmentation process in the small intestine. Throughout the digestive tract, bulk flows are thus generated, constituting the convective part of nutrient transport (by advection).

The other component of nutrient transport is diffusion. It is especially important in the small intestine, where the remaining particles in the digestate will finally reach the molecular size. Although an advection contribution still propels the digestate, the absorption of nutrients and micronutrients is mostly achieved by diffusion. First, molecules have to pass through the mucus layer (or unstirred water layer), which is a highly viscous mucin gel covering the intestinal villi lined with intestinal absorptive cells. As advection is very limited in this layer, this is done by either supramolecular (e.g. lipid mixed micelles) or molecular (e.g. peptides or saccharides) diffusion.

14.2.3 Biological Processes

Absorption is the transport across the intestinal cell membrane, which is either passive or active. Passive transport is actually diffusion, which can be either simple for lipophilic small molecules (passage down the concentration gradient across the membrane) or facilitated for hydrophilic or large molecules (binding to membrane protein that opens a channel). On the other hand, active transport is a process requiring energy, moving molecules against the concentration gradient, contrary to diffusion. Energy is supplied by ATP hydrolysis or electrochemical gradient.

14.3 DEVELOPMENT OF DIGESTION AND ABSORPTION MODELING

The early modeling efforts in the 1960s were mainly related to oral drug. The first works were compartmental, aiming to deduce drug absorption from blood or urinary concentration data (Nelson, 1961; Wagner and Nelson, 1964; Riegelman et al., 1968; Loo and Riegelman, 1968). While it is out of the scope of this chapter, it should be noted that theoretical studies within the gastrointestinal tract focused on drug release out of various dosage forms (see, for example, Higuchi, 1967; Siepmann and Siepmann, 2008). Then, derivations for drug absorption from the gastrointestinal tract were established. Because most excipients are not digestible, the modeling of digestion was not undertaken. This was done later by nutritionists and food scientists. In this section, we review the different types of model that were developed.

14.3.1 Drug Absorption

The first attempt to use a physical model of absorption in the context of the gastrointestinal tract was published in 1966 for urea, in the form of a macroscale diffusion–convection (or diffusion–advection) equation (Lifson and Hakim, 1966). However, this did not include any physicochemical parameter.

Then, a model comprising a lipid phase (for the cell membrane) was derived for drug absorption in 1970, only based on diffusion (Suzuki et al., 1970). The initial aqueous diffusion was supposed to obey Fick's first law, whereas the succeeding diffusion (within the lipid) was supposed to obey Fick's second law. A mass balance of the ionized and unionized species for drug, buffer, and water was included based on dissociation constants depending on pH. Only unionized drugs could diffuse in the lipid phase, with a distribution determined by the partition coefficient.

A significant advance was made in 1980 when a more realistic diffusion–convection–absorption equation was derived (Ni et al., 1980), with the possibility to include the above physicochemical parameters (Amidon et al., 1981). The basic equation is now known as the dispersion model (Yu et al., 1996; Huang et al., 2009):

$$\frac{\partial C}{\partial t} = \alpha \frac{\partial^2 C}{\partial x^2} - \beta \frac{\partial C}{\partial x} - \gamma C \tag{14.1}$$

where C is the drug concentration, t is the time, x is the longitudinal distance in the intestine represented as a tube, $\alpha = D$ is the longitudinal diffusion coefficient, β is the longitudinal velocity, equal to $Q/(\pi R^2)$, and γ is the absorption rate constant, equal to $2P_e/R$, where Q is the flow rate, R is the radius of the intestinal lumen, and P_e is the apparent permeability coefficient. Note that the symbols are harmonized in the equations, so each symbol will be explicated only once in the text.

Starting in the 1980s and completed in 1995, the work in the research group of Gordon Amidon established a steady-state model based on mass balances (Amidon et al., 1995; Yu et al., 1996). It enabled the construction of a biopharmaceutics classification system, based on dimensionless numbers, which are:

- The dose number $Do = \frac{C_0}{C_s} = \frac{M_0/V_0}{C_s}$,
- The dissolution number $Dn = \frac{t_{res}}{t_{diss}} = \frac{\pi R^2 L/Q}{r_0^2 \rho/(3DC_s)}$,
- The absorption number $An = \frac{t_{res}}{t_{abs}} = \frac{\pi R^2 L/Q}{R/(2P_e)} = \frac{2P_e \pi RL}{Q}$

where M_0 is the starting drug mass (dose), V_0 the starting dilution volume, C_S is the solubility of the drug, L is the tube length, r_0 is the starting radius of the drug particle, and ρ is the density of the particle. Note that the dose number is a concentration ratio, while the dissolution and the absorption numbers are time ratios, relative to the intestinal residence time t_{res}.

The dimensionless mass balance equations for particle radius and drug concentration are:

$$\frac{dr^*}{dx^*} = -\frac{Dn(1 - C^*)}{3r^*}, \tag{14.2}$$

$$\frac{dC^*}{dx^*} = Dn\left(1 - C^*\right)r^* Do - 2AnC^* \tag{14.3}$$

where $C^* = C/C_S$, $r^* = r/r_0$, and $x^* = x/L$.

Compared to the dispersion model, we see that other parameters are included, namely drug solubility, drug particle size and density. Solubility and permeability can be made pH-dependent.

Finally, another type of approach was used in the 1990s, based on compartments, resulting in the CAT (compartmental absorption and transit) model (Yu et al., 1996; Yu and Amidon, 1999). One of its advantages is to describe the whole gastrointestinal tract. The transit is defined as:

$$\frac{dM_s}{dt} = -K_s M_s, \tag{14.4}$$

$$\frac{dM_n}{dt} = K_t M_{n-1} - K_t M_n \quad \text{for } n = 1, 2, \ldots, 7, \tag{14.5}$$

$$\frac{dM_c}{dt} = K_t M_7 \tag{14.6}$$

for stomach, small intestine (7 compartments), and colon, respectively. K_s and K_t are the rate constants for gastric emptying and small intestinal transit. Absorption is assumed to occur only in the small intestine, which is divided into seven compartments such that

$$\frac{dM_a}{dt} = K_a \sum_{n=1}^{7} M_n \tag{14.7}$$

where K_a is the absorption rate constant.

Nowadays, simulation programs are available, using mainly dispersion or compartmental models. They include the parameters of the dimensionless numbers, and others related to release, degradation, food effect (fasted or fed physiology), active transport, metabolism, etc. (Huang et al., 2009).

14.3.2 Animal Feed Digestion and Absorption

Although there were early attempts to model feed digestion and absorption in pigs (Tomassone and Laplace, 1973; Usry et al., 1991), a complete description of a compartmental model was published only in the 1990s, around the same time as the CAT model (Bastianelli et al., 1996). Hydrolysis of all nutrients (proteins, lipids, carbohydrates) was accounted for and it was assumed to take place only in the intestines. Absorption of hydrolysis products and minerals was also assumed to take place only in the intestines. Secretion of endogenous proteins, lipids, and minerals was taken into account. A fermentation model was used for the colon. This simulation was further refined by Strathe et al. in 2008, with the description of hydrolysis and absorption as saturable processes (enzymatic reaction and active transport, respectively). An inhibitory effect of dietary fiber on proteolytic enzymes was implemented. Notably, the exogenous and endogenous proteins were treated separately, as they follow different hydrolysis and absorption kinetics. Note that this last point was developed in an intermediate model of the same nature but focusing only on protein digestion and absorption (Rivest et al., 2000). Recently, a similar model was applied to the digestion and absorption of proteins from milk gels, with the goal to understand the physiologic role of gel structure (Le Feunteun et al., 2014). Most parameters of the previous works were accounted for, except for hydrolysis which was not represented. The stomach was divided into two compartments. In the first compartment, milk clotting due to acidity and aggregation depending on protein processing (native vs. denatured) was included. In the second compartment, endogenous secretions and gastric emptying were simulated. Blood amino acid concentration resulting from intestinal absorption was successfully matched for mini-pig data, shown to be highly dependent on the gel structure.

An engineering approach based on chemical reactors was also developed. It initially focused on colonic hydrolysis and fermentation, represented by enzymes-catalyzed and microbes-autocatalyzed reactions, respectively (Penry and Jumars, 1987). These were modeled by Michaelis–Menten equations. Later, the same approach was applied to hydrolysis and absorption in the small intestine (Jumars, 2000a, 2000b). More recently, a generalization was conducted, using an advection–reaction equation, thus implementing a convective term (Logan et al., 2002). Its form is

$$\frac{\partial n}{\partial t} = -\vartheta \frac{\partial n}{\partial x} - \nu(x, t, n) \tag{14.8}$$

where n is the nutrient concentration, ϑ is the velocity, and ν is the consumption rate (by hydrolysis and absorption), set to obey either first-order, Michaelis–Menten, or sigmoid kinetics. Furthermore, this model was extended to saccular organs, e.g. the stomach (Logan et al., 2003).

A recent effort has been proposed to implement most processes in the small intestine, including pulsed flow, water proportion, secretion of digestive juices, brush border enzymatic reactions, and indirectly accounting for processes in the stomach (Taghipoor et al., 2012, 2014). The transport equation is

$$\frac{d^2x}{dt^2} = \tau\left(1 - \frac{1}{c}\frac{dx}{dt}\right)\frac{c_0 + c_1 R}{a + bx} - \frac{K_{visco}}{[W(t)]}\frac{dx}{dt} \tag{14.9}$$

where τ is the mean effect of the pulses by unit time, c is the mean velocity of the waves, and $[W]$ is the water mass proportion. The other constants are the parameters of the model, estimated from biological data. Hydrolysis and absorption were supposed to obey first-order kinetics. A distinction was made between non-solubilized and solubilized species.

Note that all these macroscale models represented flow using advection or a mass balance, and none of them included diffusion. Most of them were developed in the pig, because large sets of physiologic data are available. Some equivalent models exist in the ruminant, but are less transposable to the human, as the rumen system considerably differs from the human stomach.

14.4 MICROSCALE MODELING OF FOOD DIGESTION AND ABSORPTION

Currently, there is no complete model of food digestion and absorption in humans. The efforts consist in the description of specific processes, usually within one organ. These models are generally food-oriented, accounting for detailed structures at the microscale.

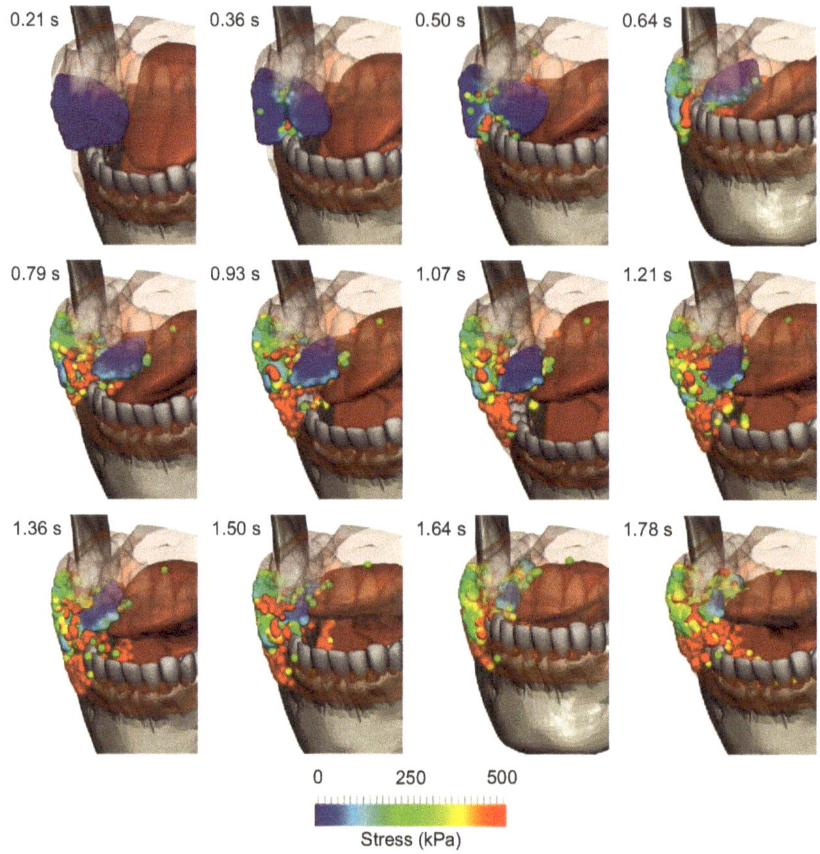

FIGURE 14.1 Simulation of agar gel chewing, displaying the breakdown into particles and their von Mises stress. Reproduced from Harrison et al. (2014a) with permission from John Wiley and Sons.

14.4.1 Mastication

The first step of food digestion is oral processing. Models exist concerning sensory features, which are important but only indirectly related to digestion (Taylor, 2002; Chen, 2009, 2014). Other crucial aspects are the saliva incorporation and the breakdown of solid foods into small pieces, essential for further processing. Early models were developed to predict the particle size reduction during food mastication (Lucas and Luke, 1983; Prinz and Lucas, 1997) or the maximum force during artificial (non-food) biting (Koolstra et al., 1988). Recently, the use of finite element or smoothed particle hydrodynamics methods allowed realistic simulations of food mastication (Röhrle and Pullan, 2007; Harrison et al., 2014a; Zhang and Hui, 2015). An example is given in Fig. 14.1, where the mastication of a 5% agar gel was simulated by smoothed particle hydrodynamics (SPH). The von Mises stress for each particle was calculated, and the resulting fragment size distribution was compared to experiments. Fig. 14.2 shows that the predictability depends on the system studied (2% vs. 5%

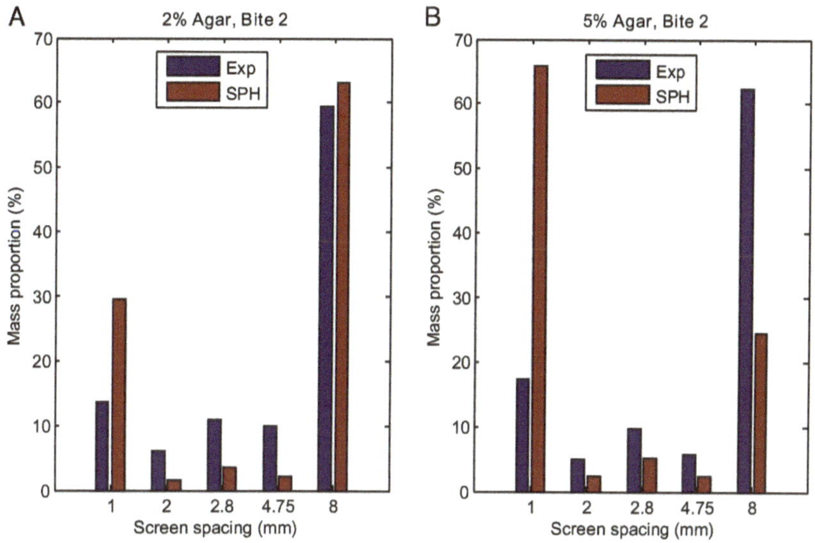

FIGURE 14.2 Experimental and simulated fragment size distributions of agar gels after two chewing cycles. Reproduced from Harrison et al. (2014a) with permission from John Wiley and Sons.

agar gels). Such simulations can reach a high degree of detail. The most achieved simulation already includes the mechanical effect of saliva (Harrison and Cleary, 2014), and further efforts could account for the release of taste, aroma, and the catalytic role of enzymes in saliva (Harrison et al., 2014b).

14.4.2 Gastric Flow and Mixing

Similarly to the oral cavity, the human stomach was only modeled for mechanical purposes so far. The first simulation, presented in 2004 by the research group of James Brasseur, was based on a lattice Boltzmann method to characterize the velocity and pressure fields in a realistic 2D reconstruction of the stomach (Pal et al., 2004, 2007). Later, a similar approach was undertaken in 3D, using a computational fluid dynamics solver to simulate the velocity and pressure fields (Ferrua and Singh, 2010; Ferrua et al., 2014). Different types of fluid were simulated, namely Newtonian and shear thinning fluids. The resulting velocity fields are shown in Fig. 14.3 to illustrate the role of food rheology on the laminar flow and recirculation flow (vortices). In parallel, a simulation based on a finite-volume method was developed for the prediction of the velocity fields in a 2D idealized geometry of the antrum, where the secretion and mixing of pepsin were included, but not its catalytic action (Kozu et al., 2010). Note that all these models were only developed in the case of liquid food, so no breakdown could be investigated. In the future, the catalytic effect of enzymes as well as the behavior of solid food should be included for further realism.

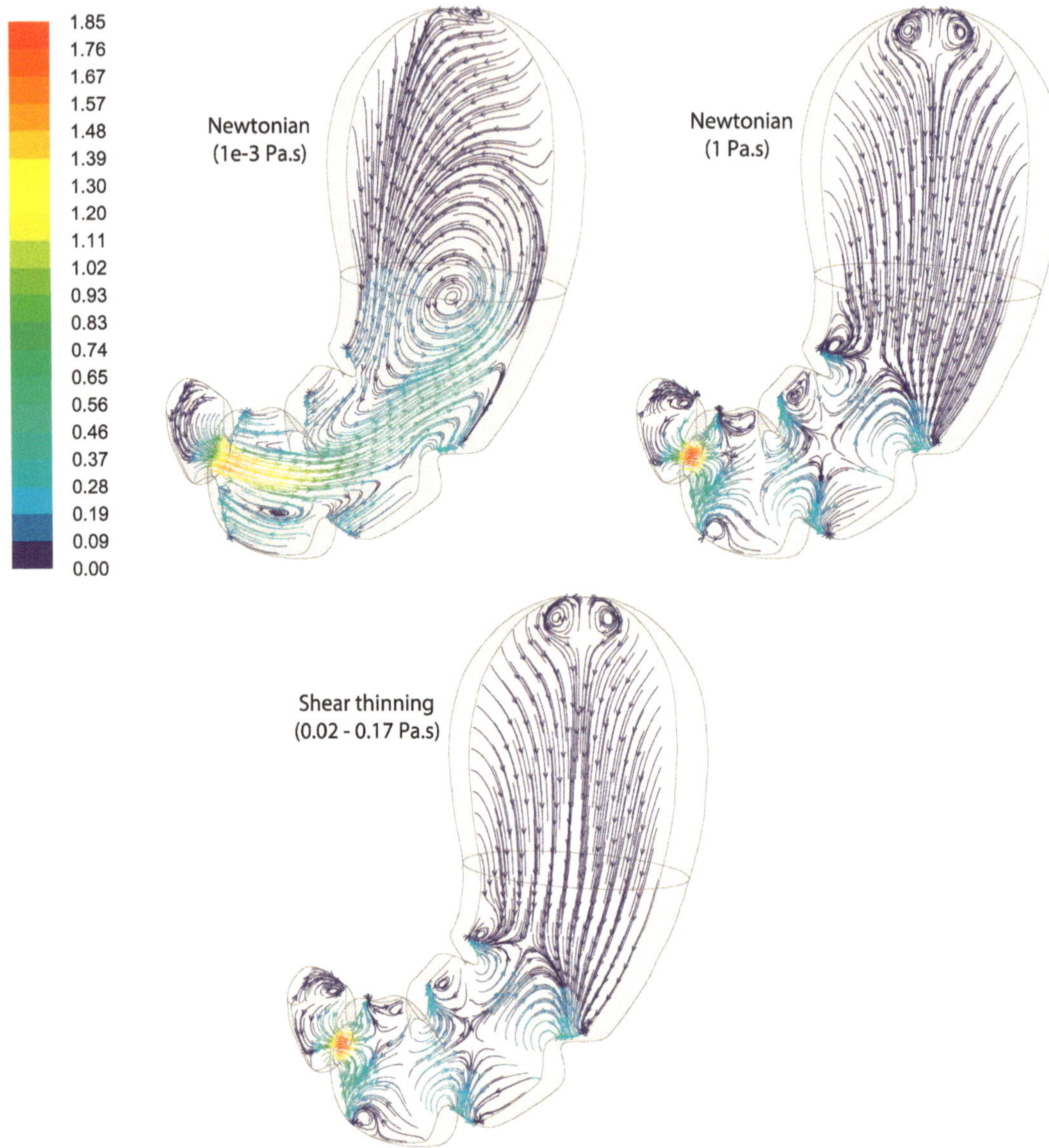

FIGURE 14.3 Simulated velocity field in the stomach for liquids with different rheological properties, colored by velocity magnitude ($cm\,s^{-1}$). Reproduced from Ferrua et al. (2014) with permission from Elsevier.

14.4.3 Intestinal Digestion and Absorption

Several approaches have been adopted to investigate the small intestine numerically. First, the research group of James Brasseur used a multiscale lattice Boltzmann method to resolve both the macroscale phenomena that occur in the intestinal lumen and the microscale phenomena

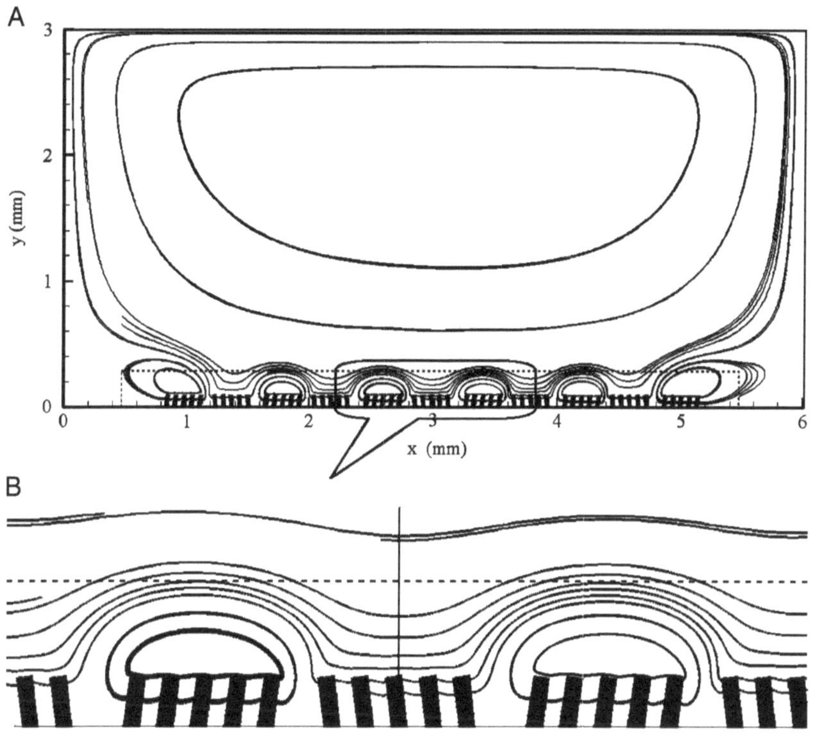

FIGURE 14.4 Simulated instantaneous streamlines in the intestinal lumen (A), with an expanded view near the villi (B). Reproduced from Wang et al. (2010b) with permission from Springer.

that occur at the villi, in a 2D idealized geometry (Wang et al., 2010a, 2010b). A luminal bulk flow, a pendular motion for the villi, and absorption of nutrients were modeled at steady state. The results indicated a coupling between the advection in the lumen and the diffusion in the unstirred water layer, both resulting in vortices that enhance absorption. These findings are illustrated in Fig. 14.4.

Later, the microscale part of the model was explored dynamically by another research group (Lentle et al., 2013; Lim et al., 2015). The simulation was focused around the villi, with alternative hypotheses: no luminal bulk flow, a relaxation/contraction motion for the villi. Similar results were reported, showing the formation of vortices around the villi, highlighting the important role of villi motion for optimal absorption. The flow and absorption patterns during a relaxation/contraction cycle are presented in Fig. 14.5.

Another approach was initiated in the research group of Julian McClements, for the hydrolysis and release of lipid from emulsion droplet (Li and McClements, 2010). Trying to elucidate the role of different physicochemical characteristics of the system, these authors proposed a zeroth order kinetics equation of general form

$$\frac{dm}{dt} = kA(t) \tag{14.10}$$

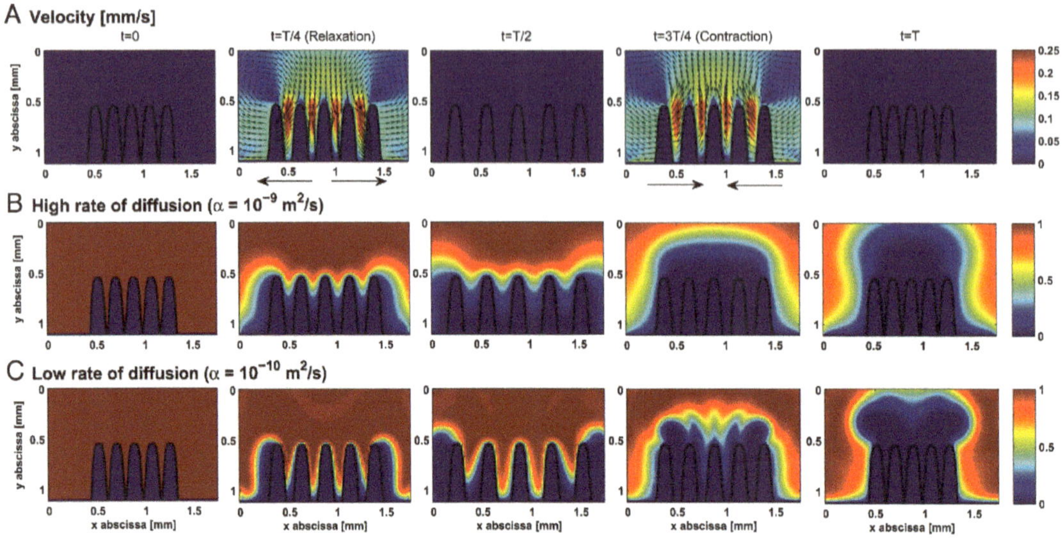

FIGURE 14.5 Simulated velocities and concentrations of a tracer nutrient near the intestinal villi. Reproduced from Lentle et al. (2013) with permission from John Wiley and Sons.

where m is the mass of lipid released from the droplet, k is the mass transfer coefficient, and A is the total surface area of the droplets. This equation can be solved analytically to give (Marze and Choimet, 2012; Gaucel et al., 2015):

$$\frac{\phi}{\phi_{max}} = 1 - \max\left\{0, \left(1 - \frac{kMt}{nr_0\rho_0}\right)^3\right\} \tag{14.11}$$

where ϕ and ϕ_{max} are the proportion and maximum proportion of lipid released from the droplet, M is the triglyceride molar mass (equal to $3M_S$ in the case of a triglyceride with three identical carbon chains, where M_S is the molar mass of the solubilized digestion products), r_0 and ρ_0 are the initial droplet radius and density, and n is a coefficient depending on the type of measurement. Note there was an error in the original calculation of the solution of Eq. (14.10) (Li and McClements, 2010), but the correct form was derived afterwards (Marze and Choimet, 2012; Gaucel et al., 2015). Eq. (14.11) was used to analyze the digestion of emulsified triglyceride, which is hydrolyzed to yield two free fatty acids and one monoglyceride, further solubilized in bile salt micelles. Depending on the technique used to evaluate the hydrolysis of triglyceride, only free fatty acids (pH-stat) or all digestion products (HPLC) are measured. The interpretation should then be made with $n = 2$ or $n = 3$, respectively. Eq. (14.10) can be modified to express the temporal derivative of the droplet radius, which can be compared to measurements (Marze and Choimet, 2012). Note that k is an effective coefficient, as it represents both hydrolysis and solubilization, but can be used to interpret hydrolysis alone or solubilization alone, depending on the experimental data. A first order kinetics equation was also proposed by Marze and Choimet (2012) in the form

$$\frac{dm}{dt} = \frac{kA(t)(m_\infty - m)}{V_T} \tag{14.12}$$

where m_∞ is the maximum mass of digestion products that can be solubilized, and V_T is the total volume of the emulsion sample. This equation can be solved with various hypotheses concerning $(m_\infty - m)$, either that solubility is infinite, or governed by the Kelvin equation, which is

$$m = m_\infty \exp\left(\frac{2\gamma M/3}{\rho N_A k_B T_K r}\right) \tag{14.13}$$

and can be approximated by

$$m \approx m_\infty \left(1 + \frac{2\gamma M/3}{\rho N_A k_B T_K r}\right) \tag{14.14}$$

where γ is the interfacial tension, N_A is the Avogadro constant, k_B is the Boltzmann constant, and T_K is the temperature in Kelvin. Note that the interfacial tension can be further decomposed using interfacial viscoelastic properties. When using the Kelvin equation, analytical solutions exist only for the temporal derivative of the droplet radius. In all cases, it was shown that the first order form was in better agreement with experimental data than the zeroth order form was, especially when the interfacial viscoelastic properties were accounted for (Marze and Choimet, 2012). This finding is illustrated in Fig. 14.6, where the experimental and simulated droplet mean diameters are plotted as a function of intestinal digestion time.

Recently, increasingly complex models for starch digestion in the gastrointestinal tract were developed (Moxon et al., 2016). The most complete model was in the form of advection–reaction–absorption equations:

$$\frac{\partial S}{\partial t} = \gamma S_S - \vartheta \frac{\partial S}{\partial x} - \frac{V_{max}}{K_m + S} S, \tag{14.15}$$

$$\frac{\partial G}{\partial t} = -\vartheta \frac{\partial G}{\partial x} + \frac{V_{max}}{K_m + S} S - \frac{2fK}{R} G \tag{14.16}$$

where S and S_S are the starch concentrations in the small intestine and in the stomach, γ is the decay constant for stomach emptying, V_{max} is the maximum hydrolysis rate, K_m is the Michaelis constant, G is the glucose concentration in the small intestine, f represents the increase in absorptive surface area due to the folds of the intestinal wall, and K is the mass transfer coefficient for absorption, expressed as $K = 1.62(\frac{\vartheta D^2}{2RL})^{1/3}$, with all other parameters already defined previously, the diffusion coefficient D being calculated using the Stokes–Einstein equation.

14.4.4 Multi-Agents Simulation

This type of modeling was first developed in ecology and social sciences, to investigate the interactions between various actors. This was applied to swarm, migration, or predation be-

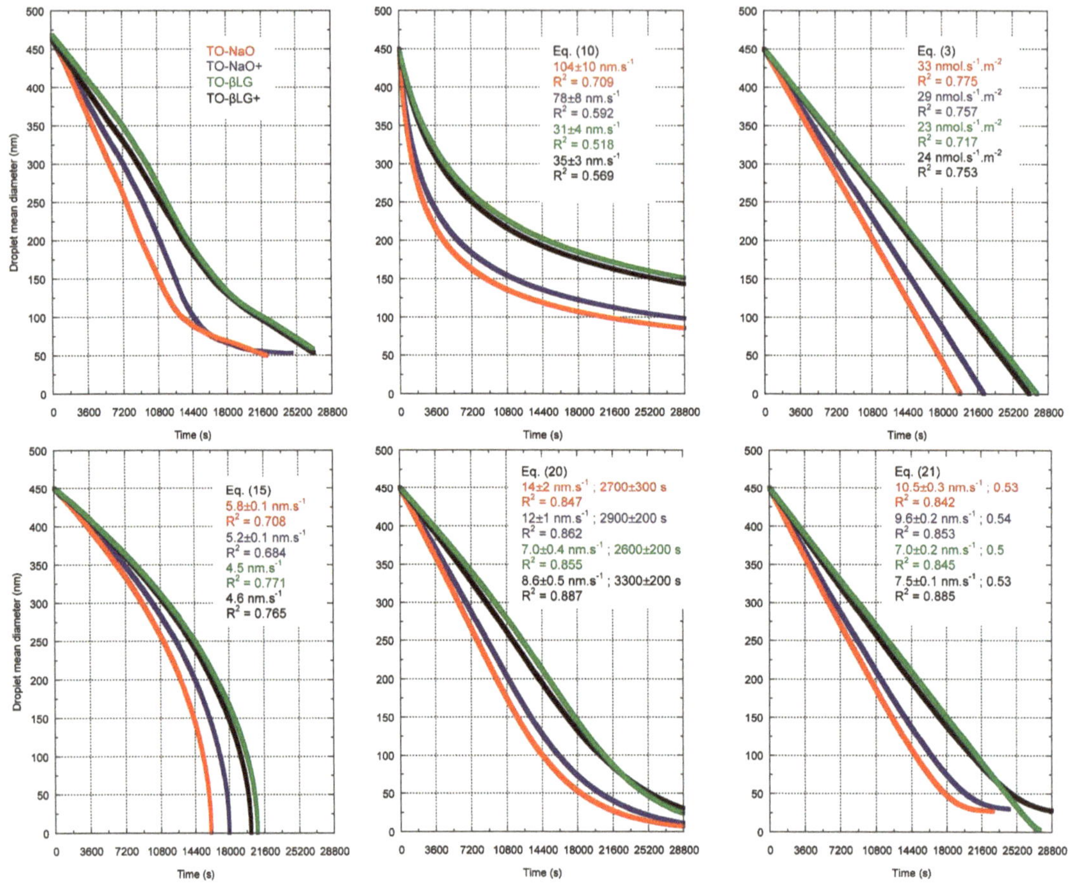

FIGURE 14.6 Mean droplet diameter during in vitro intestinal digestion of various emulsions. The graph on the top-left reports the experimental results. The other graphs report the simulated results using various models. The simple first and zeroth order kinetics correspond to graphs on the top. The first order kinetics integrating the Kelvin equation, the interfacial viscosity or elasticity correspond to graphs on the bottom, from left to right, respectively. Reproduced from Marze and Choimet (2012) with permission from The Royal Society of Chemistry.

haviors, but also to entire organizations such as ant or bee colony. Then, it was enlarged to humans for the study of crowd behavior, as well as cultural, political, social, or economical interactions. In these fields, the applications are abundant, such as diffusion of ideas or values, decision-making processes, business or industrial organizations (Bonabeau, 2002). More recently, it was developed at the microscale for biology, chemistry, and physics (Hunt et al., 2009). Whatever the application is, the basic principle is the use of different agents (particles) that have specific behaviors (properties), and interaction rules. As the type or number of agents changes, the response of the whole system is expected to change, in a deterministic or in a stochastic manner, owing to the possibility of emergent phenomena.

Recently, the use of a 2D multi-agents simulation was proposed to study lipid digestion (Marze, 2014, 2015). Each type of molecule was represented as a particle of given mass, with specific physicochemical properties and interaction rules. The model consisted in a square droplet containing triglyceride and lipophilic micronutrient particles, and lined with interfacial particles (emulsifier stabilizing the interface). It was immersed in a digestive fluid aqueous environment containing a single type of particle, representing enzymes or enzymes/bile salts.

All the simulation rules are outlined in Fig. 14.7. Briefly, all the lipophilic particles were set to diffuse with a relative diffusion coefficient based on the Stokes–Einstein equation, thus depending on their molar mass and on the lipid phase viscosity. The digestive fluid particles (IDF in Fig. 14.7) were set to diffuse faster using position randomization, reflecting their much higher diffusion coefficient, and the possible advection of the whole droplet relative to the aqueous phase. Interfacial particles were set to remain located at the interface, where hydrolysis and solubilization generally occur. When a contact was established between a digestive fluid particle and an interfacial particle, the first was either removed or kept in the simulation, and the second was activated. In this state, a contact with a lipophilic particle resulted in hydrolysis or solubilization, the interfacial particle returning to the inactivated state. This rule was set to account for the interfacial competition between digestive molecules (enzymes, bile salts) and digestion products. The contact rules for hydrolysis and solubilization were derived from experimental data on lipolysis rate and solubilization ratio (mass of solubilizate per mass of bile salt) respectively, the number of contacts required being inversely proportional to these quantities. As seen in Fig. 14.7, triglyceride is hydrolyzed to a fatty acid and a diglyceride, the latter being hydrolyzed to a fatty acid and a monoglyceride. Only fatty acids, monoglycerides, and lipophilic micronutrients can be solubilized in bile salts. Note that the solubilization or micellization concepts are mostly used in pharmacology, while the concept of bioaccessibility (percentage of (micro-)nutrient transferred in bile salt micelles) is preferred in food science, but all refer to the same notion. The droplet size was set to decrease in proportion to the number of solubilized fatty acid and monoglyceride. The simulations where the digestive particles were removed or kept were called the saturated bile or the recycled bile cases, respectively. These represented the *in vitro* or the *in vivo* cases, respectively. The simulations were stopped when there was either no more digestive particles or no more lipids. Pictures in Fig. 14.8 show the evolution of a typical simulation in the saturated bile case.

The main finding reported in the first study (Marze, 2014) was the generic relationship between vitamin bioaccessibility and fatty acid bioaccessibility (Fig. 14.9), even though the solubilization enhancement by digestion products was not taken into account. This aspect, as well as other ones, was accounted for in a second study (Marze, 2015). First, the incorporation of digestion products into bile salt micelle resulted in a higher final vitamin A bioaccessibility. The final bioaccessibility of vitamin A was also increased by its ability to remain at the interface (in the form of retinol). On the other hand, the addition of a gastric step only had a minor effect. Triglyceride mixtures as well as triglyceride–limonene mixtures were also investigated as the oil phase, using the Kendall–Monroe equation to calculate the mixture viscosity. The final bioaccessibilities and their kinetics were significantly affected compared to the pure

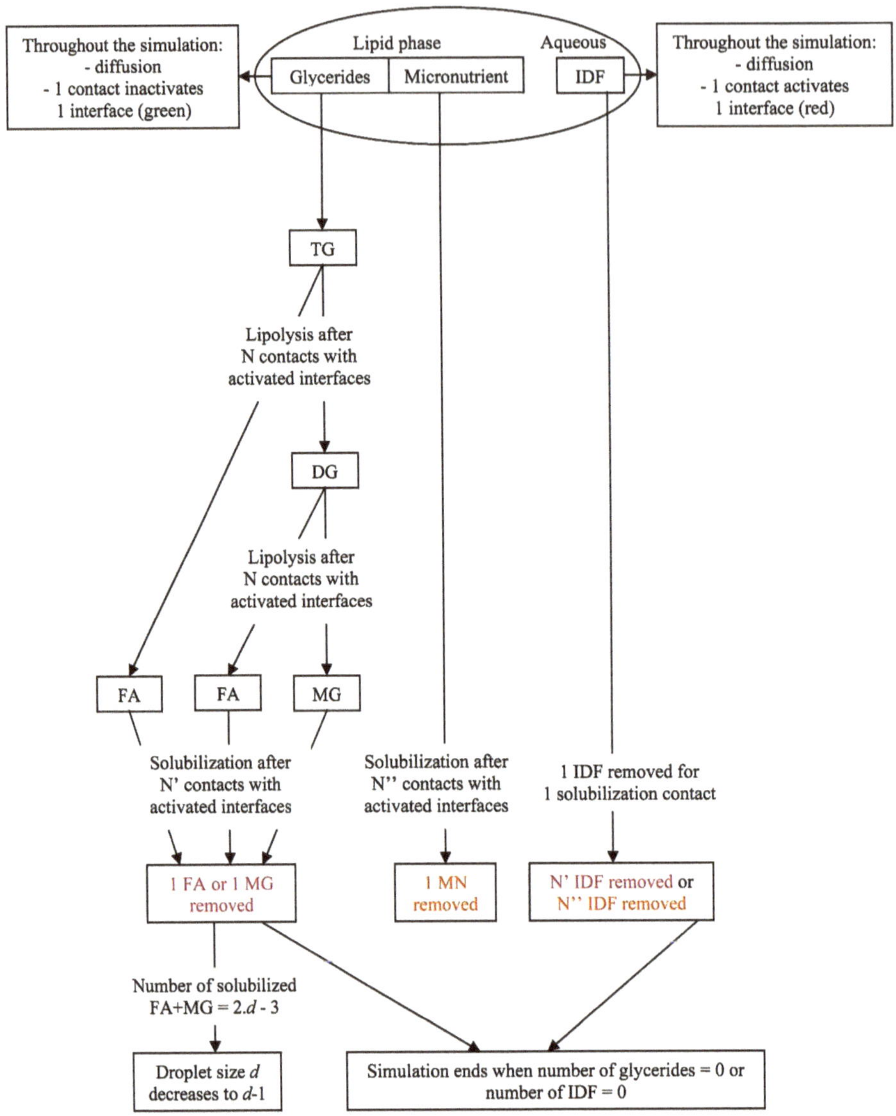

FIGURE 14.7 Diagram showing the rules applied to simulate intestinal lipid digestion. IDF, intestinal digestive fluid; TG, triglyceride; DG, diglyceride; MG, monoglyceride; FA, fatty acid; MN, micronutrient. Reproduced from Marze (2014) with permission from The Royal Society of Chemistry.

compounds. The results for the triglyceride–limonene mixtures are presented in Fig. 14.10. For a given digestion duration, it shows that the more limonene in the mixture, the lower the fatty acid and the vitamin A bioaccessibilities are. Although not quantitative in terms of time, this approach is interesting to compare the effects of various physiological, structural, and formulation parameters on lipid digestion.

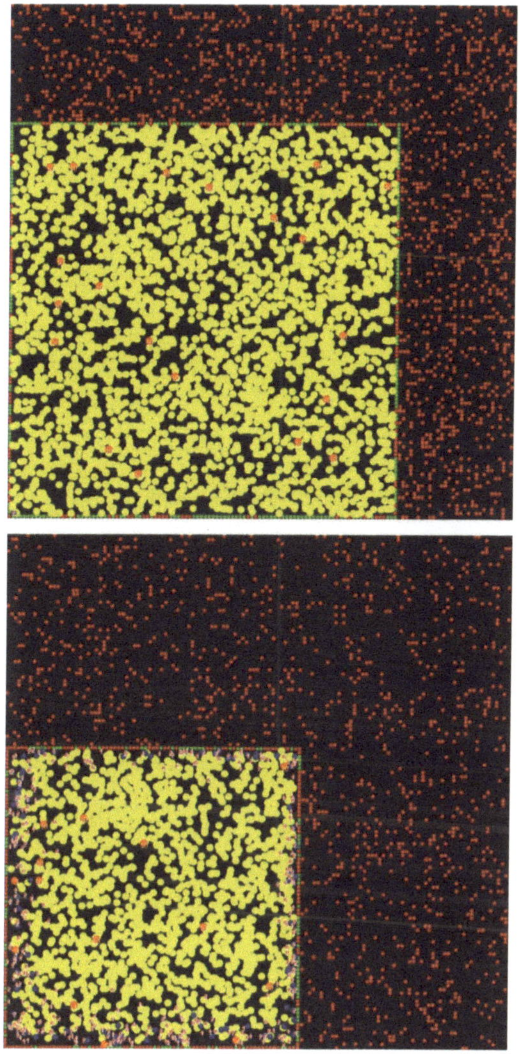

FIGURE 14.8 Pictures from a typical simulation of intestinal lipid digestion in the saturated bile case: (top) first step of the simulation, (bottom) step at which half of the bile was saturated. TG, yellow; MN, orange; IDF, red; interface, green; DG, blue; MG and FA, pink. The numbers represent the number of contacts needed to solubilize the products in bile. Reproduced from Marze (2014) with permission from The Royal Society of Chemistry.

14.5 CONCLUSION

Modeling of food digestion developed in parallel to modeling of oral drug absorption, yet in a scarcer way. Thus, this field is more fragmented, most models being established for a specific nutrient, a specific process, or a specific part of the digestive tract. Nevertheless, de-

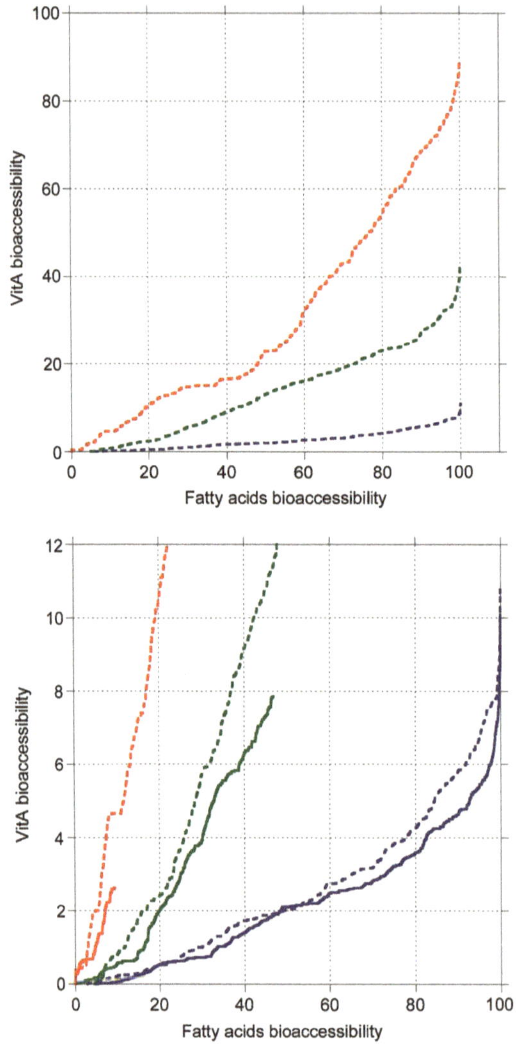

FIGURE 14.9 Results from various simulations of intestinal lipid digestion, plotted as vitamin A bioaccessibility vs. fatty acid bioaccessibility. Trieicosapentaenoin and tridocosahexaenoin, red; triolein, green; tricaprylin, blue. Dashed line, recycled bile; full line, saturated bile. Reproduced from Marze (2014) with permission from The Royal Society of Chemistry.

scriptions tend to be much more detailed. For example, the geometries of the oral cavity and of the stomach are reproduced with high fidelity to investigate mechanical aspects. Another example is the inclusion of the villi for the study of intestinal flow, mixing, and absorption. Overall, these microscale models are very useful to highlight the parameters and mechanisms controlling specific processes. As already initiated recently, some future models will try to include most nutrients, compartments, and processes. Nevertheless, the ultimate goal

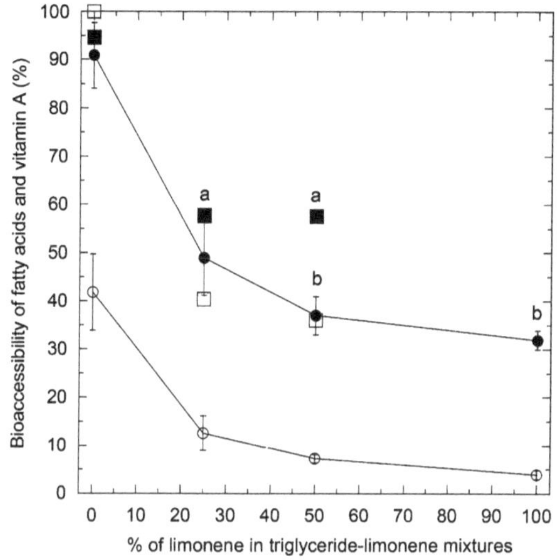

FIGURE 14.10 Results from various simulations of intestinal lipid digestion where the lipid phase is a mixture of triglyceride and limonene, plotted as the bioaccessibility of fatty acids (square) and vitamin A (circle) for the same digestion duration, with tricaprylin (empty symbol) or triolein (full symbol). Reproduced from Marze (2015) with permission from The Royal Society of Chemistry.

to simulate digestion comprehensively is still far ahead, as food is a complex mixture of nutrients with multiscale structures. Some of their interactions occurring during digestion are not precisely known. Moreover, the succession of processes throughout the digestive tract likely plays an important role and needs to be accounted for. Finally, micronutrients such as minerals and vitamins were scarcely included in the models, although they play an essential metabolic role.

References

Ahmed, A.B., Gilja, O.H., Hausken, T., Gregersen, H., Matre, K., 2009. Strain measurement during antral contractions by ultrasound strain rate imaging: influence of erythromycin. Neurogastroenterol. Motil. 21, 170–179.

Amidon, G.E., Ho, N.F.H., French, A.B., Higuchi, W.I., 1981. Predicted absorption rates with simultaneous bulk fluid flow in the intestinal tract. J. Theor. Biol. 89, 195–210.

Amidon, G.L., Lennernäs, H., Shah, V.P., Crison, J.R., 1995. A theoretical basis for a biopharmaceutic drug classification: the correlation of in vitro drug product dissolution and in vivo bioavailability. Pharm. Res. 12, 413–420.

Anderson, D.J., 1956. Measurement of stress in mastication. J. Dent. Res. 35, 664–670.

Bastianelli, D., Sauvant, D., Rerat, A., 1996. Mathematical modeling of digestion and nutrient absorption in pigs. J. Anim. Sci. 74, 1873–1887.

Bennink, R., Peeters, M., Van den Maegdenbergh, V., Geypens, B., Rutgeerts, P., De Roo, M., Mortelmans, L., 1999. Evaluation of small-bowel transit for solid and liquid test meal in healthy men and women. Eur. J. Nucl. Med. 26, 1560–1566.

Bloch, H., 1987. Man's curiosity about food digestion: an historical overview. J. Natl. Med. Assoc. 79, 1223–1227.

Bonabeau, E., 2002. Agent-based modeling: methods and techniques for simulating human systems. Proc. Natl. Acad. Sci. USA 99, 7280–7287.

Camilleri, M., Zinsmeister, A.R., Greydanus, M.P., Brown, M.L., Proano, M., 1991. Towards a less costly but accurate test of gastric emptying and small bowel transit. Dig. Dis. Sci. 36, 609–615.

Chen, J., 2009. Food oral processing – a review. Food Hydrocoll. 23, 1–25.

Chen, J., 2014. Food oral processing: some important underpinning principles of eating and sensory perception. Food Struct. 1, 91–105.

Degen, L.P., Phillips, S.F., 1996. Variability of gastrointestinal transit in healthy women and men. Gut 39, 299–305.

Dejak, B., Młotkowski, A., Romanowicz, M., 2003. Finite element analysis of stresses in molars during clenching and mastication. J. Prosthet. Dent. 90, 591–597.

de Loubens, C., Lentle, R.G., Love, R.J., Hulls, C., Janssen, P.W.M., 2013. Fluid mechanical consequences of pendular activity, segmentation and pyloric outflow in the proximal duodenum of the rat and the guinea pig. J. R. Soc. Interface 10, 20130027.

Desipio, J., Friedenberg, F.K., Korimilli, A., Richter, J.E., Parkman, H.P., Fisher, R.S., 2007. High-resolution solid-state manometry of the antropyloroduodenal region. Neurogastroenterol. Motil. 19, 188–195.

Ehrlein, H., Stockmann, A., 1998. Absorption of nutrients is only slightly reduced by supplementing enteral formulas with viscous fiber in miniature pigs. J. Nutr. 128, 2446–2455.

Ferrua, M.J., Singh, R.P., 2010. Modeling the fluid dynamics in a human stomach to gain insight of food digestion. J. Food Sci. 75, R151–R162.

Ferrua, M.J., Xue, Z., Singh, R.P., 2014. On the kinematics and efficiency of advective mixing during gastric digestion – a numerical analysis. J. Biomech. 47, 3664–3673.

Gaucel, S., Trelea, I.C., Le Feunteun, S., 2015. Comment on new mathematical model for interpreting pH-stat digestion profiles: impact of lipid droplet characteristics on in vitro digestibility. J. Agric. Food Chem. 63, 10352–10353.

Graff, J., Brinch, K., Madsen, J.L., 2001. Gastrointestinal mean transit times in young and middle-aged healthy subjects. Clin. Physiol. 21, 253–259.

Harrison, S.M., Cleary, P.W., 2014. Towards modelling of fluid flow and food breakage by the teeth in the oral cavity using smoothed particle hydrodynamics (SPH). Eur. Food Res. Technol. 238, 185–215.

Harrison, S.M., Eyres, G., Cleary, P.W., Sinnott, M.D., Delahunty, C., Lundin, L., 2014a. Computational modeling of food oral breakdown using smoothed particle hydrodynamics. J. Texture Stud. 45, 97–109.

Harrison, S.M., Cleary, P.W., Eyres, G., Sinnott, M.D., Lundin, L., 2014b. Challenges in computational modelling of food breakdown and flavour release. Food Funct. 5, 2792–2805.

Haruma, K., Wiste, J.A., Camilleri, M., 1994. Effect of octreotide on gastrointestinal pressure profiles in health and in functional and organic gastrointestinal disorders. Gut 35, 1064–1069.

Hausken, T., Mundt, M., Samsom, M., 2002. Low antroduodenal pressure gradients are responsible for gastric emptying of a low-caloric liquid meal in humans. Neurogastroenterol. Motil. 14, 97–105.

Higuchi, W.I., 1967. Diffusional models useful in biopharmaceutics – drug release rate processes. J. Pharm. Sci. 56, 315–324.

Hiiemae, K., 2004. Mechanisms of food reduction, transport and deglutition: how the texture of food affects feeding behavior. J. Texture Stud. 35, 171–200.

Huang, W., Lee, S.L., Yu, L.X., 2009. Mechanistic approaches to predicting oral drug absorption. AAPS J. 11, 217–224.

Hunt, C.A., Ropella, G.E.P., Lam, T.N., Tang, J., Kim, S.H.J., Engelberg, J.A., Sheikh-Bahaei, S., 2009. At the biological modeling and simulation frontier. Pharm. Res. 26, 2369–2400.

Hveem, K., Sun, W.M., Hebbard, G., Horowitz, M., Doran, S., Dent, J., 2001. Relationship between ultrasonically detected phasic antral contractions and antral pressure. Am. J. Physiol. 281, G95–G101.

Indireshkumar, K., Brasseur, J.G., Faas, H., Hebbard, G.S., Kunz, P., Dent, J., Feinle, C., Li, M., Boesiger, P., Fried, M., Schwizer, W., 2000. Relative contributions of "pressure pump" and "peristaltic pump" to gastric emptying. Am. J. Physiol. 278, G604–G616.

Johnson, L.R., 2014. Gastrointestinal Physiology, 8th edition. Elsevier Mosby, Philadelphia.

Jumars, P.A., 2000a. Animal guts as ideal chemical reactors: maximizing absorption rates. Am. Nat. 155, 527–543.

Jumars, P.A., 2000b. Animal guts as nonideal chemical reactors: partial mixing and axial variation in absorption kinetics. Am. Nat. 155, 544–555.

Koolstra, J.H., Van Eijden, T.M.G.J., Weijs, W.A., Naeije, M., 1988. A three-dimensional mathematical model of the human masticatory system predicting maximum possible bite forces. J. Biomech. 21, 563–576.

Kozu, H., Kobayashi, I., Nakajima, M., Uemura, K., Sato, S., Ichikawa, S., 2010. Analysis of flow phenomena in gastric contents induced by human gastric peristalsis using CFD. Food Biophys. 5, 330–336.

Kwiatek, M.A., Menne, D., Steingoetter, A., Goetze, O., Forras-Kaufman, Z., Kaufman, E., Fruehauf, H., Boesiger, P., Fried, M., Schwizer, W., Fox, M.R., 2009. Effect of meal volume and calorie load on postprandial gastric function and emptying: studies under physiological conditions by combined fiber-optic pressure measurement and MRI. Am. J. Physiol. 297, G894–G901.

Le Feunteun, S., Barbe, F., Remond, D., Menard, O., Le Gouar, Y., Dupont, D., Laroche, B., 2014. Food Bioprocess Technol. 7, 1099–1113.

Lentle, R.G., Janssen, P.W.M., De Loubens, C., Lim, Y.F., Hulls, C., Chambers, P., 2013. Mucosal microfolds augment mixing at the wall of the distal ileum of the brushtail possum. Neurogastroenterol. Motil. 25, 881–e700.

Li, Y., McClements, D.J., 2010. New mathematical model for interpreting pH-stat digestion profiles: impact of lipid droplet characteristics on in vitro digestibility. J. Agric. Food Chem. 58, 8085–8092.

Lifson, N., Hakim, A.A., 1966. Simple diffusive–convective model for intestinal absorption of a nonelectrolyte (urea). Am. J. Physiol. 211, 1137–1146.

Lim, Y.F., De Loubens, C., Love, R.J., Lentle, R.G., Janssen, P.W.M., 2015. Flow and mixing by small intestine villi. Food Funct. 6, 1787–1795.

Logan, J.D., Joern, A., Wolesensky, W., 2002. Location, time, and temperature dependence of digestion in simple animal tracts. J. Theor. Biol. 216, 5–18.

Logan, J.D., Joern, A., Wolesensky, W., 2003. Chemical reactor models of optimal digestion efficiency with constant foraging costs. Ecol. Model. 168, 25–38.

Loo, J.C.K., Riegelman, S., 1968. New method for calculating the intrinsic absorption rate of drugs. J. Pharm. Sci. 57, 918–928.

Lucas, P.W., Luke, D.A., 1983. Computer simulation of the breakdown of carrot particles during human mastication. Arch. Oral Biol. 28, 821–826.

Marze, S., Choimet, M., 2012. In vitro digestion of emulsions: mechanistic and experimental models. Soft Matter 8, 10982–10993.

Marze, S., 2014. A coarse-grained simulation to study the digestion and bioaccessibility of lipophilic nutrients and micronutrients in emulsion. Food Funct. 5, 129–139.

Marze, S., 2015. Refining in silico simulation to study digestion parameters affecting the bioaccessibility of lipophilic nutrients and micronutrients. Food Funct. 6, 115–124.

Moxon, T.E., Gouseti, O., Bakalis, S., 2016. In silico modelling of mass transfer & absorption in the human gut. J. Food Eng. 176, 110–120.

Nelson, E., 1961. Kinetics of drug absorption, distribution, metabolism, and excretion. J. Pharm. Sci. 50, 181–192.

Ni, P.F., Ho, N.F.H., Fox, J.L., Leuenberger, H., Higuchi, W.I., 1980. Theoretical model studies of intestinal drug absorption V. Non-steady-state fluid flow and absorption. Int. J. Pharm. 5, 33–47.

Ono, T., Hori, K., Nokubi, T., 2004. Pattern of tongue pressure on hard palate during swallowing. Dysphagia 19, 259–264.

Pal, A., Indireshkumar, K., Schwizer, W., Abrahamsson, B., Fried, M., Brasseur, J.G., 2004. Gastric flow and mixing studied using computer simulation. Proc. R. Soc. B 271, 2587–2594.

Pal, A., Brasseur, J.G., Abrahamsson, B., 2007. A stomach road or "Magenstrasse" for gastric emptying. J. Biomech. 40, 1202–1210.

Palmer, J.B., Rudin, N.J., Lara, G., Crompton, A.W., 1992. Coordination of mastication and swallowing. Dysphagia 7, 187–200.

Penry, D.L., Jumars, P.A., 1987. Modeling animal guts as chemical reactors. Am. Nat. 129, 69–96.

Prinz, J.F., Lucas, P.W., 1997. An optimization model for mastication and swallowing in mammals. Proc. R. Soc. B 264, 1715–1721.

Read, N.W., Al-Janabi, M.N., Holgate, A.M., Barber, D.C., Edwards, C.A., 1986. Simultaneous measurement of gastric emptying, small bowel residence and colonic filling of a solid meal by the use of the gamma camera. Gut 27, 300–308.

Riegelman, S., Loo, J.C.K., Rowland, M., 1968. Shortcomings in pharmacokinetic analysis by conceiving the body to exhibit properties of a single compartment. J. Pharm. Sci. 57, 117–123.

Rivest, J., Bernier, J.F., Pomar, C., 2000. A dynamic model of protein digestion in the small intestine of pigs. J. Anim. Sci. 78, 328–340.

Röhrle, O., Pullan, A.J., 2007. Three-dimensional finite element modelling of muscle forces during mastication. J. Biomech. 40, 3363–3372.

Shama, F., Sherman, P., 1973. Identification of stimuli controlling the sensory evaluation of viscosity II Oral methods. J. Texture Stud. 4, 111–118.

Siepmann, J., Siepmann, F., 2008. Mathematical modeling of drug delivery. Int. J. Pharm. 364, 328–343.

Strathe, A.B., Danfær, A., Chwalibog, A., 2008. A dynamic model of digestion and absorption in pigs. Anim. Feed Sci. Technol. 143, 328–371.

Suzuki, A., Higuchi, W.I., Ho, N.F.H., 1970. Theoretical model studies of drug absorption and transport in the gastrointestinal tract I. J. Pharm. Sci. 59, 644–651.

Taghipoor, M., Lescoat, P., Licois, J.R., Georgelin, C., Barles, G., 2012. Mathematical modeling of transport and degradation of feedstuffs in the small intestine. J. Theor. Biol. 294, 114–121.

Taghipoor, M., Barles, G., Georgelin, C., Licois, J.R., Lescoat, P., 2014. Math. Biosci. 258, 101–112.

Taylor, A.J., 2002. Release and transport of flavors in vivo: physicochemical, physiological, and perceptual considerations. Compr. Rev. Food Sci. Food Saf. 1, 45–57.

Tomassone, R., Laplace, J.P., 1973. Utilization of compartment systems for studying the physiology of digestion. Ann. Biol. Anim. Biochim. Biophys. 13, 747–753.

Usry, J.L., Turner, L.W., Stahly, T.S., Bridges, T.C., Gates, R.S., 1991. GI tract simulation model of the growing pig. Trans. ASABE 34, 1879–1890.

Wagner, J.G., Nelson, E., 1964. Kinetic analysis of blood levels and urinary excretion in the absorptive phase after single doses of drug. J. Pharm. Sci. 53, 1392–1403.

Wang, Y., Brasseur, J.G., Banco, G.G., Webb, A.G., Ailiani, A.C., Neuberger, T., 2010a. A multiscale lattice Boltzmann model of macro- to micro-scale transport, with applications to gut function. Philos. Trans. R. Soc. A 368, 2863–2880.

Wang, Y., Brasseur, J.G., Banco, G.G., Webb, A.G., Ailiani, A.C., Neuberger, T., 2010b. Development of a Lattice–Boltzmann method for multiscale transport and absorption with application to intestinal function. In: De, S., Guilak, F., Mofrad, M.R.K. (Eds.), Computational Modeling in Biomechanics. Springer, London.

Yu, L.X., Lipka, E., Crison, J.R., Amidon, G.L., 1996. Transport approaches to the biopharmaceutical design of oral drug delivery systems: prediction of intestinal absorption. Adv. Drug Deliv. Rev. 19, 359–376.

Yu, L.X., Amidon, G.L., 1999. A compartmental absorption and transit model for estimating oral drug absorption. Int. J. Pharm. 186, 119–125.

Zhang, X., Hui, M., 2015. The research and development on simulation of oral cavity food chewing system. Adv. J. Food Sci. Technol. 7, 553–557.

Index

A
Absorption, 34, 116, 178, 179, 183, 186, 187, 202, 245, 247, 353, 354, 356, 358, 359, 363, 365, 370
Advection, 89, 157, 160, 161, 168, 180, 181, 238, 312, 322, 332, 356, 359, 363, 365, 367
Airway surface liquid, 116, 118, 135, 137, 150
Alterations, 77–79, 82, 83, 92, 106
Analyte, 46, 51–53, 55–57, 71
Analyte concentration, 50, 52, 54–57, 71
Anterograde, 336, 337, 339, 341, 344, 347
Antibodies, 11, 53, 55, 56
Applications, engineering, 312, 328, 332, 333
Aqueous phase, 22, 24–27, 32, 34, 35, 367
AR, *see* aspect ratio
Arterial tissues, 77, 79, 82, 86–89, 99, 101
Aspect ratio (AR), 45, 53, 58, 60, 155, 156, 161
Assays, 44, 47–49, 55–57
Axisymmetric case, 95
Axon, 336–349
Axon terminals, 335, 336, 338, 339, 349

B
Bioaccessibility, 367, 368, 370
Biochemical assays, 44, 52, 53, 55, 65, 70, 71
Biochemical environment, 78, 79, 84, 85, 89–92, 106
Biochemical reactions, 43, 44, 46, 47, 50, 53, 71
Bioconvection, 195–199, 202
Biodegradation, 20, 22, 24, 29
Biofilm, 25, 27, 29, 31–35, 192
Biofilm formation, 25, 29, 31, 32, 34
Bioheat term, 262, 273, 275, 276, 305
Biological membranes, 1, 2, 208, 237
Biological tissue, 247, 261–264, 266, 268, 270, 277, 279, 293, 294, 312, 313, 322
Biomembranes, 209, 211, 221, 227, 232
Biomolecular transport, 44, 71
Bioremediation, 20
Biosurfactants, 19, 20, 22–24, 26–28, 32, 35
Blood, 243, 244, 247, 248, 253, 261, 262, 264–270, 272, 277, 279, 356
Blood flow, 89, 243, 244, 247, 248, 250–257, 270
Blood perfusion, 254, 255, 266, 267
Blood velocity, 262, 269–272
Boundary, 84, 122, 123, 147, 161, 181, 203, 248, 257, 277–279, 294, 313, 321, 326, 328, 329, 331
Boundary conditions, 44, 60, 61, 65, 89, 90, 119, 120, 122, 123, 132, 165, 166, 252, 253, 257, 277, 278, 287, 288, 292, 293, 295, 328–331
  generalized, 279, 280
  homogeneous, 282, 283, 294
  non-homogeneous, 277, 279, 280, 282, 287–289, 293
  periodic, 120, 123, 124, 214
Boutons, 336–350
  passant, 335, 336, 339
  proximal, 339, 341–343, 346–349
Branches, 324, 325, 339, 340

C
Cell bodies, 29, 30, 174–177, 202
Cell concentrations, 180, 184–189, 191–193, 195, 196, 200
Cell culture, 42–44, 60
Cell densities, 59, 64, 65, 67, 179, 181, 184, 185, 188, 189, 198
Cell monolayer, 46, 49, 52, 58–63, 65, 67, 69, 312
Cell problems, 312, 319–321, 323, 324, 332
Cell surface, 22, 23, 25, 26, 29, 59, 67, 174
Cell wall materials, 156, 160, 161, 165–168
Cells, 4, 26–32, 42–44, 48, 49, 57–60, 62–69, 77, 78, 155, 156, 172–175, 177, 178, 180–186, 188–197, 199–203, 208, 209, 221, 222
  biofilm, 32
  biological, 22, 78, 82, 218
  bottom-heavy, 193, 194
  ciliated, 114, 115
  ciliated epithelium, 120, 129, 134, 136
  motile, 26
  planktonic, 27
  smooth muscle, 80, 82, 97, 106
  transport of, 190, 202
Channel heights, 50, 52, 57, 59, 62, 69
CHARMM, 2, 3
Chemical concentration, 180, 181, 202
Chemistry, 90, 366, 368–371
Chemotaxis, 26, 176, 180–182, 189, 190, 203
CIMs (curvature-inducing molecules), 227, 228, 237–239
Coarse-grain simulations, 7, 9, 11
COC (cyclic olefin copolymer), 62, 67–69
Collagen, 80, 83, 88, 247
Collagen fibers, 80, 83, 86, 87, 95, 96, 100, 103, 108
Collagen volume fractions, 94, 95, 99, 103, 104
Compartments, 55, 271, 340–342, 358, 359, 370
Competitive immunoassays, 54–56, 69
Computational domain, 119, 120, 222, 235
Computational models, 53, 55, 56, 61, 64–68, 71, 226, 243
Computations, 93, 118, 121, 129–136, 147, 149, 298, 323, 325, 326

Concentration, saturated DCV, 340, 343, 346
Concentration distribution, 161, 165–167, 176
Concentration of ligand, 47, 48
Concentrations of DCVs, 341, 345, 348
Configuration, 62, 84–86, 92, 94, 95, 188, 199, 212
Consistency, 141–143, 145, 146
Constituents, 7, 78, 79, 83, 87, 103, 104, 106
Contact, physical, 21, 22
Continuum model of laser heating, 244
Convection, 46, 50–52, 119, 120, 126, 130, 131, 147, 156, 158, 201, 266, 356, 357
Convective transport, 43, 49, 57
Cooling, 172, 300–304
Cox–Merz rule, 141, 142, 144–146
Crude oil, 19, 21, 28, 30, 32, 34
Culture devices, microfluidic cell, 44, 49, 58, 60, 62–64, 66, 70, 71
Culture medium, 43, 46, 59–62, 65, 66
Curvature, 33, 222, 223, 229–231, 233, 237, 244
  spontaneous, 222, 223, 227–233, 235–238
Curvature-inducing molecules, *see* CIMs
Curves, 144–146, 189, 190, 228
Cycles, 149, 207, 214, 361
Cyclic olefin copolymer (COC), 62, 67–69
Cystic fibrosis, 113–115, 138, 143, 148, 149, 151
Cytoplasmic streaming, 155, 156, 158, 160, 162

D

DCV circulation, 338, 339, 343, 347–350
DCV concentration, 340, 341, 347
DCV fluxes, 340–343, 347
DCV production, 338, 339, 341, 343–347, 350
DCV production rates, 338, 345, 347, 350
DCV transport, 335–339, 343, 349
Dense core vesicles (DCVs), 335–350
Density
  inter-molecular cross-links, 94, 95, 103, 104
  laser power, 262, 300–304
Depth, 172, 177, 178, 180, 183–187, 189, 190, 195–197, 199, 202, 257, 293
Diffusion, 46, 51–55, 66, 71, 82, 89, 130, 131, 156, 160, 161, 168, 180, 234, 238, 356, 357, 359
Diffusion coefficient, 34, 46, 55, 59, 122, 160, 161, 365, 367
Diffusion constant, 216, 217
Diffusion dynamics, 157, 160, 161
Diffusion equation, 46, 50, 51, 119, 120, 126, 130, 276
  dual-phase lag bio-heat, 302
Diffusivity, 61, 180, 182, 188, 189
Digestion, 353, 358, 360, 364, 371
Digestive tract, 353, 354, 356, 369, 371
Dimensionless numbers, 49, 57, 357, 358
Direction, 26, 29, 114, 156, 174–183, 192, 193, 197, 200, 201, 214, 250, 280, 294, 340, 347
Dispersion model, 357
Distal bouton, 337, 339–341, 343–347
Domains, 14, 53, 54, 58, 59, 61, 119, 121, 123, 213, 214, 216, 217, 235, 236, 313, 315, 316, 321, 322, 329, 331, 332
  biological, 262–264
DPL (dual-phase lag), 261, 262, 272, 276, 289, 292, 293
Droplets, 30, 32, 35, 197, 364, 367
Drug absorption, 353, 356, 357
Dual-phase lag, *see* DPL
Duration, 214, 245, 246, 253, 338, 347
Dynamics, 4, 6, 31, 156, 195, 197, 198, 209, 214, 215, 224, 227, 236, 311, 316, 332

E

ECM (extracellular matrix), 20, 29, 31, 32, 34, 78, 83, 262
Elastin, 80, 83, 94–96, 103, 104
Endocytosis, 11, 221, 222, 227, 233, 237, 239
Endothelial cells, 43, 67, 80, 82, 94, 98
Energy, 22, 28, 171, 172, 175, 176, 202, 211, 222, 224–226, 228–230, 232, 233, 236, 244, 245, 338, 356
Enzymes, 68, 77, 105, 354, 361, 367
Eukaryotes, 6, 173, 174, 176, 177, 202, 203
Experiments, numerical, 312, 322, 323, 326, 329
Extracellular matrix, *see* ECM

F

Fatty acids, 354, 367, 368, 371
Fibers, 80, 86–88, 103, 106
Fission, 227, 236, 237
Flatlanders, 29, 30, 34
Fluid, periciliary, 114, 116, 118, 120, 137, 149, 150
Fluid domains, surrounding, 321, 332
Fluid flow, 26, 27, 35, 43, 50, 65, 70, 156, 160, 161, 163, 313, 322, 323, 325, 332
Fluid layer, periciliary, 120, 129, 135, 137, 147
Fluid motion, 41, 44, 53, 71, 72
Fluid velocity, 53, 56, 129, 225, 226, 314, 316
Fluid viscosity, periciliary, 134, 135, 147
Fluxes, 54, 57, 59, 159, 161, 165, 166, 181, 190, 340–342, 346–349
Fokker–Planck equation, 175, 177, 182, 184
Food digestion, 359, 360, 369
Formation, 5–7, 13, 24, 32, 48, 49, 171, 192, 194, 197, 222, 227, 236, 363
Fountain streaming, 155–168
Frequency, 114, 135, 137, 139, 141, 142, 146, 149
Function of depth, 184, 185, 187
Fusion, 11, 227, 232, 236, 336, 337

G

Gastric emptying, 358, 359
Gaussian, 209, 222, 223, 228, 229, 252

Geometries, 41, 45, 49, 50, 52, 53, 65, 201, 323, 324, 326, 332, 370
Gradients, chemical, 26, 181, 188, 198, 200–202
Green's function (GF), 262, 289–292, 295, 296, 298, 303
GROMOS, 2, 3
Growth, 8, 9, 13, 28, 29, 34, 64, 77, 78, 117, 156, 232, 312, 313, 315, 316, 318, 321, 322, 326, 332
Growth factors, 62, 63, 82, 83
Growth mode, 25–27, 35
  flatlander's, 30–32
Gyrotaxis, 176–178, 180, 181, 183, 194, 196, 197, 199

H
Harmonic term, 276, 278–281
Heat transfer, 243, 244, 252, 254, 265
Hepatocytes, 59, 67–69, 71
Hexadecane, 27–30
Homeostatic state, 91, 92, 96, 99–101, 103–105, 108
Hydrocarbons, 19, 25
Hydrodynamics, 22, 157, 159, 161, 162, 168, 174, 225, 232, 236, 239
Hydrolysis, 207, 354, 358, 359, 363, 364, 367
Hydrophobic molecules, 24, 25
Hydrophobicity, 25–27, 29

I
Illumination, 176–179, 183, 186, 202
Immersed boundary method, 118
Immersing fluid, 190, 195, 196, 202, 203
Infection, 114, 116, 117, 142
Instabilities, 194, 196–198, 202
Interface, 5, 24, 26, 28–33, 42, 47, 60–62, 118, 197, 225, 232–235, 313–317, 319, 323, 367
  biofilm, 32, 34
Interfacial area, 23, 24, 29, 31
Interfacial tension, 20, 28–30, 33, 34, 365
Intestine, 354–359, 362, 365
Irradiation, 245–248, 253–255, 257

L
Large amplitude oscillations (LAOs), 144
Laser, short pulse, 244, 245, 257
Laser irradiation, 246–249, 251, 253–257, 261, 263, 293, 294
  short pulse, 245, 253–255, 257
Ligand, 47, 48, 52, 55, 208
Light absorption, 178
Light intensity, 177–179, 186–188
Light source, 177–179, 247
Lipid, 2, 4–6, 9–11, 13, 14, 22, 23, 209–212, 221, 222, 224, 227, 228, 354–358, 363, 364, 367
Lipid bilayers, 4, 5, 12, 208, 211–213, 222–225, 233
Lipid membranes, 1, 3, 12
Lipid molecules, 209, 211
Lipid phases, 211, 237, 357, 371
Lipid values, 3–6
Living cells, 41, 43, 208
Local Thermal Non-Equilibrium (LTNE), 261, 262, 292, 302

M
Macroscale, 86, 311, 312, 317–320, 322, 332, 355
Macroscale description, 312, 313, 318, 322
Mass spectrometry, *see* MS
Mass transport, 42, 44, 50, 59, 65, 70–72
Mass transport equation, 51, 54
Mastication, 360
Mathematical modeling, 117, 232, 338
Mathematical models, 48, 113, 114, 137, 149, 157, 173, 227, 237, 261
MD, *see* molecular dynamics
Mean squared displacement, *see* MSD
Mechanical response, 79, 82, 87, 88
Mechanics, arterial, 78, 84–86, 89, 92, 93, 103, 106
Media lamellar units, *see* MLUs
Medium, 43, 61, 158, 160, 166, 245, 250, 262
Membrane, 1, 4–12, 25, 61, 65, 173, 211, 214, 216–218, 221–224, 226–232, 235–239, 356
  eukaryotic, 6
  inhomogeneous, 222
  prokaryotic, 8
Membrane components
  mixed, 228, 229
  separated, 228, 229
Membrane fusion, 11
Membrane models, 6, 7, 12–14
Membrane proteins, 4, 22, 25, 217, 356
Micelles, 22, 24, 28, 30, 35
Michaelis–Menten equation, 49, 52, 359
Michaelis–Menten kinetics, 48, 58, 59, 63–65, 359
Microbes, 8, 19, 20, 22–27, 29–32, 35
Microbial growth, 25, 27, 29, 33, 34
Microbial growth modes, 25
Microbial strategies, 25, 28, 35
Microchannel, 44–46, 53, 55, 58–60, 62, 64, 68
Microfluidic, 44, 46, 51, 53, 55, 56, 69–71
Microfluidic cell culture, 42, 43, 52, 53, 57, 68, 70, 71
Microfluidic cell culture models, 53, 58, 62
Microfluidic cell culture platforms, 42, 44, 65, 70–72
Microfluidic cell culture systems, 41, 44, 46, 47, 71
Microfluidic devices, 46, 49, 50, 53, 55, 57, 60, 62, 67, 70, 71
Microfluidic immunoassays, 53, 55, 56
Microfluidic systems, 41, 42, 47, 72
Micronutrients, 356, 368, 371
Microorganisms, 20, 21, 171, 173–180, 182, 190, 192, 195, 196, 199, 202, 203

Microscale, 20, 31, 33, 42, 45, 79, 312, 316, 318, 319, 329, 332, 359, 362, 366
Microscale domain, 313, 316, 318, 319
Microscale growth, 316, 321, 332, 333
MLUs (media lamellar units), 96, 101, 104, 105, 108
Model geometries, 53, 71
Model parameters, 56, 57, 79, 100, 342, 343
Modeling biomolecular transport, 42, 71
Modeling domain, 54, 55, 58, 65, 68
Models
  compartmental, 358
  dynamic, 183
  minimal lipid, 3
  photokinesis, 179, 184, 185
  phototaxis, 183, 184, 186
  power-law, 141, 143, 145
  reduced, 203
  two-concentration, 349
  two-step, 261, 266
  vascular, 243, 244
Models of chemotaxis, 203
Molecular dynamics (MD), 1, 3, 12, 14, 208, 211, 227
Molecular dynamics simulations, 1, 2, 10, 12, 13
Molecules, 1, 2, 4, 6, 11, 14, 56, 70, 78, 82–84, 90, 91, 160, 226–228, 237, 354, 356
  cholesterol, 5, 6
MS (mass spectrometry), 70, 71, 364
MSD (mean squared displacement), 216
Mucins, 118, 119, 129–131
Mucins ratio, cross-linked, 120, 121, 134, 135
Mucociliary clearance, 114, 116–118, 135–137, 149
Mucociliary transport, 135, 148, 149
Mucus, 113, 114, 117, 118, 134, 135, 137–139, 141–150
Mucus layer, 114, 129, 135, 137, 147, 148, 356
Mucus samples, 141, 143–146, 148
Multicomponent membranes, 14, 228, 229, 236
Multiscale, 79, 86, 311–313, 333
Mustacich, 162, 163, 165

N
Navier–Stokes equations, 44, 45, 50, 51, 68, 118, 131, 225, 226, 231–233, 238
Neurons, 337, 340, 349, 350
Neuropeptides, 335–338
Non-dimensionalization, 49–52
Numerical approximations, 322, 329–331
Numerical modeling, 58, 65, 71, 243, 255–257
Numerical models, 60, 151, 254–257
Numerical simulations, 60, 68, 69, 109, 114, 135, 137, 145, 161, 162, 199, 200, 238, 325, 328
Numerical values, 263, 265–267, 274, 276
Nutrient concentration, 58, 64, 314, 315, 328, 330, 331, 359
Nutrient consumption, 58, 64, 67, 313, 315, 329, 332
Nutrient transport, 156, 312, 314, 316, 320, 322, 332, 356
Nutrients, 31, 43, 46, 48, 57, 58, 64–66, 77, 82, 155, 156, 202, 312, 313, 315, 316, 322, 353–356, 369–371

O
Oil biodegradation, 20, 22, 23, 27–30, 35
Oil compounds, 22–24, 27, 32, 35
Oil surface, 20, 29–32
Oily phase, 20, 22, 25, 27, 29–32, 34, 35
Oily substrates, 19, 20, 22, 25, 26, 28, 32, 33
Oxygen, 43, 46, 48, 51, 52, 58–62, 65, 67–71, 77, 82, 171, 172, 176, 180, 188, 199
  diffusion coefficient of, 59, 61, 65
Oxygen concentrations, 59–62, 65, 67, 69, 70, 188, 201
Oxygen flux, 61, 65
Oxygen gradients, 180, 188, 189
Oxygen levels, 60, 68, 69, 172, 197, 200
Oxygen pressure drop (OPD), 62, 69
Oxygenator, 60, 61, 69
  internal, 60, 61, 69
Oxytaxis, 180, 197, 199, 202

P
Particles, 2, 30, 131, 138, 160, 182, 191, 208, 209, 211, 356, 357, 360, 366, 367
  interfacial, 367
  solvent, 209–211, 213
Pathologies, 77, 78, 116, 138, 141, 142
Patients, 116, 117, 141–143, 149
Péclet numbers, 57, 160, 165–167
Periodic conditions, 123–126
Permission, 185, 189, 193–195, 199–201, 237, 360–364, 366, 368–371
Phantoms, 247–249, 253, 254, 257, 258
Phase field, 222, 232–234
Phase field formulation, 232, 233
Phase field models, 222, 233, 235, 238
Phase separation, 227, 228, 236, 355
Phenomenological model, 186
Phospholipid membranes, 12, 13
Phospholipids, 5, 6, 8, 9, 12, 23
Photosynthesis, 172, 173, 175, 202
Phytoplankton, 172, 173, 176, 177
Pi, radius relationship, 101–105
Plant cells, 155, 156, 160
Plants, 21, 155–157, 172, 173, 312
Plots of tissue temperature, 301–304
Plumes, 194, 198–200
Pollen tube growth, 156, 167
Pollen tubes, 155–168
  apical region of, 160, 161, 165–167
  fast tip-growing, 156–158

fast tip-growth of, 155–157, 167, 168
growing, 168
shank region of, 161, 163, 164
Porous medium, 20, 27, 68, 198, 199, 201, 202, 261, 268, 270, 277, 279, 312, 321, 329, 332
Porous membrane, 65, 66
Porous tissue domain, 314, 315, 332
Power-law index, 139, 142, 143, 145, 146
Prokaryotes, 8, 10, 171, 173, 174, 176, 177, 202
Proteins, 4–8, 11, 13, 23, 25, 31, 41, 43, 46, 208, 209, 218, 221, 222, 236, 237, 354, 358
Pulses, 198, 199, 244, 245, 250, 331, 359

R
RBCs, *see* red blood cells
Receptors, 12, 46–48, 52, 54, 83
Recovery stroke, 129, 130, 135, 137, 148, 149, 174
Red blood cells (RBCs), 221, 224
Reference configuration, 85–87, 96, 107, 108
Remodeling, 81–83, 85, 91, 92, 96, 101–105, 109
Remodeling activation, 103
Remodeling laws, 79, 85, 91, 92, 98, 104
Remodeling mechanisms, 84, 88, 91, 92, 96, 98, 105
Remodeling problem, 90–92, 98, 99
Research group, 357, 361–363
Resident DCVs, 339–343, 349
Resident state, 340–342, 345, 346, 348, 349
Reynolds number, 45, 50, 216, 217, 224, 225
Rhamnolipids, 23, 27, 28
Room temperature, 158, 252–254

S
Samples, 55, 141, 143, 144, 148
Scattering, 178, 179, 183, 186, 187, 198, 202, 203
Scenarios, 102–105, 203
Shear rate, 139, 143–147, 150, 151, 355
Shear stresses, 43, 61, 65, 67–69, 139
Simulations, 1, 3–14, 33, 60, 63, 68, 100, 134, 135, 161, 211, 214, 216, 217, 360, 361, 367, 369
molecular, 1, 5, 8, 9, 14
Slides, 163, 165, 166, 197
SMCs (smooth muscle cells), 80, 82, 83, 88, 94, 97, 98, 101, 106
Smoluchowski equation, 182, 190, 200, 203
Smooth muscle cells, *see* SMCs
Smoothed particle hydrodynamics (SPH), 360
Snapshots, 134, 136, 199, 200, 214
Soft tissues, 247, 248, 333
Solubility, 32, 354, 357, 365
Soma, 335–339, 341, 343–347, 349, 350
Spatial distribution of cells, 192, 194
Species, 28, 29, 35, 46, 52, 89, 160, 161, 166
SPH, *see* smoothed particle hydrodynamics
Sputum, 116, 141, 142, 151
Stiffness, 87, 222, 227–231, 235–238
Stimuli, external, 174, 175, 181, 182
Stomach, 354, 355, 358, 359, 361, 362, 365, 370
Stroke, effective, 129, 130, 137, 149
Structural features, 85, 90, 91, 94, 96, 99, 319
Substrate, 21, 22, 25, 48, 49, 52, 222, 354
Subsurface temperature, 255–257
Surface temperature, 248, 251, 253–255
transient, 254–256
Surface tension, 28, 226, 234, 235
Suspended growth, 25, 34, 35
Suspension, 175, 177, 178, 181, 183–185, 188, 189, 192, 194–203
bacterial, 200, 201
flowing, 181
shallow, 181, 183, 189, 190, 197
Swimmer, 192, 208, 211, 212, 214–217
active, 215–217
passive, 215–217
Symmetry, 102, 167, 177, 323, 328
Synapses, 337–339

T
Taxis function, 178, 186, 187
Temperature, 22, 43, 44, 87, 143, 211, 213, 244–248, 253, 254, 257, 261, 262, 267, 279, 293, 354, 365
Temperature distribution, 304
Temperature solution, 283, 287, 289
Tension, 26, 34, 228
Terminals, 335, 337–339, 345–347, 349, 350
type III, 338, 339, 343, 345–347, 349, 350
Therapies, 117
Thermal dose, 246, 247
Thermal therapy, 246, 304
Thylakoid membranes, 7, 8
Tip-growth, fast, 155, 158, 160, 164, 165, 168
Tissue, 72, 78–80, 83–85, 92, 106, 243–246, 249, 257, 261, 262, 265, 266, 277, 278, 304, 312–314, 328, 329, 331–333
human, 245–247
target, 244–247
Tissue domains, 313, 329, 332
Tissue mechanics, 78
Tissue phantoms, 247, 249, 254, 255, 257
optical, 247
vascularized synthetic, 250, 254, 257
Tissue temperature, 247, 276, 278–280, 300–304
Torque, photo-reactive, 179, 184–186
Torque balance, 177, 179, 181, 186, 198, 203
Tortuosity, 312, 322, 325, 326, 328
Trachea, 113–115, 134, 135
Transition, 139, 147, 211, 213, 214, 217, 239, 325

Transport
active, 22, 356, 358
transmembrane, 22, 23, 25
Transport phenomena, 77, 78, 82, 84, 313
Transport problems, 51, 78, 85, 89, 90, 92, 102, 106, 320
Transport processes, 78, 79, 82, 326, 332, 333
Transport-remodeling problems, 102
Triglyceride, 354, 364, 367, 368, 371
Turgor pressure, 156, 157, 163, 164, 166, 168
Type III boutons, 335, 337–339, 342, 343, 345–350

U

Uniqueness, 127, 128, 234, 319

V

Velocity, 33, 45, 50, 56, 119, 130, 131, 135, 147–150, 159, 160, 175, 177–181, 209, 216, 217, 359, 361
sliding, 161, 165, 166
Velocity component, 159
Velocity field, 44, 46, 51, 131, 132, 134, 162–165, 231, 361
Vesicles, 10, 156, 160, 161, 167, 208, 221, 222, 224, 232, 236–238, 336, 339, 343
multicomponent, 229, 236
Vessels, 81, 244, 247, 249–251, 261, 262, 265, 270, 272
Viscosity, 44, 114, 119, 120, 127, 130, 131, 134, 138, 141, 143–147, 151, 217, 225, 234, 247, 355
shear, 146

W

Water interface, 20, 24–27, 29–32, 34, 35
Water surface, 178, 186, 189, 190
Weakly nonlinear stability (WNS), 198

FSC
www.fsc.org
MIX
Papier | Fördert
gute Waldnutzung
FSC® C083411

Kolibri 360 GmbH
Ferdinand-Jühlke-Straße 7
99095 Erfurt, Deutschland
produktsicherheit@kolibri360.de